Springer Series in Reliability Engineering

Series Editor

Today's modern systems have become increasingly complex to design and build, while the demand for reliability and cost effective development continues. Reliability is one of the most important attributes in all these systems, including aerospace applications, real-time control, medical applications, defense systems, human decision-making, and home-security products. Growing international competition has increased the need for all designers, managers, practitioners, scientists and engineers to ensure a level of reliability of their product before release at the lowest cost. The interest in reliability has been growing in recent years and this trend will continue during the next decade and beyond.

The Springer Series in Reliability Engineering publishes books, monographs and edited volumes in important areas of current theoretical research development in reliability and in areas that attempt to bridge the gap between theory and application in areas of interest to practitioners in industry, laboratories, business, and government.

Now with 100 volumes!

Indexed in Scopus and EI Compendex

Interested authors should contact the series editor, Hoang Pham, Department of Industrial and Systems Engineering, Rutgers University, Piscataway, NJ 08854, USA. Email: **hopham@soe.rutgers.edu, or Anthony Doyle, Executive Editor, Springer, London. Email:anthony.doyle@springer.com.**

* * *

Fuqing Yuan

Essentials of Reliability Engineering

Fundamentals and Applications

Springer

Fuqing Yuan
Department of Technology and Safety
UiT The Arctic University of Norway
Tromsø, Norway

ISSN 1614-7839 ISSN 2196-999X (electronic)
Springer Series in Reliability Engineering
ISBN 978-3-032-15543-6 ISBN 978-3-032-15544-3 (eBook)
https://doi.org/10.1007/978-3-032-15544-3

This Springer imprint is published by the registered company Springer Nature Switzerland AG
The registered company address is: Gewerbestrasse 11, 6330 Cham, Switzerland

To Jennie and Hsiping
To Jinmei

Preface

Over the past two decades of advocacy in reliability engineering, I have witnessed reliability's evolution from a little-known topic to a well-known subject. Today, reliability plays an increasingly prominent role across a wide range of industries. In electronic industries, the reliability team is an indispensable group to support product design. In oil and gas industry, reliability teams are critical to ensure safe operation and maintenance optimization. In automobile industry, reliability is involved in design and warranty data analysis. Academically, the importance of reliability is also rising. Many universities now offer reliability engineering as an important course for senior undergraduate and graduate students, reflecting the growing demand from industry.

Reliability engineering is a multidisciplinary subject that involves knowledge from science, engineering, economics, management, and others. The quantitative methods employed in reliability are heavily dependent on probability and statistical techniques. For instance, life data analysis, which utilizes statistical distributions, often appears almost indistinguishable from applied statistics. As a result, some researchers consider reliability to be a branch of applied statistics. Conversely, certain qualitative approaches in reliability such as those used in reliability growth originate from management disciplines, and some reliability researchers are therefore affiliated with management departments. Compared with well-established fields such as electrical engineering, mechanical engineering, or computer science, reliability engineering lacks a clear identity.

Reliability engineering continues to evolve to adapt to emerging technologies and new industrial requirements. In recent years, reliability experiments have become one focus area of reliability, driven by the practical needs of industry. Designing effective reliability experiments requires a thorough understanding of failure mechanisms, which often involve principles from physics, chemistry, and other scientific disciplines. In the domain of methodology, artificial intelligence (AI) has also significantly impacted reliability. Machine learning algorithms are increasingly applied to reliability tasks, such as data analysis and failure diagnosis. The AI method is even dominating in several major reliability journals. This interdisciplinary mixture with other subjects makes reliability increasingly diverse, posing challenges to unify all reliability activities as a cohesive subject.

This book presents both traditional reliability methods, such as life data analysis and system reliability evaluation, and modern approaches, including failure physics and simulation-based techniques. The introduction of simulation method is very necessary as it relies less on assumptions while very effective. Moreover, it is a less discussed topic in traditional reliability books. This book is written to be understandable to readers without advanced backgrounds in mathematics or specific engineering fields. Moreover, based on my teaching experience in reliability, most students can apply methods such as the Weibull model to analyze data, but few really understand the underlying principles. Consequently, they cannot adequately interpret the results they obtained when it is applied to real industry problems. This book, therefore, explores these fundamentals, explaining the reasoning behind the methods and including key mathematical derivations.

Reliability engineering involves a substantial amount of mathematics, and I have personally devoted much effort to the mathematical aspects of my research. However, mathematics is not the goal; it is a tool. In real industrial applications, overly complex methods may not help solve problems; instead, they may only make them more complicated. Therefore, this book focuses on methods that are practical and effective rather than those that are the most mathematically advanced. To keep the book self-contained, some fundamental and essential mathematical concepts are included to help readers understand the underlying principles and, if desired, develop their own models. Furthermore, since several methods discussed in this book require computational implementation, basic programming is necessary to perform the computing.

Finally, I would like to express my gratitude to all my students who have attended my lectures and provided valuable feedback over the past ten years. I am also deeply grateful to several of my teachers in reliability engineering for introducing and guiding me in this fascinating field.

Tromsø, Norway
2025, Oct 10

Fuqing Yuan

Competing Interests The author has no competing interests to declare that are relevant to the content of this manuscript.

Contents

About the Author

Fuqing Yuan was a software engineer. In 2004, he accidentally encountered quality/reliability topics and, unfortunately, he found them fascinating. Since then, he has devoted himself to this field. During this period, he obtained both his master's and Ph.D. degrees. Currently, he works as an Associate Professor at the University of Tromsø in Norway. Throughout his research on reliability, he has explored probability, statistics, optimization, numerical analysis, simulation, materials science, physics, signal processing, condition monitoring, sensor development, and machine learning. This multidisciplinary research has given him a broad and diverse knowledge base. This book represents part of the outcome of his work in the field. History is a continuous process, and reliability remains a persistent challenge in industry. It is his hope that this book will help students and researchers discover more effective and efficient methods for addressing reliability problems.

Chapter 1
Introduction and Concepts

Developing an *immortal* product has long been a human aspiration. However, the frequent accidents caused by engineering systems remind us that this dream remains a formidable challenge. We must live with inherently unreliable systems and learn to manage their limitations to gain the greatest benefit. This chapter first outlines the historical development of reliability engineering and highlights several notable practical applications. It then introduces key concepts and definitions fundamental to the field.

1.1 Introduction

The reliability issue came to concern during World War II when military system became complex. In US army, it was reported as much as 50 percent of the equipment and spare parts in storage was unserviceable before being used during the war in pacific, and only about one third of navy electronic equipment can operate properly after the war in 1950 [1]. The unreliability of the military equipment was the major driving force for the development of reliability. Particularly, the failure problem of vacuum tube, like transistors nowadays, was especially concerned as it was a fundamental unit for the military equipment, for example in the radar, rockets, and the later guided missile system [1, 2]. In 1950, an Ad Hoc Group on Reliability of Electronic Equipment that was closely connected to the US military was established to address the failure in electronics including the tube [3]. This group was upgraded to Advisory Group on Reliability of Electronic Equipment (AGREE) under the department of defense of US in 1952 [3, 4]. AGREE was a group supported by US army, navy and air force to address reliability especially. The AGREE played a vital role in the development of reliability at that time. The AGREE laid a foundation for the reliability as a subject [5]. Other than military purpose, commercial airlines also emphasized reliability during this period [1]. The Aeronautical Radio, Inc. (ARINC) which initially worked on the tube problem in the

F. Yuan, *Essentials of Reliability Engineering*, Springer Series in Reliability Engineering, https://doi.org/10.1007/978-3-032-15544-3_1

communication and navigation equipment in the aircraft, but later on the ARINC also worked for the US Navy at the beginning of 1950s [6]. Additionally, some earlier work can be related to the reliability activities in parallel with the above-mentioned. For example, the Poisson distribution—whose limiting case leads to the widely used exponential distribution in reliability—was studied from a mathematical perspective by Erlang in 1909 [7]. In 1939, Weibull introduced a statistical distribution, later known as the Weibull distribution, to model the life distribution of materials, which is an important distribution used in reliability engineering later [8, 9]. In short, early reliability work focused mainly on electronics, and reliability was at that time regarded as a problem rather than a subject.

In the 1960s, problematic vacuum tubes—which had triggered many reliability concerns—were replaced by solid-state semiconductor devices with much higher reliability [6]. Nevertheless, the importance of reliability did not fade, as systems continued to experience failures. During this period, the electronic unit gained a remarkable improvement in reliability, which was estimated to have failure rate about five to six order of magnitude due to the use of the integrated circuity [6]. In this period, the physics of failure entered the field of reliability, employing advanced instruments to investigate failure mechanisms such as electronic components became increasingly miniaturized. In this period, the nowadays well-known faut tree analysis was introduced for the development of missile systems and for the design of new commercial airplane [10]. NASA was also joining the efforts to solve the reliability issue. The Failure Modes, Effects and Criticality Analysis (FMECA) were also used for failure analysis for spacecraft in this period [6]. It is noteworthy that during the 1960s, the then widely used Weibull analyses were improved and promoted for use by aircraft companies [11]. In 1964, ARINC published a book titled "Reliability Engineering" [12], where reliability was first recognized as an engineering discipline.

Today, reliability has become a subject applied across a wide range of industries, beyond the military and aerospace sectors mentioned above. The new industries such as the computer, telecommunication such as the mobile phone, automotive industries such as autonomous cars, and the traditional heavy industries such as oil and gas, paper mills, steel companies, mining etc. are all concerning the issues of reliability. Achieving high reliability offers a significant competitive edge in an intensely contested market. The wider range of industries concerning reliability drive the progress of reliability further and make the subject more diverse, deeper and broader. The reliability methods now cover the top-level data analysis to the lower level of physical analysis. From another perspective, reliability tends from small scale, for example only one component, one system, to a large scale, e.g., the whole telecommunication network system, the whole oil & gas plant, the whole pipeline system. Reliability analysis methods have also become increasingly sophisticated, evolving from simple probability models to multivariate and mixed distributions, and even incorporating deep learning approaches from machine learning.

Nowadays, reliability engineering as a subject is interdisciplinary involving knowledge of mathematics, especially probability and statistics for uncertainty analysis, physics& chemistry for failure analysis, computer science for simulation.

Reliability methods manifest differently depending on the specific context. In life data analysis, reliability is an application of statistics. In software engineering, reliability concerns factors such as bug-free operation, code quality, and maintainability. For electronic units, reliability involves thermal design and electromagnetic compatibility/interference considerations. Mechanical engineering addresses reliability through stress–strength analysis and vibration analysis. In structural engineering, for systems such as ships or buildings, reliability focuses on mechanical analysis of structures and the uncertainties of material properties. In ship design, reliability is closely related to stability and maintainability. While reliability exhibits diverse facets across different engineering domains, there are fundamental principles common to all. This book focuses on these shared aspects of reliability methods.

1.2 Counter-Failure Methods

Let us begin with a personal experience. After one summer, my daughter's bicycle, left outdoors, became rusty and discolored. The chain and chainring no longer fit properly, as shown in Fig. 1.1. Exposure to rain and the surrounding environment damaged the new bicycle due to the lack of protection. This example illustrates that engineering systems—even modern ones—can be highly vulnerable to natural environmental conditions.

Engineering systems operate in certain environments, especially harsh environments, would have degradation, as the example in Fig. 1.1. The knowledge from physics tells us only under Kelvin 0°, corresponding to −273.11 °C, the solid structure is not dynamic, i.e., no kinetic energy. At all higher temperatures, the atomic would be dynamic and some atoms would escape from the constraint of its neighbors to other position. If two materials border each other, as shown in the left of Fig. 1.2, the two materials tend to mix, as demonstrated in right of Fig. 1.2. If the

Fig. 1.1 Bicycle corrosion due to environmental exposure

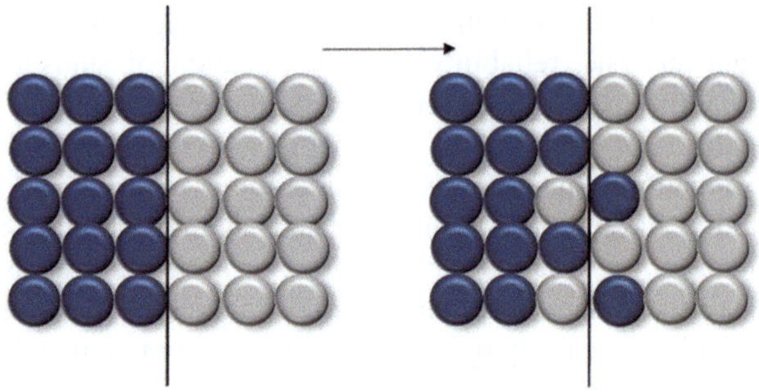

Fig. 1.2 Atomic diffusion

atomic mixture occurred, named diffusion in physics, that is not desirable, failure would be resulted consequently.

Corrosion and diffusion are two fundamental mechanisms that can lead to failure. Most of the physical world is inherently dynamic, and if systems are not properly designed, failures are inevitable—sooner or later. The following sections present examples of countermeasures to control and prevent failure, including discussions of the concepts of "Six Sigma" and maintenance strategies.

1.2.1 Maintenance

Maintenance is all actions necessary for retaining an item in or restoring it to a specified condition [13]. Maintenance is a very effective and efficient measurement to counter failure in industry. For example, for transportation system such as railway, the maintenance is critical to ensure the normal operation. The railway system is subject to regular inspection to ensure the railway can be operated reliably, as after some operating period, rail track would become irregular, rail faster and sleeper might disfunction, the overhead line would wear out.

Maintenance can be generally classified as preventive maintenance and corrective maintenance. Preventive as its name suggests is to take action to prevent the failure occurrence. Corrective maintenance acts after the failure occurred. When planning maintenance, for some noncritical component, it can adopt corrective maintenance to save cost, which is called "run to failure". While for most industries, such as oil and gas, it is more common to have both maintenance policies. For maintenance, a key point is the maintenance point, which is the time planned to perform the maintenance activity. For instance, if we determine the performance cannot be tolerated under certain level, such as the horizontal dashed line shown in Fig. 1.3, then the circle point is the maintenance point.

As the result of maintenance, the maintenance effectiveness has three well-known categories:

I. **As good as new**: The performance of machine has been restored as a new system after the maintenance.

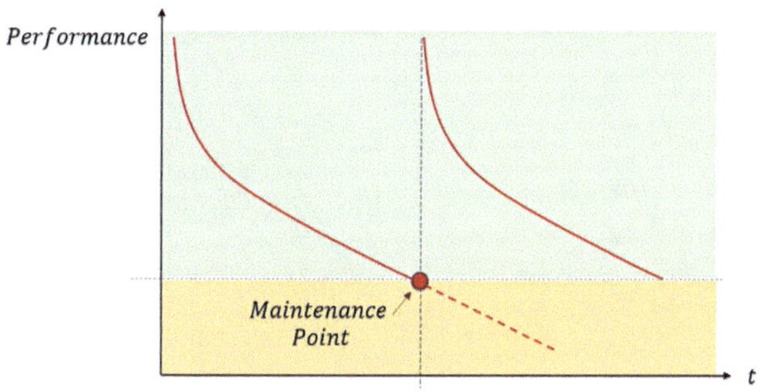

Fig. 1.3 Maintenance point

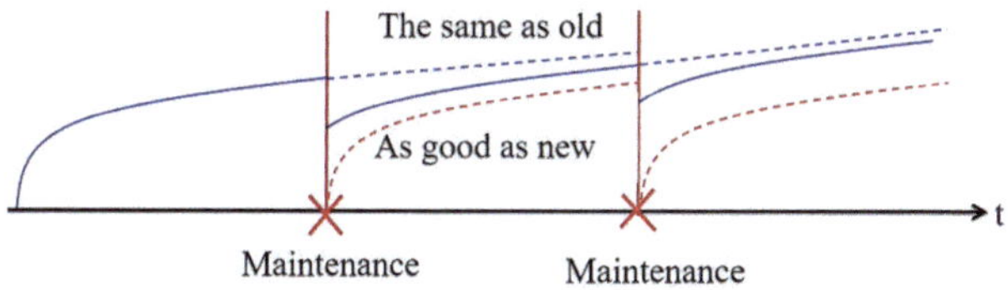

Fig. 1.4 Maintenance effectiveness

II. **As bad as old** or **the same as old**: Failure has been removed, but the failure rate or performance of the machine remains unchanged after maintenance.
III. **Imperfect repair**. The performance of the machine is between the Categories I and II.

Figure 1.4 demonstrates the three categories of maintenance effectiveness, where the solid line is the failure rate after imperfect repair.

Maintenance is a subject highly relevant to reliability. The performance shown in Fig. 1.3 can be reliability. In practice, the maintenance policy can be based on the reliability level of the system and thus reliability plays a central role for the maintenance activities.

1.2.2 Design Modification

Another measure to prevent failure is to improve the design. It is an effective approach if it is feasible both from technical and from economic perspective. The designer, manufacturer and maybe the user might meet to investigate the root cause of the failure. If the cause is found to be technological error, design modification will be introduced to prevent failure recurrence. Take aircraft as an example. Boeing airplanes are generally safe passenger aircraft that enable us to travel around the world, although accidents caused by technical errors occasionally occur. For instance, in 2013, a Boeing 787 Dreamliner was reported to have dangerous smoke and burning in the rear of the cabin [14, 15]. Investigation found the smoke was from a fire inside a lithium-ion battery, which was short circuited. Despite no casualty was reported, the error was serious, and catastrophic consequence could

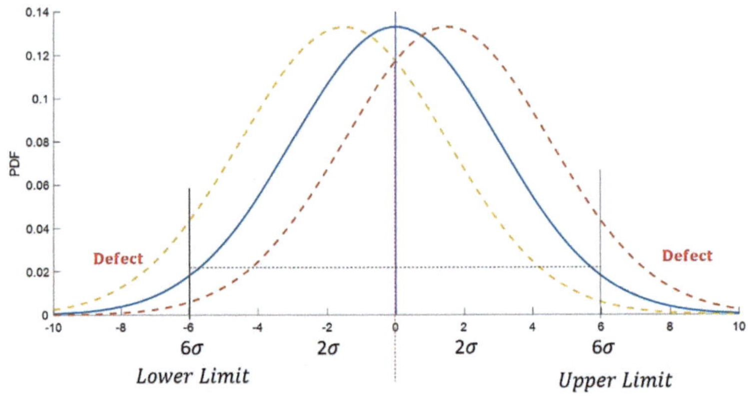

Fig. 1.5 Concept of six sigma

result. The Boeing 787 was the latest commercial airplane by Boeing with advanced technology, and the accident was surely unexpected. As aviation components were strictly tested and heavily regulated before joining the service, such events are thus very rare. The lithium-ion battery was soon redesigned by adding insulation among the cells in the battery to minimize the risk of short circuit once fire occurs again. In commercial aviation, even minor failures are unacceptable. Ongoing improvements in design are a key strategy for preventing such failures.

Adopting a no-failure policy comes at an extremely high cost. Only a few special industries like commercial aviation, military and nuclear power industry practice it. More civic industries can tolerate some failure/defect to some extent. In the industrial production, there is a well-known concept named "Six Sigma". The concept defines the product is with a probability of 99.99966% without defect, suggesting around 3.4 defects per million products. The concept is claimed by many manufacturers as the target of defect rate in their products. The Six Sigma was adopted its name from statistics, assuring the upper and lower limit of specification of product are within $\pm 6\sigma$, where the name "Six Sigma" obtained. Shown from Fig. 1.5, the probability of defect is the sum of the two defect areas locating outside of the 6σ region.

Standardizing the normal distribution to $N(0, 1)$, the sum of the two defect areas is then

$$\Phi(-6) + 1 - \Phi(6) = 1.97 \times 10^{-9}$$

If considering the mean with uncertainty, common in practice, defining a shift of mean with a $\pm 1.5\sigma$ for 6σ, the defect area is then computed by the distribution depicted by any one of the dashed lines in Fig. 1.5. The total defect area is then

$$\Phi(-7.5) + 1 - \Phi(4.5) = \Phi(-4.5) + 1 - \Phi(7.5) = 3.4 \times 10^{-6}$$

suggesting a 3.4 defect per million, which is the target of Six Sigma for the product quality. The Six Sigma is a topic in quality engineering, but is also discussed in reliability engineering.

1.3 Reliability Assessment

If the reliability of a system can be determined accurately, both failure prevention and maintenance optimization are significantly simplified. Reliability assessment evaluates the system's reliability level, which can generally be performed using three approaches: Life Data Analysis, Experimental Analysis, and Low-Level Chemical/ Physical Failure Model Analysis.

1.3.1 Life Data Analysis

Collecting failure data from field or other sources, then analyzing them to evaluate the reliability is a popular approach in the reliability practice. In reliability engineering, a major work is data analysis. The data can be life data, for example time to failure data, maintenance data, technical configuration data, manufacture data or operation data. Life data is the data about the system life. It contains information about when the system fails. Life data is a major data source for reliability data analysis if there are available. In practice, life data normally is not present in raw data, and it should be extracted from other data. In the computer management system, data is more often shown as records such as the date-time when the system started to work, the time when it shut down. The life data is then extracted from such kinds of raw data. Knowledge of informatic about database SQL language, computer programming language such Matlab for preprocessing might require, as the amount of data records can reach the million level. Manually sorting it is impossible. Considerable efforts are needed to extract the life data from these raw data. Some challenges could be confronted, for example as:

- Data amount is huge. Raw data size could reach 100 GB level. SQL language is perhaps needed to extract needed data if the data is structured, i.e., presenting in a form of table.
- Data format is not consistent. For example, the date format might be in "yyyy-mm-dd" or "dd-mm-yyyy" or other form. SQL or some other advanced language such as Matlab might be used to format the data into a unique style. Reliability engineers should check the data format, otherwise the analysis will be based on wrong data set.

Fig. 1.6 Reliability data analysis procedure

- Data record error. The engineer might type a number erroneously. For example, the age of a man should be positive, while the database might record a negative value. If possible, write some programming code to screen such data. Eliminating the erroneous data is tedious and time-consuming, while these efforts are very necessary.

Dealing with huge amounts of failure data is a challenge but it is exciting work for reliability engineers. While another extreme situation, unfortunately is more common, is data lacking. The lack of data is usually due either to no failures occurring or to failures not being recorded. Many industries such as wind power and some oil & gas plants store life data in maintenance databases. The maintenance database might only store the data at system/subsystem level, not at the component level. If our interest is of component, the data is then unavailable. For example, at a wind power farm, for maintenance activities report, the engineer only describes failure at subsystem level or only concerns bigger component, such as turbine, the lubrication system, transformer, the hydraulic pump. The lower-level parts are not seen reported. The data lack will result in insufficient data size for life data analysis, while data amount is important for an accurate reliability analysis. To solve this problem, international organizations and many industries have established failure data collection system. ISO has a standard ISO 14224 for reliability data collection of oil and gas facilities [16]. Some military systems have their own standards to collect reliability data but are not available to public. Civil aviation also has its own system for reliability data collection. The airlines collect failure data and send the failure data to the manufacturer for analysis. Attributing to these data collection systems, after each accident, the aviation industry can have very accurate data released. The manufacturer of airplane aggregates data from the airline and thus has a larger data amount, which is beneficial for the reliability analysis. The Norwegian oil industry has a database named OREDA collects data from oil and gas companies for more than 40 years. The reliability analysis results are published as handbook for offshore and onshore equipment. After sufficient data is available, life data analysis can be performed to evaluate the reliability level more accurately. A general procedure for reliability data analysis is summarized as Fig. 1.6.

In classical reliability engineering, Life Data Analysis plays a major role; accordingly, the first four chapters of this book focus on this subject.

Life Data Analysis Approaches

Life data analysis mainly uses probability and statistics methods, fitting the observed data to a probability model, statistic model, or using the union/intersection of probability of the event such as in the faulty tree, event tree analysis. Of these methods, statistical analysis is one of the most widely used.

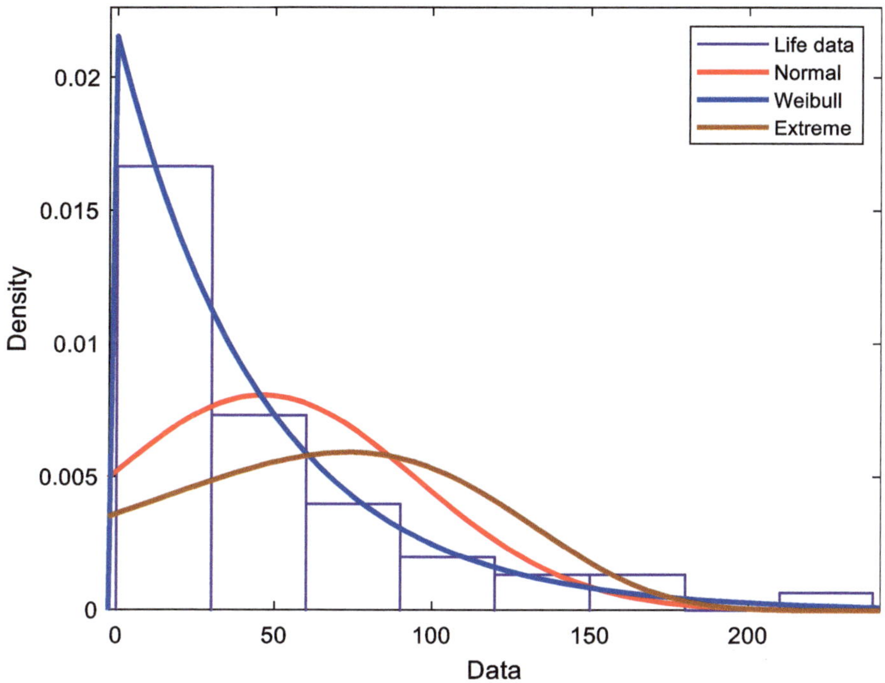

Fig. 1.7 Data fitting using various distributions

- **Using Classical Statistics**: As shown in Fig. 1.7, this approach fits the failure data to a specified statistical distribution such as Normal, Weibull, Exponential, Gamma distribution. A critical task of this approach is the parameter estimation given available life data, which is essentially distribution analysis.
- **Bayesian Approach**: The Bayesian approach is an alternative approach to classical statistics, as shown in Fig. 1.8. Bayesian approach considers the parameter in the distribution as variable. Numerical method is usually used for computation. Bayesian approach is advantageous when the model is complex, for which the classical statistic is hard to obtain analytical solution. Bayesian approach employs simulation to obtain solutions, which is a major drawback of this approach. Simulation engages in heavy computation and takes longer time. Chapter 10 of this book discusses the simulation method in detail.
- **Machine Learning**: Machine learning has become a popular data analysis approach in the past decade. It relies on a group of basic mathematical functions to fit observed data. The parameters in the basic functions do not have physical explanation on the data. Fitted functions will change with the data flexibly, due to which the approach is merely data driven. As the use of a large number of basic functions, the model is very complex, and training algorithm is required to obtain model parameters. Figure 1.9 illustrates an example to demonstrate how the machine learning model can fit the data flexibly. All machine learning methods

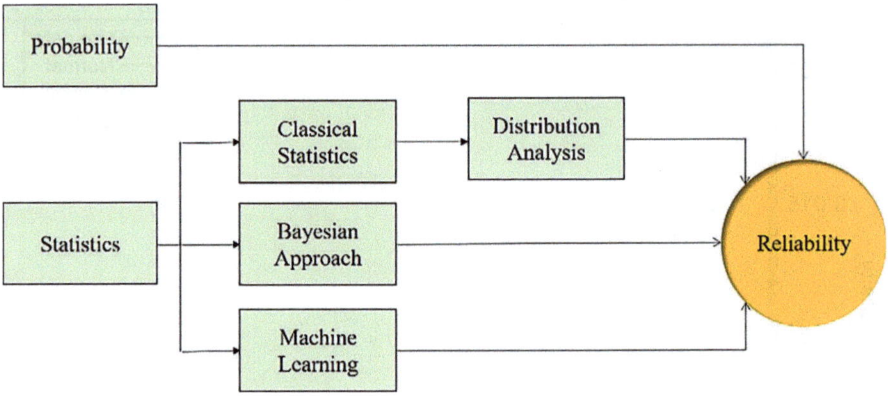

Fig. 1.8 Life data analysis approaches

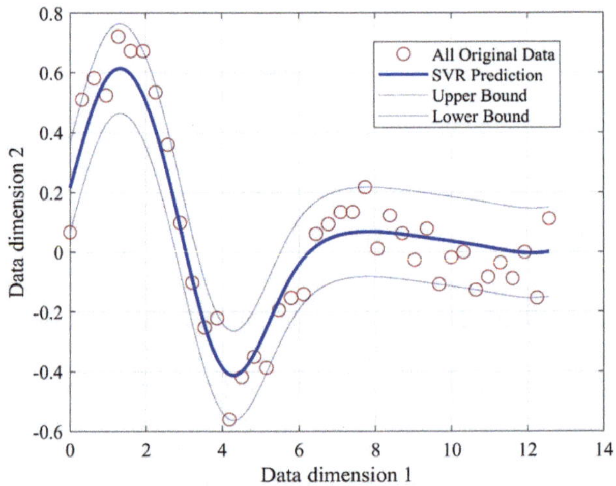

Fig. 1.9 Machine learning model fits the data

such as artificial neural network, support vector machine, hidden Markov train, kNN, and random forest can be used to reliability data analysis by taking advantage of their flexible fitting. Machine learning for data analysis is a broad topic. This book does not cover machine learning methods.

1.3.2 Reliability Evaluation by Experiment

Life data provides valuable information for reliability analysis. However, such data is not available for products that are still in the design phase or in the early stages of service, since failures have not yet occurred. Performing life tests at the lab is an alternative approach for reliability analysis for new products. A common practice,

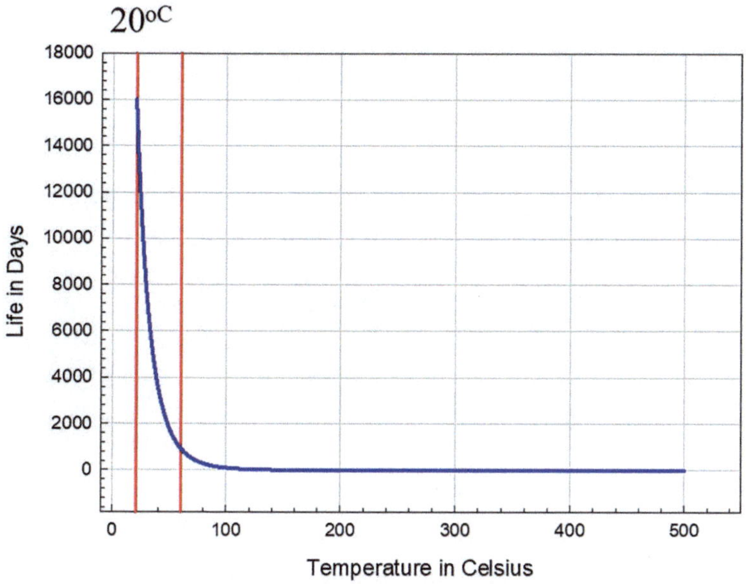

Fig. 1.10 Life of product drops with temperature

often followed by many electronics companies, is to select a certain number of products for test and then observe whether any of them fail and, if so, how many failures occur during the test. The test can simulate the operating condition of the real field by adjusting the humidity, temperature, gas concentration, vibration, or radiation level etc. For example, spacecraft, whether at the system level or for individual components, undergo such test prior to launch to ensure safe operation in space.

For system life tests, long lifetime systems might not fail during the test period. For example, a hard drive can have an average life up to 50–100 years. Observing failure time for such product is not realistic. In reliability we can employ accelerated life tests to shorten the system life to a range within our observational time. The accelerated life test achieves the aim by giving higher stress on the product by adjusting parameters of the operational condition, mainly humidity and temperature. Then applying accelerated life model to transform the life under the accelerated condition to that in the real operating condition. Figure 1.10 illustrates an example on how the life drops with the operational condition, where for the product at 20 °C with life 44 years will have life only 3.6 days at temperature 180 °C. Observational time within days is feasible to be conducted at laboratory. In practice, increasing the operating temperature to shorten product life is a common approach for electronic device life tests.

Environmental chamber is an apparatus designed to perform the acceleration life test. The chamber can adjust the temperature, for example at a range from −50 °C to 300 °C, humidity from 20% to 98% etc. Figure 1.11 shows a chamber in our lab. The large hole on the left is used to route the wire out for data reading.

Fig. 1.11 Environmental chamber for life test

Other than life test, there are some other tests that are conducted for reliability. For electronic devices, other reliability tests are for example:

(a) Falling Test. To test if the device will be damaged if it drops from 2 m or 3 m height.
(b) EMC test. Test if the device can function normally in the given electromagnetic wave radiation environment.
(c) Vibration test. Test if the device can function when the device is operating at given level of vibration.

Chapter 12 of this book will provide a detailed discussion on experiments and accelerated life tests.

1.3.3 Reliability Analysis by Using Physical Model

Engineering failures, when not caused by human error, ultimately stem from mechanical, electrical, electronic, dysfunctions. For example, in the case of a bridge collapse due to overload, a mathematical model can be developed that incorporates factors such as load, material strength, wind speed, and temperature to predict the time to failure. In this book, such an approach is referred to as a physical model. Some industries prefer failure prediction using physical models, which are expected to be more accurate and convincing than traditional life data analysis. One limitation of life data analysis is for example: when asked why it fails within 10 h, it is difficult to answer from a purely physical perspective, but it can be readily addressed if the failure time is derived from a physical model.

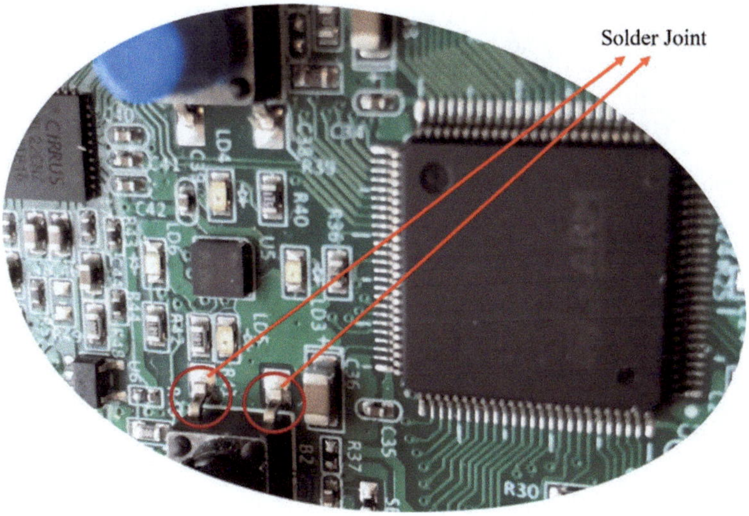

Fig. 1.12 Solder joint on a MCU board

However, developing physical models is not a simple task. It depends on the component, root cause, failure mode, failure mechanisms. For example, when a cellphone encounters black screen error, the root causes could be

- LED panel damaged
- Control unit failed
- Battery damaged
- Main processor failed
- Connection lost
- etc.

For each cause, for example, if the control unit failed, it might be the chip damaged, solder lose, resistor burned, capacitor damaged. If solder connection loses, which is a common problem in electronic devices, as shown in Fig. 1.12, a mechanical model can be developed by addressing connection stress and connection strength. Thereafter to use this model we can find out how the connection strength would decrease after each stress cycle. The failure time can then be estimated from the number of experienced cycles.

Model development is a challenging process because it requires both a solid understanding of the underlying physics of failure and an appropriate mathematical formula to represent the failure mechanism. In practice, identifying a suitable physical model for a specific type of failure is often difficult, and in many cases, we fail to find the model that can adequately predict the failure time. The situation becomes even more complex when multiple failure mechanisms are linked to the same root cause. Consequently, for a given type of failure, capturing all possible root causes and developing corresponding physical models demands considerable effort.

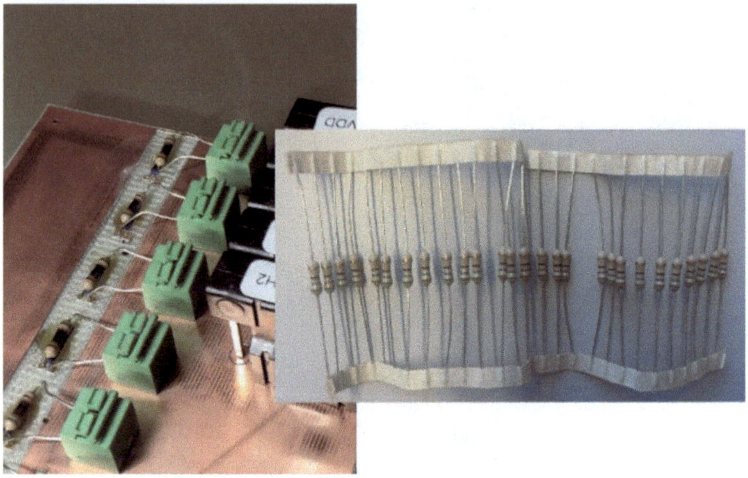

Fig. 1.13 Burning of resistors

The physical model is often complex in form and may, in some cases, involve mathematical formulas that do not have a closed-form expression. To illustrate, we present a physical model for a simple type of failure: the burning of a resistor, as shown in Fig. 1.13. In this example, thermal physics is applied to model the lifetime of a resistor subjected to an overstressed voltage. After appropriate simplifications, the physical model can be expressed as follows.

$$t = \frac{kS}{Cm} \ln \frac{V^2}{V^2 + \rho_0 kwT_0 S - \rho_0 kwTS}$$

k: Thermal conductivity in J/(s.cm^2.°C)
w: Ratio of length to cross-sectional area of wire
ρ_0: Resistivity
V: Applied voltage
T: Temperature
T_0: Initial ambient temperature
C: Heat capacitance
S: Surface area of resistor contacting with air
m: Mass of the resistor

Disregarding the physical meaning of each parameter, the purpose of presenting this model is merely to illustrate the complexity of physical models. Even in this simplified case, the model involves nine parameters, and once uncertainty is considered, each parameter is treated as a random variable. Consequently, the applicability of the model to accurately predict failure time is significantly compromised by this uncertainty. A more detailed discussion of approaches to address uncertainty in physical models is provided in Chap. 7 of this book.

Developing a practical physical model is often a time-consuming task. For engineering systems containing thousands of components—which is typical in modern applications—it is infeasible to develop physical models for every failure mode and failure type. From this perspective, reliability evaluation based on life data collected from the field is a much more cost-effective way. In practice, physical models are generally considered only for critical failures.

1.4 Reliability Practice in Industries

Beyond military systems in the 1940s and 1950s, many other industries have also emphasized reliability efforts to support product design, optimize maintenance, conduct experiments, and analyze failure data. Practitioners in these fields, often with multidisciplinary engineering backgrounds, have been a strong driving force in implementing reliability practices and promoting the development of new reliability methods and models. This section presents examples of activities undertaken in different industries.

1.4.1 Reliability in Electronic Company

The role of reliability in electronics companies is particularly prominent. Within such organizations, reliability engineers may work in an independent reliability group or be embedded within other teams, such as thermal design, circuit design, or software development. Modern electronic devices are highly complex, typically consisting of numerous components and subsystems, including processors, memory, batteries and control units, integrated into a compact and portable case. Ensuring the reliability of a system composed of so many interacting units is a considerable challenge.

Although the mean life of many electronic components is typically very long—for instance, certain chip-type units can achieve lifetimes of up to 50,000 years under favorable operating conditions [17]—failures can still arise from issues such as poor connections or mismatches in thermal expansion among different units. These weaknesses can ultimately compromise the overall reliability of the product. To ensure that a product meets its reliability goals, electronics companies typically engage in a wide range of reliability-related activities, for example:

- Define the reliability target. For example, set a target of reliability that less than 3% product will be returned for repair after putting in the market.
- Controlling the component reliability supplied from third part. The supplier may provide some documents about: the expected life of their product, how this life was assessed, and how the life test experiment was carried out. From the document, the reliability engineer should find out if the component can reach

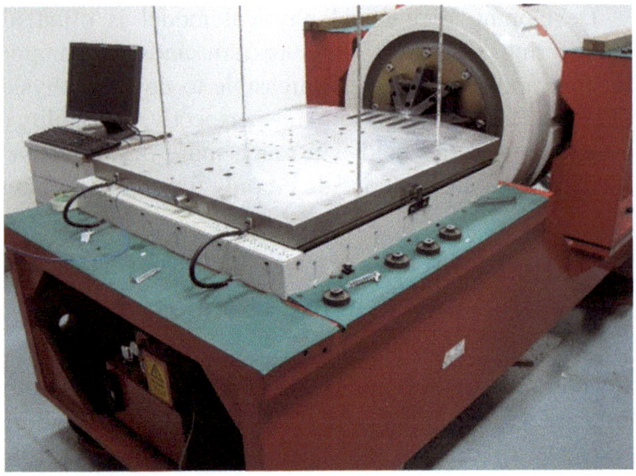

Fig. 1.14 A vibration test platform

his requirement and if sufficient test have been carried out by the supplier. For some critical components, reliability engineers, other than obtaining reliability from the document, could perform their own experiment to assess the reliability level.

- Perform falling test, aiming to assess the survivability from accidental drop for a new product.
- Perform vibration test (as shown in Fig. 1.14), and electromagnetic inference test to check if the standard has been met and if the product can run safely under certain operating environment.
- Perform life test. When prototypes are available, pick up a certain number of prototypes to run them to failure and assess the failure rate. In this task, reliability engineers should determine the size of products for tests. Conduct test over a defined period, for example, 6 months. Record the number of failures, failure time and analyze the test result to evaluate if the reliability target has been met.
- Perform acceleration life test. Design and carry acceleration test to evaluate the life of a new product/ component, or to evaluate if new types of connection are strong enough. This type of experimental test is a popular and significant work in the electronic industry.

1.4.2 Reliability Activities in Aviation

Reliability in the aviation is a successful practice. The catastrophic consequences of aviation accidents make the demand of reliability for this industry extremely high. The aviation is thus heavily regulated and documentation is required with almost all of its activities, from its design, manufacturing, operation and maintenance [18]. The heavy regulation, stringent documentation requirement leads to a high cost, but it ensures the reliability level. The accident rate per million departure is 2.05 in 2022, and the number of fatal accidents in Europe is even 0 for the year 2022 [19]. The reliability activities for a commercial aircraft are involved throughout all the life cycle:

- During the design/development phase, the reliability program identifies and defines the reliability requirements. Reliability indices are selected for example as mean flight hour between failure, mean time between failure, mean time between unscheduled removal. Safety index, availability index, maintainability index and durability index that related to reliability are all defined systematically [20]. The reliability definition is for both the whole aircraft and for its constituent components. Once the index of the whole system is defined, reliability engineer then allocates the system reliability to each of the individual components. Techniques such as system reliability method according to reliability block diagram, ARINC Apportionment Technique, Equal apportionment technique can be used to accomplish the allocation task [21]. After the design, the reliability engineer verifies if the target reliability level can be met, which is a challenge work as no failure data available at that stage. Techniques using physics of failure, reliability prediction methods such as reliability block diagram, part count analysis, Markov Chain, Monte Carlo simulation etc. can be used to predict reliability. The fault tree analysis, failure modes and effects analysis are also important tools used. The reliability work in this stage is vital and is a tedious procedure often with iterative efforts. For these activities, reliability engineers must have a thorough understanding of the new design and be familiar with the underlying mechanisms in order to develop an accurate reliability model.
- Reliability activities are also present in the manufacturing stage to ensure the designed reliability level is achieved in the manufactured components. Quality assurance in the aviation industry is governed by aircraft manufacturers such as Boeing and Airbus, in accordance with established standards and regulations. The afore discussed Six Sigma is also one of the methods employed.
- Reliability activities continue to play a crucial role even after the aircraft has been delivered to the airline regarding maintenance. During the design phase, the maintenance is already a consideration. Design out of maintenance prefers a design with easier and ergonomic maintenance. The condition monitoring system is integrated into the design to monitor the operation to avoid failure and accident. The well-known concept of reliability-centered maintenance originated from the aviation industry. When in service, reliability activities in aviation maintenance involves data collection, setting alert levels, data representation, data analysis [18].

1.5 Basic Concepts and Definitions

In this section, we introduce the fundamental concepts and definitions related to reliability, including failure, failure rate, MTBF, and MTTF. Since no standard definitions exist for these terms, the definitions adopted here reflect the most accepted conventions.

1.5.1 Reliability and Probability

An intuitive understanding of reliability could be the degree of a system can be relied on or can be trusted. Reliability can be defined either in a qualitative way such as low, medium, high, or in a quantitative way using number. The quantitative way sounds more scientific. Adopting probability to define reliability was shown at the beginning of the reliability subject [22], while it was controversial when the failure time is considered as random [6].

Definition by Rowe in 1952 [22]

> a reliable valve is characterized by a very high probability that it will operate normally when taken from stock and installed in equipment for which it was intended, and by an extremely low probability that it will fail during subsequent operation in that equipment for some definite period of time.

Definition adopted by Ricard.R.Carhart from Rand Corporation in 1953 [1]

> The reliability of a given component or system is the probability that it will perform its required function under given conditions for a specified operating time.

Definition given by ARIC in 1955 [23].

> The reliability of an electronic product is the probability that the product will give satisfactory performance for a given period of time when used in the manner and for the purpose intended.

A wide accepted one is as defined by ASQ (American Society for Quality) [24]

> The **probability** that an item can perform a **required function** without failure under **stated conditions** for a **specified interval**.

The influential US military standards defined in a similar way as above but with a slight difference [13]. In those definitions, reliability is measured in terms of probability. The probability is how likely an event to occur. Probability is numerical and measurable, and it is defined by a percentage ranged from 0 to 100, or from 0 to 1. The reliability is therefore following the same range, if it adopts probability as its measurement. Reliability can also be measured using metrics other than probability.

It is natural to consider the probability as the reliability. We illustrate a practical example to demonstrate it. An IT service company has business of repairing laptops. They have collected a large dataset documenting the failure times of the laptops. It found the percentage of failed laptops differs considerably with brands. Within 3 years, the best laptop has 12% failed, the worst has 25.6% failed, a value double than the best, as shown in Fig. 1.15. In a way of reliability thinking, the reliability of the brands varies.

In this example, the repair company might not be aware of the concept of reliability. They just roughly calculate the percentage of failed laptops, i.e.

Fig. 1.15 Percentage of failure of laptops

	Brand	% of failed	% of survived
1	A	15.6%	84.4%
2	B	15.7%	84.3%
3	C	16.8%	83.2%
4	D	17.4%	..
5	E	18.3%	..
6	F	21.5%	..
7	G	23.3%	..
8	H	23.5%	76.5%
9	I	25.6%	74.4%

$$\widehat{P} = (\mathrm{N}_{Failed}/\mathrm{N}_{Total}) \times 100\% \tag{1.1}$$

It can be proved mathematically that this percentage can be considered as the probability of failure for the three-year period. Assumes implicitly the probability of failure exists and unique. And further, each individual laptop from the same brand has the same probability of failure, despite their probability of failure essentially varies individually. Ignoring the difference is a way to simplify the problem. Suppose the real probability of failure is P which ranges from 0 to 1. Within this specific time T, suppose each individual laptop is independent from each other, i.e., a failure from one laptop will not affect the failure from another one. Define an event as the failing of the laptop and the event has the probability P. For a total number of N_{Total} laptops from a same brand, if the laptop failed, we count it 1, otherwise 0, i.e.

$$N(i) = \begin{cases} 1; Failed\ within\ T \\ 0 : Not\ Failed\ within\ T \end{cases} \tag{1.2}$$

The number of failures is then

$$N_{Failuire} = \sum_{i=1}^{N_{Total}} N(i) \tag{1.3}$$

For each individual laptop, the failure event can be modeled as a Bernoulli random variable, representing a single trial with success (failure) or non-success (no failure). Equation (1.3) therefore corresponds to the sum of Bernoulli random variables. The distribution of this sum can be derived, which under independence assumptions follows a Binomial distribution. Suppose a failure pattern is given as follows:

$$\begin{aligned} &N(1) = 1; \\ &N(2) = 0; \\ &N(3) = 1; \\ &\ldots \\ &N(i) = 0 \\ &\ldots \\ &N(N_{Total}) = 1; \end{aligned} \tag{1.4}$$

Since the failure probability is P, the probability of no failure is then $1 - P$. As each individual laptop is independent, the joint probability for Eq. (1.4) is then

$$P_{joint} = P * (1 - P) * P * \ldots * (1 - P) * \ldots * P. \tag{1.5}$$

Rewrite the Eq. (1.5) as

$$P_{joint} = P^{N_{Failed}} * (1 - P)^{N_{Not\ Failed}} \tag{1.6}$$

Given the observed number of failures is $N_{Failure}$, the possibility to have the permutation of (1.4) is then $1/\binom{N_{Total}}{N_{Failure}}$. The $\binom{N_{Total}}{N_{Failure}}$ is the combination of N_{Total} choosing $N_{Failure}$. We can find the

$$\frac{1}{\binom{N_{Total}}{N_{Failure}}} * \frac{N_{Failure}}{N_{Total}} \approx P^{N_{Failed}} * (1 - P)^{N_{Not\ Failed}} \tag{1.7}$$

And the probability of occurrence of number of $N_{Failure}$ is

$$\binom{N_{Total}}{N_{Failure}} P^{N_{Failed}} * (1 - P)^{N_{Not\ Failed}} \tag{1.8}$$

The number of failures $N_{Failure}$ is a random variable, essentially the Eq. (1.8) is the density function of Binomial distribution. The expectation of the $N_{Failure}$ is then [25]

$$E(N_{Failure}) = N_{Total} * P \tag{1.9}$$

From Eq. (1.9), it is obvious that

$$\frac{E(N_{Failure})}{N_{Total}} = P \tag{1.10}$$

If the really observed number of failures is approximated as the $E(N_{Failure})$, Eq. (1.1) is actually equal to Eq. (1.10), meaning the $\widehat{P}$ in Eq. (1.1) can be used as an estimate of the probability of failure. Since the reliability is complementary event to the event of probability of failure, then reliability is

$$\widehat{R} = 1 - \widehat{P} \tag{1.11}$$

The variance of random number $N_{Failure}$ is [25].

$$Var(N_{Failure}) = N_{Total} * P * (1 - P) \tag{1.12}$$

Then it is easy to find the variance of P for Eq. (1.10) is

$$Var\left(\widehat{P}\right) = \frac{P * (1 - P)}{N_{Total}} \tag{1.13}$$

According to Eq. (1.11), the $Var\left(\widehat{R}\right) = Var\left(1 - \widehat{P}\right) = Var(1) + Var\left(-\widehat{P}\right) = 0 + (-1)^2 Var\left(\widehat{P}\right) = Var\left(\widehat{P}\right)$, the reliability $\widehat{R}$ has then been proved to have the same variance with $\widehat{P}$. Eq. (1.13) suggests a larger number of laptops N_{Total} on test, a higher reliable estimate, which is consistent with our experience. Figure 1.16 demonstrates an example where reliability is estimated as 0.844. The standard deviation varies with sample size N_{Total} = 10, 50, 100, 3000 as the 0.1147, 0.0513, 0.0363, 0.0066 respectively according to Eq. (1.13). From Fig. 1.16, it shows when the sample size increases, the distribution will concentrate around the estimated 0.844 more and more, suggesting a higher sample size implies a higher reliable estimate.

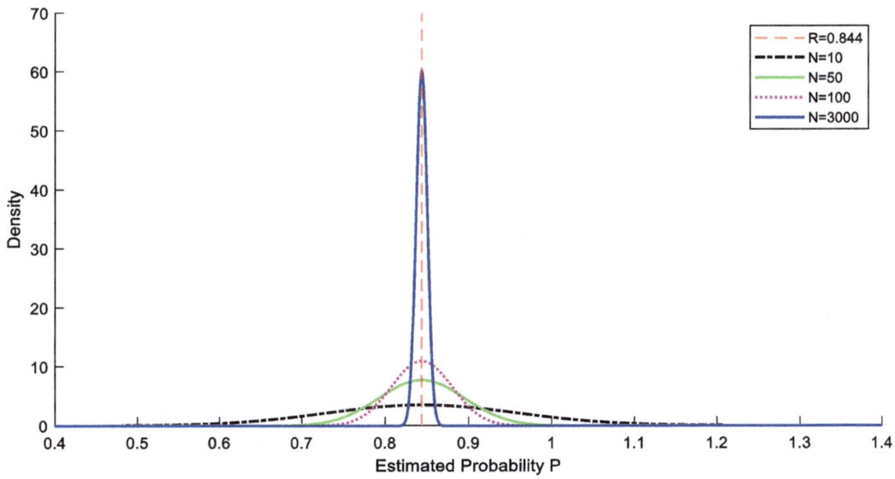

Fig. 1.16 Standard deviation with increasing sample size

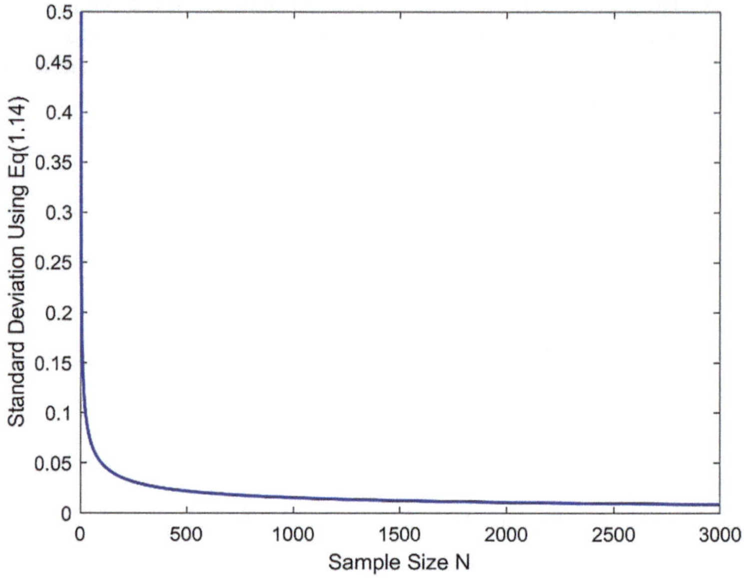

Fig. 1.17 Standard deviation with increasing sample size

Following Eq. (1.13), we have one further investigation: Since the P is ranged within $[0, 1]$, the $P * (1 - P)$ is with a maximal value $\frac{1}{4}$ at $P = 1/2$, which can be readily found from $P * (1 - P) = \frac{1}{4} - \left(P - \frac{1}{2}\right)^2 \leq \frac{1}{4}$. The variance of $\widehat{P}$ is at its maximum as

$$Max\left[Var\left(\widehat{P}\right)\right] = \frac{1}{4 * N_{Total}} \tag{1.14}$$

The standard deviation derived from Eq. (1.14) is plotted for various sample sizes in Fig. 1.17, where it shows the standard deviation drops and approaches to 0.

Equation (1.14) can be used to calculate the maximal number of samples for test, given a certain tolerable error, or to control the estimation error using Eq. (1.1). Equation (1.14) is the upper limit of variance given the sample size N_{Total}, i.e. the worst-case scenario. The real variance should not be lower than this.

Example 1.1

A PC company is planning to test the reliability of a to-be released PC. They use the Eq. (1.11) to evaluate the reliability. It requires at most 10% of standard deviation of test results for reliability. How many PCs should the company propose to safeguard a less than 10% standard deviation?

(continued)

Example 1.1 (continued)

Solution: The aim is to tolerate at most 10% of estimation standard deviation. According to Eq. (1.14), the standard deviation should be

$$\sqrt{\frac{1}{4 * N_{Total}}} \leq 10\%$$

Then the

$$N_{Total} \geq 25.$$

A minimum of 25 PCs is required for the test.

Let us come back the reliability. We still use the above discussed case to explain the reliability definition. The example considers reliability over a 3-year period. In our experience, for longer periods such as 5, 10 years, reliability should decrease certainly, since the likelihood of failure increases with prolonged operation. Thus, reliability is time dependent. Reliability addresses the time dependency as "**specified interval**" in the definition. The **specified interval** can be temporal as calendar time for this PC case. It can also be flight hours for aircraft, travelled distance for automobiles.

Back to the "**probability**" in the definition, using probability to quantify reliability can be easily understood as probability has been widely accepted by engineering. Many engineering problems can be formulated into probability or statistical problems. The theoretical foundation of probability is solid, and it has been well developed. Nowadays, in reliability engineering, reliability value implicitly means probability. However, using probability has shortcomings when coming to decision making. The usage of probability arises some dilemmas. Practitioners from industries often ask: should I shut down the machine when the reliability is 80%? Will failure occur in the next hour when reliability is 40%? The probability cannot help making decisions for such questions, which is a limitation of using probability. Moreover, in practice, the "time-dependent" has limitation. Sometimes we just intend to compare the reliability level for products regardless of time. Reliability using probability associated with time will complicate the problem. Therefore, in some applications, it is necessary to use other measures for reliability.

1.5.2 Failure

The definition of failure is context sensitive. For example, an excellent student got B grade in an exam. He considers it as failure as he expected a higher-grade A. Another student also got B Grade. But he didn't consider it a failure, as he expected grade D

Fig. 1.18 Failure of resistors

or E. Definition of failure depends on the original expectation. One cannot absolutely define an event as failure or no failure without considering the background. There are several definitions on failure related to reliability:

- The inability, because of defect(s), of an item, product, or service to perform its required functions as needed (ASQ, American Society for Quality [24]).
- The event, or inoperable state, in which any item or part of an item does not, or would not, perform as previously specified (MIL-STD-721C [13]).
- Failure can be any incident or condition that causes an industrial plant, manufactured product, process, material, or service to degrade or become unsuitable or unable to perform its intended function or purpose safely, reliably, and cost-effectively (Witherell 1994 [26]).

These definitions are similar, but all include the "expected" as "needed", "specified" or "intended". Let's have engineering example. Figure 1.18 shows a resistor burned. It can be often considered having failed as the obvious burning mark was shown and the smoke was smelled. However, when measuring the resistance, it found the resistance does not change so much, decreasing only around 5%. From this point, one might not define it as failure, as the electronics might still be able to function. Failure is therefore context-dependent, varying with the perspective from which it is defined.

1.5.3 MTBF and MTTF

Related to the definition of failure, we can introduce another concept MTBF (Mean Time Between Failure). MTBF is a popular concept of reliability in industries. It is single valued and easily understood. For example, a MTBF of an electronic unit is 12,000 hours, the MTBF of pump is 3000 hours. The MTBF of an aircraft is 1200 hours. The MTBF can be illustrated as a stochastic process as shown in Fig. 1.19.

As suggested by its definition, the calculation of MTBF is as [24]

$$MTBF = \frac{\sum \Delta t_i}{n} \tag{1.15}$$

The Δt_i is the operating time between two consecutive failures. The n is the total number of failures.

Example 1.2
A machine failure record: the first failure occurred 1 day after installation. The 2nd failure occurred after 2 days of the 1st failure, 3rd failure occurred 5 days after 2nd failure. Calculate the MTBF.

Solution: According to Eq. (1.15), the

$$MTBF = \frac{1 + 2 + 5}{3} = 3 \text{ days}$$

The MTBF is 3 days.

It should be noted that the concept of MTBF applies only to repairable items. The "Between" in the definition means repair activities involved in the process. For non-repairable system, a corresponding concept is MTTF (Mean Time To Failure). MTTF defines the average life of non-repairable items from the same population, as illustrated in Fig. 1.20. The item is "run to failure", i.e., no repair is considered. An elaborated definition of MTTF is as:

- The total number of life units of an item divided by the total number of failures within that population, during a particular measurement interval under stated conditions (MIL-STD-721C [13]).

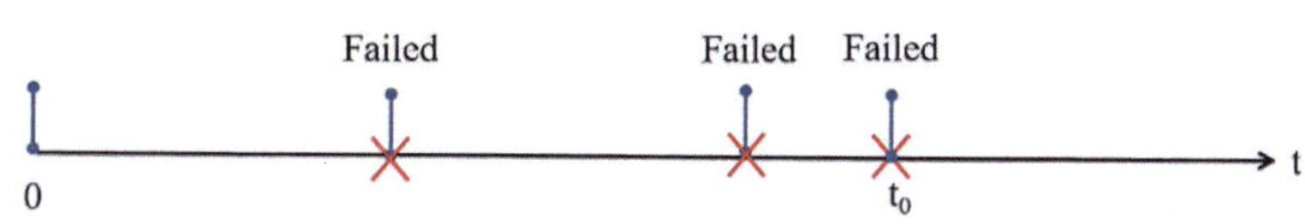

Fig. 1.19 A failure process

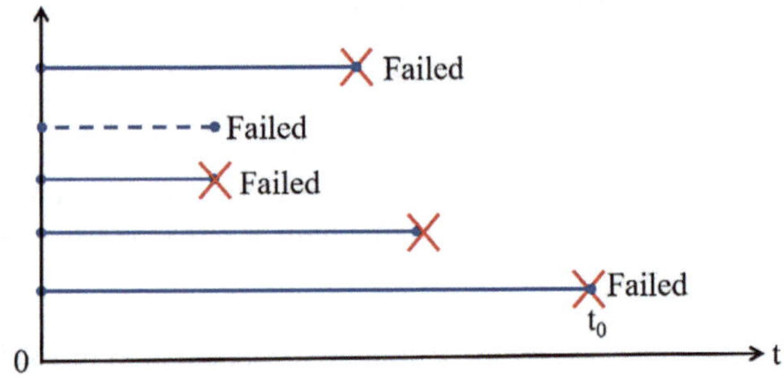

Fig. 1.20 Illustration of unrepairable item failure

Table 1.1 Time to failure data (in seconds)

	Stress		
Life	15 V	15.5 V	16.7 V
1	17.9	21.0	14.6
2	21.8	22.6	14.6
3	19.4	21.8	13.1
4	17.9	14.8	18.5
5	28.7	15.6	11.5
6	21.0	13.2	15.4
7	29.5	30.3	27.0

In this definition, the life unit may be in terms of flight hours, mileage, load, or other measures that quantify the component's operational life. For the "run to failure" test, mathematically, the MTTF can be calculated as

$$MTTF = \frac{\sum t_i}{n} \tag{1.16}$$

The n is the size of the population. The $\sum t_i$ is the total life for all population. The t_i is the life (operating time until to failure) for the ith sample. The "run to failure" is common in life test experiment. The PC example illustrated in Sect. 1.5.1 is one of such cases.

Example 1.3
Total 7 resistors are running the "run to failure" experiment. Life is tabulated in Table 1.1. What is the MTTF at stress 15V?

Solution: According to Eq. (1.16), the total life is

$$17.9 + 21.8 + 19.4 + 17.9 + 28.7 + 21.0 + 29.5 = 156.2s$$

$$MTTF = \frac{156.2}{7} = 22.3143 \text{ s}$$

The MTTF at 15 V is 22.3143 seconds.

Table 1.2 Specification of a HDD [27]

Reliability	Specification
Error rate	1 per 10^{15}
At 40 °C Nr of load-unload cycles	6×10^5
MTBF (in million hours)	2.0
Warranty (in years)	5

Data source: HGST Data Sheet Ultrastar 5K3000, 2012 [27]

MTTF and MTBF are different as described in the definition. It concerns if the item is repairable. In practice, for an item, it is not distinct between repairable or non-repairable. It depends on the requirements. For example, the Hard Driver Disk (HDD) is repairable physically. However, in industries, HDD is rarely repaired due to the high repair cost. It becomes non-repairable economically. It is also worth noting that some industries use MTBF as if they were MTTF, even for items that cannot be repaired. For example, in the specification of some electronic units, the MTBF is sometimes used as "mean life", even most electronic units are non-repairable. Table 1.2 shows the specification of an HDD. It uses MTBF, while the correct term could be "MTTF" since practically HDD will, most probably, not be repaired.

1.5.4 Failure Rate

Failure rate is another term frequently used in industry to describe reliability. Outside the field of reliability, in popular science discussions, the failure rate may refer to the "probability of failure", implying the percentage of failed product from a population. In reliability engineering, failure rate is still a concept related to probability, but not the "probability of failure". Reliability engineering defines the failure rate as the "rate of deterioration" referring to ROCOF, meaning Rate Of OCcurrence Of Failures [28]. In this book, by convention, the failure rate is implicitly defined as ROCOF which is defined as:

- The total number of failures within an item population, divided by the total number of life units expended by that population, during a particular measurement interval under stated conditions (MIL-STD-721C [13]).
- The number of failures per unit time (for equal time intervals, ASQ [24])

The rate of deterioration is defined as $\frac{dQ}{Q}$ where Q is quantity for general [5]. For a special case, given the rate as constant velocity as v, then

$$\frac{dQ}{Q} = -v \tag{1.17}$$

Integrate both sides of Eq. (1.17), then

$$\ln Q = -vt + C \tag{1.18}$$

where C is constant, the (1.17) is equivalent to

$$Q(t) = Ae^{-vt} \tag{1.19}$$

If the Q is the percentage of survived product from a population, then $Q(t) = R(t)$ with $R\ (t = 0) = 1$. The $A = 1$ and the Eq. (1.19) is rewritten as

$$R(t) = e^{-vt} \tag{1.20}$$

which is the reliability function of Exponential distribution. The v is the failure rate of the distribution. From this derivation, the failure rate finds the physical meaning as rate of deterioration. Failure rate v is usually denoted by lambda λ, and $\lambda(t)$ when it is time dependent. When the failure rate is constant, the failure rate is the reciprocal of the MTBF or MTTF.

$$\lambda = \frac{1}{MTBF} \tag{1.21}$$

or

$$\lambda = \frac{1}{MTTF} \tag{1.22}$$

While the conversion between failure rate and MTBF or MTTF is simple, it applies only when the failure rate is constant. We will revisit this topic in Chap. 4 for more detailed discussion.

Example 1.4
We randomly selected 5 non-repairable items to run to failure, the failure times in days are.

$$100, 220, 1000, 1200, 2000$$

What is the failure rate?

Solution: The MTTF is

$$MTTF = \frac{100 + 220 + 1000 + 1200 + 2000}{5} = 904 \text{ days}$$

According to (1.22),

$$\lambda = \frac{1}{904} = 0.0011/\text{days}$$

The failure rate is 0.0011 per day.

Table 1.3 Specification of a Type of Resistor [29]

Reliability	Model 1	Model 2	Model 3
Operating voltage	50V	75V	150 V
Failure rate	$\leq 2 \times 10^{-9}/h$	$\leq 2 \times 10^{-9}/h$	$\leq 2 \times 10^{-9}/h$

Data Source: BC Components Product Data Sheet BC08 [29]

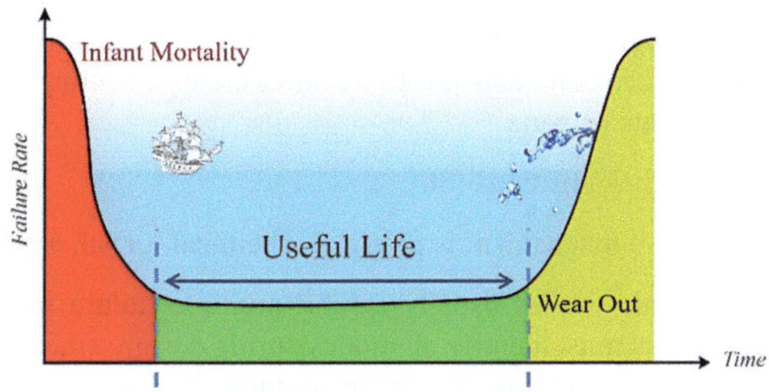

Fig. 1.21 Conceptual failure rate of a product in its life cycle

Example 1.4 illustrates the definition of failure rate. In electronic industries, the failure rate is usually used as the reliability index in the specification. Table 1.3 shows an example of resistor.

Failure rate is a simple, single-valued reliability metric when failure rate is constant. One can compare the reliability level just by comparing their failure rate values. It is also necessary to note that MTBF/MTTF or the failure rate is not the life that an item will be expected. It is just the mean life of the population. Most products will fail before the MTBF/MTTF. It is important to keep this in mind when selecting components. This topic will be discussed further in Chap. 4.

Failure rate refers to rate of deterioration. For a life cycle of product, this rate could differ at its life stage. At the end of the life cycle, the product is prone to failure due to aging such as the airplane, while in the beginning, the rate could be high due to weak connection, damage from the transposition or installation error. The time between them is then normal, no aging, no installation error. This hypothesis is consistent with our intuition. Analogous to a huma being, an infant gets ill easily and the elder also falls ill easily, while the age between them is less prone to being ill. Thus, the failure rate is lower in the midst but higher in both ends, like a bath. In reliability, it is dubbed as a bathtub curve, as shown in Fig. 1.21. The earlier stage is named infant mortality stage; the middle stage is useful life stage and the last is wear out stage. A product should pass the infant stage before entering into service, and the wear out stage should have more maintenance.

Bathtub curve gives a tip for new system design. Products should have a longer period in their useful life stage, and items still in the infant stage should not reach the customer. The bathtub concept started from the beginning of reliability for then electronics in 1950s [5], and have widely distributed in the reliability society. However, this is somewhat controversial, as it is primarily a general empirical conclusion. Few cases in the literature have verified systems strictly exhibiting a bathtub-shaped failure rate.

1.5.5 Failure Mode and Failure Mechanism

Failure Mode and failure mechanism are often mentioned in reliability society. For example, when we perform FMECA (Failure mode, effects, and criticality analysis) or FMEA (Failure mode and effects analysis) analysis, we should list the failure Mode and failure mechanism. Failure mode is defined as [30]:

- Symptom by which a failure is observed; e.g., Open, short, drift, functional faults for electronic; and brittle fracture, creep, buckling, fatigue for mechanical components or parts.

Failure mekanisme is [30]:

- Failure mechanism is physical, chemical, or other process resulting in a failure.

Technical systems usually exhibit multiple failure modes, including hidden ones. Engineers must identify all modes during design, since unidentified modes can cause serious accidents, as seen in the automotive industry. Table 1.4 shows an example of failure modes for a cellphone.

Failure mechanisms examine failures from a lower-level, more fundamental perspective. It requires good understanding of the failure of physics or chemistry. The failure mechanisms leading to one failure mode could be multiple. "Broken phone shell" failure mode could be from the failure mechanism of extremely high mechanic force applied on the shell. The applied force exceeds the tensile strength of the shell material; or it could be due to the thermal expansion, when the thermal coefficients of the material inside the phone vary significantly. Under a rising temperature, the expansion of the material mismatches and stress induced, and that stress exceeds the tensile strength of the shell material, leading to a broken shell consequence.

1.5.6 Data Censoring

Not all the "run to failure" experiments can have an observed failure time. In many cases, waiting for a failure is impractical, so the experiment is terminated before

Table 1.4 Failure modes of cellphone

Component	Failure mode
Keyboard	Not respond to actions
Battery	Does not function
	Not able to be charged
	Low durability
	Explosion
Phone Shell	Broken
	Deformation
	Obvious external scratches
…	…

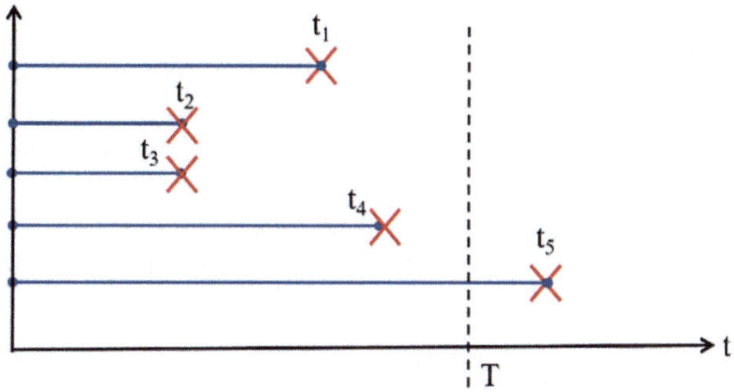

Fig. 1.22 Right censored data

failure occurs. The failure time obtained is therefore censored. In general, concerning censoring, four categories of failure data can be:

- Complete life data. All failure times are obtained for the samples on test. The above described in Fig. 1.20 shows this situation.
- Right censored data. Some failures have not occurred during the observed period. It could occur after the observed time or the failure will never occur. Right censored data is common in industries. For the PC example, before the PC was released to the market, developers take certain number of PCs to run to failure. The test lasts 180 days. Most PCs will not be observed failure during the 180 days. The life data collected is thus censored. As illustrated in Fig. 1.22, where at time T, the observation is terminated, the t_5 has not been observed at the time T.

The MTTF calculation Eq. (1.16) for right censored situation can be modified to

$$MTTF = \frac{Total\ Life}{r} \tag{1.19}$$

Total life is the sum of the lives who run to failure or the censored time if it has not failed. For case shown in Fig. 1.22, Total life is

$$Total\ Life = t_1 + t_2 + t_3 + t_4 + T$$

The r is the number of faiures observed. In Fig. 1.22, r is 4. We will discuss it further in Chap. 4 where we will have the statistical derivation of Eq. (1.19).

Example 1.5

Same as in Example 1.4, we randomly selected 5 non-repairable items to run to failure. The observation is terminated at days 1500. The failure times in days are.

(continued)

Example 1.5 (continued)

$$100, 220, 1000, 1200$$

The last item has not been observed failure. What is the failure rate?

Solution: The MTTF is

$$MTTF = \frac{100 + 220 + 1000 + 1200 + 1500}{4} = 1005 \text{ days}$$

According to Eq. (1.19),

$$\lambda = \frac{1}{1005} = 0.000995/\text{days}$$

The failure rate is 0.000995 per day.

The right censored data loses exact failure time for item failed after the observed time. It thus loses information, and it is less accurate than the complete life data. However, in some situations, we have to stop the observation as the failure occurrence is rare, and continuing the experiment is too expensive.

- Left censored data. As shown in Fig. 1.23, one starts to observe at time T_0. The failure times after T_0 are observed. The situation before T_0 are unknown. Some might have already failed before the observation. For example, when doing maintenance, the engineer checks the machines at certain time intervals. When they check the machine, some machines have already run for some period, and some could have already failed but unknown. Some operating and failure times before T_0 are lost. The failure data collected is left censored.
- Interval censored data. Only certain time intervals are watched. As shown in Fig. 1.24, only the failures between time T_0 and T_1 are observed. The other failure times are unknown. It could happen in the periodic inspection. For example, engineer checks the system every 10 hours, and each inspection takes 1 hour.

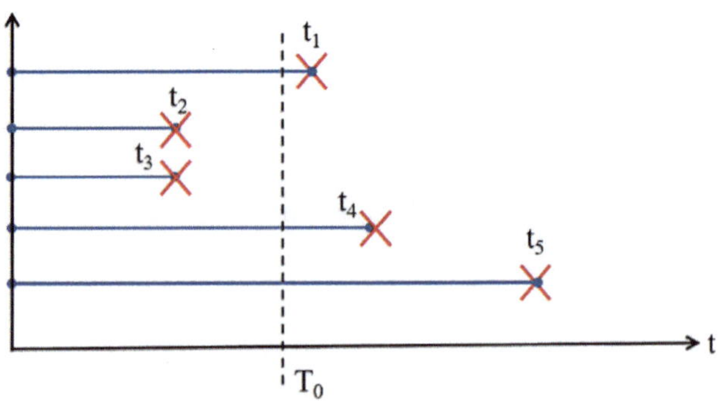

Fig. 1.23 Left censored data

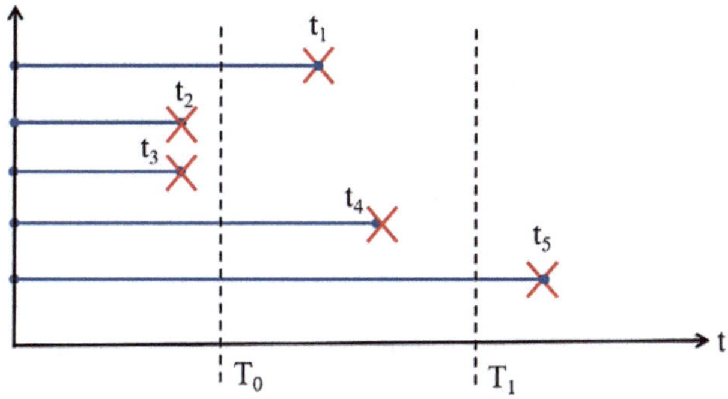

Fig. 1.24 Interval censored data

Only the failure occurred within the 1 hour, the exact failure time can be recorded. The other failure times are censored. The data obtained are thus interval censored.

1.6 Reliability Standards

The reliability standards cover data collection and analysis, sample method, reliability evaluation, system reliability modeling, reliability prediction, reliability test, issued by organization such as ISO (International Standard Organization), IEC (International Electrotechnical Commission), US Military and some specific industries such as nuclear and oil & gas. The following list some of those standards related to reliability.

For data collection and analysis, e.g.

- *ISO 6527 (1982) Nuclear power plants - Reliability data exchange - General guidelines*. This guideline provides templates for data exchange related to reliability. It also provides failure rate calculation formulas.
- *ISO 7385 (1983) Nuclear power plants - Guidelines to ensure quality of collected data on reliability*. This standard establishes guidelines to ensure the accuracy and quality of availability and reliability data collected in the context of nuclear power plant operations.
- *IEC 60319 (1999) Presentation and specification of reliability data for electronic components*. This standard provides guidelines for data collection and presentation for reliability of electronic components.
- *IEC 60300–3-2 (2004) Dependability management - Part 3: Application guide - Section 2: Collection of dependability data from the field*. This standard provides guidelines for data collection for reliability, maintainability, availability and maintenance analysis.
- *ISO 14224 (2006) Petroleum and natural gas industries - Collection and exchange of reliability and maintenance data for equipment*. It is standard for oil & gas industry. It provides a standard format for equipment in petroleum, natural gas and petrochemical industries related to reliability and maintenance.

For product design regarding reliability, e.g.

- *IEC 61164 (1995) Reliability growth - Statistical tests and estimation methods.* The standard provides models and methods for reliability growth assessment based on failure data. It also includes formulas for point estimation, confidence intervals, and methods for goodness-of-fit test.
- *IEC 61014 (2003) Programmes for reliability growth.* Provides guidelines to improve product reliability.
- *IEC 62429 (2007) Reliability growth - Stress testing for early failures in unique complex systems.* It provides guidance on reliability growth during the final test or acceptance test of new complex systems.

The US military handbooks and standards play an important role in reliability engineering. Some industrial practitioners follow these military standards in their reliability work. Some salient standards are:

- MIL-HDBK-108 (1960) Sampling procedures and tables for life and reliability test (Based on Exponential Distribution).
- MIL-HDBK-721c (1981) Definition of terms for reliability and maintainability. The definitions of the terms are widely used in reliability engineering. Our book follows some of these definitions.
- MIL-HDBK-756b (1981) Reliability modeling and prediction. A standard describes the method to calculate system reliability.
- MIL-STD-105D (1989) Sampling procedure and tables for inspection by attributes.
- MIL-HDBK-217F (1991) Reliability prediction of electronic equipment. An influential standard for electronic industries until now. The reliability prediction method described by this standard is still widely used in the electronic industry.
- MIL-HDBK-781A (1996) Handbook for reliability test methods, plans, and environments for engineering, development qualification, and production. A comprehensive handbook for reliability test. An excellent introduction on reliability experiments for beginners.
- MIL-HDBK-338b (1998) Electronic reliability design handbook. A comprehensive handbook with 1046 pages for electronic system design. It covers almost every aspect in reliability: hardware, software, life data analysis, reliability experiments and experimental data analysis.
- MIL-HDBK-690d (2005) Failure rate sampling plans and procedures. The standards address sampling methods and procedures with detailed explanations presented in clear and accessible language. Illustrative examples are included to demonstrate the numerical calculations.

There are many other handbooks, standards related to reliability for specific industries. The issue of the reliability standards in many industries reflects the importance of reliability.

1.7 Reliability Analysis Using Matlab

The use of software for computation is indispensable in reliability analysis. Many reliability models and methods depend on software, as manual computation is impractical or impossible. Matlab is general computing software, easy to learn and easy to code with, for both engineering and scientific applications. It provides plenty of built-in functions and toolboxes or apps to use. Upon statistical functions provided by Matlab, as most reliability analysis methods are essentially statistical analysis, we can write our own script code for our purpose. The following are a summary of functions, toolbox we can use for reliability analysis.

- Graphical distribution fitting tool. Without requiring any programming coding, this tool provides a software type of toolbox for distribution analysis. It supports most statistical distribution used in reliability such as Exponential, Weibull, Normal, Lognormal. Provided with your data to the Matlab, the toolbox can perform parameter estimation, graphic plot, and goodness of fit test.
- Statistical functions. For reliability analysis, coding using Matlab is necessary for calculation. Matlab provides a wide range of built-in statistical functions, which include the standard function such as mean, variance, and functions for statistical distribution. Relevant to reliability, the Matlab provides functions: random number generating, distribution parameter estimation, CDF computation etc. Table 1.5 lists the functions related to Weibull distribution. Similar functions are provided for other statistical distribution. The use of these built-in functions can save significant time for coding.

Problems and Questions *Please select the best answer for question 1–3:*

1. A reliability statement contains the following four parts:
   (a) Probability, parameters, environment, condition
   (b) Function, failure rate, useful time, likelihood
   (c) Condition, failure rate, MTBF, failure
   (d) Probability, time, function, condition
2. Which of following statements is not in reliability definition

Table 1.5 Functions for Weibull distribution

wblcdf()	Weibull cumulative distribution function. We can use it to calculate the probability of failure for time of interest.
wblfit()	Parameter estimation function. To calculate parameter values and confidence interval of the estimated parameters.
wblinv()	The inverse function of Weibull CDF. Given probability of failure, we can calculate the time corresponding to a given probability.
wbllike()	Calculate the Weibull negative log-likelihood for given data and parameters. The function is seldom used in basic reliability analysis. But it is useful when reliability analysis requires advanced statistic method.

(a) Probability
(b) Stated Condition
(c) Useful life
(d) Specified period of time
(e) Required Function

3. MTBF would be used instead of MTTF if

(a) The product is repairable
(b) The product is discarded
(c) The product is non-repairable
(d) The failure occurs only once

4. A sample of five units from a population is tested to failure.

Unit number 1 failed at 162 hours
Unit number 2 failed at 157 hours
Unit number 3 failed at 146 hours
Unit number 4 failed at 173 hours
Unit number 5 failed at 155 hours

What is MTBF or MTTF of this product?

5. A system has failure times (calendar time in days) as

$$177\ 242\ 293\ 336\ 368\ 395\ 410$$

The repair times are assumed negligible. This means that the system is assumed to be functioning again almost immediately after a failure is encountered. What is MTBF or MTTF of this system?

6. A PC company runs a life test experiment for their new PCs. The experiment runs 180 days. Total 24 PCs run the life test. At the end of the test, only 3 failed with failure times (calendar time in days) as

$$3, 150, 170$$

What is MTBF or MTTF of the PCs?

7. Fourteen of a batch of 500 computer chips fail the final screening test. Estimate the failure probability and why it is? What is the error of the estimate of the failure probability?

Mathematic Review Questions

8. $\int_0^1 (x^4 - 3\sqrt{x})dx = ?$
9. $\int_1^2 (\sqrt{x} + \frac{2}{x})dx = ?$
10. Find the first derivative of $x^3 + 5x^2 - 9x + 2 = ?$
11. Find the first derivative of $\frac{x^3}{x^3+5} = ?$
12. A function is of form

$$f(x) = x^2(1-x)^{n-2}$$

Please find the x_0 that maximizes $f(x)$?

13. Occurrence of Event A is 0.95, Event B is 0.90. if the A and B is independent, what is the $A \cap B = ?$ (Both A and B will occur).

References

1. Carhart RR (1953) A survey of the current status of the electronic reliability problem. The Rand Corporation
2. Defence Technical Information Center (1960) Proceedings 6th joint military-industry guided missile reliability. Defence Documentation Center
3. Spiegel J, Bennett EM (1960) military system reliability: Department of Defense Contributions. IRE Transactions on Reliability and Quality Control RQC-9(3):1–8
4. Barlow RE, Proschan F, Hunter LC (1996) Mathematical theory of reliability. SIAM, Philadelphia
5. AGREE. Equipment (1957) Reliability of military electronic equipment. Assistant Secretary of Defence
6. Knight CR (1991) Four decades of reliability progress. In: Annual reliability and maintainability symposium. 1991 proceedings, pp 156–160
7. Erlang AK (1909) The theory of probabilities and telephone conversations. Nyt Tidsskrift for Matematik B 20:33–39
8. Weibull W (1951) A statistical distribution function of wide applicability. J Appl Mech 18(3): 293–297
9. W. Weibull, A statistical theory of the strength of materials. Generalstabens litografiska anstalts förlag, 1939
10. Ericson CA II (1999) Fault tree analysis-a history. In: Proceedings of the 17th international system safety conference, Orlando
11. Abernethy RB (2004) The new Weibull handbook. Dr. Robert Abernethy
12. Von Alven WH (1964) Reliability engineering. Prentice-Hall
13. Department of Defense (1981) MIL-STD-721C definitions of terms for reliability and maintainability
14. Drew C, Mouawad J. Safety board gives new details on boeing battery fire, but not cause. NYtimes
15. Kolly JM, Panagiotou J, Czech B (2013) The investigation of a lithium-ion battery fire onboard a Boeing 787 by the US National Transportation safety board. Safety Research Corporation of America, Dothan, AL, pp 1–18
16. ISO (1991) ISO 14224: petroleum, petrochemical and natural gas industries - collection and exchange of reliability and maintenance data for equipment. Department of Defense, USA
17. Low H (2015) How to improve system availability and minimize down time with Hercules MCUs. Texas Instruments
18. Kinnison HA, Siddiqui T (2013) Aviation maintenance management, 2nd edn. Ma Graw Hill
19. International Civil Aviation Organization (2023) Safety report. International Civil Aviation Organization, Montréal
20. Zio E, Fan M, Zeng Z, Kang R (2019) Application of reliability technologies in civil aviation: lessons learnt and perspectives. Chin J Aeronaut 32(1):143–158
21. U. Department of Defense (1998) MIL-HDBK-338B electronic reliability design handbook
22. Rowe EG (1952) Technique of trustworthy valves. Proc IRE 40(10):1166–1177
23. Knight CR, Jervis ER, Herd GR (1955) The definition of terms of interest in the study of reliability. IRE Trans Reliab Qual Control PGRQC-5:34–56
24. Benbow DW, Broome HW (2008) The certified reliability engineer handbook. ASQ Quality Press, Milwaukee, WI

25. Kendall MG, Stuart A, Ord JK, Arnold SF, O'Hagan A (1994) Kendall's advanced theory of statistics, 6th edn. Edward Arnold and Halsted Press, London
26. Witherell CE (1994) Mechanical failure avoidance : strategies and techniques. McGraw-Hill, New York
27. HGST Wester Digital Company(2012) 3.5-inch enterprise CoolSpin hard disk drives, USA
28. Yeh L (1997) The rate of occurrence of failures. J Appl Probab 34(1):234–247
29. BC Components (2002) Data sheet professional flat chip resistors
30. Alessandro B (2004) Reliability engineering theory and practise. Springer, Berlin Heidelberg

Chapter 2
Reliability Mathematics

Probability and statistics are the primary mathematical approaches applied in the field of reliability. This chapter begins by introducing the fundamental concepts of probability. Then, starting from empirical data distributions, it presents several important theoretical statistical distributions. Finally, key statistical properties and categories of distribution parameters relevant to reliability are discussed.

2.1 The First Probability

Physically, the failure of a machine is deterministic resulting from deterministic factors, and the event "Fail" and "Not Fail" will not occur simultaneously. From these aspects, the introduction of probability to the occurrence of failure seems counter intuitive. People with little or no background in probability sometimes express confusion about its meaning. For example, if we say that a machine has an 80% chance of failing and a 20% chance of not failing, they may find it puzzling and ask whether the machine will fail or not. This is a fundamental question of probability. How to interpret the seemingly paradoxical probability in a physical world? Let us take the famous coin-tossing as an example. Coin has two sides. When we toss it, the outcome will be either this side or another side, as shown in Fig. 2.1.

For each tossing, the outcome is unknown. It is a consequence of mechanical interaction with our hand, ambient conditions such as wind direction, speed, even the strength of electromagnetic, and some other factors. These factors are hard to measure, or to control, and thus the outcome is unpredictable. Then we leave the question of trying to predict the outcome for each toss and assess the frequency of the occurrence of each side only, since it is doable by experiments. By repeating to toss a coin and calculate the frequency for one side, the plot of the frequency against the number of experiments is shown in Fig. 2.2, the frequency of one side tends to be stabilized to a certain value which is deterministic.

F. Yuan, *Essentials of Reliability Engineering*, Springer Series in Reliability Engineering, https://doi.org/10.1007/978-3-032-15544-3_2

Fig. 2.1 Two sides of a coin

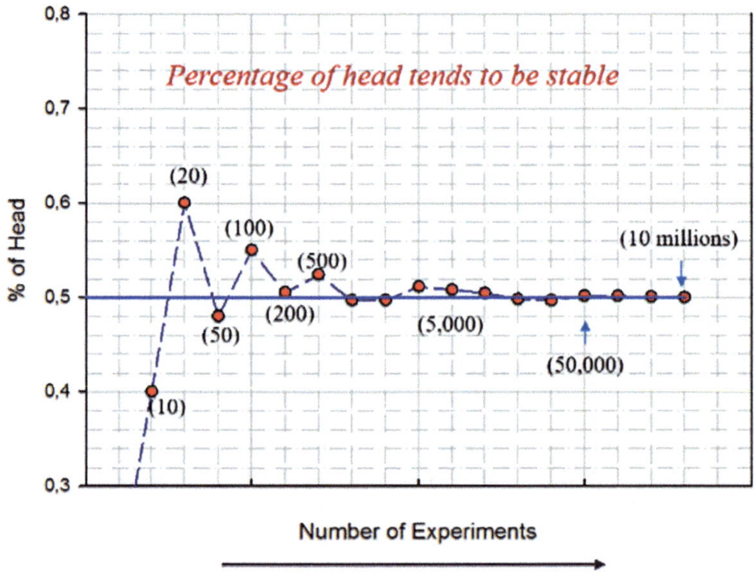

Fig. 2.2 Tendency of frequency

This stabilized value is then the probability. For a failure event, if the possibility of failure is fixed, the failure frequency will exhibit a similar pattern, chaotic in the beginning and stabilizing later, as in Fig. 2.2. Probability can then be defined as the frequency of an event when the number of experiment n is infinite:

$$n \rightarrow \infty$$

It will be assumed throughout this book, unless otherwise specified, to denote the probability as P. The probability 0 implies failure will certainly not occur, while the 1 implies it will certainly occur. The occurrence of coin showing head or tail side is named event in probability. The probability must satisfy:

(a) $0 \leq P(A_i) \leq 1$ for all event A_i
(b) $\sum P(A_i) = 1$. The sum of the probability for all possible events is 1.

In the case of failure, there are only two possible events: fail or not fail. The sum of the probability of the two events is unity. If the probability of failure is $P(A)$, the probability of no failure is then $1 - P(A)$. In reliability engineering, failure is essentially deterministic. However, as the available knowledge to determine the failure is not sufficient, we then consider the unknown as uncertain and therefore probability is used.

2.2 Combination of Events

The occurrence of failure is an event in probability. Each event is associated with a probability, if it exists. What of interest, for example, if the probability of unit A fails is $P(A)$, unit B fails is $P(B)$, what is the probability of both will occur? Or if any constituent unit failure will lead to the whole system failure, what is the probability of the whole system failure?

2.2.1 Intersection of Events

The probability of both A and B occurring is an intersection of events. In probability language, it is written as $P(A \cap B)$. Conventionally, it can illustrate the event and probability as shown in shown in Fig. 2.3 with diagram. The use of sets for illustration of probability can refer to the measure theory.

When event A and B are mutually independent, the probability

$$P(A \cap B) = P(A) * P(B) \tag{2.1}$$

Equation (2.1) is held only when event A and B are independent. Example 2.1 illustrates Eq. (2.1).

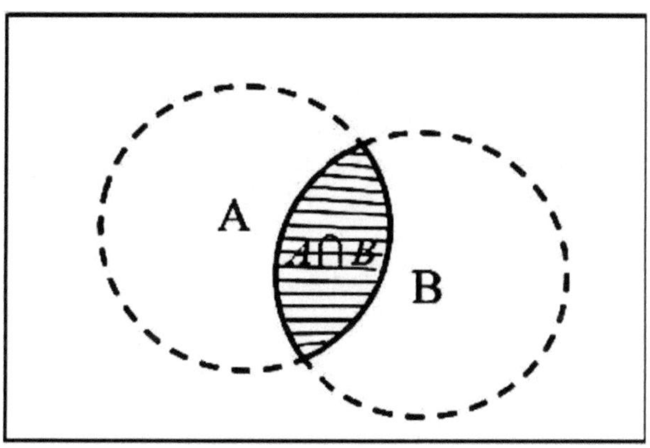

Fig. 2.3 Intersection of events

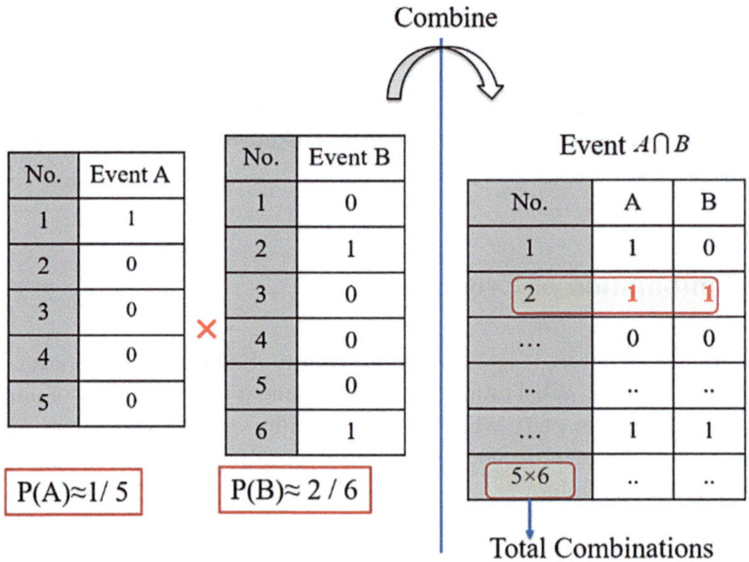

Fig. 2.4 Combination of events A and B

Example 2.1
Event A occurs in 1 out of 5 trials. Event B occurs in 2 out of 6 trials, as shown in Fig. 2.4. When they are mutually independent, the occurrence of A will not affect B. What is $P(A \cap B)$?

Solution: When the A and B is mutually independent, the $A \cap B$ is the event of A = 1 and B = 1. The total combination of A and B is 5*6 = 30. The combination of A = 1 and B = 1 is 2 out the 30. The probability of $A \cap B$ is thus

$$P(A \cap B) = \frac{2}{30} = \frac{1}{5} * \frac{2}{6} = P(A) * P(B)$$

Obviously, the $P(A \cap B) = P(A) * P(B)$.

However, when we say if event A occurs, the event B will also occur, i.e., if A = 1, the B will be 1 as well. This is a dependent case. The combination will be different. The total combination remains 30. But the number of combinations of A = 1 and B = 1 will be 6. The

$$P(A \cap B) = \frac{6}{30} \neq \frac{1}{5} * \frac{2}{6} \neq P(A) * P(B)$$

(continued)

Example 2.1 (continued)
For another dependent scenario, if we say if A occurs, the B will occur; and if B occurs, the A will occur, i.e., either A or B is one, the other one will change to 1, then the total number of combinations for A = 1 and B = 1 is 14. The

$$P(A \cap B) = \frac{14}{30} \neq \frac{1}{5} * \frac{2}{6} \neq P(A) * P(B)$$

Example 2.1 demonstrates the impact of the dependency between the events. The last two scenarios are dependent situations, where we can see Eq. (2.1) does not work.

2.2.2 *Union of Events*

Either A occurring or B occurring leads to system failure. This is a union of events. In probability theory, it is denoted as $P(A \cup B)$. Diagrams regarding sets as shown in Fig. 2.5 are used to illustrate the union event.

The probability of union is

$$P(A \cup B) = P(A) + P(B) - P(A \cap B) \tag{2.2}$$

For mutually independent A and B, the $P(A \cap B) = P(A) * P(B)$. The second term on the right hand is necessary for an overlapping scenario.

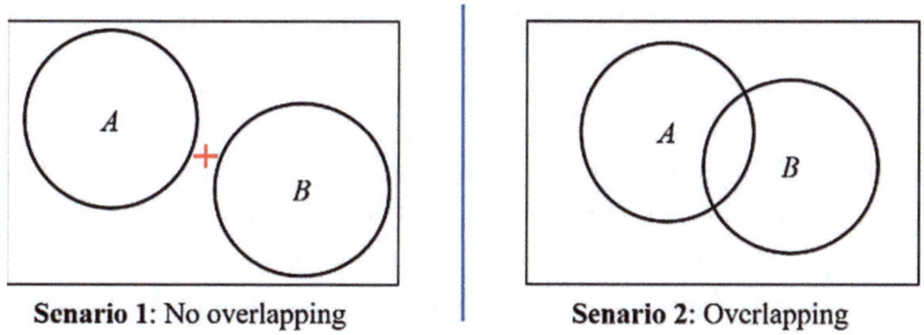

Fig. 2.5 Union of events

Example 2.2
An airplane has two engines. The airplane is considered to have failed if the failure of any single engine occurs. Assume the two engines are mutually independent. The probability of failure for one engine is 0.99. What is the probability of airplane failure?

Solution: It is a union event problem. The

$$P(A \cup B) = P(A) + P(B) - P(A \cap B) = 0.99 + 0.99 - 0.99 * 0.99 = 0.9999$$

The probability of failure is 0.9999.

If without the second term $P(A \cap B)$, a mistake can be made commonly, the $P(A \cup B) = P(A) + P(B) = 1.98 > 1$. Obviously, it is incorrect.

2.2.3 *Conditional Probability*

Occurrence of event A will cause the occurrence of event B with a certain probability. A short-circuit failure on the mainboard may cause the burning of chips onboard. A rainstorm may induce traffic accidents. They are examples of conditional probability. The probability of Event B occurs conditional to condition A is written as

$$P(B/A) = \frac{P(A \cap B)}{P(A)} \tag{2.3}$$

The $P(A \cap B)$ represents the probability of both A and B will occur. Equation (2.3) is abstract and we can take an example to interpret it.

Example 2.3
A box contains red colored balls n_r and blue colored balls n_b. Balls inside are mixed without any regular pattern. We pick up one ball two times without putting back as demonstrated in Fig. 2.6. The event of interest is what is the probability of: red (Event A) was picked for the first time and the blue (Event B) was picked for the second time?

Solution: The probability of picking up a red one for the first time is $P(A) = \frac{n_r}{n_r + n_b}$. For the second time, the probability of picking up blue one is $P(B/A) = \frac{n_b}{n_r + n_b - 1}$, conditioned on the event A of red was picked for the first time. Then the probability of: for the first time red was picked and for the second time the blue was selected is $\frac{n_r}{n_r + n_b} \cdot \frac{n_b}{n_r + n_b - 1}$, for which we can write it as $P(A).\ P(B/A) = P(A \cap B)$.

In this case, the $P(A).P(B) = \frac{n_r}{n_r + n_b} \cdot \frac{n_b}{n_r + n_b}$. It shows the $P(A).\ P(B/A) = P(A \cap B) \neq P(A).\ P(B)$, as. Event A and Event B are dependent.

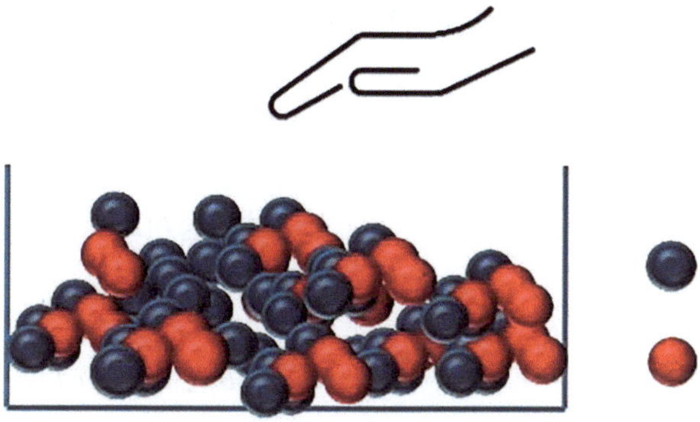

Fig. 2.6 Ball picking

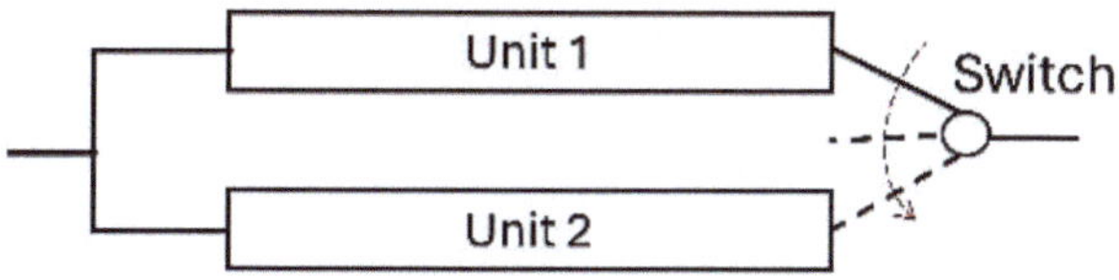

Fig. 2.7 Backup unit

As shown in Example 2.3, it is very common to use the

$$P(A \cap B) = P(A).P(B/A) \tag{2.4}$$

In practice, we should be careful to distinguish the conditional probability *P(B/A)* from the joint probability P(A∩B) since their expressions are similar.

Example 2.4
The primary cooling unit in a nuclear power plant has reliability 0.997. There is also a backup cooling unit to substitute for the primary unit when it fails as shown in Fig. 2.7. The reliability of the backup unit is 0.920. What is the reliability of the whole cooling system. Assume two cooling unit independent from each other [1].

Solution: The failure of primary cooling unit is *P(A)*. In this case, there are two scenarios where the cooling system is reliable. Scenario 1 assumes the primary never failed; Scenario 2 assumes the primary failed but the backup works. The two scenarios are exclusive. The scenario 1 is with probability 0.997. The scenario 2 is conditional probability, it is conditioned the primary failed and it will start to work. Applying Eq. (2.4), the probability is (1–0.997) *0.920 = 0.00275, where *P(B/A)* = 0.920. For the whole cooling system, it is 0.997 + 0.00275 = 0.9998 since the two scenarios are exclusive.

2.2.4 Important Properties

Assume all events mutually independently. It can find from $P(A \cap B) = P(A) * P(B)$, the

$$P(A \cap B) \leq P(A) \ or \ P(B) \tag{2.5}$$

and

$$P(A \cup B) \geq P(A) \ or \ P(B) \tag{2.6}$$

For Eq. (2.5) can be readily derived since $P(A) \leq 1$ or $P(B) \leq 1$. For the proof of Eq. (2.6), according to Eq. (2.2),

$$P(A \cup B) = P(A) + P(B) - P(A)P(B) = P(A)[1 - P(B)] + P(B) \geq P(B)$$
$$(\text{Since } 1 - \text{P(B)} \leq 1, \text{then P(A)}[1 - \text{P(B)}] \geq 0)$$

And similarly,

$$P(A \cup B) = P(A) + P(B) - P(A)P(B) = P(B)[1 - P(A)] + P(A) \geq P(A)$$

Equation (2.6) is thus proved.

2.3 Failure Rate Function

The airplane tends to fail more frequently when its life approaches the designed life. An old laptop is expected to malfunction more than a brand new one. Failure rate can capture this nature to measure the "tendency" of the failure. In Chap. 1, the definition of failure rate was discussed, this chapter illustrates it further from a statistical perspective.

2.3.1 Imperial Failure Rate Function

We illustrate the failure rate with a empirical example. There is total 10 products on the test. They failed at year

0.1 0.3 1.2 1.3 1.5 2.1 2.3 2.4 2.7 3.2

By grouping them by year, the number of failures from year 1 to year 3 is 2, 3, 4 respectively. Figure 2.8 illustrates the failure process for the 3 years.

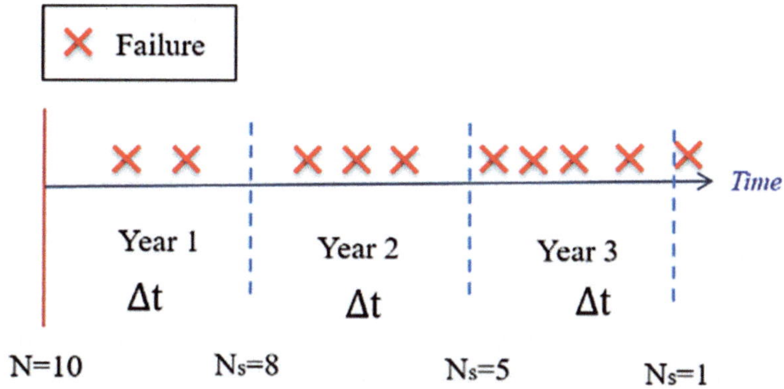

Fig. 2.8 Failure process

Fig. 2.9 Probability of failure against time

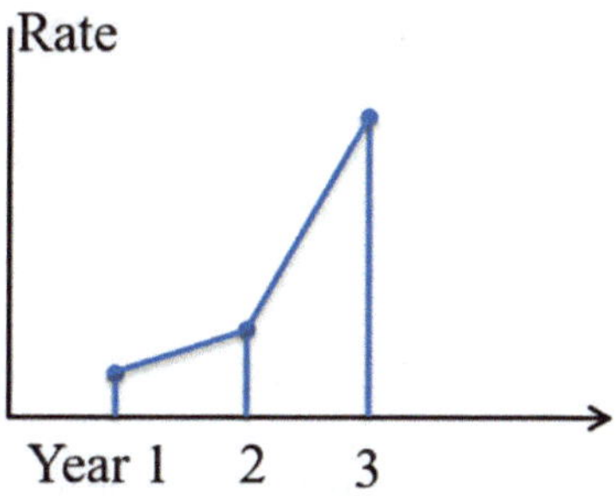

In Fig. 2.8, the N_s denotes the number of survived products and let N_f denote the number of failed in this time interval, then

Year 1: $N_s = 10$; $N_f = 2$; $\frac{N_f}{N_s} = 2/10 = 0.2$
Year 2: $N_s = 8$; $N_f = 3$; $\frac{N_f}{N_s} = 3/8 = 0.375$
Year 3: $N_s = 5$; $N_f = 4$; $\frac{N_f}{N_s} = 4/5 = 0.80$

Initially, the probability of failure $\frac{N_f}{N_s} = 0.2$. When time runs to the 2nd year, 8 remains, and the probability of failure is 0.375. For the 3rd year, it is 0.80. The probability of failure tends to increase, as illustrated in Fig. 2.9.

We then divide the probability of failure by the time interval Δt

$$r = \frac{N_f}{N_s} / \Delta t \tag{2.7}$$

One can imagine if we have more products involved in the test, we can group the lives in month, days or hours etc. Then we can expect to obtain a much smoother curve than Fig. 2.9. Equation (2.7) can then model by a mathematical function of time. The r is thus written as $r(t)$. This is the failure rate function in reliability engineering, which is defined as a real-valued, continuous function. This function

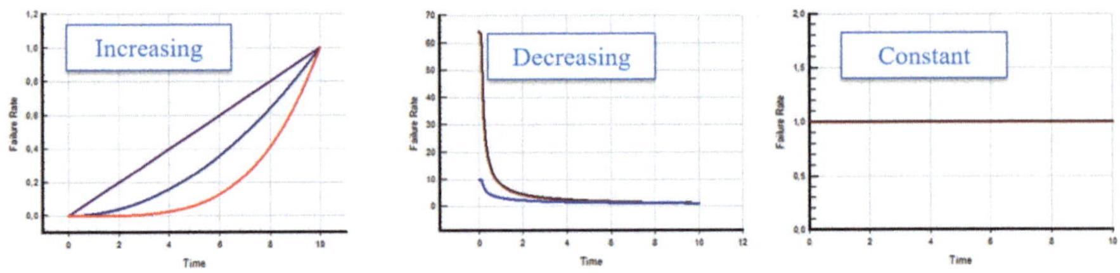

Fig. 2.10 Monotone failure rate

can be constant, increasing, decreasing, or even not monotone. Figure 2.10 shows the three monotone failure rates, while Fig. 2.9 is an empirical failure function.

2.4 Probability Density Function

We might be interested in how the data is distributed according to the time. Suppose the time to failure data for air conditioning system are (in hours) (Data adapted from Ref. [2])

23 261 87 7 120 14 62 47 225 71 246 21 42 20 5 12 120 11 3 14 71 11 14 11 16 90 1 16 52 95

Grouping the data at a time interval of 30 hours, each time slot contains certain number of failures as in Table 2.1. The frequency is the percentage of the failure occurred in this time interval.

A figure can be plotted as the frequency against the time, as shown in Fig. 2.11, which is essentially histogram plot. This failure frequency, as we discussed in Chap. 1, is an approximate failure probability.

If we further divide the frequency by the time interval, then a probability density in terms of time is obtained, which is shown in Fig. 2.12.

If one uses a mathematical function $f(t)$ to fit the plot in Fig. 2.12, this mathematical function is then a Probability Density Function (PDF). And imagine when the number of data is infinite and time interval tends to $\Delta t \to 0$. The $f(t)$ can be continuous. This continuous function is the PDF discussed in probability and statistics books. Some common PDF such as Exponential distribution is

$$f(t) = \lambda e^{-\lambda t} \tag{2.8}$$

PDF of Weill distribution

$$f(t) = \frac{\beta}{\alpha}\left(\frac{t}{\alpha}\right)^{\beta-1} e^{-\left(\frac{t}{\alpha}\right)^{\beta}} \tag{2.9}$$

Table 2.1 Division of data

No.	1	2	3	4	5	6	7	8	9	10
Range	1–30	31–60	61–90	91–120	121–150	151–180	181–210	211–240	241–270	271–300
No. of failures	16	3	5	3	0	0	0	1	2	0
Frequency	16/30 = 0.53	0.1	0.17	0.1	0	0	0	0.03	0.067	0
f_i	0.53/30 = 0.018	0.0033	0.0057	0.0033	0	0	0	0.001	0.0022	0

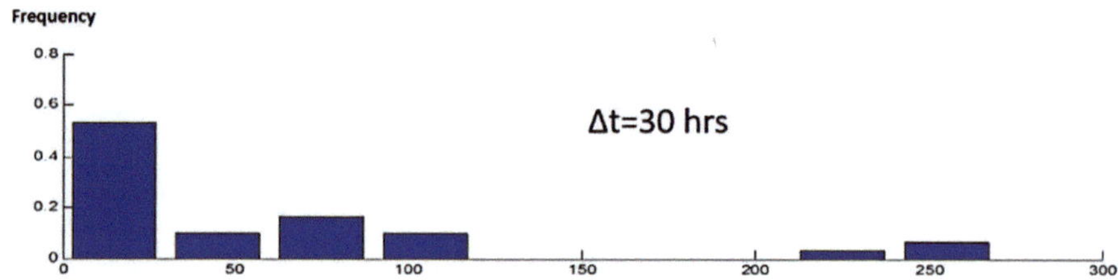

Fig. 2.11 Failure drequency

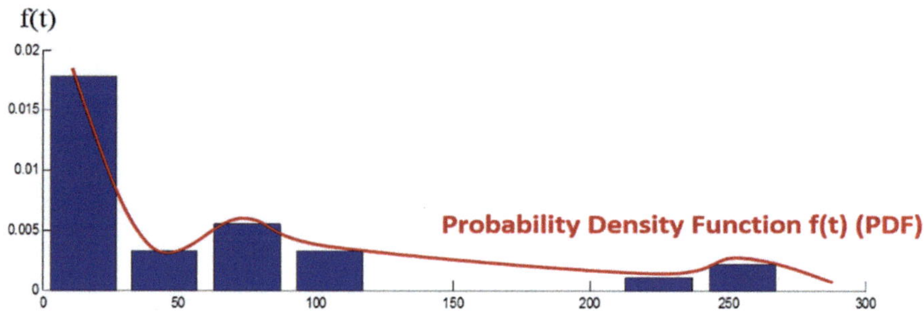

Fig. 2.12 Frequency and probability density against Time

Since $f(t)$ is the probability in the interval is divided by the time interval Δt, then the probability of failure is $\sum_{\Delta t \to 0} f(t_i)\Delta t$, which is equivalent to

$$F(t) = \int_{-\infty}^{t} f(\tau)d\tau \tag{2.10}$$

Equation (2.10) is Cumulative Distribution Function (CDF), used to obtain the probability of failure for some time interval. When the distribution is only positive valued, Eq. (2.10) is rewritten as

$$F(t) = \int_{0}^{t} f(\tau)d\tau \tag{2.11}$$

For reliability engineering, two special time points are $F(t = 0) = 0$, and the $F(t = +\infty) = 1$. It implies at time $t = 0$, the probability of failure is 0, and the machine will surely fail when time continues to an infinite period $+\infty$. The $F(t)$ is monotonically increasing with t. It makes sense, as the probability of failure should always increase, as shown in Fig. 2.13.

Since the $f(t)$ is defined as the probability divided by Δt, when the $\Delta t < 1$, the density function $f(t)$ could be ≥ 1. Figure 2.14 shows the PDF for a Normal distribution with mean 50 and standard deviation 0.1, of which PDF exhibits bigger than

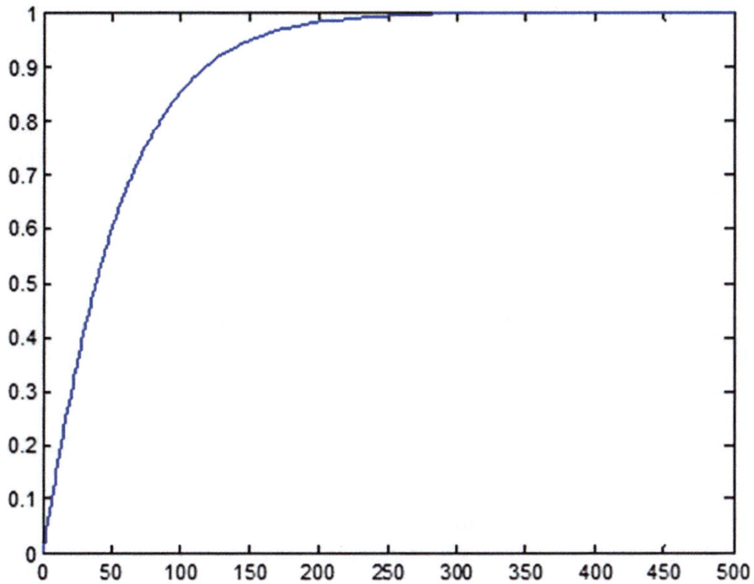

Fig. 2.13 Plot of CDF

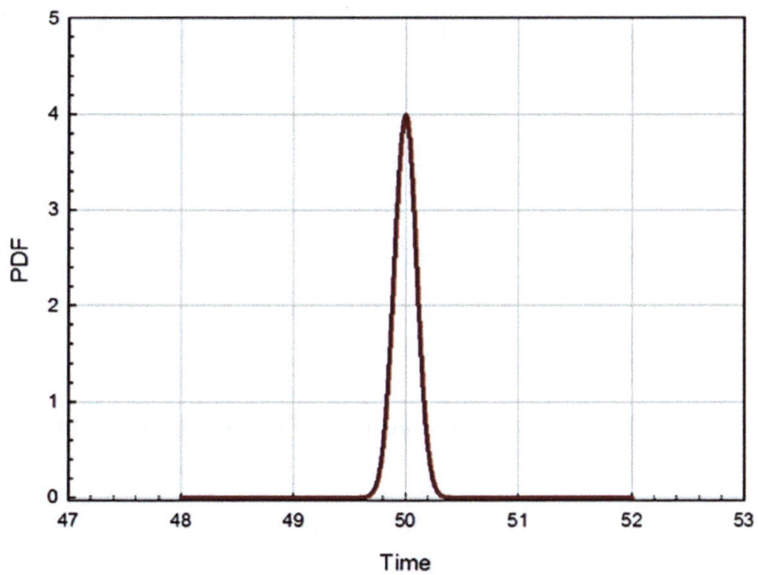

Fig. 2.14 PDF of a Normal distribution

1. The demonstration of Fig. 2.14 is necessary, as many misunderstood the PDF is within range 0 to 1, which is wrong.

Example 2.5
Failure of an electronic system follows a probability density function (PDF) as (t in hour)

$$f(t) = \lambda e^{-\lambda t}; \lambda = 1, t > 0$$

What is the probability of the item will fail within the first 1 hour?
Solution: The probability of $t = [0, 1]$ is

$$F(t=1) = \int_0^1 \lambda e^{-\lambda\tau} d\tau = \left[-e^{-\lambda\tau}\right]_0^1 = 1 - e^{-1} \approx 0.6321$$

The probability of failure within the first 1 hour is 0.6321.

2.5 Reliability, Failure Rate and PDF

Both reliability and failure rate can characterize the reliability level. They can essentially be converted from each other uniquely. Equation (2.7) can be rewritten as

$$r = \frac{\frac{N_f}{N_s}}{\Delta t} = \frac{\left(\frac{N_f}{N}\right)/\Delta t}{\frac{N_s}{N}} \tag{2.12}$$

When $N \to \infty$, the $\left(\frac{N_f}{N}\right)/\Delta t \to f(t)$ and $\frac{N_s}{N} \to R(t)$, Eq. (2.12) is rewritten as

$$r(t) = \frac{f(t)}{R(t)} \tag{2.13}$$

Equation (2.13) tells the relationship among failure rate, reliability and PDF. The $r(t)$ can be rewritten as

$$r(t) = \frac{f(t)}{R(t)} = \frac{F'(t)}{R(t)} = \frac{(1 - R(t))'}{R(t)} = -\frac{R'(t)}{R(t)}$$

Integrate both sides $\int_0^t r(\tau)d\tau = -\int_0^t \frac{R'(\tau)}{R(\tau)} d\tau$. The $\int_0^t \frac{R'(\tau)}{R(\tau)} d\tau = \ln R(t) - \ln R(0)$. The reliability $R(0) = 1$. Then the $\int_0^t \frac{R'(\tau)}{R(\tau)} d\tau = \ln R(t)$. Therefore $\ln R(t) = -\int_0^t r(\tau)d\tau$. We obtain

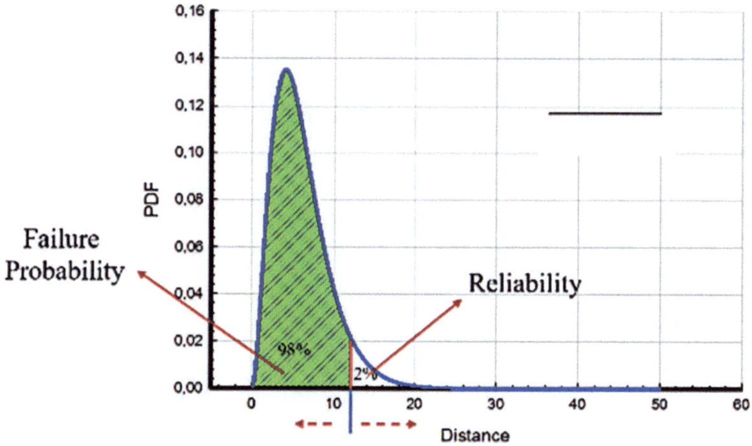

Fig. 2.15 Reliability function from PDF

Table 2.2 Summary of CDF, PDF, reliability and failure rate

$F(t) = \int_{-\infty}^{t} f(\tau)d\tau$
$R(t) = \int_{t}^{+\infty} f(\tau)d\tau$
$r(t) = \frac{f(t)}{R(t)} = \frac{f(t)}{1 - \int_{-\infty}^{t} f(\tau)d\tau}$
$R(t) = e^{-\int_0^t r(\tau)d\tau}$
$F(t) = 1 - R(t)$

$$R(t) = e^{-\int_0^t r(\tau)d\tau} \tag{2.14}$$

One can derive the reliability function from failure rate function using Eq. (2.14), which is a popular method in reliability calculation.

Alternatively, the reliability function can be derived from the PDF. The shadowed region right of time t in Fig. 2.15 is the probability of failure, the right side of figure is obviously the reliability function. The integration region of the reliability is from t to $+\infty$, that means the failure will occur after the time t, as shown in following Eq. (2.15)

$$R(t) = \int_t^{+\infty} f(\tau)d\tau \tag{2.15}$$

We summarize the relationship among CDF, PDF, failure rate function and reliability function in Table 2.2.

Example 2.6
Failure of an electronic system follows a probability density function (PDF) with constant failure rate λ. What is the reliability function, PDF and CDF.

Solution: The reliability at time t is

$$R(t) = e^{-\int_0^t r(\tau)d\tau} = e^{-\int_0^t \lambda d\tau} = e^{-\lambda t}$$

The CDF is $F(t) = 1 - R(t) = 1 - e^{-\lambda t}$. From Eq. (2.13), the PDF is

$$f(t) = r(t) * R(t) = \lambda.e^{-\lambda t}$$

2.6 Typical Statistical Distributions

Section 2.4 describes the probability density function, any real valued continuous function $f(t)$ with $\int_{-\infty}^{+\infty} f(\tau)d\tau = 1$. For life data analysis, the $f(\tau)$ should be positively defined since failure time is positive. This section discusses some typical distributions, either discrete or continuous distribution in brief. Some important distributions for reliability will be further discussed in coming chapters.

2.6.1 Binomial Distribution

The Binomial distribution is discretely distributed along integer 0, 1, 2, 3, ... when the events have only two consequences: failing/functional, or head/tail as in the coin tossing example. Suppose we want to draw a sample size n from a box. In this box, percentage θ is blue and percentage $1 - \theta$ is red, as shown on Fig. 2.16. What is of interest is the probability of number of x blue balls can be drawn from the box, where there are total number of n balls, either red or blue.

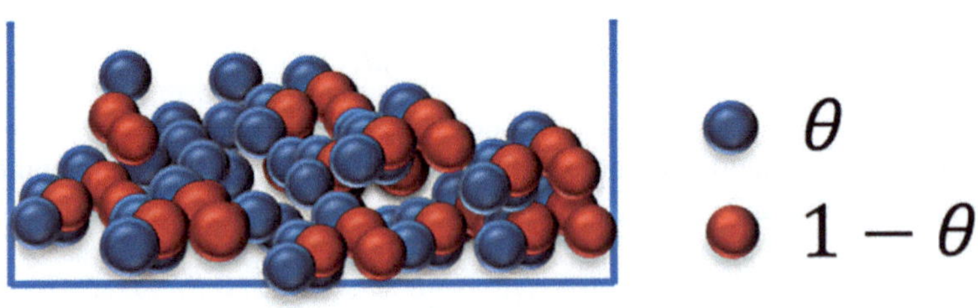

Fig. 2.16 Binomial distribution example

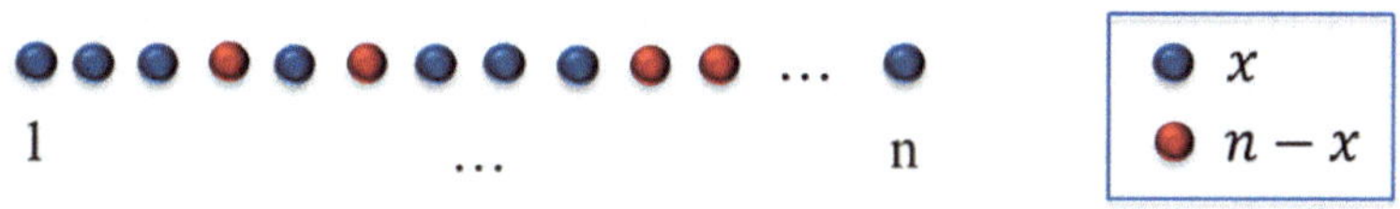

Fig. 2.17 One round of number *n* drawing

For one round of number *n* drawings (ball will be put back after each drawing), as shown in Fig. 2.17, the probability of blue one picked is θ, the red one picked is $1 - \theta$.

Each drawing is independent. The joint probability of *n* drawing is then

$$P = \theta\theta\theta(1-\theta)\theta(1-\theta) \tag{2.16}$$

Equation (2.16) is rewritten as

$$P = \theta^x(1-\theta)^{n-x} \tag{2.17}$$

There are number of $\binom{n}{x}$ different sequences as in Fig. 2.17 of *x* blue and $n - x$ red. Each sequence is exclusive. The probability for number of *x* from *n* is then

$$b(x) = \binom{n}{x}\theta^x(1-\theta)^{n-x} \tag{2.18}$$

The $\binom{n}{x}$ denotes choosing *x* from *n*, which defined as

$$\binom{n}{x} = \frac{n!}{(n-x)!x!}; x = 0, 1, \ldots, n \tag{2.19}$$

Equation (2.18) is the probability density function of the Binomial distribution. Figure 2.18 shows a plot of the Binomial distribution, in which the p is the θ in Eq. (2.18).

The Table 2.3 shows the mean and variance of Binomial distribution.

Example 2.7

Total number of 100 ICs (Integrated Circuit units) are tested in the lab. At the completion of the test it is observed two of the units failed during the test. What is the failure probability of the IC?

Solution: According to Table 2.3, the failure probability is estimated by

$$\widehat{P} = \frac{x}{n} = \frac{2}{100} = 0.02$$

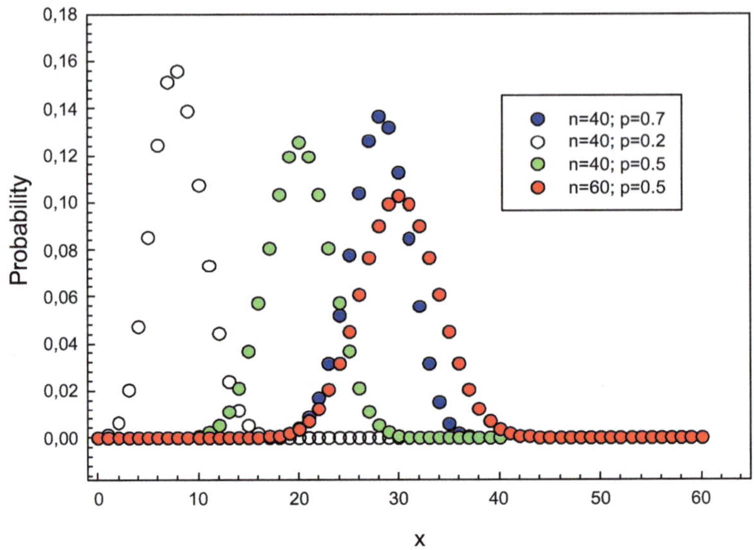

Fig. 2.18 PDF of Binomial distribution

Table 2.3 Properties of Binomial distribution

Mean	$n\theta$
Variance	$n\theta(1-\theta)$
MLE (Maximum Likelihood Estimator)	$\widehat{\theta}=\frac{x}{n}$

2.7 Poisson Distribution

Poisson distribution is a special case of Binomial distribution with n → ∞. It is widely used in applications such as

- the number of phone calls received at an exchange or call center in an hour;
- the number of customers arriving at tax office per day;
- the number of students enter the library per day;
- the number of failures of a complex system per year.

Let $\lambda = n\theta$ and $\theta = \lambda/n$ substitute it into Binomial distribution. Then

$$\lim_{n\to\infty} b(x)=\binom{n}{x}\theta^x(1-\theta)^{n-x}=\frac{n!}{x!(n-x)!}\frac{\lambda^x}{n^x}\frac{(1-\lambda/n)^n}{(1-\mu/n)^x}$$

$$=\frac{n(n-1)\ldots(n-x+1)}{x!}\frac{\lambda^x}{n^x}\frac{(1-\lambda/n)^n}{(1-\lambda/n)^x}=1\left(1-\frac{1}{n}\right)\ldots\left(1-\frac{x-1}{n}\right)\frac{\lambda^x}{x!}\frac{\left(1-\frac{\lambda}{n}\right)^n}{\left(1-\frac{\lambda}{n}\right)^x}=\frac{\lambda^x}{x!}e^{-\lambda}$$

i.e.

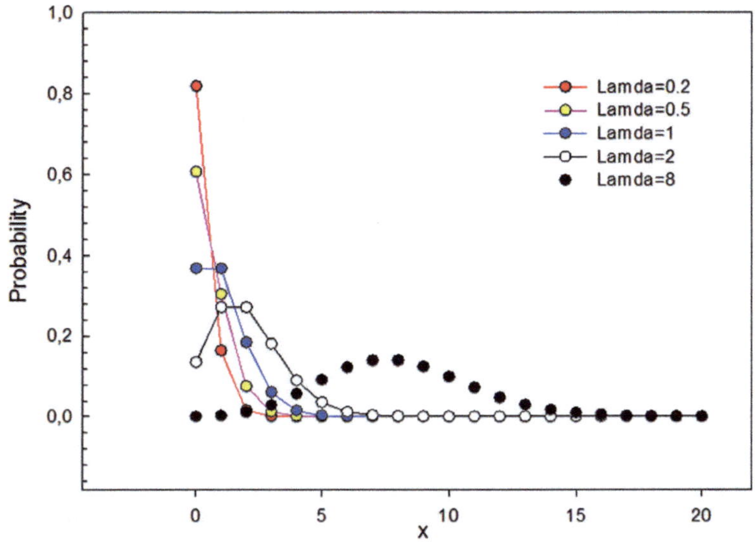

Fig. 2.19 PDF of Poisson distribution

$$\lim_{n \to \infty} b(x) = \frac{\lambda^x}{x!} e^{-\lambda} \tag{2.20}$$

To differentiate the $b(x)$ in the Binomial distribution, we rewrite Eq. (2.20) as

$$p(x) = \frac{\lambda^x}{x!} e^{-\lambda} \tag{2.21}$$

Equation (2.21) is the PDF of the Poisson distribution. Figure 2.19 plots the PDF for various λ.

For a simplified calculation on the Binomial distribution, the Poisson can approximate the Binomial distribution for a larger n and smaller θ as

$$b(x) \cong \frac{(n\theta)^x}{x!} e^{-n\theta} \tag{2.22}$$

Table 2.4 exhibits the approximation of using Eq. (2.22).

Some statistical properties of Poisson distribution are shown in Table 2.5.

Table 2.4 Poisson approximates Binomial ($\theta = 0.01$; n = 100))

x	$b(x)$	p(x)
0	0,366,032	0,367,879
1	0,36,973	0,367,879
2	0,184,865	0,18,394
3	0,060999	0,061313
4	0,014942	0,015328
5	0,002898	0,003066
6	0,000463	0,000511

Table 2.5 Properties of Poisson distribution

Mean	λ
Variance	λ

Example 2.8

The number of failures of a system follows Poisson distribution. The average number of failures per hour is 1.2.

Question 1: What is the probability of exactly three failures in 3 hours?

Solution: Mean number of failures in 3 hours is $1.2 \times 3 = 3.6$ ($\lambda = 3.6$). So the probability of exactly three failures in 3 hours is

$$\frac{e^{-3.6}(3.6)^3}{3!} = 0.212$$

Question 2: What is the probability of at least two failures in 2 hours?

Solution: Mean number of failures in 2 hours is $1.2 \times 2 = 2.4$ ($\lambda = 2.4$).

$$P(x \geq 2) = 1 - P(x = 0) - P(x = 1) = 1 - \frac{e^{-2.4}(2.4)^0}{0!} - \frac{e^{-2.4}(2.4)^1}{1!} = 1 - 0.091 - 0.218 = 0.691$$

The probability of at least two failures in 2 hours is 0.691.

2.7.1 *Normal Distribution*

Normal distribution is the most common distribution in statistics. It has broad applications in engineering. For example, the noise in the signal is considered as Normal distribution. The parameter drift in the electronic device also follows Normal distribution. In the regression analysis, the residual is assumed following Normal distribution. Normal distribution is also widely used for statistical hypothesis test. In reliability engineering, the Normal distribution is seldom used to fit the life data but is often used to access the performance of estimators, or to obtain the

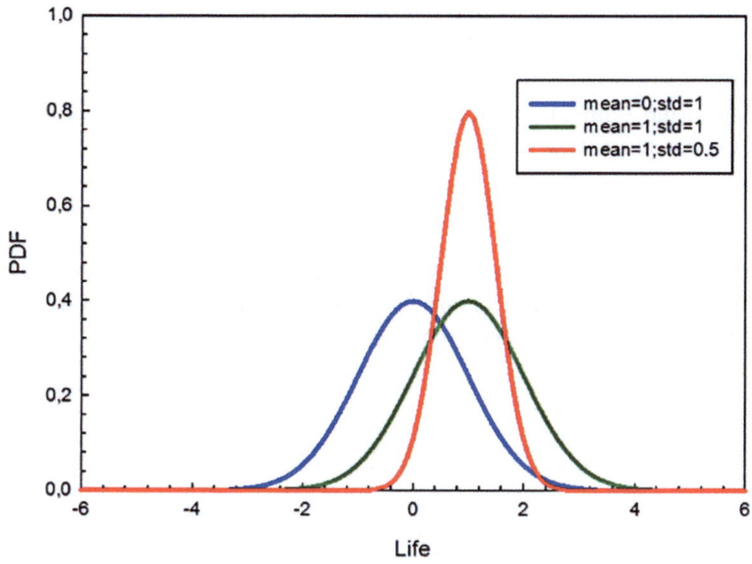

Fig. 2.20 PDF of Normal distribution

Table 2.6 Properties of Normal distribution

Mean	μ
Variance	σ^2
MLE (maximum likelihood estimator)	$\widehat{\mu} = \frac{\sum x_i}{n}$; $\widehat{\sigma^2} = \frac{\sum (x_i - \widehat{\mu})^2}{n}$

confidence interval of estimator. The probability density function of Normal distribution is

$$f(t) = \frac{1}{\sigma\sqrt{2\pi}} e^{-\frac{1}{2}\left(\frac{t-\mu}{\sigma}\right)^2} \tag{2.23}$$

The two parameters, mean μ and standard deviation σ, are two typical parameters in statistics. The PDF of Normal distribution is symmetric with one peak as shown in Fig. 2.20. The peak point corresponds to the mean μ.

The statistical properties of Normal distribution are shown in Table 2.6. The $\overline{x}$ denotes the mean of samples.

The standard Normal distribution is defined as $\mu = 0$, $\sigma = 1$. The corresponding PDF is then

$$f(z) = \frac{1}{\sqrt{2\pi}} e^{-\frac{1}{2}z^2} \tag{2.24}$$

The CDF is

$$\Phi(z) = \int_{-\infty}^{z} f(t)dt \tag{2.25}$$

According to the symmetry of the Normal distribution PDF function, the Normal distribution has properties

$$\Phi(-z) = 1 - \Phi(z) \tag{2.26}$$

It is worth noting that the previously discussed Binomial distribution is also symmetric. Furthermore, when the number of trials is large and the probability of success is not too close to 0 or 1, the Binomial distribution can be approximated by Normal distribution as

$$z = \frac{x - n\theta}{\sqrt{n\theta(1-\theta)}} \tag{2.27}$$

Returning to the Normal distribution, we can rewrite Eq. (2.25) in quantile as shown in many statistics textbook as follows

$$\Phi(z_\alpha) = \int_{-\infty}^{z_\alpha} \frac{1}{\sqrt{2\pi}} e^{-\frac{1}{2}t^2} dt = \alpha \tag{2.28}$$

The z_α denotes the z value where the CDF value is α. One salient property of z_α is

$$z_\alpha = -z_{1-\alpha} \tag{2.29}$$

which is attributed from the symmetry of the PDF. Table 2.7 shows some quantile values of standard Normal distribution. More values can be found from the Normal tables in statistics book [1].

Figure 2.21 demonstrates the symmetry of Normal distribution.

When the random variable, let us say t, follows nonstandard Normal distribution, we can use the following to standardize it

$$z = \frac{t - \mu}{\sigma} \tag{2.30}$$

Table 2.7 Important quantile values

α	0.02	0.025	0.05	0.1	0.9	0.95	0.975	0.98
z_α	−2.05375	−1.95996	−1.64485	−1.28155	1.28155	1.64485	1.95996	2.05375

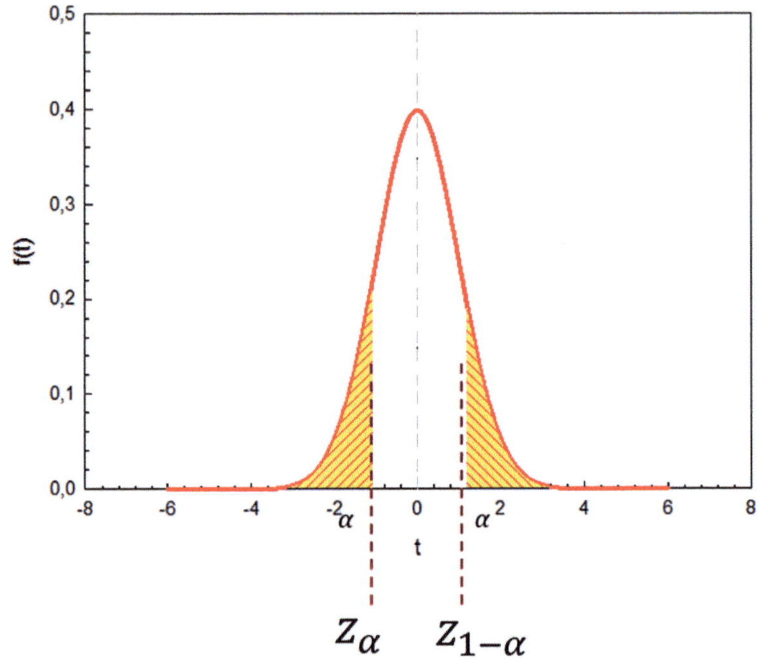

Fig. 2.21 Symmetry of PDF of Normal distribution

Example 2.9
Length of a product follows Normal distribution. What is the probability of the length of the product within 3σ from its mean (σ is standard deviation)?

Solution: It means $z = \frac{|t-\mu|}{\sigma} \leq \frac{3\sigma}{\sigma} = 3$, so $-3 \leq z \leq 3$.

$$P(-3 \leq z \leq 3) = \Phi(3) - \Phi(-3)$$

The $\Phi(-3) = 0.0013$. According to symmetry of the PDF in Normal distribution, the

$$\Phi(3) - \Phi(-3) = 1 - 2\Phi(-3) = 1 - 0.0026 = 0.9974$$

It is at a probability of 99.74% of the length within 3σ.

2.7.2 *Exponential Distribution*

The Exponential distribution is an important distribution in the reliability. This chapter describes it in brief. A further discussion will be presented in Chap. 4. The PDF of the one parameter Exponential is

$$f(t) = \lambda e^{-\lambda t}; t \geq 0; \lambda > 0 \tag{2.31}$$

The PDF of two parameter Exponential is

$$f(t) = \lambda e^{-\lambda(t-\gamma)}; t \geq 0; \lambda > 0 \tag{2.32}$$

where γ is location parameter. The description of location parameter refers to Sect. 2.7.1. Two parameters exponential distribution is not so often used in reliability. Our focus will be on the one parameter Exponential in this book.

The CDF function of Eq. (2.31) is

$$F(t) = 1 - e^{-\lambda t} \tag{2.33}$$

Table 2.8 exhibits some major statistical properties of Exponential distribution.

2.7.3 Weibull Distribution

Weibull distribution is another important distribution for reliability. The PDF of the Weibull distribution is

$$f(t) = \frac{\beta}{\alpha}\left(\frac{t}{\alpha}\right)^{\beta-1} e^{-\left(\frac{t}{\alpha}\right)^{\beta}} \tag{2.34}$$

where α is scale parameter, β is the shape parameter. The description of shape and scale parameter refers to Sect. 2.7. The CDF of Weill distribution is

$$F(t) = 1 - e^{-\left(\frac{t}{\alpha}\right)^{\beta}} \tag{2.35}$$

Table 2.9 exhibits the major properties of this distribution.

The $\Gamma(t)$ called Gamma function is defined as

$$\Gamma(t) = \int_0^{+\infty} x^{t-1} e^{-x} dx \tag{2.36}$$

Weibull distribution is another important distribution widely used in reliability distribution. Chapter 5 of this book will address this distribution specifically.

Table 2.8 Properties of Exponential distribution

Mean	$1/\lambda$
Variance	$1/\lambda^2$
MLE	$\widehat{\lambda} = 1/\frac{\sum x_i}{n}$

Table 2.9 Properties of Weibull distribution

Mean	$\alpha\Gamma(1 + 1/\beta)$
Variance	$\alpha^2\left[\Gamma\left(1+\frac{2}{\beta}\right) - \left(\Gamma\left(1+\frac{1}{\beta}\right)\right)^2\right]$
MLE	No analytical solution

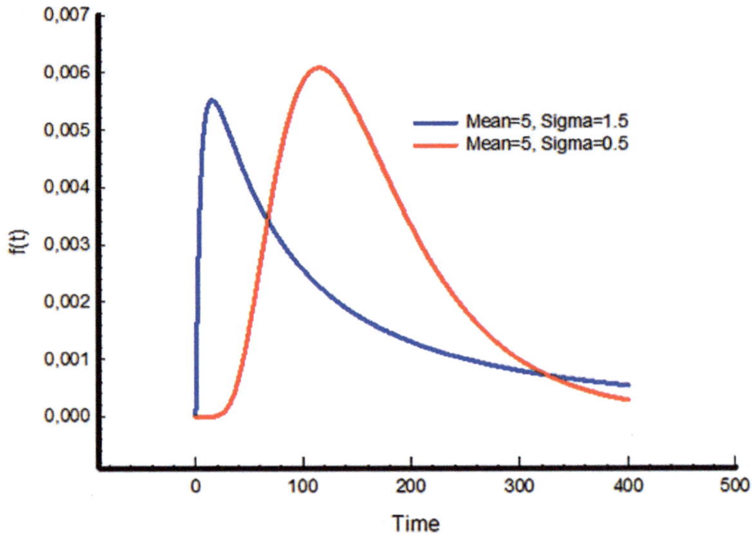

Fig. 2.22 PDF of Lognormal distribution

2.7.4 Lognormal Distribution

Normal distribution is valued within a range of infinity covering the negative region. This is not meaningful for strictly positive data, for example, time measurements. The Normal distribution thus does not fit. For positive-valued time *t*, we can replace the *t* by ln(t) to ensure the positivity. The range of ln(t) is from $-\infty$ to $-\infty$. Normal distribution can hence again be used. The PDF of Lognormal distribution is

$$f(\mathrm{t}) = \frac{1}{\sigma\sqrt{2\pi}} e^{-\frac{1}{2}\left(\frac{\ln t - \mu}{\sigma}\right)^2} \quad t > 0 \tag{2.37}$$

The plot of the PDF is shown in Fig. 2.22. The Lognormal is not symmetric anymore, which is different than that of Normal distribution.

Table 2.10 exhibits the major properties of the Lognormal distribution.

Table 2.10 Properties of Lognormal distribution

Mean	$e^{(\mu+\sigma^2/2)}$
Variance	$\left(e^{\sigma^2}-1\right)e^{(2\mu+\sigma^2)}$
MLE	$\widehat{\mu}=\frac{1}{n}\sum \ln t_i, \widehat{\sigma^2}=\frac{1}{n}\sum(\ln t_i-\widehat{\mu})^2$ Or $\widehat{\sigma^2}=\frac{1}{n-1}\sum(\ln t_i-\widehat{\mu})^2$

Example 2.10

Times to failure of a mining equipment are [3]:

2 3 3 4 4 4 10 12 15 16 18 18 23 24 24 26 27 28 29 30 32 32 35 36
44 50 51 54 54 55 60 65 68 72 76 89 90 112 114 134 141 158 164 211

Assume time to failure follows Lognormal distribution. Find out the parameters, plot the pdf function, and compare the plot with Weibull distribution?

Solution: According to Table 2.10, the parameters for the Lognormal

$$\widehat{\mu}=\frac{1}{n}\sum \ln t_i=3.4506$$

$$\widehat{\sigma}=\sqrt{\frac{1}{n-1}\sum(\ln t_i-\widehat{\mu})^2}=1.1698$$

The parameters for Weibull are known as $\widehat{\alpha}=54$; $\widehat{\beta}=1.07$. We plot the two PDFs by comparing with the empirical PDF as in Fig. 2.23.

The PDF shape of Lognormal and Weibull is similar. It implies for the same data sets, both distributions can fit them.

2.7.5 *Gamma Distribution*

A system will fail until it has experienced a certain number of shocks. For example, the keys in keyboard will fail after it has been hit to certain amount of limit. If the average number of hits at each time unit is constant, the life of the keyboard then follows Gamma distribution. Gamma distribution is a very flexible and general distribution, and it plays a significant role in engineering. It can be used to fit failure data as well. A three parameters Gamma distribution is with PDF of

$$f(x)=\frac{\beta}{\theta\Gamma(k)}\left(\frac{x}{\theta}\right)^{k\beta-1}e^{-\left(\frac{x}{\theta}\right)^{\beta}};x>0;k>0;\theta>0;\beta>0 \tag{2.38}$$

Equation (2.38) is called generalized Gamma function of which the exponential, Weibull etc. is a special case of it, as shown in Table 2.11.

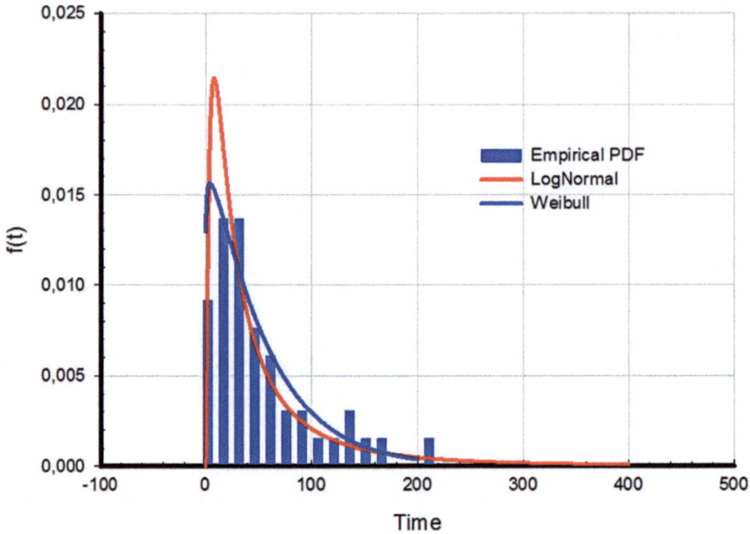

Fig. 2.23 PDF of Lognormal and Weibull

Table 2.11 Generalization of Gamma function

Parameters of (2.38)	Equivalent distribution
$k = 1$	Two parameters Weibull distribution
$k = 1, \beta = 1$	One parameter Exponential distribution
$k = \frac{n}{2}, \beta = 1, \theta = 2$	Chi-squared (χ^2) distribution with degree of n

Table 2.12 Properties of Gamma distribution

Mean	$k\theta$
Variance	$k\theta^2$

Equation (2.38) is very general and its maximum likelihood estimators are complex. In engineering, letting $\beta = 1$ in Eq. (2.38), a simpler but more popular Gamma distribution can be derived with the PDF as

$$f(\mathrm{x}) = \frac{x^{k-1} e^{-\frac{x}{\theta}}}{\theta^k \Gamma(k)}; x > 0, k > 0, \theta > 0 \tag{2.39}$$

This distribution Eq. (2.39) has close relation with Exponential distribution. We will discuss it further in Chap. 4. The summary of Gamma distribution's properties corresponding to Eq. (2.39) is shown in Table 2.12.

2.7.6 Erlang Distribution

Erlang distribution is a special case of Gamma distribution with the k in Eq. (2.38) is integer. Since $\Gamma(k) = (k - 1)!$, Eq. (2.39) is then rewritten as

$$f(x) = \frac{x^{k-1} e^{-\frac{x}{\theta}}}{\theta^k (k-1)!}; x > 0, k = 1, 2, 3, \ldots, \theta > 0 \tag{2.40}$$

The mean and variance hold the same as in the Table 2.12, as it is a special case of Eq. (2.39) for discrete k as integer. Let $\lambda = \frac{1}{\theta}$. The CDF of the Erlang distribution is

$$F(x) = 1 - \sum_{m=0}^{m=k-1} \frac{(\lambda x)^m}{m!} e^{-\lambda x} \tag{2.41}$$

It is of the form of the sum of Poisson distributions. The Erlang distribution can be used, for instance, to model the waiting time in the queueing, to model the number of failures for a given certain time period for repairable system.

2.7.7 Extreme Value Distribution

Extreme Value distribution is the limiting distribution from the extremum of some samples, assuming they belong to the same arbitrary distribution. For example, the maximum mean daily flows of flood follow Extreme Value distribution. An example of reliability: the deepest pit threatens the pipe most and the depth of the deepest pit along the pipe can model with Extreme Value distribution [4]. Extreme Value distribution has several forms, the Type I (Gumbel) is most relevant to reliability. The PDF of Gumbel distribution is

$$f(x) = \frac{1}{\beta} e^{\frac{x-\mu}{\beta}} . e^{\left[-e^{\frac{x-\mu}{\beta}}\right]}; x \in (-\infty, +\infty), \beta > 0 \tag{2.42}$$

The CDF of Eq. (2.42) is

$$F(x) = 1 - e^{\left[-e^{\frac{x-\mu}{\beta}}\right]} \tag{2.43}$$

Of importance of the reliability, the Weibull distribution can be converted into an Extreme Value Distribution. The cumulative probability function of Weibull distribution is

Table 2.13 Properties of Extreme distribution [5]

Mean	$\mu + 0.5772\beta$
Variance	$\frac{\pi^2}{6}\beta^2$

$$F(t) = 1 - \exp\left(-(t/\alpha)^{\beta}\right) \tag{2.44}$$

Equation (2.44) converts into an extreme value distribution by let $t' = \ln t$

$$F(t') = 1 - \exp\left(-\exp\left(\frac{t' - \mu}{\sigma}\right)\right) \tag{2.45}$$

The parameters in the two distributions corresponds as

$$\alpha = e^{\mu}; \beta = 1/\sigma \tag{2.46}$$

Same as for the Weibull distribution, the Gumbel Extreme distribution has no closed form estimator in its maximum likelihood estimation. It relies on numerical approach to obtain solutions. The mean and variance of Extreme distribution is shown in Table 2.13.

2.7.8 Chi-Squared Distribution

Chi-squared distribution is a complex distribution that is used often for hypothesis tests in statistics. In reliability, it is seldom used to model the failure data but often used for goodness of fit tests. The distribution is defined as the sum of number of k independent squared standard Normal distribution. The n is the freedom degree that is the sole parameter of this distribution.

$$Z_1^2 + Z_2^2 + Z_3^2 + \ldots + Z_n^2 \sim \chi_n^2 \tag{2.47}$$

The PDF of this distribution is given by following equation and one PDF is plotted as shown in Fig. 2.24.

$$f(x) = \frac{2^{-n/2} x^{n/2-1} e^{-x/2}}{\Gamma(n/2)} \tag{2.48}$$

The PDF is asymmetric, thus Eq. (2.29) used to simplify the calculation for the Normal distribution does not hold for the Chi-Squared. The quantile value and the CDF of Chi-squared distribution is a complex expression that will not be discussed in this book. One can refer to textbook for more detail [1].

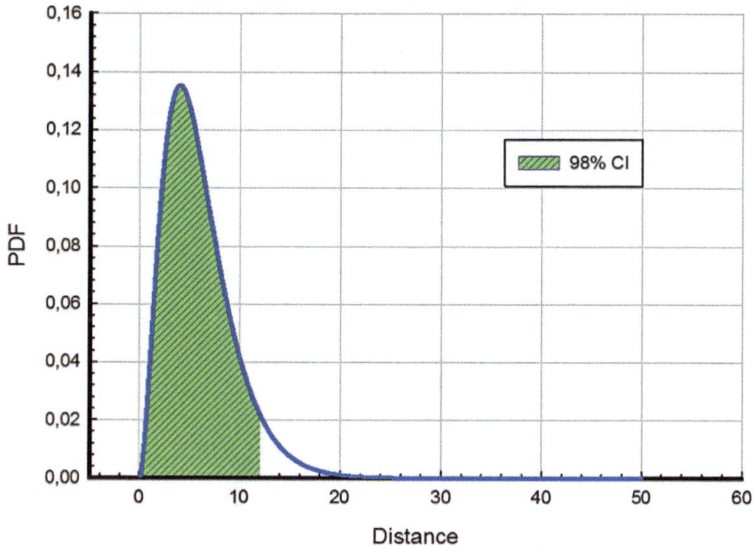

Fig. 2.24 PDF of Chi-squared distribution

2.7.9 Student Distribution

Student distribution is similar to the Normal distribution with a symmetric PDF. The student distribution is also a widely used distribution for hypothesis test. Student distribution is defined as

$$T = \frac{Z}{\sqrt{\frac{V}{n}}} \tag{2.49}$$

where Z denotes standard Normal distribution, V denotes Chi-squared distribution with degree of freedom n. The sole parameter of this distribution is the degree of freedom n. The PDF function of it is given by

$$f(x) = \frac{\Gamma\lfloor(n+1)/2\rfloor}{\sqrt{n\pi}\Gamma(n/2)}\left(1 + x^2/n\right)^{-(n+1)/2} \tag{2.50}$$

The following Fig. 2.25 demonstrates the symmetric property of this distribution.

The quantile value of student distribution can be found in most statistics textbooks. One popular application of student distribution is the mean and standard deviation of any distributions can be constructed as a student distribution, if the independent samples are from the same distribution, i.e.

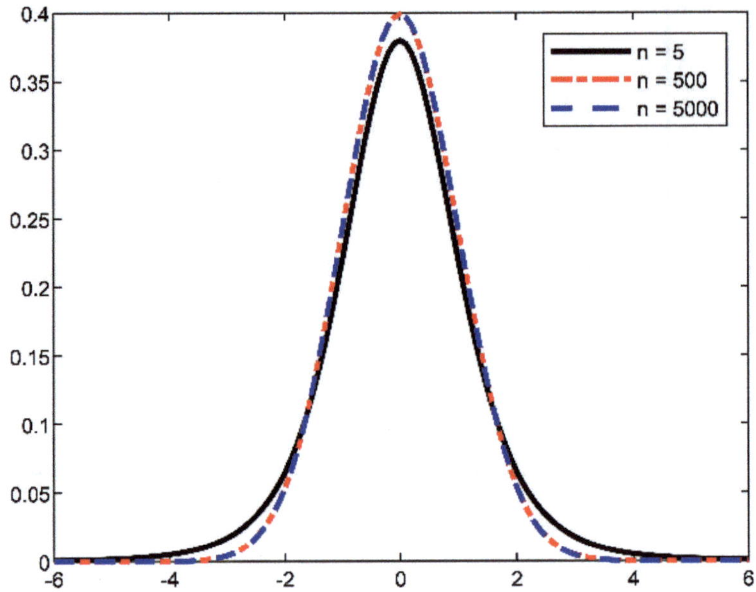

Fig. 2.25 PDF of student distribution

$$\frac{\overline{x}-\mu}{\sqrt{s/n}} \sim t \tag{2.51}$$

where $\overline{x}$ is the average of the samples and s is the standard deviation $s = \sqrt{\frac{\sum (x_i - \overline{x})^2}{n-1}}$.

2.7.10 *F Distribution*

F distribution, also named Fisher-Snedecor distribution, is derived from Chi-squared distribution. It is defined as

$$F = \frac{U/m}{V/n} \tag{2.52}$$

where U is Chi-squared distribution with degree m, V is Chi-squared distribution with degree n. The U and V is independent. The PDF function of the F distribution is rather complex as

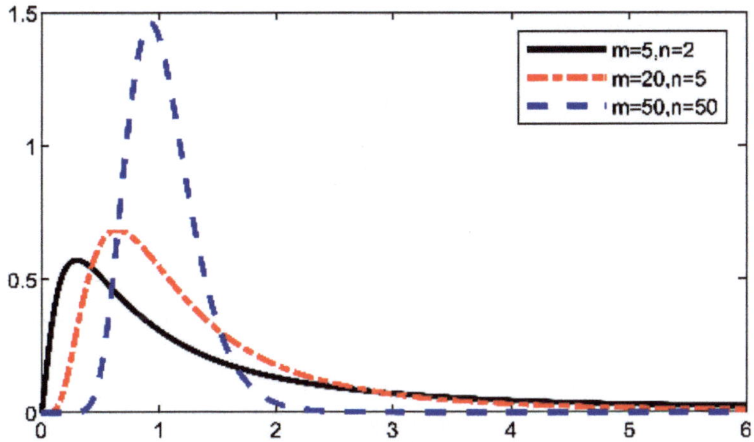

Fig. 2.26 PDF of F distribution

$$f(x) = \frac{1}{B\left(\frac{m}{2}, \frac{n}{2}\right)} (m/n)^{m/2} x^{\frac{m}{2}-1} \left(1 + \frac{m}{n}x\right)^{-(m+n)/2} \tag{2.53}$$

where $B(.)$ is beta function $B\left(\frac{m}{2}, \frac{n}{2}\right) = \frac{\Gamma\left(\frac{m}{2}\right)\Gamma\left(\frac{n}{2}\right)}{\Gamma\left(\frac{m+n}{2}\right)}$. The PDF of this distribution is asymmetric, as shown in Fig. 2.26.

The quantile value of F distribution can be found in F table. The F-distribution is widely used in regression analysis for testing the significance of the parameters.

2.8 Location, Shape and Scale Parameters

Parameters in statistical distributions vary significantly. But many distributions can have their parameters classified as location, shape and scale parameters.

2.8.1 Location Parameter

For example, in failure time data analysis, at the time when the time starts to count, the products have already run for certain period t_0. It implies the life data should be added a t_0. In the statistical distribution, the t_0 is a kind of location parameter. In the plot of PDF and CDF, the location parameter shifts the plot left or right without changing the shape and scale of the plot, as shown in Fig. 2.27.

The γ in Fig. 2.27 is the location parameter in three parameters Weibull distribution. One can refer to the two-parameter Weibull distribution of Eq. (2.34) to compare the three-parameter Weibull as follows

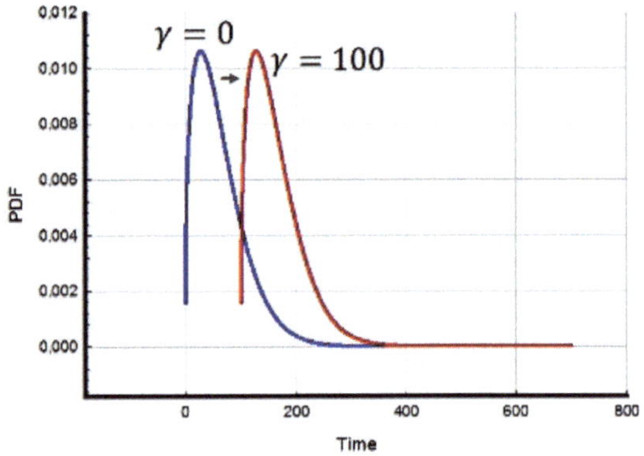

Fig. 2.27 Location parameter

$$f(t) = \frac{\beta}{\alpha}\left(\frac{t-\gamma}{\alpha}\right)^{\beta-1} e^{-\left(\frac{t-\gamma}{\alpha}\right)^{\beta}} \tag{2.54}$$

In mathematical terms, if the PDF function can be rewritten in a form $f(t - \gamma)$, then the γ is location parameter. All distribution can have a location. For the Exponential distribution in Eq. (2.31), the PDF can be modified to $f(t - \gamma) = \lambda e^{-\lambda(t - \gamma)}$ by adding the location parameter, which is the two parameter Exponential distribution. In reliability engineering, the location parameter is rarely used. When location parameter is present, the parameter estimation will be complicated, and one should avoid using this parameter if not necessary.

2.8.2 *Scale Parameter*

The α in the Weibull distribution as shown in Eq. (2.54) and the standard deviation in the Normal distribution are scale parameters. In the case of failure time analysis, scale parameter scales up/down the failure time. For example, when the time unit is changed from hours to days, the scale parameter will scale accordingly. Figure 2.28 shows a PDF plot for a Weibull distribution. The red (lower) one was estimated with scale parameter 42 hours, when the failure time was recorded in hours. For the same life data, if we use days instead of hours, all the life data are then divided by 24 to convert to days. The scale parameter accordingly becomes 42/24 when estimated by maximum likelihood method, while the shape of the distribution plot remains the same, as shown in Fig. 2.28. In mathematical terms, if the PDF can be written in form of $\frac{1}{\alpha} f\left(\frac{t}{\alpha}\right)$, then the α is scale parameter. The failure rate parameter in the Exponential distribution is scale parameter, since obviously its PDF is of the form $\frac{1}{\alpha} f\left(\frac{t}{\alpha}\right)$.

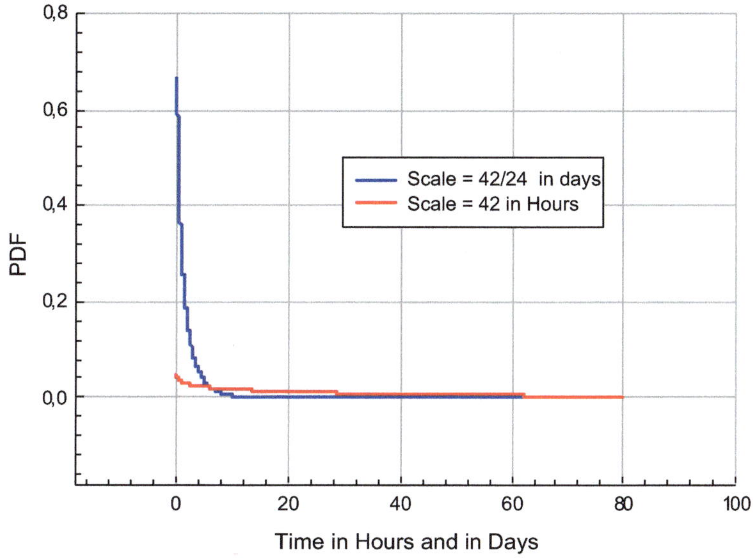

Fig. 2.28 Scale parameter

2.8.3 Shape Parameter

Shape parameter characterizes the general shape of the distribution. The β in the Weibull distribution of Eq. (2.34) and Gamma function of Eq. (2.38) are shape parameters. For the Weibull distribution, the shape parameter affects the monotonicity of the failure rate function and the shape of the PDF. Figure 2.29 plots the PDFs for various Weibull distributions with various shape parameters. The shapes of these distributions exhibit significant difference.

2.9 Mean, Median and Mode

The mean is the statistical average of the distribution. The median is the point at which the probability is 0.5. The mode is the point of the PDF function with maximum value. Given a PDF $f(t)$, the mean of the distribution is

$$\mu = \int_{0}^{+\infty} tf(t)dt \tag{2.55}$$

Figure 2.30 demonstrates the mean of the Normal and Weibull distribution.

The median of a distribution is the T_{median} at which the probability is 0.5

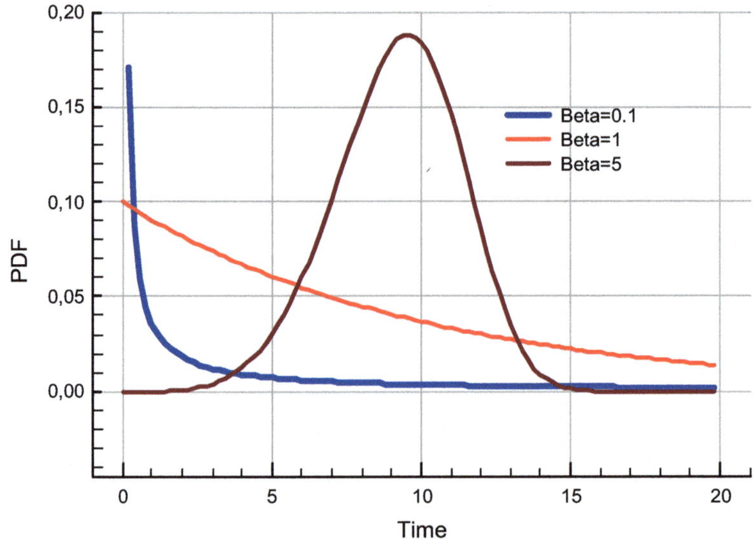

Fig. 2.29 Shape parameter

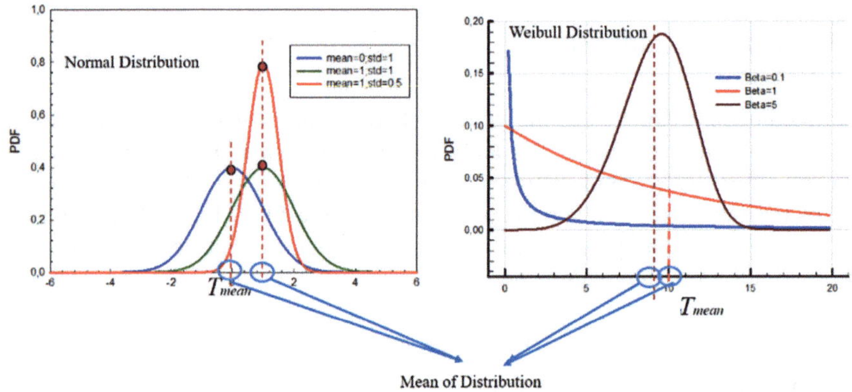

Fig. 2.30 Mean of Normal and Weibull distributions

$$\int_0^{T_{median}} f(t)dt = 0.5 \tag{2.56}$$

Figure 2.31 demonstrates the median for Normal and Weibull distribution.

The mode is at which the PDF has maximum value. For Normal distribution, it has only one maximum. For Weibull distribution, one maximum exists when the shape parameter is greater than 1, as demonstrated in Fig. 2.32.

The mean, median and mode are equal only for Normal distribution, as shown in the Left figure of Fig. 2.33. For Weibull distribution, they vary, as demonstrated by

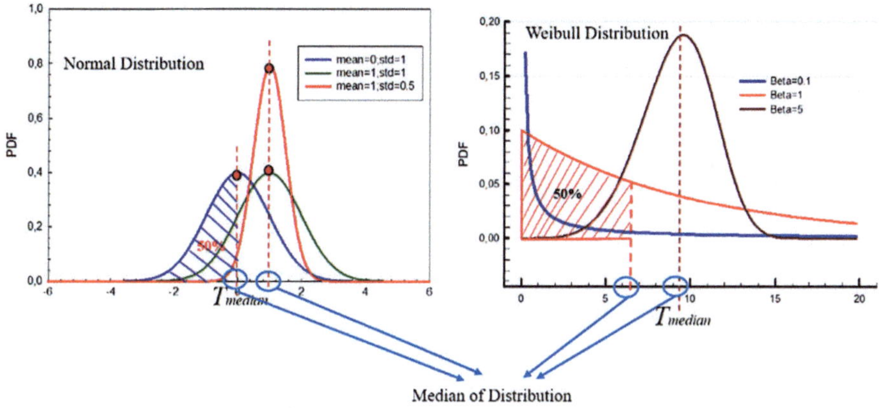

Fig. 2.31 Median of a distribution

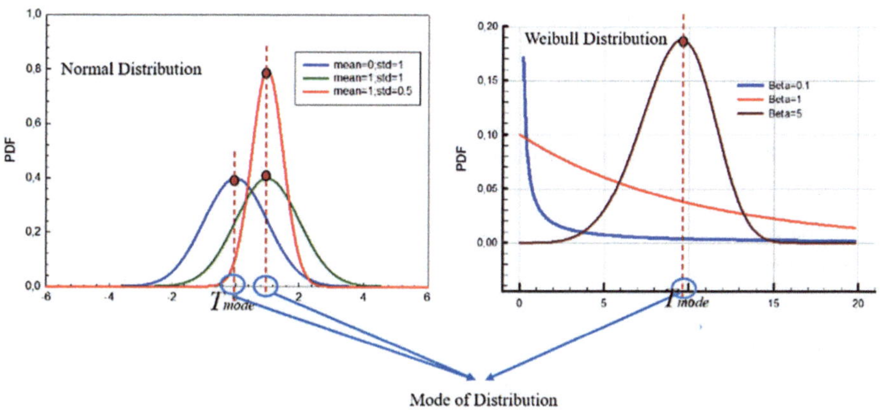

Fig. 2.32 Mode of a distribution

the right figure of Fig. 2.33, where shape parameter is 5 and 2 respectively. The mean, median and mode locate at different points of the time on the x-axis.

Problems and Questions

1. Which one of the following statements is wrong

   (a) Reliability of compressor after 3 years is still 0.9
   (b) The failure rate of a system is 1.1 per hour
   (c) The failure rate of the pump is 0.4 per year
   (d) Reliability of a system with redundancy can reach 1.05

2. A company has warranted its computers for 1 year of service. The computers have a constant failure rate and have a MTTF of 10 years. About what percentage of the computers are functioning at the end of the warranty period?

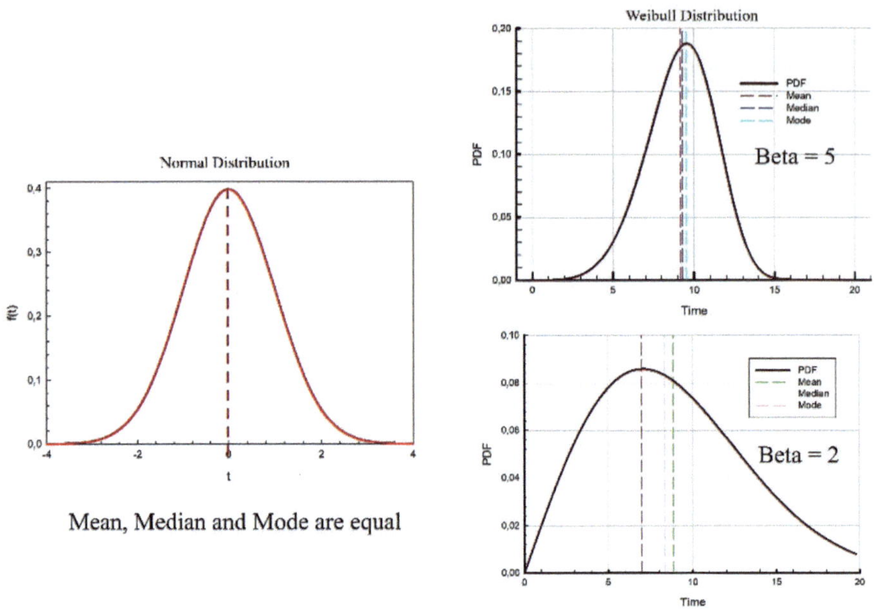

Fig. 2.33 Mean, median and mode for distributions

(a) 90.5%
(b) 20.2%
(c) 1.9%
(d) 99.9%
(e) None of the above

3. Which one of the statements is correct

   (a) Reliability can be greater than 1.
   (b) Probability density function (PDF) can be greater than 1.
   (c) Failure rate cannot be greater than 1.
   (d) Cumulative density function (CDF) can be greater than 1.

4. A system suffers two failure modes. The two failure modes are **dependent** on each other. The probability of failure mode one is 0.8. The probability of failure mode two is 0.7. What is the probability of both failure modes occur?

   (a) 0.70
   (b) 0.56
   (c) 0.72
   (d) Not enough information is given to determine the answer.

5. If we want to fit the life data where life are positive, which one distribution is not able to use:

   (a) Two parameter Weibull distribution
   (b) One parameter Exponential distribution
   (c) Lognormal distribution
   (d) Normal distribution

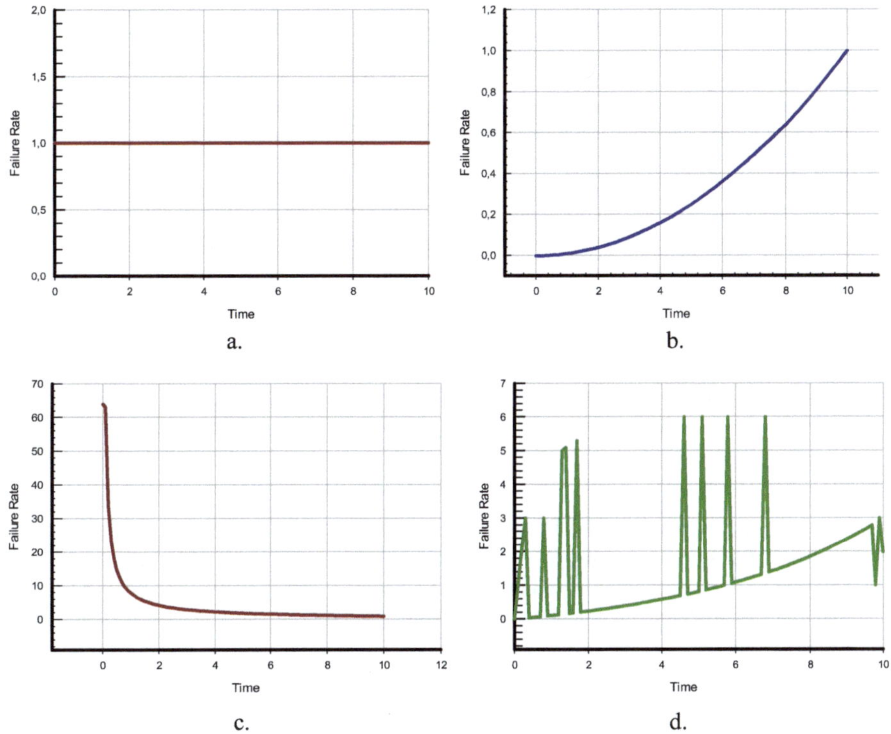

Fig. 2.34 Four types of failure rate: (**a**) Constant Failure Rate; (**b**) Increasing Failure Rate; (**c**) Decreasing Failure Rate; (**d**) Non-monotonic Failure Rate

6. Which one of the following statements is **correct**?

   (a) Probability density function (PDF) value cannot be over 1.
   (b) Reliability at infinite time ($t = \infty$) will be 1 (reliability is defined between 0 and 1).
   (c) Cumulative density function (CDF) at infinite time ($t = \infty$) will be 0.
   (d) Cumulative density function value cannot be over 1 (if probability of failure is defined between 0 and 1).

7. A mining equipment's time to failure follows Weibull distribution with shape parameter 1 and scale parameter 100 hours. Find the approximate reliability of the product at 100 hours.

   (a) 0.50
   (b) 0.75
   (c) 0.37
   (d) 0.99
   (e) 0.89
   (f) None of the above

8. Which failure rate could be when the shape parameter in Weibull distribution is less than one as shown in Fig. 2.34?
9. Ten unrepairable products failed within 100 hours at times

$$1;\ \ 14; 25;\ \ 37;\ \ 48;\ \ 59;\ \ 65;\ \ 78;\ \ 85;\ \ 99$$

(a) Plot the empirical PDF (group the data by each 10 hours)
(b) Plot empirical failure rate (group the data by each 10 hours)
(c) Is the failure rate increasing, deceasing or constant? Which distribution can fit this data?

10. If the failure rate of a system is

$$r(t) = t^2; t > 0$$

(a) What is the reliability function?
(b) What is CDF (Cumulative Distribution Function) $F(t)$ and its PDF function $f(t)$?
(c) For this distribution, is the failure rate increasing or decreasing with time t?

11. A distribution with failure rate:

$$r(t) = at^{\beta}; a > 0; \beta > 0; t \in [0; +\infty\ ]$$

The a and β is parameters. Please determine the monotonicity of the failure rate function for various a and β.

12. One parameter Exponential distribution (PDF $f(t) = \lambda \exp(-\lambda t)$) is the simplest but an important distribution in reliability engineering. Suppose we have recorded lifetimes for identical items as

20.3 59.6 74.0 99.0 25.3 31.4 42.2 1.8 63.2 3.6 28.4 32.6

(a) What is the failure rate λ?
(b) Estimate the reliability for a service time of 10 hours.

13. If the CDF for times to failure for a system is

$$F(t) = 1 - \frac{100}{(t+10)^2}; t \geq 0$$

(a) What is reliability function R(t)?
(b) What is the failure rate as a function of time? Does the failure rate increase or decrease with time?
(c) Suppose the system cannot run with reliability under 95%. What is maximum service time for the system?

14. The life of a mining equipment has been known following Weibull distribution, the parameter estimated is known as (α is scale parameter, β is shape parameter)

$$\alpha = 1000\ hrs; \beta = 2$$

(a) What is the maximum operating hours, when it requires the reliability to be at least 95%?
(b) The mining company has 100 such equipment. They are identical. How many equipment are expected to be survived after 500 hours?
(c) If the scale parameter is $\alpha = 1000$ hrs, the shape parameter is unknown. What should the shape parameter be, in order to guarantee the reliability of 98% for 500 hours?

15. If the reliability function of a system is

$$R(t) = e^{-t^3}; t \geq 0$$

(a) What is CDF (Cumulative Distribution Function) F(t) and its PDF function f(t)?
(b) What is the failure rate function r(t)?
(c) For this distribution, is the failure rate increasing with time t, or decreasing with time t? For a system with this failure rate, is the system deteriorating with time t (e.g. more vulnerable to failure with time t)?

16. The time to failure of a component is assumed to be uniformly distributed over (*a*, *b*). The probability density is thus

$$f(t) = \frac{1}{b-a}\ for\ a < t \leq b$$

Derive the corresponding reliability function *R*(*t*) and failure rate function *r*(*t*). Draw a sketch of *r*(*t*).

17. Suppose that we have observed the random sample $x_1, x_2, x_3, \ldots, x_n$. The data follows Normal distribution

$$f(x) = \frac{1}{\sqrt{2\pi\theta}} e^{-\frac{(x-\mu)^2}{2\theta}}$$

Find the maximum likelihood estimators for μ and θ.

18. The times to failure are 7, 12, 19, 29, 41, and 67 hours for an unrepairable system. Suppose TTFs follow Rayleigh distribution with PDF

$$f(t) = \frac{t}{\theta^2} e^{-t^2/(2\theta^2)}, t > 0$$

(a) Estimate the parameter θ using maximum likelihood method.
(b) If it requires the reliability of the unit to be 95%, what is the unit's maximum service time?

19. The PDF for the time-to-failure of an appliance is (*t* is in years)

$$f(t) = \frac{32}{(t+4)^3}, t > 0$$

(a) Find the reliability function *R*(*t*).
(b) Find the failure rate $\lambda(t)$.
(c) Find the MTTF. (hint: $MTTF = \int tf(t)dt$).

20. A device has a constant failure rate of $\lambda = 1$/hour.

(a) What is the probability that it will fail during the first 1 hour of operation?
(b) If that the device has been successfully operated for 1 hour. What is the probability that it will fail during the next 1 hour of operation (Conditional probability)?

21. The length of a batch of products follows Normal distribution. What is the probability of the length of the product within 2σ from its mean (σ is std)?
22. A random variable x follows Normal distribution with mean $\mu = 1$ and standard deviation $\sigma = 2$,

    (a) Please convert the x to a standardized Normal variable.
    (b) Find out the probability for $x \leq 0$.

References

1. Johnson RA, Bhattacharyya GK (2010) Statistics: principles and methods. Wiley
2. Proschan F (1963) Theoretical explanation of observed decreasing failure rate. Technometrics 5(3):375–383
3. Kumar U, Klefsjö B, Granholm S (1989) Reliability investigation for a fleet of load haul dump machines in a Swedish mine. Reliab Eng Syst Safe 26(4):341–361
4. Jarrah A, Bigerelle M, Guillemot G, Najjar D, Iost A, Nianga JM (2011) A generic statistical methodology to predict the maximum pit depth of a localized corrosion process. Corros Sci 53(8):2453–2467
5. Krishnamoorthy K (2006) Handbook of statistical distributions with applications. Chapman & Hall/CRC, Boca Raton

Chapter 3
Essential Elements of Statistics for Reliability

Many methods are used in reliability engineering for both quantitative and qualitative analysis, among which statistical methods are undoubtedly among the most important. This chapter presents key statistical concepts and methods that are essential for reliability analysis. Statistics, as a fundamental branch of mathematics, is well developed with mature theories. However, applying statistical methods to reliability problems without a proper understanding of their fundamentals and limitations can be risky. Misconceptions and misuse of statistical methods are common in state-of-the-art applications of reliability engineering. This hchapter introduces statistics from a practical perspective for reliability. It begins with parameter estimation, including discussions on confidence interval derivation, consistency, and asymptotic normality. It then addresses parameter bias, methods for goodness-of-fit test, and approaches for approximating cumulative distribution functions.

3.1 Maximum Likelihood Estimation

Reliability models are commonly related to probability or statistical distribution. The unknown parameter in the probability or statistical model should be estimated to be used for practical application. We firstly assume, usually implicitly, that the dedicated probability or statistical model can fit the data. Parameter estimation is a procedure trying to fit the data to the model. There are many parameter estimation methods we can employ from statistics, for example maximum likelihood estimation, moment estimation and Bayesian estimation. Among them, the Maximum Likelihood Estimation (MLE) is the dominant one. Given a group of observed data, the MLE assumes the most likely parameter that can fit the data as the best one. This principle seems intuitive, but it has been verified theoretically plausible. Imaging a gambling game as in Fig. 3.1, the probability of winning is shown on each box. One reasonable selection is to choose the box with the highest probability,

F. Yuan, *Essentials of Reliability Engineering*, Springer Series in Reliability Engineering, https://doi.org/10.1007/978-3-032-15544-3_3

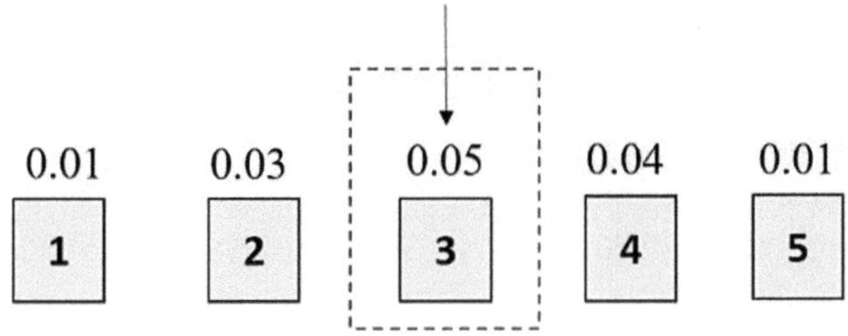

Fig. 3.1 Emulating of maximum likelihood estimation

meaning to choose option 3 in Fig. 3.1, since 0.05 is the highest probability, meaning it has the highest chance to win.

Back to the statistical issue, to employ the principle introduced in Fig. 3.1, we should find the number on the box. The number is called likelihood now. If the reliability model is a statistical distribution and suppose the observed failure times are $t_1, t_2, \ldots, t_i, \ldots t_n$ and they are independent and identically distributed with PDF $f(t_i)$. The independent and identically distributed, abbreviated as i.i.d, is often referred in our book. The i.i.d implies any two events have no statistical dependency. For example, the occurrence of event A does not affect that of event B. The assumption is very strong, and despite most practical cases doesn't meet it, we still assume it as this assumption can simplify the problem significantly. For the independent lifetime $t_1, t_2, \ldots, t_i, \ldots t_n$, the likelihood of this series is its joint distribution, which is defined as follows and we can find it in many statistics textbooks.

$$L(\boldsymbol{\theta}/t) = \prod_{i=1}^{n} f(t_i) \tag{3.1}$$

We can explain the likelihood formula of Eq. (3.1) in a more understandable way. As we discussed in Chap. 2, the $f(t_i)$ can be considered as the probability of failure occurring at time t_i in some way, although not exactly.

The probability of failure at time t_i is actually $f(t_i)\Delta t$. The Δt is an extremely small value as close as zero. The probability for the lifetime series $t_1, t_2, \ldots, t_i, \ldots t_n$ is then $f(t_1)\Delta t.\,f(t_2)\Delta t.\,f(t_3)\Delta t\ldots f(t_n)\Delta t$ where the multiplication like $f(t_1)\Delta t.\,f(t_2)\Delta t$ for joint probability is due to the independence assumption. If no independence assumption, the calculation will be very complex. The Δt is just a constant and it does not carry any information of the parameter. And it is obvious the $f(t_1).\,f(t_2).\,f(t_3)\ldots f(t_n)$ is proportional to $f(t_1)\Delta t.\,f(t_2)\Delta t.\,f(t_3)\Delta t\ldots f(t_n)\Delta t$. In mathematical terms, we can write it as:

$$f(t_1).f(t_2).f(t_3)\ldots f(t_n) \propto f(t_1)\Delta t.f(t_2)\Delta t.f(t_3)\Delta t\ldots f(t_n)\Delta t$$

We can just ignore the Δt since it can be considered as constant, which results to Eq. (3.1). Equation (3.1) function is called likelihood function. The use of likelihood is to differentiate it from the probability since likelihood is not probability as we discussed above. According to the maximum probability principle, as illustrated in Fig. 3.1, this corresponds to maximum likelihood in this case, the best parameter $\boldsymbol{\theta}$ is the parameter corresponding to the maximum value of $L(\boldsymbol{\theta}/t)$, which is the Maximum Likelihood Estimation (MLE) method.

We can use a numerical example to demonstrate the MLE. Suppose the time to failure is as follows

2.5 5.9 3.8 1.2 3.76 1.25 0.24 1.01 6.4 1.8

Suppose the data follows the Exponential distribution with PDF $f(t_i) = \lambda \exp(-\lambda t)$. The $L(\boldsymbol{\theta}/t)$ values from Eq. (3.1) for λ from 0.1 to 0.8 is as Table 3.1.

According to the MLE principle, the λ with the highest likelihood is the best. In Table 3.1, we can choose the $\lambda = 0.4$ as the best, since it has the highest likelihood value L as 1.5164E-09. From Table 3.1, we observe the $L(\boldsymbol{\theta}/t)$ is of extremely small magnitude. For ease of operation, we take the logarithm of Eq. (3.1) as

$$\ln L(\boldsymbol{\theta}/t) = \sum_{i=1}^{n} \ln f(t_i) \tag{3.2}$$

The optimal value λ corresponding to the maximum value of Eq. (3.2) is same as the optimal value λ in Eq. (3.1), as shown in Fig. 3.2, while Eq. (3.2) is more tractable.

From the plot of figure, we can find there is obvious a maximum point existing. We should find this point using some methods. Surely using the plot of the likelihood as shown Fig. 3.2 is not a good method. If the PDF is a continuous function, suitable for most reliability case, we can obtain the maximum point by taking the first derivative of function (3.2) and equal it to zero as follows

$$\frac{\partial \ln L}{\partial \boldsymbol{\theta}} = 0 \tag{3.3}$$

The solution to Eq. (3.3) is the maximum likelihood estimate, which is considered as the best parameter for the failure time $t_1, t_2, \ldots, t_i, \ldots t_n$. There has another issue regarding the applicability of using Eq. (3.3). The solution to Eq. (3.3) could be a minimum point instead of the desired maximum point. We will resolve this issue later by discussing the consistency of the MLE.

3.1.1 Feasibility of MLE

This section proves the solution to Eq. (3.3) is the maximum point of the ML function. We take the general Exponential family distribution as example. Exponential family represents a large number of distributions. Exponential, Normal

Table 3.1 Numerical values of likelihood function

λ	0.1	0.2	0.3	0.4	0.5	0.6	0.7	0.8
L	6.1667E-12	3.8941E-10	1.3848E-09	1.5164E-09	8.7092E-10	3.3254E-10	9.5801E-11	2.2457E-11
$\ln L$	−25.8119	−21.6664	−20.3977	−20.3069	−20.8615	−21.8243	−23.0687	−24.5194

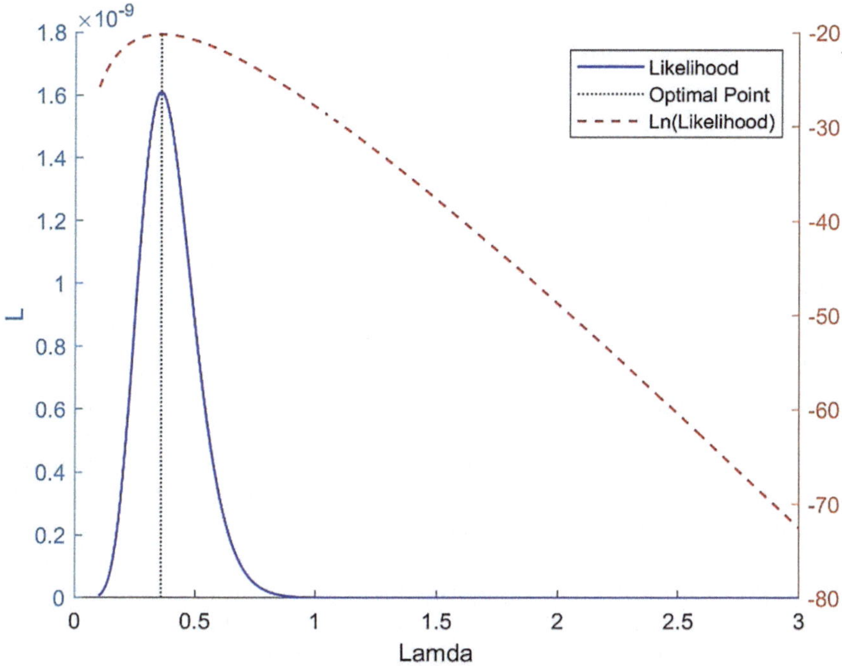

Fig. 3.2 Likelihood function plot

distribution belong to the exponential family. The Exponential family distribution is defined of a form of PDF as follows:

$$f(\boldsymbol{t};\boldsymbol{\theta}) = b(\boldsymbol{t})e^{\boldsymbol{\theta}' T(\boldsymbol{t}) - \boldsymbol{c}(\boldsymbol{\theta})} \tag{3.4}$$

The Exponential family has some convenient statistical properties. Appendix 3A presents two of them. The log likelihood function is

$$\ln L(\boldsymbol{\theta}/\boldsymbol{t}) = \sum_{i=1}^{n} \{\ln b(\boldsymbol{t}_i) + \boldsymbol{\theta}' T(\boldsymbol{t}_i) - \boldsymbol{c}(\boldsymbol{\theta})\} \tag{3.5}$$

The first derivative is not hard to be obtained as follows and let it equal to zero

$$\frac{\partial \ln L}{\partial \boldsymbol{\theta}} = \sum_{i=1}^{n} T(\boldsymbol{t}_i) - n\frac{\partial \boldsymbol{c}}{\partial \boldsymbol{\theta}} = 0 \tag{3.6}$$

And the

$$\frac{\partial^2 \ln L}{\partial \boldsymbol{\theta}^2} = -n\frac{\partial^2 \boldsymbol{c}}{\partial \boldsymbol{\theta}^2} = -nVar(T(\boldsymbol{t})) < 0 \tag{3.7}$$

From Appendix 3A, it has been known the $\frac{\partial^2 \boldsymbol{c}}{\partial \boldsymbol{\theta}^2} = var(T(\boldsymbol{t})) > 0$, therefore, the $\frac{\partial^2 \ln L}{\partial \boldsymbol{\theta}^2} < 0$, which guarantees the solution from Eq. (3.6) is the maximum point. The Exponential distribution, Normal distribution, multi-Normal distribution which of interest in this book belong to the Exponential family. For the Exponential distribution with PDF $f(t) = \lambda \exp(-\lambda t)$. Corresponding to the general form Eq. (3.4), the $b(x) = 1$, the $\boldsymbol{\theta} = \lambda$, $T(\boldsymbol{t}) = t$, $\boldsymbol{c}(\boldsymbol{\theta}) = -\ln \lambda$. Since the $\boldsymbol{c}(\boldsymbol{\theta}) = -\ln \lambda$, according to Eq. (3.7), the $\frac{\partial^2 \boldsymbol{c}}{\partial \boldsymbol{\theta}^2} = \frac{1}{\lambda^2}$, which is same as given in the Table 2.8 in Chap. 2.

The abovementioned description proved the $\ln L(\boldsymbol{\theta}/t) = \sum_{i=1}^{n} f(t_i)$ has the maximum point for likelihood function of the Exponential family, while it cannot prove the estimator obtained from the maximum likelihood is close to the true parameter. Consistency is a concept that ensures the estimated parameter converges to the true parameter when the number of sample size increases. Investigating the consistency of MLE is necessary as it is a necessary condition of using MLE. An estimator is far from the true parameter is useless. The concept of consistency is less familiar in reliability engineering. As estimator is a random number, it is certainly desirable that the randomness can be reduced by increasing the sample size. This is the consistency.

It can find the MLE, not limited to Exponential family, asymptotically approaches to the true parameters when $n \longrightarrow \infty$ with probability 1. According to the Jensen inequality [1], by using the inequality $E(\ln(.)) < \ln E(.)$. The $\boldsymbol{\theta}$ is the true parameters of $f(\boldsymbol{t}; \boldsymbol{\theta})$ and $\widehat{\boldsymbol{\theta}}$ is estimators of $\boldsymbol{\theta}$. Then

$$E\left(\ln\left(\frac{f\left(\boldsymbol{t};\widehat{\boldsymbol{\theta}}\right)}{f(\boldsymbol{t};\boldsymbol{\theta})}\right)\right) < \ln E\left(\frac{f\left(\boldsymbol{t};\widehat{\boldsymbol{\theta}}\right)}{f(\boldsymbol{t};\boldsymbol{\theta})}\right) \tag{3.8}$$

Since $E\left(\frac{f\left(\boldsymbol{t};\widehat{\boldsymbol{\theta}}\right)}{f(\boldsymbol{t};\boldsymbol{\theta})}\right) = \int \frac{f\left(\boldsymbol{t};\widehat{\boldsymbol{\theta}}\right)}{f(\boldsymbol{t};\boldsymbol{\theta})} f(\boldsymbol{t};\boldsymbol{\theta})dt = \int f\left(\boldsymbol{t};\widehat{\boldsymbol{\theta}}\right)dt = 1$, from Eq. (3.8), the

$$E\left(\ln\left(\frac{f\left(\boldsymbol{t};\widehat{\boldsymbol{\theta}}\right)}{f(\boldsymbol{t};\boldsymbol{\theta})}\right)\right) \le \ln E\left(\frac{f\left(\boldsymbol{t};\widehat{\boldsymbol{\theta}}\right)}{f(\boldsymbol{t};\boldsymbol{\theta})}\right) = 0 \tag{3.9}$$

i.e.

$$E\left(\ln\left(f\left(\boldsymbol{t};\widehat{\boldsymbol{\theta}}\right)\right)\right) - E(\ln(f(\boldsymbol{t};\boldsymbol{\theta}))) < 0$$

Since

$$\frac{\sum_{i=1}^{n} f\left(t;\widehat{\theta}\right)}{n} \xrightarrow{n\to\infty} E\left(\ln\left(f\left(t;\widehat{\theta}\right)\right)\right);$$

$$\frac{\sum_{i=1}^{n} f(t;\theta)}{n} \xrightarrow{n\to\infty} E(\ln(f(t;\theta)))$$

and let $\widehat{\theta}$ **is** $\theta \pm \delta$, where δ is arbitrarily small, then

$$\frac{\sum_{i=1}^{n} f(t;\theta\pm\delta)}{n} - \frac{\sum_{i=1}^{n} f(t;\theta)}{n} \xrightarrow{n\to\infty} E\left(\ln\left(f\left(t;\widehat{\theta}\right)\right)\right) - E(\ln(f(t;\theta))) < 0 \quad (3.10)$$

Equation (3.10) suggests there is a local maximum within interval $[\theta - \delta, \theta + \delta]$. The local parameter is written as $\widehat{\theta}$ that can be obtained from $\frac{\partial \ln L}{\partial \theta} = 0$. Equation (3.10) thus guarantees a local maximum existing for MLE and thus the $\frac{\partial \ln L}{\partial \theta} = 0$ have solution with probability 1 when $n \longrightarrow \infty$. The $\widehat{\theta}$ is a value between $\theta - \delta$ and $\theta + \delta$. It is not hard to find the error $\left|\theta - \widehat{\theta}\right| < 2\delta$. Since δ is arbitrary and can be extremely small, the $\widehat{\theta}$ can be arbitrarily close to the true θ. Conclusively, the MLE can be arbitrarily close to the true parameters when $n \longrightarrow \infty$. Figure 3.3 demonstrates a simulation result for an example when the sample size n from 5 to 1000, where it can be seen the estimator tends to reach the real parameter when n increases. The consistency ensures the plausibility of using the maximum likelihood estimation. The consistency of MLE can also be proved by other approach such as using Taylor expansion other than the above discussed method.

3.1.2 Fisher Information

The maximum likelihood estimation is the most acceptable parameter estimating method in statistics. Other than the above-mentioned consistency of the point estimation, the variance of estimator can be derived from the likelihood function, which can offer a great advantage for inference on the estimator. Fisher information matrix can provide the estimator variance in a simple way. Fisher information is a milestone in statistics and informatics. Many hypothesis-test and statistical inference in statistics is based on it. The derivation of the Fisher information matrix requires advanced statistics knowledge. This chapter provides a simple inference of it. One can ignore the inference and go to the conclusion directly if it is necessary.

Definition 3.1 Fisher Information. For a continuous pdf function $f(t;\theta)$ with number of m parameters, written as vector $\theta_{m\times 1}$. Define a vector of random variable as

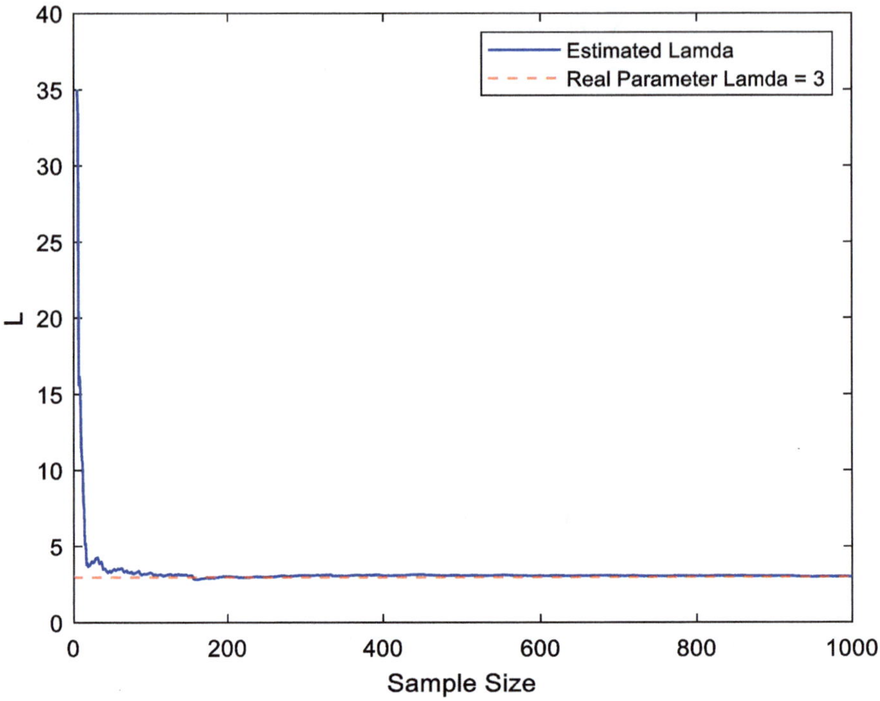

Fig. 3.3 MLE approaches to true parameter

$$S_{\theta}(t) = \left(\frac{\partial \ln f(\boldsymbol{t};\boldsymbol{\theta})}{\partial \theta_1}; \quad \ldots \quad \frac{\partial \ln f(\boldsymbol{t};\boldsymbol{\theta})}{\partial \theta_m} \right)_{m \times 1} \tag{3.11}$$

The Fisher information is defined as

$$\boldsymbol{I}(\boldsymbol{\theta}) = E(S_{\theta}(t) S_{\theta}{}'(t))_{m \times m} \tag{3.12}$$

It is well known $\int f(t;\boldsymbol{\theta})dt = 1$. Then $\int \frac{\partial f(t;\boldsymbol{\theta})}{\partial \theta_i} dt = 0$. Write the $\frac{\partial f(t;\boldsymbol{\theta})}{\partial \boldsymbol{\theta}} = f'(t;\boldsymbol{\theta})$. Therefore

$$E(S_{\theta}(t))_{m \times 1} = E_{\theta}\left[\frac{f'(t;\boldsymbol{\theta})}{f(t;\boldsymbol{\theta})}\right] = \int f'(t;\boldsymbol{\theta})dt = \mathbf{0}_{m \times 1}$$

Therefore, we can find the

$$\boldsymbol{I}(\boldsymbol{\theta}) = E(S_{\theta}(t) S_{\theta}{}'(t))_{m \times m} = E(S_{\theta}(t) S_{\theta}{}'(t))_{m \times m} - E(S_{\theta}(t))_{m \times 1} . E(S_{\theta}{}'(t))_{1 \times m} = Var_{\theta}(S_{\theta}(t)) \tag{3.13}$$

The $S_{\boldsymbol{\theta}}(\boldsymbol{t})$ is the first derivative of function $\ln f(\boldsymbol{t};\boldsymbol{\theta})$. Larger derivative means a higher sensitivity at the true parameter $\boldsymbol{\theta}$, suggesting the $\boldsymbol{\theta}$ more distinct, i.e. containing more information.

For Exponential distribution family distribution, the fisher information matrix must exist.

Example 3.1

Derive the Fisher Information Matrix for exponential distribution with PDF $f(t) = \lambda \exp(-\lambda t)$.

Solution: $\ln f(t;\boldsymbol{\theta}) = \ln \lambda - \lambda t$. Then

$$\frac{\partial \ln f(t;\boldsymbol{\theta})}{\partial \theta_1} = \frac{1}{\lambda} - t$$

$$E(S(t)S'(t))_{m\times m} = E\left\{\left(\frac{1}{\lambda} - t\right)\left(\frac{1}{\lambda} - t\right)\right\} = E\left\{\frac{1}{\lambda^2} - 2\frac{1}{\lambda}t + t^2\right\}$$

$$E(t^2) = \frac{2}{\lambda^2}; E(t^2) = \frac{1}{\lambda};$$

$$E(S(t)S'(t))_{m\times m} = \frac{1}{\lambda^2} - 2\frac{1}{\lambda}\frac{1}{\lambda} + \frac{2}{\lambda^2} = \frac{1}{\lambda^2}$$

The Fisher information matrix is $\frac{1}{\lambda^2}$.

Find the information using (3.12) or (3.13) is complicated. For continuous PDF, a much easy method is to use the formula as follows

$$\boldsymbol{I}(\boldsymbol{\theta}) = -E_{\boldsymbol{\theta}}\left[\frac{\partial^2 \ln f(t;\boldsymbol{\theta})}{\partial \boldsymbol{\theta}^2}\right] \tag{3.14}$$

Proof:

$$\frac{\partial^2 \ln f(t;\boldsymbol{\theta})}{\partial \theta_i \partial \theta_j} = \frac{\frac{\partial^2 \partial^2 f(t;\boldsymbol{\theta})}{\partial \theta_i \partial \theta_j}}{f(t;\boldsymbol{\theta})} - \frac{\frac{\partial f(t;\boldsymbol{\theta})}{\partial \theta_i}}{f(t;\boldsymbol{\theta})} \cdot \frac{\frac{\partial f(t;\boldsymbol{\theta})}{\partial \theta_j}}{f(t;\boldsymbol{\theta})} \tag{3.15}$$

Take expectation of both side of Eq. (3.15).

$$E_{\boldsymbol{\theta}}\left[\frac{\partial^2 \ln f(t;\boldsymbol{\theta})}{\partial \theta_i \partial \theta_j}\right] = E_{\boldsymbol{\theta}}\left[\frac{\frac{\partial^2 f(t;\boldsymbol{\theta})}{\partial \theta_i \partial \theta_j}}{f(t;\boldsymbol{\theta})}\right] - E_{\boldsymbol{\theta}}\left[\frac{\frac{\partial f(t;\boldsymbol{\theta})}{\partial \theta_i}}{f(t;\boldsymbol{\theta})} \cdot \frac{\frac{\partial f(t;\boldsymbol{\theta})}{\partial \theta_j}}{f(t;\boldsymbol{\theta})}\right]$$

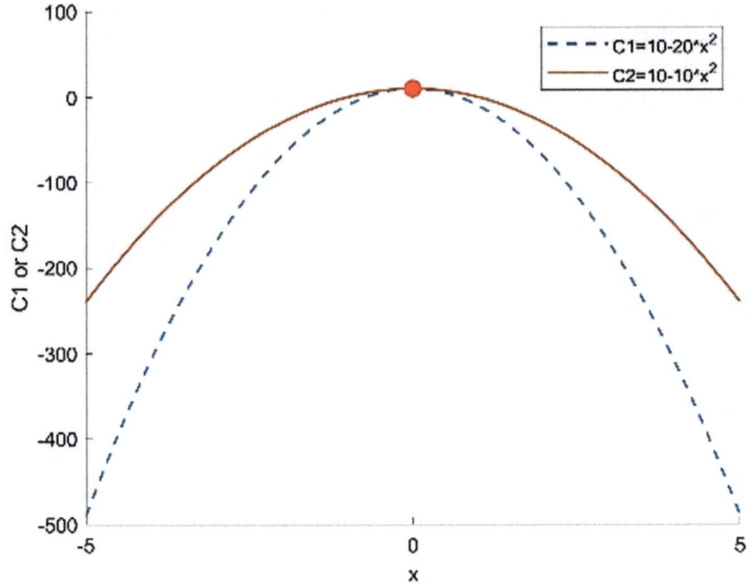

Fig. 3.4 Second derivative and information

It is not hard to find $E_{\boldsymbol{\theta}}\left[\frac{\frac{\partial^2 f(t;\boldsymbol{\theta})}{\partial\theta_i\partial\theta_j}}{f(t;\boldsymbol{\theta})}\right] = \frac{\partial^2}{\partial\theta_i\partial\theta_j}\int f(t;\boldsymbol{\theta})dt = 0$ since $\int f(t;\boldsymbol{\theta})dt = 1$. According to the definition of $\boldsymbol{I}(\boldsymbol{\theta})$ shown in Eq. (3.12), the $E_{\boldsymbol{\theta}}\left[\frac{\frac{\partial f(t;\boldsymbol{\theta})}{\partial\theta_i}}{f(t;\boldsymbol{\theta})}\cdot\frac{\frac{\partial f(t;\boldsymbol{\theta})}{\partial\theta_j}}{f(t;\boldsymbol{\theta})}\right] = I_{ij}$. According to Eq. (3.15), $I_{ij}(\boldsymbol{\theta}) = -E_{\boldsymbol{\theta}}\left[\frac{\partial^2 \ln f(t;\boldsymbol{\theta})}{\partial\theta_i\partial\theta_j}\right]$. Equation (3.14) is proved. The validity of Eq. (3.14) relies on certain regularity conditions. Since most continuous PDF functions satisfy these conditions, this book omits a discussion of them.

To help understanding the relationship between the so-called information and the second derivative shown in Eq. (3.14), we can use an intuitive demonstration to explain it empirically. Geographically, the second derivative represents the degree of curvature, higher degree of curvature, higher the absolute second derivative value. For example, as demonstrated in Fig. 3.4, the curvature at the point marked in the red circle has higher 2nd derivative for C1 than that for C2, and the C1 is less dispersed. The absolute of second derivative of C1 is 20, for C2 is 10. Then from information perspective, the information contains in C1 at the top point (marked in red dot) is more concentrated than that in C2.

Example 3.2
Derive the Fisher Information Matrix using Eq. (3.14) for exponential distribution with PDF $f(t) = \lambda \exp(-\lambda t)$.

Solution: $\ln f(t;\boldsymbol{\theta}) = \ln \lambda - \lambda t$. Then $\frac{\partial \ln f(t;\boldsymbol{\theta})}{\partial \theta_1} = \frac{1}{\lambda} - t$. Then

$$\frac{\partial^2 \ln f(t;\boldsymbol{\theta})}{\partial \boldsymbol{\theta}^2} = -\frac{1}{\lambda^2}.$$

According to Eq. (3.4), the $\boldsymbol{I}(\boldsymbol{\theta}) = -E_{\boldsymbol{\theta}}\left[\frac{\partial^2 \ln f(t;\boldsymbol{\theta})}{\partial \boldsymbol{\theta}^2}\right] = \frac{1}{\lambda^2}$, which is the same as from the Example 3.1.

The Fisher information leads to the Cramer-Rao or Information Inequality which is a fundamental inequality for parameter estimation and for the reliability data analysis.

Theorem 3.1 Suppose the estimator of function $g(\boldsymbol{\theta}_{m\times 1})$, written as $g(\boldsymbol{\theta})$ for simplicity, is the statistics $\widehat{\boldsymbol{\theta}}(\boldsymbol{t})$ and the $\widehat{\boldsymbol{\theta}}(\boldsymbol{t})$ is an unbiased estimator of $g(\boldsymbol{\theta})$. Let $\Delta_{m\times 1} = \frac{\partial g(\boldsymbol{\theta})}{\partial \boldsymbol{\theta}}$. The variance $Var_{\boldsymbol{\theta}}\left(\widehat{\boldsymbol{\theta}}(\boldsymbol{t})\right)$ has a lower bound

$$Var_{\boldsymbol{\theta}}\left(\widehat{\boldsymbol{\theta}}(\boldsymbol{t})\right) \geq \Delta_{m\times 1}{}' I^{-1} \Delta_{m\times 1} \tag{3.16}$$

Formula (3.16) is named Fisher information inequality or Crammer-Rao bound or C-R bound [2].

Proof The $Cov\left(\widehat{\boldsymbol{\theta}}(\boldsymbol{t}), S_{\boldsymbol{\theta}}(t)\right) = E\left(\widehat{\boldsymbol{\theta}}(\boldsymbol{t}) S_{\boldsymbol{\theta}}(t)\right) - E\left(\widehat{\boldsymbol{\theta}}(\boldsymbol{t})\right) E(S_{\boldsymbol{\theta}}(t)) = E\left(\widehat{\boldsymbol{\theta}}(\boldsymbol{t}) S_{\boldsymbol{\theta}}(t)\right)$. The $S_{\boldsymbol{\theta}}(t)$ is defined in Eq. (3.11). The $\widehat{\boldsymbol{\theta}}(\boldsymbol{t})$ is defined as an unbiased estimator of $(\boldsymbol{\theta})$, meaning

$$E\left(\widehat{\boldsymbol{\theta}}(\boldsymbol{t})\right) = \int \widehat{\boldsymbol{\theta}}(\boldsymbol{t}) f(\boldsymbol{t};\boldsymbol{\theta}) d\boldsymbol{\theta} = g(\boldsymbol{\theta}) \tag{3.17}$$

Differentiate Eq. (3.17) on both side in respect to $\boldsymbol{\theta}$, we obtain

$$\int \widehat{\boldsymbol{\theta}}(\boldsymbol{t}) \frac{\partial f(\boldsymbol{t};\boldsymbol{\theta})}{\partial \boldsymbol{\theta}} d\boldsymbol{t} = E\left(\widehat{\boldsymbol{\theta}}(\boldsymbol{t}) S_{\boldsymbol{\theta}}(t)\right) = \frac{\partial g(\boldsymbol{t};\boldsymbol{\theta})}{\partial \boldsymbol{\theta}} = \Delta_{m\times 1} \tag{3.18}$$

Then

$$Cov\left(\widehat{\boldsymbol{\theta}}(\boldsymbol{t}), S_{\boldsymbol{\theta}}(t)\right) = \Delta_{m\times 1} \tag{3.19}$$

Propose a quantity $Var_{\boldsymbol{\theta}}\left(\widehat{\boldsymbol{\theta}}(\boldsymbol{x}) - \boldsymbol{c}'S_{\boldsymbol{\theta}}(x)\right)$, where $\boldsymbol{c}' = (c_1, c_2, \ldots, c_m)$, real valued coefficients. Then according to the definition of variance with dependent variables, we obtain

$$Var_{\boldsymbol{\theta}}\left(\widehat{\boldsymbol{\theta}}(t) - \sum_{i=1}^{m} c_i S_i(t)\right) = Var_{\boldsymbol{\theta}}\left(\widehat{\boldsymbol{\theta}}(\boldsymbol{t})\right) + \boldsymbol{c}'Var_{\boldsymbol{\theta}}(S_{\boldsymbol{\theta}}(t)S_{\boldsymbol{\theta}}(t)')\boldsymbol{c} - 2Cov\left(\widehat{\boldsymbol{\theta}}(t), \boldsymbol{c}'S_{\boldsymbol{\theta}}(t)\right) \tag{3.20}$$

According to Eq. (3.13), the $Var_{\boldsymbol{\theta}}(S_{\boldsymbol{\theta}}(\boldsymbol{t})S_{\boldsymbol{\theta}}(\boldsymbol{t})') = I(\boldsymbol{\theta})$, and according to Eq. (3.19), $Cov\left(\widehat{\boldsymbol{\theta}}(\boldsymbol{t}), \boldsymbol{c}'S_{\boldsymbol{\theta}}(t)\right) = \boldsymbol{c}'\Delta_{m\times 1}$, Eq. (3.20) is rearranged as

$$Var_{\boldsymbol{\theta}}\left(\widehat{\boldsymbol{\theta}}(\boldsymbol{t}) - \sum_{i=1}^{m} c_i S_i(\boldsymbol{t})\right) = Var_{\boldsymbol{\theta}}\left(\widehat{\boldsymbol{\theta}}(\boldsymbol{t})\right) + \boldsymbol{c}'I(\boldsymbol{\theta})\boldsymbol{c} - 2\boldsymbol{c}'\Delta_{m\times 1} \tag{3.21}$$

Since variance is a nonnegative value, the right hand of Eq. (3.21) is then

$$Var_{\boldsymbol{\theta}}\left(\widehat{\boldsymbol{\theta}}(\boldsymbol{t})\right) + \boldsymbol{c}'I(\boldsymbol{\theta})\boldsymbol{c} - 2\boldsymbol{c}'\Delta_{m\times 1} \geq 0$$

i.e.

$$Var_{\boldsymbol{\theta}}\left(\widehat{\boldsymbol{\theta}}(\boldsymbol{t})\right) \geq 2\boldsymbol{c}'\Delta_{m\times 1} - \boldsymbol{c}'I(\boldsymbol{\theta})\boldsymbol{c} \tag{3.22}$$

The $2\boldsymbol{c}'\Delta_{m\times 1} - \boldsymbol{c}'I(\boldsymbol{\theta})\boldsymbol{c}$ is with extremal value at point

$$\boldsymbol{c}^* = I(\boldsymbol{\theta})^{-1}\Delta_{m\times 1} \tag{3.23}$$

which is obtained by find the solution $\boldsymbol{c}^*$ from $\frac{\partial}{\partial c} = 0$. Substitute the $\boldsymbol{c}^*$ into (3.22), Eq. (3.22) is rewritten as

$$Var_{\boldsymbol{\theta}}\left(\widehat{\boldsymbol{\theta}}(\boldsymbol{t})\right) \geq 2\boldsymbol{c}'\Delta_{m\times 1} - \boldsymbol{c}'I(\boldsymbol{\theta})\boldsymbol{c} = 2\Delta_{m\times 1}{}'I(\boldsymbol{\theta})^{-1}\Delta_{m\times 1}$$
$$- \Delta_{m\times 1}{}'I(\boldsymbol{\theta})^{-1}I(\boldsymbol{\theta})I(\boldsymbol{\theta})^{-1}\Delta_{m\times 1} = \Delta_{m\times 1}{}'I(\boldsymbol{\theta})^{-1}\Delta_{m\times 1}.$$

Equation (3.16) thus follows.

In practical situation, it is very common to let the general $g(\boldsymbol{\theta}) = \boldsymbol{\theta}$. The **C-R** bound is then

$$Var_{\boldsymbol{\theta}}\left(\widehat{\boldsymbol{\theta}}(\boldsymbol{t})\right) \geq I^{-1} \tag{3.24}$$

Example 3.3

The Binomial distribution is with PDF $b(x) = \binom{n}{x}\theta^x(1-\theta)^{n-x}$. The expectation of the distribution is $E(x) = n\theta$, variance is $n\theta(1-\theta)$. An estimator is $\widehat{\theta} = \frac{x}{n}$. Derive the C-R lower bound of the estimator.

Solution: Firstly, we shall check the bias of $\widehat{\theta} = \frac{x}{n}$. The $E\left(\widehat{\theta}\right) = \frac{E(x)}{n} = \frac{n\theta}{n} = \theta$. The estimator is unbiased. The Fisher information can be derived from the second derivative of the $\ln b(x)$.

$$\ln b(x) = \ln\binom{n}{x} + x\ln\theta + (n-x)\ln(1-\theta)$$

$$\frac{\partial^2 \ln f(t;\boldsymbol{\theta})}{\partial \boldsymbol{\theta}^2} = -\frac{x}{\theta^2} - \frac{n-x}{(1-\theta)^2}$$

Then $E\left(\frac{\partial^2 \ln f(t;\boldsymbol{\theta})}{\partial \boldsymbol{\theta}^2}\right) = -\frac{E(x)}{\theta^2} - \frac{n-E(x)}{(1-\theta)^2} = -\frac{n}{\theta} - \frac{n}{1-\theta} = -\frac{n}{\theta(1-\theta)}$. The $I(\theta) = \frac{n}{\theta(1-\theta)}$. Since $E\left(\widehat{\theta}\right) = \theta$, we can obtain $g(\theta) = \theta$, $\Delta_{m\times 1} = 1$, $I^{-1} = \frac{\theta(1-\theta)}{n}$, corresponding to the C-R inequality (3.16). Then

$$Var_{\boldsymbol{\theta}}\left(\widehat{\boldsymbol{\theta}}(\boldsymbol{t})\right) \geq \Delta_{m\times 1}{}' I^{-1} \Delta_{m\times 1} = \frac{\theta(1-\theta)}{n}$$

The real variance of the estimator $\frac{x}{n}$ is $Var_{\boldsymbol{\theta}}\left(\widehat{\boldsymbol{\theta}}(\boldsymbol{t}) = \frac{x}{n}\right) = \frac{Var(x)}{n^2} = \frac{n\theta(1-\theta)}{n^2} = \frac{\theta(1-\theta)}{n}$, which is the same as the C-R lower bound. The estimator has reached the C-R lower bound. But it is worth mentioning that not all estimators can reach its C-R lower bound.

Example 3.4

Derive the C-R variance bound for unbiased estimator of Exponential distribution with PDF $f(t) = \lambda\exp(-\lambda t)$.

Solution: From Example 3.1, we obtained the Fisher information matrix for Exponential distribution is

$$I(\lambda) = \frac{1}{\lambda^2}$$

(continued)

Example 3.4 (continued)
According to Eq. (3.24), the C-R lower variance bound for the unbiased estimator of Exponential is then

$$Var\left(\widehat{\lambda}\right) \geq \lambda^2$$

3.1.3 C-R Lower Variance Bound of MLE

The Fisher information discussed above is defined upon the $\ln f(t;\boldsymbol{\theta})$. For n independent observation data t_i; $i = 1, 2, 3, \ldots n$, we can have a log likelihood function as follows as the likelihood function in MLE.

$$\ln L(\boldsymbol{\theta}/\boldsymbol{t}) = \sum_{i=1}^{n} \ln f(t_i)$$

From the definition of the Fisher information matrix, it is easy to find for the n i.i.d data, the

$$\boldsymbol{I_n}(\boldsymbol{\theta}) = n\boldsymbol{I}(\boldsymbol{\theta}) \tag{3.25}$$

Equation (3.25) suggests the overall information from the i.i.d data is just the addition of individual data. For the i.i.d data, the Fisher information matrix $\boldsymbol{I_n}(\boldsymbol{\theta})$ is from the $\ln L(\boldsymbol{\theta}/\boldsymbol{t}) = \sum_{i=1}^{n} \ln f(t_i)$ by

$$\boldsymbol{I_n}\left(\theta_{ij}\right) = -\frac{\partial^2 \ln L(\boldsymbol{\theta}/\boldsymbol{t})}{\partial\theta_i\theta_j} \tag{3.26}$$

The C-R bound of variance of unbiased estimators from Eqs. (3.16) and (3.24) still applies as

$$Var_{\boldsymbol{\theta}}\left(\widehat{\boldsymbol{\theta}}(\boldsymbol{t})\right) \geq \Delta_{m\times 1}{}' I_n{}^{-1} \Delta_{m\times 1} \tag{3.27}$$

In statistics, use Eqs. (3.26) and (3.27) to obtain the variance is common as it can derive from MLE likelihood function.

Example 3.5
Derive the C-R variance bound for unbiased estimator of Exponential distribution with PDF $f(t) = \lambda \exp(-\lambda t)$ with n independent observations t_i; $i = 1, 2, 3, \ldots n$.

Solution: According to Eq. (3.27),

$$-\ln L(\lambda) = -\ln\left\{\lambda e^{-\lambda t_1}\lambda e^{-\lambda t_2}\ldots\lambda e^{-\lambda t_n}\right\} = -n\ln\lambda + \lambda\sum_{i=1}^{i=n} t_i$$

The second derivative of the log likelihood function is

$$\boldsymbol{I}_n = \frac{\partial^2 - \ln L(\lambda)}{\partial\lambda\partial\lambda} = n/\lambda^2$$

The Fisher information is then $\boldsymbol{I}_n = \frac{\partial^2 - \ln L(\lambda)}{\partial\lambda\partial\lambda} = n/\lambda^2$. The C-R lower bound of variance is then

$$Var\left(\widehat{\lambda}\right) \geq \boldsymbol{I}_n{}^{-1} = \lambda^2/n$$

Derivation of the C-R bound is simple and requires only the explicit expression of PDF. The C-R bound is a universal bound for all unbiased estimator, regardless of the estimation method. In reliability data analysis, when the variance of the estimator is not explicit, or the variance from the estimator is too complex to be evaluated, we usually use the C-R bound to approximate the variance of estimators, albeit not exactly in some cases.

3.1.4 Asymptotic Normality of MLE

Asymptotic Normality is an important property of an estimator which lays the foundation of many statistical inference and statistical test. We have proved the MLE estimator, written as $\widehat{\theta}_n$ approach to the true parameter θ_0 when $n \to \infty$. Another important property of MLE is its normality when $n \to \infty$.

Theorem 3.2 The maximum likelihood estimator is asymptotically approaching to a Normal distribution when $n \to \infty$.

$$\left(\widehat{\boldsymbol{\theta}}_n - \boldsymbol{\theta}_0\right) \xrightarrow{n\to\infty} N\left(0, I_n{}^{-1}\right)$$

Proof [2] Denoting the

$$\mathcal{L}_n=\sum_{i=1}^{n}\ln f(t_i;\boldsymbol{\theta});\ \mathcal{L}_n{}'=\sum_{i=1}^{n}\frac{\partial\ln f(t_i;\boldsymbol{\theta})}{\partial\boldsymbol{\theta}};\ \mathcal{L}_n{}''=\sum_{i=1}^{n}\frac{\partial^2\ln f(t_i;\boldsymbol{\theta})}{\partial\boldsymbol{\theta}^2}$$

According to the mean value theorem, for a continuous function $f(x)$ on the closed interval $[a,b]$, there exists a value $c\in(a,b)$ which lets the first derivative of $f^1(c)=\frac{f(b)-f(a)}{b-a}$. Let $\tilde{\theta}$ is a value between $\widehat{\theta}_n$ and θ_0, then

$$\mathcal{L}_n{}''\left(\tilde{\boldsymbol{\theta}}\right)=\frac{\mathcal{L}_n{}'\left(\widehat{\boldsymbol{\theta}}_n\right)-\mathcal{L}_n{}'(\boldsymbol{\theta}_0)}{\widehat{\boldsymbol{\theta}}_n-\boldsymbol{\theta}_0}\tag{3.28}$$

According to the definition of MLE, the $\mathcal{L}_n{}'\left(\widehat{\boldsymbol{\theta}}_n\right)=0$. Rearrange Eq. (3.28) as

$$\widehat{\boldsymbol{\theta}}_n-\boldsymbol{\theta}_0=-\frac{\mathcal{L}_n{}'(\boldsymbol{\theta}_0)}{\mathcal{L}_n{}''\left(\tilde{\boldsymbol{\theta}}\right)}\tag{3.29}$$

In the derivation of Fisher information, we have known

$$S_{\boldsymbol{\theta}}(t)=\left(\frac{\partial\ln f(t;\boldsymbol{\theta})}{\partial\theta_1};\quad\cdots\quad\frac{\partial\ln f(t;\boldsymbol{\theta})}{\partial\theta_m}\right)_{m\times 1}$$

Therefore, the $\mathcal{L}_n{}'(\boldsymbol{\theta}_0)=\sum_{i=1}^{n}S(t_i)$, for each $S(t_i)$, $E(S(t_i))=0$, $Var(S(t_i))=I(\boldsymbol{\theta})$, the

$$E(\mathcal{L}_n{}'(\boldsymbol{\theta}_0))=0;\quad Var(\mathcal{L}_n{}'(\boldsymbol{\theta}_0))=nI(\boldsymbol{\theta})\tag{3.30}$$

According to the central limit theorem,

$$\mathcal{L}_n{}'(\boldsymbol{\theta}_0)\xrightarrow{n\to\infty}N(0,nI(\boldsymbol{\theta}))\tag{3.31}$$

For the $\mathcal{L}_n{}''\left(\tilde{\boldsymbol{\theta}}\right)$ in Eq. (3.29), when $n\to\infty$, $\widehat{\boldsymbol{\theta}}_n\ {}^{n\to\infty}\to\boldsymbol{\theta}_0$, then $\mathcal{L}_n{}''\left(\tilde{\boldsymbol{\theta}}\right)\ {}^{n\to\infty}\to\mathcal{L}_n{}''(\boldsymbol{\theta}_0)$. While $\mathcal{L}_n{}''(\boldsymbol{\theta}_0)=-nI(\boldsymbol{\theta}_0)$ according to Eq. (3.25), therefore,

$$\mathcal{L}_n{}''\left(\tilde{\boldsymbol{\theta}}\right)\xrightarrow{n\to\infty}-nI(\boldsymbol{\theta}_0)\tag{3.32}$$

According to the Slutsky's theorem, and letting $nI(\boldsymbol{\theta})$ denoted by I_n, Eq. (3.29) is rewritten to

$$\widehat{\boldsymbol{\theta}}_n - \boldsymbol{\theta}_0 = -\frac{\mathcal{L}_n{}'(\boldsymbol{\theta}_0)}{\mathcal{L}_n{}''(\tilde{\boldsymbol{\theta}})} \xrightarrow{n\to\infty} \frac{N(0, I_n(\boldsymbol{\theta}))}{I_n(\boldsymbol{\theta})} = N\left(0, I_n{}^{-1}\right)$$

i.e.

$$\left(\widehat{\boldsymbol{\theta}}_n - \boldsymbol{\theta}_0\right) \xrightarrow{n\to\infty} N\left(0, I_n{}^{-1}\right) \tag{3.33}$$

Or equivalently,

$$\widehat{\boldsymbol{\theta}}_n \xrightarrow{n\to\infty} N\left(\boldsymbol{\theta}_0, I_n{}^{-1}\right) \tag{3.34}$$

The normality of MLE does not help the parameter estimation, while it plays a significant role for the parameter inference, e.g. test the significance of parameter, finding the confidence interval of estimate.

Example 3.6
Derive the asymptotic normality of maximum likelihood estimator for distribution with PDF $f(t) = \lambda \exp(-\lambda t)$ with n independent observations t_i; $i = 1, 2, 3, \ldots n$.

Solution: According to Eq. (3.33), and the $I_n = n/\lambda^2$ from Example 3.5, the $I_n{}^{-1} = \lambda^2/n$, the MLE of the exponential distribution follows Normal distribution asymptotically as

$$\widehat{\lambda} \sim N\left(\lambda, \lambda^2/n\right)$$

Simulation is used to valid the efficiency of Eq. (3.34). For the Exponential distribution discussed in Example 3.6, an unbiased estimate of λ is as follows, see Eq. (4.19) in Chap. 4.

$$\widehat{\lambda} = \frac{n-1}{\sum t_i}$$

The real $\lambda = 0.1$ in this simulation and the unbiased $\widehat{\lambda}$ should follow distribution as

$$\widehat{\lambda} \sim N\left(\lambda, \frac{\lambda^2}{n}\right) = N\left(0.1, \frac{0.01}{n}\right)$$

For small sample size, such as $n = 5$, for this case, the normality is not clear, as the distribution is skewed. When the sample size increases, the normality becomes clearer and the $N\left(\lambda, \frac{\lambda^2}{n}\right)$ can fit the data roughly, for example $n = 50$. In Fig. 3.5,

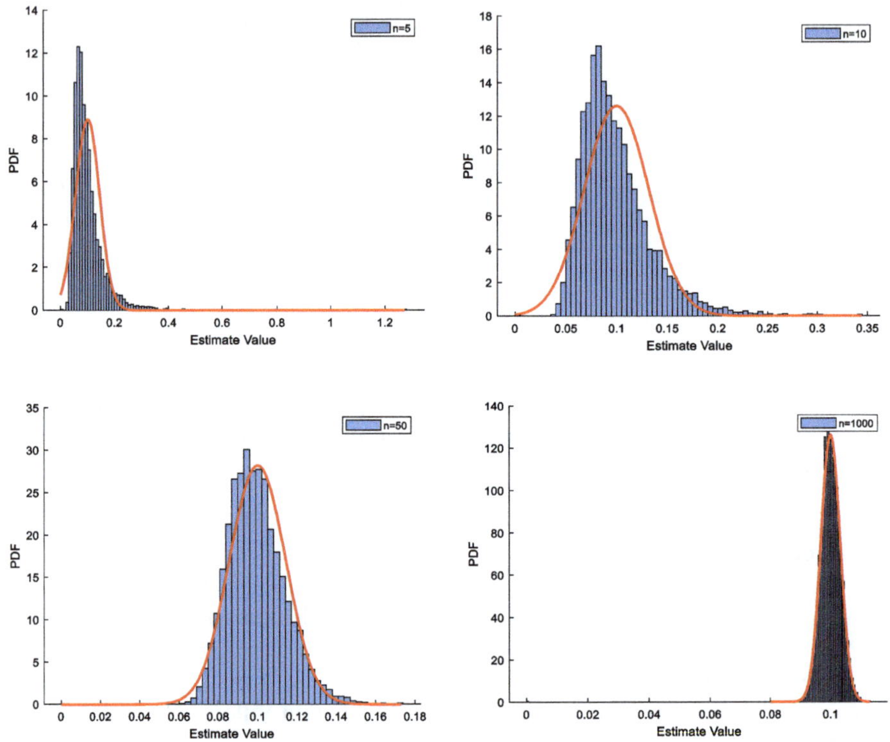

Fig. 3.5 Distribution of estimates for various sample sizes

Table 3.2 Variance with sample size

n	Simulated Variance	C-R Variance	Kurtosis
5	0.0033	0.002	33.58
10	0.0012	0.001	7.4348
50	2.0743×10^{-4}	2.0000×10^{-4}	4.1481
1000	1.2×10^{-5}	1.0×10^{-5}	/

the continuous line is the plot of $N\left(\lambda, \frac{\lambda^2}{n}\right)$. When the sample size is big, the $N\left(\lambda, \frac{\lambda^2}{n}\right)$ can fit the empirical estimates very well, as when $n = 1000$ shown in the last figure in Fig. 3.5. Conclusively, the $N\left(\lambda, \frac{\lambda^2}{n}\right)$ holds when the sample size becomes bigger. The simulation results agree with Eqs. (3.33) and (3.34) well, i.e. the normality holds when sample size becomes big. Table 3.2 illustrates the variance obtained from the simulation and the C-R variance. The variance obtained from simulation shows higher variance than the C-R one for all the sample size n, which is expected. The C-R one is minimum and the real one should be more than it, as the C-R bound defined.

3.2 Moment Estimation Method

Moment method is an alternative parameter estimation method to maximum likelihood estimation. In reliability data analysis, moment method is not commonly used, but for some complex distribution, the moment method can be simpler than maximum likelihood method.

For a random variable, the definition of moment is as

$$\mu_r = E(t^r); \tag{3.35}$$

The r is the order of the moment as $r = 1, 2, 3, \ldots$. For given observation data t_1, t_2, …, t_i, …, t_n where the exact PDF of the data is unknown, Eq. (3.35) is approximated by the empirical moment defined as

$$m_r = \frac{\sum_{i=1}^{n} t_i^r}{n}; \tag{3.36}$$

It is straightforward to have

$$m_r \xrightarrow{n \to \infty} \mu_r = E(t^r) \tag{3.37}$$

Example 3.7

Derive a moment estimator for distribution with PDF $f(t) = \lambda \exp(-\lambda t)$ with n independent observations t_i; $i = 1, 2, 3, \ldots n$.

Solution: It is known the mean of the Exponential is $E(t) = 1/\lambda$. It means the first order of moment $\mu_{r=1} = 1/\lambda$.

$$\lambda = 1/\mu_{r=1}$$

As we discussed, the $\mu_{r=1}$ can be approximated by $m_{r=1} = \frac{\sum_{i=1}^{n} t_i}{n}$. Then let

$$\lambda = \frac{1}{\mu_{r=1}} \sim \frac{1}{\frac{\sum_{i=1}^{n} t_i}{n}} = \frac{n}{\sum_{i=1}^{n} t_i}$$

The moment estimator $\widehat{\lambda}$ can be obtained when t_i is available.

Or alternatively, we know the $Var(t) = 1/\lambda^2$. And the $Var(t) = E(t^2) - E^2(t) = \mu_2 - \mu_1{}^2$. Therefore,

(continued)

Example 3.7 (continued)

$$1/\lambda^2 = \mu_2 - \mu_1^{\,2}$$
$$\lambda^2 = 1/(\mu_2 - \mu_1^{\,2})$$

Therefore,

$$\lambda = \sqrt{1/(\mu_2 - \mu_1^{\,2})}$$

As $\mu_1 \sim m_1 = \frac{\sum_{i=1}^{n} t_i}{n}$; $\mu_2 \sim m_2 = \frac{\sum_{i=1}^{n} t_i^2}{n}$, the estimator is then

$$\widehat{\lambda} = \sqrt{1/\left(\frac{\sum_{i=1}^{n} t_i^2}{n} - \left(\frac{\sum_{i=1}^{n} t_i}{n}\right)^2\right)}$$

Let the two estimators derived in this example as $\lambda_1 = \frac{1}{\mu_1}$ and $\lambda_2 = \sqrt{1/(\mu_2 - \mu_1^{\,2})}$. To check the difference between them, let

$$\lambda_1^{\,2} = \frac{1}{\mu_1^{\,2}}; \lambda_2^{\,2} = 1/(\mu_2 - \mu_1^{\,2})$$

Obviously, the $\mu_1^{\,2} \neq \mu_2 - \mu_1^{\,2}$, implying $\lambda_1 \neq \lambda_2$, suggesting for a same parameter, the estimator values are different. This is drawback of the moment method.

In general, the parameter to be estimated can be written as a function of moment as follows,

$$\theta = g(\mu_1, \mu_2, \mu_3, \ldots) \tag{3.38}$$

For example, in Example 3.7, the $\lambda = 1/\mu_{r\,=\,1}$, the $g(\mu_1, \mu_2, \mu_3, \ldots) = 1/\mu_1$. For $\lambda = \sqrt{1/(\mu_2 - \mu_1^{\,2})}$, the $g(\mu_1, \mu_2, \mu_3, \ldots) = \sqrt{1/(\mu_2 - \mu_1^{\,2})}$.

Theorem 3.3 When the $m_r \overset{n\to\infty}{\longrightarrow} \mu_r$, the $g(m_1, m_2, m_3, \ldots) \overset{n\to\infty}{\longrightarrow} g(\mu_1, \mu_2, \mu_3, \ldots)$ given the function $g(.)$ is continuous, which is the consistency of the moment method. Without providing proof in this chapter, we use consistency as

$$g(m_1, m_2, m_3, \ldots) \overset{n\to\infty}{\longrightarrow} g(\mu_1, \mu_2, \mu_3, \ldots) \tag{3.39}$$

Same as the maximum likelihood method, the moment method is also asymptotic normality, which is a desirable property.

Theorem 3.4 Suppose the observed data are i.i.d as t_1, t_2, …. , t_i, …t_n. The parameters are denoted as vector $\boldsymbol{\theta} = (\theta_1, \theta_2, \theta_3, \ldots \theta_m)'$, and the parameter $\theta_j = g_j(\mu_1, \mu_2, \ldots, \mu_k); j = 1, 2, 3, \ldots m$. Suppose the $g_j(.)$ has a continuous and partial fist derivative and the first derivative is written as $\frac{\partial g_j}{\partial \mu_i}; j = 1, 2, \ldots, m; i = 1, 2, \ldots, k$.

Deonte the $\boldsymbol{G} = \begin{pmatrix} \frac{\partial g_1}{\partial \mu_1} & \frac{\partial g_1}{\partial \mu_2} & \cdots \frac{\partial g_1}{\partial \mu_k} \\ \cdots & \cdots & \cdots \\ \frac{\partial g_m}{\partial \mu_1} & \frac{\partial g_m}{\partial \mu_2} & \cdots \frac{\partial g_m}{\partial \mu_k} \end{pmatrix}_{m \times k}$. If we let $m_r \sim \mu_r$, the moment estimator is

$$\widehat{\boldsymbol{\theta}} = (g_1(m_1, m_2, \ldots, m_k), g_2(m_1, m_2, \ldots, m_k), \ldots g_m(m_1, m_2, \ldots, m_k)). \tag{3.40}$$

Then the $\widehat{\boldsymbol{\theta}}$ asymptotically follows Normal distribution [3]

$$\sqrt{n}\left(\widehat{\boldsymbol{\theta}}_n - \boldsymbol{\theta}_0\right) \xrightarrow{n \to \infty} N(0, \boldsymbol{G\Sigma G'}) \tag{3.41}$$

where $\boldsymbol{\Sigma}_{ij} = \mu_{i+j} - \mu_i \mu_j$. The Normal distribution in this book is referred to $N(\mu, \sigma^2)$, where the μ is mean and the second parameter is variance σ^2 [3].

The property of normality is not trivial as it is important for parameter inference. The proof of Eq. (3.41) is as follows.

Proof Given a random vector $\boldsymbol{m} = (m_1, m_2, \ldots, m_k)_{1 \times k}$ and $m_r \xrightarrow{n \to \infty} \mu_r$ for $r = 1, 2, 3, \ldots k$. The vector follows multivariate normal distribution when $n \to \infty$, according to the multivariate central limit theorem that is written as

$$\sqrt{n}\{(m_1, m_2, \ldots, m_k) - (\mu_1, \mu_2, \ldots, \mu_k)\} \xrightarrow{n \to \infty} \boldsymbol{N}(\boldsymbol{0}, \boldsymbol{\Sigma})$$

The element of $\boldsymbol{\Sigma}_{ij} = Cov(t^i, t^j)$ and the $\boldsymbol{\Sigma}$ is $k \times k$ matrix which is defined as

$$\begin{aligned} Cov\left(t^i, t^j\right) &= E\left\{\left(t^i - E\left(t^i\right)\right)\left(t^j - E\left(t^j\right)\right)\right\} = E\left\{\left(t^i - \mu_i\right)\left(t^j - \mu_j\right)\right\} \\ &= E\left(t^i t^j\right) - \mu_i \mu_j = E\left(t^{i+j}\right) - \mu_i \mu_j = \mu_{i+j} - \mu_i \mu_j \end{aligned}$$

For a single parameter $\theta_j = g_j(\mu_1, \mu_2, \ldots, \mu_k)$, its moment estimator is $\widehat{\theta}_j = g_j(m_1, m_2, \ldots, m_k)$. The

$$\widehat{\theta}_j - \theta_j = g_j(m_1, m_2, \ldots, m_k) - g_j(\mu_1, \mu_2, \ldots, \mu_k) \tag{3.42}$$

Let the $g_j(.)$ function as $g_j(v_1, v_2, \ldots, v_k)$ where v_i denotes a variable. According to the Taylor expansion, the $g_j(v_1, v_2, \ldots, v_k)$ can be expanded at the point $\boldsymbol{\mu} = (\mu_1, \mu_2, \ldots, \mu_k)_{1 \times k}$ as

$$g_j(m_1, m_2, \ldots, m_k) = g_j(\mu_1, \mu_2, \ldots, \mu_k) + \frac{\partial g_j(\boldsymbol{\mu})}{\partial \boldsymbol{v}}(\boldsymbol{m} - \boldsymbol{\mu})' + o(\boldsymbol{m} - \boldsymbol{\mu}) \tag{3.43}$$

As the $o(\boldsymbol{m} - \boldsymbol{\mu}) \rightarrow \mathbf{0}$ when $n \rightarrow \infty$, rearrange Eq. (3.43) as

$$g_j(m_1, m_2, \ldots, m_k) - g_j(\mu_1, \mu_2, \ldots, \mu_k) = \frac{\partial g_j(\boldsymbol{\mu})}{\partial \boldsymbol{v}}(\boldsymbol{m} - \boldsymbol{\mu})'$$

where $\frac{\partial g_j}{\partial \boldsymbol{v}} = \frac{\partial g_j(\boldsymbol{\mu})}{\partial \boldsymbol{v}} = \left(\frac{\partial g_j(\boldsymbol{\mu})}{\partial v_1}, \frac{\partial g_j(\boldsymbol{\mu})}{\partial v_2}, \ldots, \frac{\partial g_j(\boldsymbol{\mu})}{\partial v_k}\right)_{1 \times k}$. According to Eq. (3.42), the

$$\widehat{\theta}_j - \theta_j = g_j(m_1, m_2, \ldots, m_k) - g_j(\mu_1, \mu_2, \ldots, \mu_k) = \frac{\partial g_j(\boldsymbol{\mu})}{\partial \boldsymbol{v}}(\boldsymbol{m} - \boldsymbol{\mu})'$$

Since $\sqrt{n}(\boldsymbol{m} - \boldsymbol{\mu})_{1 \times k} \sim \boldsymbol{N}(\mathbf{0}_{1 \times k}, \boldsymbol{\Sigma})$, then $\sqrt{n}(\boldsymbol{m} - \boldsymbol{\mu})'_{1 \times k} \sim \boldsymbol{N}(\mathbf{0}_{k \times 1}, \boldsymbol{\Sigma})$.

The asymptotic normality for $g_j(m_1, m_2, \ldots, m_k) - g_j(\mu_1, \mu_2, \ldots, \mu_k)$ is then

$$\sqrt{n}\left(\widehat{\theta}_j - \theta_j\right) \sim \boldsymbol{N}\left(0, \left(\frac{\partial \boldsymbol{g}_j}{\partial \boldsymbol{v}}\right)\boldsymbol{\Sigma}\left(\frac{\partial \boldsymbol{g}_j}{\partial \boldsymbol{v}}\right)'\right) \tag{3.44}$$

For all parameter vector $\boldsymbol{\theta} = (\theta_1, \theta_2, \theta_3, \ldots \theta_m)'$, extending Eq. (3.44) to a parameter vector, let

$$\boldsymbol{G} = \left(\frac{\partial \boldsymbol{g}_1}{\partial \boldsymbol{v}}; \frac{\partial \boldsymbol{g}_2}{\partial \boldsymbol{v}}; \ldots; \frac{\partial \boldsymbol{g}_m}{\partial \boldsymbol{v}}\right)_{m \times k} = \begin{pmatrix} \frac{\partial g_1}{\partial \mu_1} & \frac{\partial g_1}{\partial \mu_2} & \cdots & \frac{\partial g_1}{\partial \mu_k} \\ \vdots & \vdots & & \vdots \\ \frac{\partial g_m}{\partial \mu_1} & \frac{\partial g_m}{\partial \mu_2} & \cdots & \frac{\partial g_m}{\partial \mu_k} \end{pmatrix}_{m \times k}$$

Then the $\sqrt{n}\left(\widehat{\boldsymbol{\theta}}_n - \boldsymbol{\theta}_0\right) \xrightarrow{n \rightarrow \infty} N(0, \boldsymbol{G\Sigma G}')$, where the $\widehat{\boldsymbol{\theta}}_n$ and $\boldsymbol{\theta}_0$ are column vector $m \times 1$. Equation (3.41) has been proved. The variance of estimator $\widehat{\theta}_j$ is the diagonal value $\boldsymbol{G}_{jj}$.

Example 3.8
The Normal distribution has two parameters u, σ^2, and the $u=\mu_1$, $\sigma^2=\mu_2-\mu_1^2$. Derive the asymptotic normality for the moment estimators $\widehat{u}=m_1$. $\widehat{\sigma^2}=m_2-m_1^2$.

Solution: The parameter vector $\boldsymbol{\theta}=(\theta_1,\theta_2,\theta_3,\ldots\theta_m)'=(u,\sigma^2)'$. $u=g_1=\mu_1$; $\sigma^2=g_2=\mu_2-\mu_1^2$.

$$\frac{\partial g_1}{\partial \mu_1}=1;\ \frac{\partial g_1}{\partial \mu_2}=0$$

$$\frac{\partial g_2}{\partial \mu_1}=-2\mu_1;\ \frac{\partial g_2}{\partial \mu_2}=1$$

According to Eq. (3.41),

$$\boldsymbol{G}=\begin{pmatrix} \frac{\partial g_1}{\partial \mu_1} & \frac{\partial g_1}{\partial \mu_2} & \cdots \frac{\partial g_1}{\partial \mu_k} \\ \cdots & \cdots & \cdots \\ \frac{\partial g_m}{\partial \mu_1} & \frac{\partial g_m}{\partial \mu_2} & \cdots \frac{\partial g_m}{\partial \mu_k} \end{pmatrix}_{m\times k}=\begin{pmatrix} 1 & 0 \\ -2\mu_1 & 1 \end{pmatrix}_{2\times 2}$$

$$\boldsymbol{\Sigma}=\begin{pmatrix} \mu_2-\mu_1^2 & \mu_3-\mu_1\mu_2 \\ \mu_3-\mu_1\mu_2 & \mu_4-\mu_2^2 \end{pmatrix},$$

where $\boldsymbol{\Sigma}_{ij}=\mu_{i+j}-\mu_i\mu_j$

$$\begin{aligned}\boldsymbol{G\Sigma G}' &=\begin{pmatrix} 1 & 0 \\ -2\mu_1 & 1 \end{pmatrix}\begin{pmatrix} \mu_2-\mu_1^2 & \mu_3-\mu_1\mu_2 \\ \mu_3-\mu_1\mu_2 & \mu_4-\mu_2^2 \end{pmatrix}\begin{pmatrix} 1 & 0 \\ -2\mu_1 & 1 \end{pmatrix}' \\ &=\begin{pmatrix} \mu_2-\mu_1^2 & \mu_3-\mu_1\mu_2 \\ 2\mu_1^3-3\mu_1\mu_2+\mu_3 & -2\mu_1\mu_3+2\mu_1^2\mu_2+\mu_4-\mu_2^2 \end{pmatrix}\begin{pmatrix} 1 & -2\mu_1 \\ 0 & 1 \end{pmatrix} \\ &=\begin{pmatrix} \mu_2-\mu_1^2 & \mu_3-3\mu_1\mu_2+2\mu_1^3 \\ 2\mu_1^3-3\mu_1\mu_2+\mu_3 & -4\mu_1^4+8\mu_1^2\mu_2-4\mu_1\mu_3+\mu_4-\mu_2^2 \end{pmatrix}\end{aligned}$$

According to Theorem 3.4, the $\sqrt{n}\begin{pmatrix} \hat{u}-u \\ \widehat{\sigma^2}-\sigma^2 \end{pmatrix}\sim N(0,\boldsymbol{G\Sigma G}')$. For estimator $\widehat{u}$, $\sqrt{n}(\widehat{u}-u)=N(0,\mu_2-\mu_1^2)$. For $\widehat{\sigma^2}$, the

$$\sqrt{n}\left(\widehat{\sigma^2}-\sigma^2\right)\sim N(0,-4\mu_1^4+8\mu_1^2\mu_2-4\mu_1\mu_3+\mu_4-\mu_2^2)$$

Example 3.9
Suppose the time to failure is as follows

2.5 5.9 3.8 1.2 3.76 1.25 0.24 1.01 6.4 1.8

Suppose the data follows the Exponential distribution. Using moment method to estimate the parameter and evaluate the estimator's variance.

Solution: One moment estimator of the failure rate is $\lambda = 1/\mu_1$ and $\mu_1 = m_1$. Then $\widehat{\lambda} = \frac{n}{\sum_{i=1}^{n} t_i} = 0.3589$.

For the variance, according to the Theorem 3.4, the $\lambda = g(\mu_1) = 1/\mu_1$. The $\frac{\partial g}{\partial \mu_1} = -1/\mu_1^2$. The $\boldsymbol{\Sigma} = \mu_2 - \mu_1^2$, therefore, $\boldsymbol{G\Sigma G'} = \left(-\frac{1}{\mu_1^2}\right)(\mu_2 - \mu_1^2)\left(-\frac{1}{\mu_1^2}\right) = \frac{\mu_2 - \mu_1^2}{\mu_1^4}$. The μ_1 and μ_2 is unknown. One approach is to use the m_1 and m_2 to approximate the μ_1 and μ_2. Then

$$\boldsymbol{G\Sigma G'} \approx \frac{m_2 - m_1^2}{m_1^4} = \frac{\frac{\sum_{i=1}^{n} t_i^2}{n} - \left(\frac{\sum_{i=1}^{n} t_i}{n}\right)^2}{\left(\frac{\sum_{i=1}^{n} t_i}{n}\right)^4} = 0.0669$$

The $\boldsymbol{G\Sigma G}'$ is the variance for $\sqrt{n}\left(\widehat{\boldsymbol{\theta}}_n - \boldsymbol{\theta}_0\right)$. For estimator $\widehat{\boldsymbol{\theta}}_n$, the variance is $\boldsymbol{G\Sigma G}'/n$. Hence, the variance for the $\widehat{\lambda}$ is $\frac{0.0669}{10} = 0.00669$.

According to the Example 3.5, the C-R bound variance is λ^2/n. If we let the true λ as the $\widehat{\lambda} = \frac{n-1}{\sum_{i=1}^{n} t_i} = 0.3230$, the C-R bound is $\frac{\lambda^2}{n} = \frac{0.3230^2}{10} = 0.0104$. The moment variance and C-R bound variance show slight difference.

3.3 Least Squares Parameter Estimation

Least Squares Estimation (LSE) is used in life data analysis as well as in accelerated life data analysis. In reliability engineering, LSE is widely applied in goodness-of-fit tests. However, in the state of the art, theoretical discussions on regression are relatively limited. This section addresses the least squares method from a more theoretical perspective, focusing on estimator bias, estimator variance, and selected statistical tests.

3.3.1 Linear Regression Model

The linear regression model is of the form

$$y = \beta_0 + \beta_1 x_1 + \beta_2 x_2 + \ldots + \beta_r x_r + \varepsilon \tag{3.45}$$

where the ε is assumed to follow Normal distribution with mean 0 and variance σ^2.

When n independent y_i, associated with predictors x_i, $i = 1, 2, \ldots, r$ are observed. The regression can be written in matrix as follows, where the 1 is due to the consideration of the constant threshold.

$$\begin{bmatrix} y_1 \\ y_2 \\ \cdots \\ y_n \end{bmatrix} = \begin{bmatrix} 1 & x_{11} & x_{12} & \cdots & x_{1r} \\ 1 & x_{21} & x_{12} & \cdots & x_{2r} \\ \cdots & \cdots & \cdots & \cdots & \cdots \\ 1 & x_{n1} & x_{n2} & \cdots & x_{nr} \end{bmatrix} \begin{bmatrix} \beta_1 \\ \beta_2 \\ \cdots \\ \beta_{r+1} \end{bmatrix} + \begin{bmatrix} \varepsilon_1 \\ \varepsilon_2 \\ \cdots \\ \varepsilon_n \end{bmatrix} \tag{3.46}$$

Let $\boldsymbol{Y} = \begin{bmatrix} y_1 \\ y_2 \\ \cdots \\ y_n \end{bmatrix}, \boldsymbol{X} = \begin{bmatrix} 1 & x_{11} & x_{12} & \cdots & x_{1r} \\ 1 & x_{21} & x_{12} & \cdots & x_{2r} \\ \cdots & \cdots & \cdots & \cdots & \cdots \\ 1 & x_{n1} & x_{n2} & \cdots & x_{nr} \end{bmatrix}_{n \times p = (r+1)}, \boldsymbol{\beta} = \begin{bmatrix} \beta_1 \\ \beta_2 \\ \cdots \\ \beta_{r+1} \end{bmatrix}, \boldsymbol{\varepsilon} = \begin{bmatrix} \varepsilon_1 \\ \varepsilon_2 \\ \cdots \\ \varepsilon_n \end{bmatrix}.$

Equation (3.45) is rewritten as

$$\boldsymbol{Y} = X\boldsymbol{\beta} + \boldsymbol{\varepsilon} \tag{3.47}$$

Denote the estimator of $\boldsymbol{\beta}$ by $\widehat{\boldsymbol{\beta}}$. The fitted values $\widehat{\boldsymbol{Y}}$ is then given by

$$\widehat{\boldsymbol{Y}} = X\widehat{\boldsymbol{\beta}} \tag{3.48}$$

3.3.2 Least Squares Estimation Method

The least squares method is the method to find the optimal $\boldsymbol{\beta}$ by minimizing the residual between the observed $\boldsymbol{Y}$ and predicted $\widehat{\boldsymbol{Y}}$, written as

$$S(\boldsymbol{\beta}) = \left(\boldsymbol{Y} - \widehat{\boldsymbol{Y}}\right)' \left(\boldsymbol{Y} - \widehat{\boldsymbol{Y}}\right) \tag{3.49}$$

Or since $\widehat{\boldsymbol{Y}} = \boldsymbol{X}\widehat{\boldsymbol{\beta}}$, Eq. (3.49) is rewritten as

$$S(\boldsymbol{\beta}) = \left(\boldsymbol{Y} - \boldsymbol{X}\widehat{\boldsymbol{\beta}}\right)' \left(\boldsymbol{Y} - \boldsymbol{X}\widehat{\boldsymbol{\beta}}\right) \tag{3.50}$$

A standard method to find the optimal $\widehat{\boldsymbol{\beta}}$ from Eq. (3.50) is to take the first derivative as

$$\frac{\partial S(\boldsymbol{\beta})}{\partial \boldsymbol{\beta}} = \mathbf{0} \tag{3.51}$$

The explicit expression of Eq. (3.51) is then

$$-2X'Y + 2X'X\boldsymbol{\beta} = \mathbf{0} \tag{3.52}$$

The Estimator of $\boldsymbol{\beta}$ is then

$$\widehat{\boldsymbol{\beta}} = (X'X)^{-1} X'Y \tag{3.53}$$

The second derivative of (3.50) is

$$\frac{\partial^2 S(\boldsymbol{\beta})}{\partial \boldsymbol{\beta}^2} = 2X'X \tag{3.54}$$

Equation (3.54) is obviously a positive definite matrix when not all observations are identical, conditioned on which, the solution to Eq. (3.51) is unique and minimal.

Example 3.10

Experiments were conducted to obtain the activation energy E_a. The observed data are

x	$-\frac{1}{RT}$	−0.34642	−0.29693	−0.25981	−0.23094	−0.20785
y	ln$k$	−6.71713	−6.23993	−5.8396	−5.43758	−5.07678

The activation energy E_a is the slope of the assumed linear function $y = ax + b$. Estimate the activation energy using regression method.

Solution: The

$$X = \begin{pmatrix} 1 & -0.34642 \\ 1 & -0.29693 \\ 1 & -0.25981 \\ 1 & -0.23094 \\ 1 & -0.20785 \end{pmatrix}; Y = \begin{pmatrix} -6.71713 \\ -6.23993 \\ -5.8396 \\ -5.43758 \\ -5.07678 \end{pmatrix}$$

$$\widehat{\boldsymbol{\beta}} = (X'X)^{-1} X'Y = \begin{pmatrix} -2.7171 \\ 11.7185 \end{pmatrix}$$

The $X'X = \begin{pmatrix} 5 & -1.3420 \\ -1.3420 & 0.3722 \end{pmatrix}$; The 5*0.3722 − (−1.3420) * (−1.3420) = 0.060 > 0, implying the $X'X$ is positive definite. The $\widehat{\boldsymbol{\beta}}$ is the

(continued)

Example 3.10 (continued)
optimal value resulting to minimal residual $S(\beta) = \left(Y - \widehat{Y}\right)'\left(Y - \widehat{Y}\right)$. The discussion of the positive definite is for the demonstration. The $X'X$ is positive definite as the row vectors in X is different from each other. The activation energy E_a, corresponding to the slop a of the linear function is then $\widehat{\boldsymbol{\beta}}(1,2) = 11.7185$. The threshold b is $\widehat{\boldsymbol{\beta}}(1,1) = -2.7171$. The plot of the regression and the original data are shown in Fig. 3.6, where is seen the regression line fits the data well.

3.3.3 Expectation and Variance of LSE Estimator

The inference is of most interest to us in this section as it is very important to reliability data analysis. The statistical inference on regression is simpler than other parameter estimation methods, which is a clear advantage of LSE. The bias of the LSE can be checked firstly. By the definition of linear regression model, it is not hard to find the $E(\boldsymbol{\varepsilon}) = \mathbf{0}$, and observed Y are random with $E(\boldsymbol{Y}) = X\,\boldsymbol{\beta}$ according to Eq. (3.47).

According to Eq. (3.53), the expectation of $\widehat{\boldsymbol{\beta}}$ is [4]

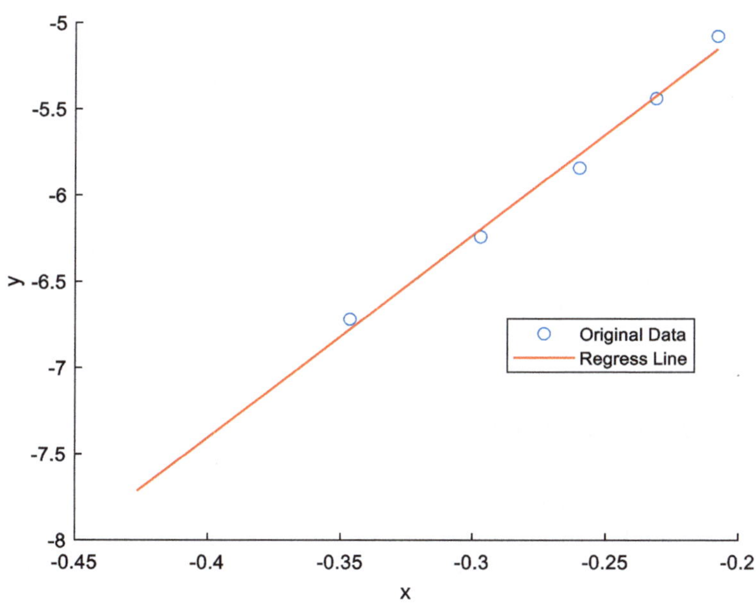

Fig. 3.6 Plot of regression for Example 3.10

$$E\left(\widehat{\boldsymbol{\beta}}\right) = E\left((X'X)^{-1}X'Y\right) = (X'X)^{-1}X'E(\mathbf{Y}) = (X'X)^{-1}X'X\boldsymbol{\beta} = \boldsymbol{\beta}$$

i.e.

$$E\left(\widehat{\boldsymbol{\beta}}\right) = \boldsymbol{\beta} \tag{3.55}$$

The LSE is thus unbiased, a highly desirable property. Another statistic of interest is the variance of $\widehat{\boldsymbol{\beta}}$. According to Eq. (3.53),

$$Var\left(\widehat{\boldsymbol{\beta}}\right) = Var\left\{(X'X)^{-1}X'\mathbf{Y}\right\}$$

The $(X'X)^{-1}X'$ is a matrix of constant and obviously, $Var(\mathbf{Y}) = \sigma^2$. Then [4]

$$\begin{aligned} Var\left(\widehat{\boldsymbol{\beta}}\right) &= \left((X'X)^{-1}X'\right)Var(\mathbf{Y})\left((X'X)^{-1}X'\right)' \\ &= \left((X'X)^{-1}X'\right)\sigma^2\left((X'X)^{-1}X'\right)' = (X'X)^{-1}X'X(X'X)^{-1}\sigma^2 \\ &= (X'X)^{-1}\sigma^2 \end{aligned} \tag{3.56}$$

The residual of regression $\boldsymbol{Y} - \widehat{\boldsymbol{Y}}$ is

$$\widehat{\boldsymbol{\varepsilon}} = Y - X\,(X'X)^{-1}X'Y \tag{3.57}$$

The $\widehat{\boldsymbol{\varepsilon}}$ is a column of $(n \times 1)$.

The $E(\widehat{\boldsymbol{\varepsilon}}) = E(Y) - X\,(X'X)^{-1}X'E(Y) = X\boldsymbol{\beta} - X\,(X'X)^{-1}X'X\boldsymbol{\beta} = \mathbf{0}$ by the $E(\boldsymbol{Y}) = X\boldsymbol{\beta}$. Thus

$$E(\widehat{\boldsymbol{\varepsilon}}) = \mathbf{0}. \tag{3.58}$$

Figure 3.7 plots the residual for the case of Example 3.10. The residual is centered around 0. The empirical mean of the residual $\widehat{\boldsymbol{\varepsilon}}$ is −2.66e-15, a number near 0, as suggested by the $E(\widehat{\boldsymbol{\varepsilon}}) = \mathbf{0}$.

3.3.4 Test Significance of Estimates

In the reliability data analysis, it is often to evaluate if some factors, e.g. the temperature can affect the Y significantly or not. In the regression, the task can be performed by statistical test to check if the corresponding coefficient β_j is zero. The hypothesis is

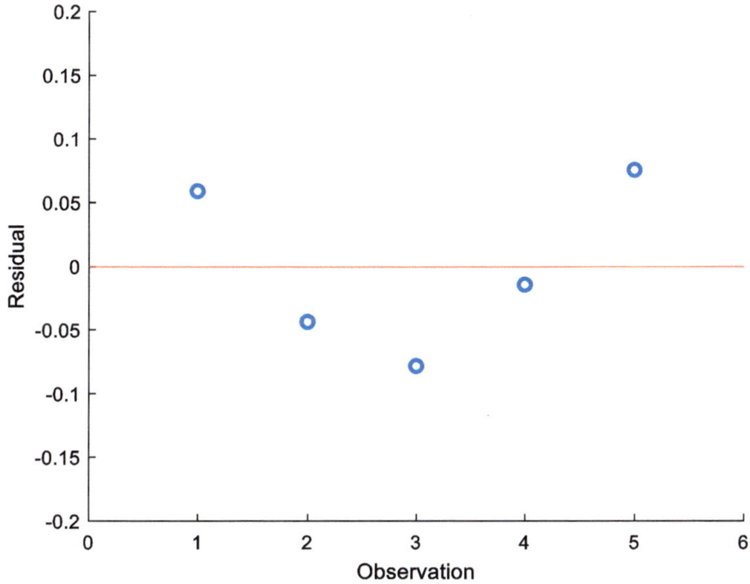

Fig. 3.7 Residual plot for Example 3.10

$$H_0 : \beta_j = 0$$

A t-test can be used. The detailed reason of using the t-test can be found in reference [5]. We have known the $E\left(\widehat{\beta}_j\right) = 0$ according to Eq. (3.55), and the variance is the jth diagonal element of $(X'X)^{-1}\sigma^2$, according to Eq. (3.56). Let the $(X'X)^{-1}$ denoted by D_{jj} for its diagonal element j. The σ^2 is unknown but it can be estimated by the standard deviation of $\widehat{\boldsymbol{\varepsilon}}$ from Eq. (3.57) as [5]

$$S_n{}^2 = \frac{1}{n-p} \sum_{i=1}^{n} (y_i - \widehat{y}_i)^2 \tag{3.59}$$

where $p = r + 1$, corresponding to the total number of parameters.

The test statistics is then [5]

$$t = \frac{\beta_j}{S_{n\sqrt{D_{jj}}}} \tag{3.60}$$

If the quantity of (3.60) is less than $t_{\frac{\alpha}{2},n-p}$ or bigger than $t_{1-\frac{\alpha}{2},n-p}$, the hypothesis $\beta_j = 0$ is then rejected, as illustrated in Fig. 3.8.

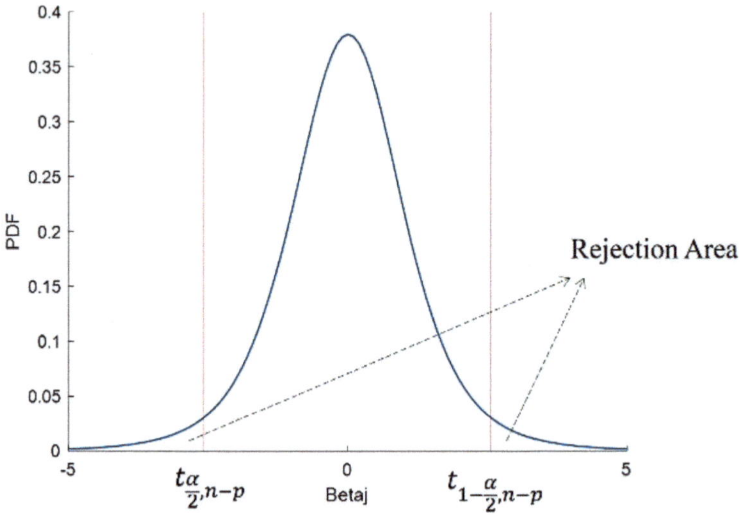

Fig. 3.8 Rejection area of t-test for LSE

Example 3.11a

Test the significance of the parameters $\widehat{\boldsymbol{\beta}}$ in Example 3.10 at significance level $\alpha = 0.05$.

Solution: The significance level $\alpha = 0.05$, The $n = 5$ as it has 5 data sets. The $r = 2$ where one is temperature, and one is the constant threshold in the regression. The $p = r + 1 = 3$.

$$t_{\frac{\alpha}{2},n-p} = t_{0.025,3} = -0.7649; t_{1-\frac{\alpha}{2},n-p} = 0.7649$$

$$S_n = \sqrt{\frac{1}{n-p}\sum_{i=1}^{n}(y_i - \widehat{y}_i)^2} = 0.0763$$

$$(X'X)^{-1} = \begin{pmatrix} 6.1811 & 22.2852 \\ 22.2852 & 83.0327 \end{pmatrix}; \widehat{\boldsymbol{\beta}} = (X'X)^{-1}X'Y = \begin{pmatrix} -2.7171 \\ 11.7185 \end{pmatrix}$$

For $\beta_1 = 0$, the threshold parameter, $D_{jj} = 6.1811$ from $(X'X)^{-1}$ for $j = 1$, the $t = \frac{\beta_j}{S_n\sqrt{D_{jj}}} = \frac{-2.7171}{0.0763*\sqrt{6.1811}} = -14.3287 < -0.7649$. The β_1 is thus significantly not 0.

For $\beta_2 = 0$, the threshold parameter, $D_{jj} = 83.0327$ for $j = 2$, the $t = \frac{\beta_j}{S_n\sqrt{D_{jj}}} = \frac{11.7185}{0.0763*\sqrt{83.0327}} = 16.8612 > 0.7649$. The β_2 is also significantly not 0.

The rejection of $\beta_j = 0; j = 1, 2$ implies both the threshold and the temperature should be in the regression model.

3.3.5 Confidence Interval of Regression

The Confidence Interval (CI) describes the degree of uncertainty of the point estimator. For purpose of reliability data analysis, the prediction $\widehat{y}$ given observed $\boldsymbol{x_0}$ and the CI of this estimator are addressed in this section. The prediction interval given point $\boldsymbol{x_0}$, without providing proof, is as follows [6]

$$\widehat{y}(\boldsymbol{x_0}) \pm t_{\frac{\alpha}{2},n-p} S_n \sqrt{1 + \boldsymbol{x_0}(X'X)^{-1}\boldsymbol{x_0}'} \tag{3.61}$$

where $\widehat{y}(\boldsymbol{x_0}) = \widehat{\beta}_0 + \widehat{\beta}_1 x_1 + \widehat{\beta}_2 x_2 + \ldots + \widehat{\beta}_r x_r$.

The CI for an estimator β_j is [5]

$$\widehat{\beta}_j \pm t_{\frac{\alpha}{2},n-p} S_{n\sqrt{D_{jj}}} \tag{3.62}$$

Example 3.11b

For the case in Example 3.10, given the point $\boldsymbol{x_0} = (1, -0.3079)$, predict the $\widehat{y}(\boldsymbol{x_0})$ and given the CI for the prediction at significance level $\alpha = 0.05$.

Solution: The prediction $\widehat{\boldsymbol{Y}} = X\widehat{\boldsymbol{\beta}} = \boldsymbol{x_0}\widehat{\boldsymbol{\beta}} = (1, -0.3079)\begin{pmatrix} -2.7171 \\ 11.7185 \end{pmatrix} = -6.3255$.

The

$$S_n = \sqrt{\frac{1}{n-p}\sum_{i=1}^{n}(y_i - \widehat{y}_i)^2} = 0.0763;\ (X'X)^{-1} = \begin{pmatrix} 6.1811 & 22.2852 \\ 22.2852 & 83.0327 \end{pmatrix}$$

The $t_{\frac{\alpha}{2},n-p} = -0.7649$. The confidence interval by $\widehat{y}(\boldsymbol{x_0}) \pm t_{\frac{\alpha}{2},n-p} S_n \sqrt{1 + \boldsymbol{x_0}(X'X)^{-1}\boldsymbol{x_0}'}$ of Eq. (3.61) is

$$\widehat{y}(\boldsymbol{x_0}) + t_{\frac{\alpha}{2},n-p} S_n \sqrt{1 + \boldsymbol{x_0}(X'X)^{-1}\boldsymbol{x_0}'} = -6.3255 - 0.7649$$
$$* 0.0763\sqrt{1 + \boldsymbol{x_0}(X'X)^{-1}\boldsymbol{x_0}'} = -6.3928$$
$$\widehat{y}(\boldsymbol{x_0}) - t_{\frac{\alpha}{2},n-p} S_n \sqrt{1 + \boldsymbol{x_0}(X'X)^{-1}\boldsymbol{x_0}'} = -6.3255 - 0.7649$$
$$* 0.0763\sqrt{1 + \boldsymbol{x_0}(X'X)^{-1}\boldsymbol{x_0}'} = -6.2582$$

Conclusively, the $\widehat{y}(\boldsymbol{x_0})$ at $\boldsymbol{x_0} = (1, -0.3079)$ is -6.3255 with interval [−6.3928, −6.2582].

3.3.6 Model Quality

Criteria to check the fitting of the regression model on the observation data is necessary as the assumption of linearity of the model could not always hold. Much work has been done to investigate this issue. The R^2, interpretated as the degree of linear model fitting the data, is a common one indicator which are included by many commercial software packages. There are several variants of R^2 definition, a simple one is [5]

$$R^2 = 1 - \frac{\sum_{i=1}^{n} (\widehat{y}_i - \overline{y})^2}{\sum_{i=1}^{n} (y_i - \overline{y})^2} = 1 - \frac{SS_{Reg}}{SS_{Total}} \tag{3.63}$$

where $\overline{y}$ is the mean of the observed y_i. An improvement of Eq. (3.63) is adjusted $\tilde{R^2}$ defined as [4]

$$\tilde{R^2} = 1 - \frac{(1 - R^2)(n - 1)}{n - p} \tag{3.64}$$

The SS_{Total} is a quantity not dependent on the linear model. Both the R^2 and $\tilde{R^2}$ is dependent on the residual. Less residual leads to a higher R^2 and $\tilde{R^2}$ value, which are both normally ranged between 0 and 1. The higher R^2 value implies higher fitting level.

For the case in Example 3.10, the $R^2 = 0.9894$, the $\tilde{R^2} = 0.9859$. In general, the fitting at this level can be considered as well fitting. The R^2 and $\tilde{R^2}$ can be used as a numerical indicator to replace the visual checking of fitting by checking the plot such as in Fig. 3.6.

3.4 Central Limit Theorem

Central Limit Theorem (CLT) is a widely used theorem that approximates the distribution of samples as a Normal distribution. For reliability data analysis, the central limit theorem can be used to approximate the distribution of failure data as a Normal distribution to simplify problems in some cases. But most often the CLT is used for statistical inference. In this book, the central limit theorem has been applied to evaluate the confidence interval of parameters in Exponential and Weibull distribution in the coming chapters.

Theorem 3.5 [1] Let x_i $(i = 1, \ldots, n)$ be i.i.d (independent and identically distributed) with $E(x_i) = u$ and $Var(x_i) = \sigma^2 < \infty$. Then, $\sqrt{n}(\overline{x} - u)$ tends to a Normal distribution $N(0, \sigma^2)$ and hence $\sqrt{n}(\overline{x} - u)/\sigma$ tends to the standard Normal distribution $N(0, 1)$.

In other words, the Central Limit Theorem (CLT) asserts that, as the sample size increases, the distribution of the sample mean approaches a Normal distribution. Larger sample size tends to have less approximation error. The CLT does not assume the sample x_i follows any distribution, only requires the samples from a same distribution, with the expectation u and a finite variance σ^2. An alternative description of the CLT is the mean of the data is approximated as a Normal distribution with mean $E(\overline{x})$ and variance $Var(\overline{x})$.

$$\overline{x} \sim N(E(\overline{x}), Var(\overline{x})) \tag{3.65}$$

i.e. the $\frac{\overline{x} - E(\overline{x})}{\sqrt{Var(\overline{x})}} \sim N(0, 1)$. Equation (3.65) is more practical than the $\sqrt{n}(\overline{x} - u)/\sigma$. It is easy to verify the $E(\overline{x}) = u$ and $Var(\overline{x}) = \frac{Var(x_1)+Var(x_2)+\ldots}{n^2} = \frac{\sigma^2}{n}$, and thus the Theorem 3.1 and Eq. (3.65) are equivalent. A notable conclusion from the $Var(\overline{x}) = \frac{\sigma^2}{n}$ is $Var(\overline{x}) = 0$ when $n \to \infty$, implying the $\overline{x} = E(\overline{x})$ when $n \to \infty$. The asymptotic normality can be demonstrated by the following example.

This example examines the asymptotic normality of samples from Exponential distribution with parameter $\lambda = 1/10$. The expectation of the distribution is $1/\lambda = 10$, the variance is $1/\lambda^2 = 100$. For the statistics mean $\overline{x}$, we plot the $\overline{x}$ and $Var(\overline{x})$ against the sample size as shown in Fig. 3.9. We can see the $\overline{x}$ tends to the expectation 10 when n becomes larger, as shown on the left of Fig. 3.9, and the $Var(\overline{x})$ tends to zero as shown on the right of Fig. 3.9. It demonstrates larger sample size, less variance in the $\overline{x}$.

The normality of the statistics can be demonstrated by plotting the histogram for various sample sizes. If the distribution is normal, the peak should center around 10 and the distribution should present as a symmetric pattern. Figure 3.10 shows when $n = 5$, the distribution of $\overline{x}$ is not symmetric. But when the sample size increases, the distribution gradually tends to be symmetrical. For $n = 50$, the distribution is roughly symmetric. For the $n = 1000$, the $\overline{x}$ exhibits an obviously

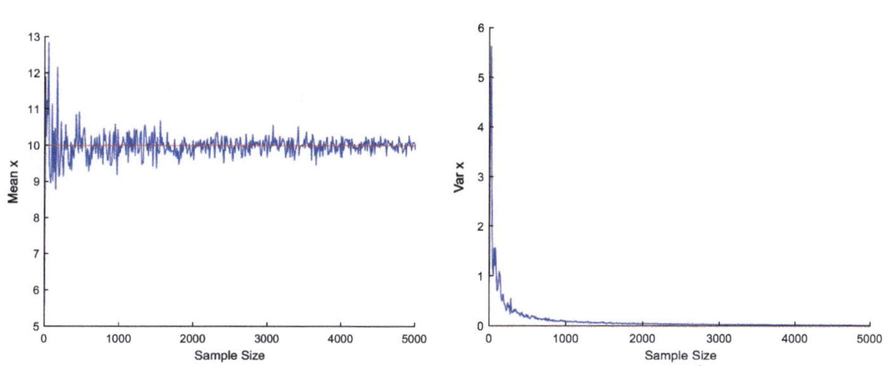

Fig. 3.9 Mean and variance for mean statistics

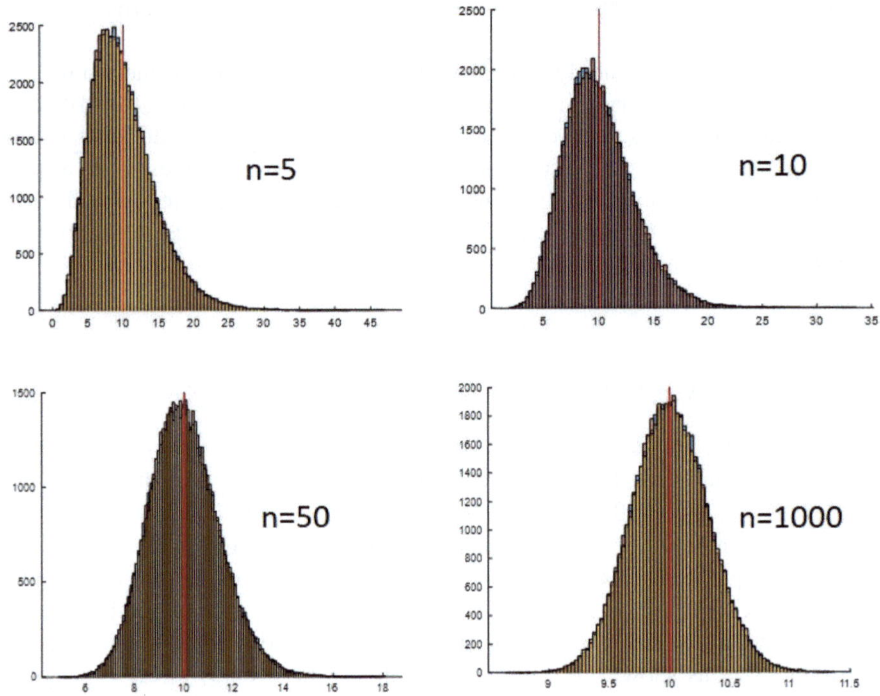

Fig. 3.10 Histogram for sample mean with various sample size

Table 3.3 Kurtosis values

Sample Size	Kurtosis	Sample Size	Kurtosis
$n = 5$	4.13	$n = 50$	3.1238
$n = 10$	3.69	$n = 1000$	3.0076

symmetricity with mean value around 10, implying the distribution of $\overline{x}$ for $n = 1000$ is very close to Normal distribution.

It is well-known the Kurtosis value of Normal distribution is 3. We can see the Kurtosis for the $\overline{x}$ for different sample sizes in Table 3.3. When the $n = 50$, the Kurtosis value is around 3 which implies it is a rough Normal distribution, as also suggested by Fig. 3.10. For $n = 1000$, the Kurtosis value is 3.0076, which is very close to 3.

This demonstration further shows the normality of the sample mean $\overline{x}$ can appear even when the sample size is not big. It can bring significant convenience in data analysis. For situations where the exact distribution of sample mean, such as for Weibull distribution, has no analytical expression, the CLT is a good approach to approximate the sample mean distribution.

The CLT discussed above are the simplest univariate. There is an extension of the CLT considering multivariate as follows.

Theorem 3.6 Given a random vector $\boldsymbol{m} = (m_1, m_2, \ldots, m_k)_{1 \times k}$ and its expectation as μ_r. Providing the $m_r \xrightarrow{n \to \infty} \mu_r$ for $r = 1, 2, 3, \ldots k$, the vector $\boldsymbol{m}$ follows multivariate Normal distribution when $n \to \infty$ as

$$\sqrt{n}\{(m_1, m_2, \ldots, m_k) - (\mu_1, \mu_2, \ldots, \mu_k)\} \xrightarrow{n \to \infty} \boldsymbol{N}(\boldsymbol{0}, \boldsymbol{\Sigma}) \tag{3.66}$$

This extension is important for reliability data analysis. The proof of the theorem is omitted in this book.

3.5 Bias of Estimator

The point estimator refers to a statistics approximating the true parameter in a statistical model. For instance, the mean of samples is an estimator of the mean parameter in the Normal distribution. The inverse of MTTF is an estimator of the failure rate parameter in Exponential distribution. As the true parameter is normally unknown, in practise we can use the estimate as the true parameter. This estimator based on observed data is a statistic of a function of the observed data. Since observed data is random, the estimator is a random variable accordingly. For example, the mean of the sample is an estimator to parameter u of Normal distribution. The following example demonstrates its randomness. Let the observed data, drawn from Normal distribution $u = 1; \sigma^2 = 16$ is

$$x_1 = 2.22; \quad x_2 = -0.97; \quad x_3 = 0.82; \quad x_4 = 0.60; \quad x_5 = 0.36$$

The estimate, written as $\widehat{u}$, is the sample mean as

$$\widehat{u} = \frac{x_1 + x_2 + x_3 + x_4 + x_5}{5} = 0.6060$$

For another observed data from the same distribution provided it is

$$x_1 = 0.09; \quad x_2 = 1; \quad x_3 = -3.91; \quad x_4 = 0.59; \quad x_5 = 6.21$$

The $\widehat{u} = 0.7960$, which differs from 0.6060. It demonstrates the randomness of the observation data leads to randomness of $\widehat{u}$. The $\widehat{u}$ follows certain distribution. Figure 3.11 shows the empirical distribution of 50,000 of estimate $\widehat{u}$.

The randomness in the estimates suggests a desirable property of estimator: an estimator is better if the expectation of the $\widehat{u}$ can equal to the true parameter u, ensuring the $\widehat{u}$ centring around the true u, which is the definition of unbiasedness of estimate, written as

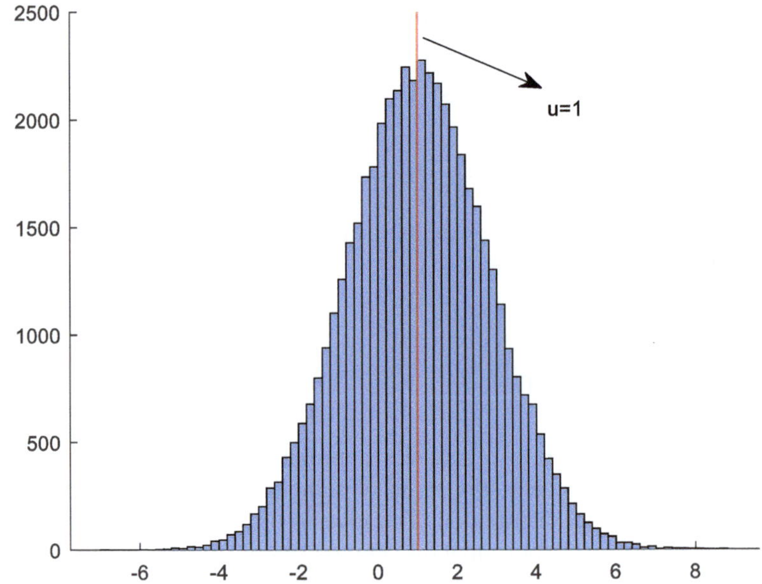

Fig. 3.11 Empirical distribution of estimates

$$E(\widehat{u}) = u \tag{3.67}$$

We have already mentioned many times of bias in the previous sections. Here we discuss it further to highlight the importance of this property as it is often ignored by reliability engineer. The $E(\widehat{u}) = E\left(\frac{x_1+x_2+x_3+x_4+x_5}{5}\right) = u$, suggesting $\frac{x_1+x_2+x_3+x_4+x_5}{5}$ is a unbiased estimator of u. In Fig. 3.11, we can see the $\widehat{u}$ centring around the true $u = 1$, suggesting it is unbiased.

Given another estimator of the u as

$$\widehat{u}_1 = \frac{x_1 + x_2 + \ldots + x_n}{n - 1} \tag{3.68}$$

The $E(\widehat{u}_1) = \frac{n}{n-1} u \neq u$, suggesting a biased estimator. The empirical distribution of $\widehat{u}_1$, shown in Fig. 3.12, presents the mean of the $\widehat{u}_1$ deviating from the true parameter $u = 1$, meaning the estimate $\widehat{u}_1$ not centering the true parameter u, suggesting it is biased.

Asymptotic Unbiasedness

Asymptotic unbiasedness defines the estimator is unbiased when the sample size is infinite. The example shown in Fig. 3.12 uses a sample size of 5, if the sample size increases to infinite, the $E(\widehat{u}_1) = \lim_{n\to\infty} \frac{n}{n-1} u = u$, meaning the estimator $\widehat{u}_1$ is asymptotically unbiased.

The bias issue is concerning when the sample size is small. For the $\widehat{u}_1 = \frac{x_1+x_2+\ldots+x_n}{n-1}$, the $E(\widehat{u}_1) = \frac{n}{n-1} u = u + \frac{1}{n-1} u$. The $\frac{1}{n-1} u$ is the biased part and

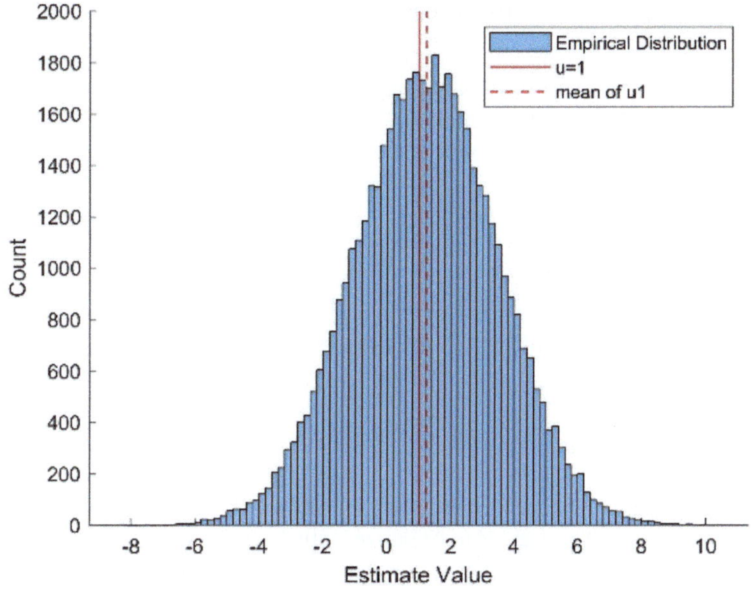

Fig. 3.12 Empirical distribution of biased estimates

decreases with sample size. For $n = 5$, the $\frac{1}{n-1}u = 0.25u$; while for $n = 1001$, $\frac{1}{n-1}u = 0.001u$, which demonstrates the bias is more significant when n is small.

Bias Correctness

Since the unbiasedness is defined as $E(\widehat{u}) = u$. In case the $E(\widehat{u}) \neq u$, we can correct the bias by introducing a constant coefficient C to let $C.E(\widehat{u}) = u$. Then,

$$C = u/E(\widehat{u}) \tag{3.69}$$

For the example of estimator $\widehat{u}_1 = \frac{x_1+x_2+\ldots+x_n}{n-1}$, the $C = (n - 1)/n$. Given the observation data as $x_1 = 0.09$; $x_2 = 1$; $x_3 = -3.91$; $x_4 = 0.59$; $x_5 = 6.21$. The $\widehat{u}_1 = \frac{x_1+x_2+\ldots+x_n}{n-1} = 0.9950$. The corrected one is then $\widehat{u}_1 = \frac{n-1}{n} * 0.9950 = 0.7960$.

Example 3.12

The Poisson distribution is with PDF as $p(x) = \frac{\lambda^x}{x!}e^{-\lambda}$. As shown in the Table 2.5 in the Chap. 2, the mean and MLE are λ and $\widehat{\lambda} = \frac{\sum x_i}{n}$ respectively. Please check the bias of the MLE.

Solution: The expectation $E\left(\widehat{\lambda}\right) = E\left(\frac{\sum x_i}{n}\right) = \frac{\sum E(x_i)}{n} = \frac{n\lambda}{n} = \lambda$. The MLE is thus unbiased.

3.6 Statistical Hypothesis Test

Other than using any subjective approach, the hypothesis test is an objective and stricter method to help decision making. This section addresses two important tests in reliability data analysis: Student t test and the likelihood ratio test.

3.6.1 *Student* t *Test*

According to Eq. (2.34) in Chap. 2, the t test is of the form

$$t = \frac{Z}{\sqrt{V/n}} \tag{3.70}$$

where Z is standard Normal distribution, V follows Chi-squared distribution with degree of freedom n. The t test is to organize statistics to be of the form of Eq. (3.70). In reliability data analysis, for example, a well-known statistic follows the t distribution is as follows

$$\frac{\overline{x} - \mu}{\sqrt{\widehat{\sigma}^2/n}} \sim t_{n-1} \tag{3.71}$$

Let the $\widehat{\sigma}$ estimated by $s = \sqrt{\frac{\sum (x_i - \overline{x})^2}{n-1}}$ and it has been known the $s^2(n - 1)/\sigma^2 \sim \chi^2(n - 1)$. And we have known the $\overline{x}$ follows Normal distribution with mean μ and variance σ^2/n, which is $\frac{\overline{x} - \mu}{\sqrt{\sigma^2/n}} \sim N(0, 1)$. It is also well-known the $\overline{x}$ and the s are independent. According to the definition of Eq. (3.70), the statistics

$$\frac{\frac{\overline{x} - \mu}{\sqrt{\sigma^2/n}}}{\sqrt{\frac{s^2(n-1)}{\sigma^2}/(n-1)}} \sim t_{n-1}$$

Since the $\frac{\frac{\overline{x} - \mu}{\sqrt{\sigma^2/n}}}{\sqrt{\frac{s^2(n-1)}{\sigma^2}/(n-1)}} = \frac{\overline{x} - \mu}{\sqrt{s^2/n}}$, Eq. (3.71) thus follows.

3.6.2 *Likelihood Ratio Function*

For reliability data analysis, other than the t test, likelihood ratio test is another popular approach used for hypothesis test and for the confidence interval derivation. A likelihood ratio is defined as

$$\lambda(t_1, t_2, \ldots, t_n) = \frac{\prod_{i=1}^{n} f\left(t; \widehat{\boldsymbol{\theta}}\right)}{\prod_{i=1}^{n} f(t; \boldsymbol{\theta_0})} \tag{3.72}$$

The $\widehat{\theta}$ is the maximum likelihood estimator, the θ_0 is true parameter. Let

$$\mathcal{L}_n(\boldsymbol{\theta}) = \sum_{i=1}^{n} \ln f(t_i; \boldsymbol{\theta}); \quad \mathcal{L}_n{}'(\boldsymbol{\theta}) = \sum_{i=1}^{n} \frac{\partial \ln f(t_i; \boldsymbol{\theta})}{\partial \boldsymbol{\theta}}; \quad \mathcal{L}_n{}''(\boldsymbol{\theta}) = \sum_{i=1}^{n} \frac{\partial^2 \ln f(t_i; \boldsymbol{\theta})}{\partial \boldsymbol{\theta}^2}$$

Expanding the $\mathcal{L}_n(\boldsymbol{\theta_0})$ at point $\mathcal{L}_n\left(\widehat{\boldsymbol{\theta}}\right)$ by Taylor expansion,

$$\mathcal{L}_n(\boldsymbol{\theta_0}) = \mathcal{L}_n\left(\widehat{\boldsymbol{\theta}}\right) + \mathcal{L}_n{}'\left(\widehat{\boldsymbol{\theta}}\right)\left(\boldsymbol{\theta_0} - \widehat{\boldsymbol{\theta}}\right) + \frac{1}{2}\left(\boldsymbol{\theta_0} - \widehat{\boldsymbol{\theta}}\right)' \mathcal{L}_n{}''(\boldsymbol{\theta_*})\left(\boldsymbol{\theta_0} - \widehat{\boldsymbol{\theta}}\right) \tag{3.73}$$

The $\boldsymbol{\theta}_*$ is between θ_0 and $\widehat{\theta}$. The $\mathcal{L}_n{}'\left(\widehat{\boldsymbol{\theta}}\right) = 0$ since the $\widehat{\boldsymbol{\theta}}$ is the maximum likelihood estimator. Equation (3.73) is then changed as

$$\mathcal{L}_n(\boldsymbol{\theta_0}) - \mathcal{L}_n\left(\widehat{\boldsymbol{\theta}}\right) = \frac{1}{2}\left(\boldsymbol{\theta_0} - \widehat{\boldsymbol{\theta}}\right)' \mathcal{L}_n{}''(\boldsymbol{\theta_*})\left(\boldsymbol{\theta_0} - \widehat{\boldsymbol{\theta}}\right) \tag{3.74}$$

When $n \to \infty$, the $\boldsymbol{\theta}_* \to \boldsymbol{\theta_0}$, $-\mathcal{L}_n{}''(\boldsymbol{\theta_*}) \to I(\boldsymbol{\theta_0})$. The right hand of Eq. (3.74) is rewritten as

$$-\frac{1}{2}\sqrt{n}\left(\boldsymbol{\theta_0} - \widehat{\boldsymbol{\theta}}\right)' I(\boldsymbol{\theta_0})\sqrt{n}\left(\boldsymbol{\theta_0} - \widehat{\boldsymbol{\theta}}\right) \tag{3.75}$$

According to the Theorem 3.2, which states the $\left(\widehat{\boldsymbol{\theta}}_n - \boldsymbol{\theta}_0\right) \xrightarrow{n \to \infty} N\left(0, I_n{}^{-1}\right)$, in this case, which is equivalent to

$$\sqrt{n}\left(\widehat{\boldsymbol{\theta}} - \boldsymbol{\theta}_0\right) \xrightarrow{n \to \infty} N\left(0, I^{-1}(\boldsymbol{\theta_0})\right) \tag{3.76}$$

Therefore, the

$$\sqrt{n} I^{1/2}(\boldsymbol{\theta}_0)\left(\widehat{\boldsymbol{\theta}} - \boldsymbol{\theta}_0\right) \xrightarrow{n \to \infty} N(0, \mathbf{1}_{\boldsymbol{k} \times \boldsymbol{k}})$$

Let the $D = \sqrt{n} I^{1/2}(\boldsymbol{\theta}_0)\left(\widehat{\boldsymbol{\theta}} - \boldsymbol{\theta}_0\right)$. The D is standard multi-variate Normal distribution. The $\sqrt{n}\left(\boldsymbol{\theta_0} - \widehat{\boldsymbol{\theta}}\right)' I(\boldsymbol{\theta_0})\sqrt{n}\left(\boldsymbol{\theta_0} - \widehat{\boldsymbol{\theta}}\right) = D'D$ hence follows $\chi^2(k)$. Recall Eq. (3.74), the

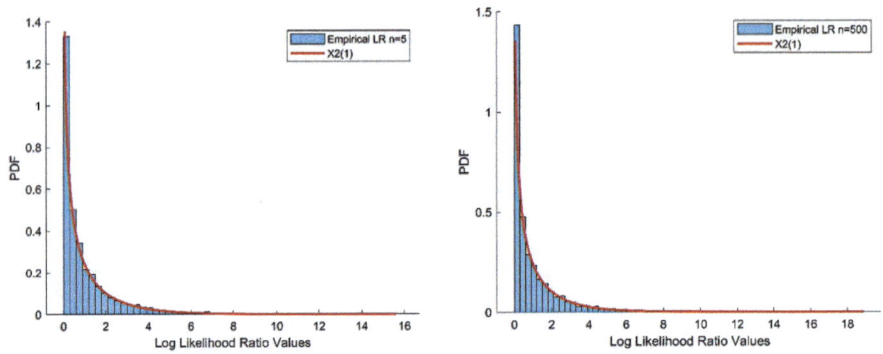

Fig. 3.13 Empirical distribution of likelihood ratio

$$2\left(\mathcal{L}_n\left(\widehat{\boldsymbol{\theta}}\right) - \mathcal{L}_n(\boldsymbol{\theta_0})\right) \sim \chi^2(k) \tag{3.77}$$

which is equivalent as $2\ \ln\ \lambda(t_1, t_2, \ldots, t_n) \sim \chi^2(k)$.

Simulation Check on Asymptotic Distribution $\chi^2(k)$

A simulation is carried to check the validation of Eq. (3.77). The random numbers were drawn from an Exponential distribution of $\lambda = 0.1$. Total 4001 groups of data were generated for sample size $n = 5$ and $n = 500$ respectively. The Exponential has only one parameter meaning $k = 1$. According to Eq. (3.77), the likelihood ratio values therefore asymptotically approach to $\chi^2(1)$ since $k = 1$. As shown in Fig. 3.13, the likelihood ratio can fit the expected distribution $\chi^2(1)$ well even when sample size is as small as 5. Equation (3.77) agrees with the simulation result plots.

Simulation Check on Asymptotic Normality of Fisher Information

The Fisher information plays a key role in the likelihood functioned related statistics. The asymptotic normality is critical to the applicability of likelihood ratio test. We conduct another simulation to check the normality for one parameter Exponential distribution. The Fisher information for the Exponential distribution is as follows

$$I(\lambda) = E\left(-\frac{\partial^2 \ln f(t;\lambda)}{\partial \lambda^2}\right) = E\left(-\frac{\partial^2(\ln \lambda + \lambda t)}{\partial \lambda^2}\right) = \frac{1}{\lambda^2}$$

According to Eq. (3.76) $\sqrt{n}\left(\widehat{\boldsymbol{\theta}}_n - \boldsymbol{\theta}_0\right) \xrightarrow{n\to\infty} N\left(0, I^{-1}(\boldsymbol{\theta_0})\right)$ and $I^{-1}(\boldsymbol{\theta_0}) = \lambda^2$, the

$$\sqrt{n}\left(\widehat{\lambda}_n - \lambda\right) \xrightarrow{n\to\infty} N\left(0, \lambda^2\right)$$

which is equivalent to

$$\widehat{\lambda}_n \xrightarrow{n\to\infty} N\left(\lambda, \frac{\lambda^2}{n}\right)$$

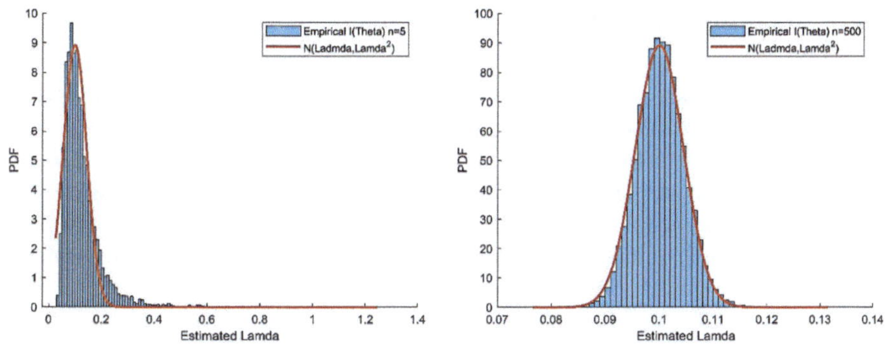

Fig. 3.14 Asymptotic normality of fisher information

The simulation runs 5000 iterations. Random numbers were drawn from Exponential distribution of $\lambda = 0.1$. As shown in Fig. 3.14, the maximum likelihood estimator $\widehat{\lambda}_n$ approaches to normality when $n \to \infty$. The normality can roughly achieve normality even when $n = 5$ as shown on the left of Fig. 3.14. For larger sample size $n = 500$, as shown on the right side of Fig. 3.14, the normality becomes obvious, as the empirical PDF shows a clear symmetrical pattern and it fits the target Normal distribution well.

Example 3.13

Use the likelihood function to derive a confident interval for Exponential distribution.

Solution: According to Eq. (3.77), the $2\left(\mathcal{L}_n(\lambda) - \mathcal{L}_n\left(\widehat{\lambda}\right)\right) \sim \chi^2(1)$. The $\widehat{\lambda}$ is the likelihood estimator.

Given significance level α, The lower bound can be obtained from following equation

$$\chi^2_{\frac{\alpha}{2}}(1) = 2\left(\mathcal{L}_n(\lambda) - \mathcal{L}_n\left(\widehat{\lambda}\right)\right).$$

Upper bound can be obtained from

$$\chi^2_{1-\frac{\alpha}{2}}(1) = 2\left(\mathcal{L}_n(\lambda) - \mathcal{L}_n\left(\widehat{\lambda}\right)\right).$$

To obtain the lower and upper bounds using the above two equations, numerical methods are required. Chapter 3 of this book provides numerical examples of applying the likelihood ratio function to construct confidence intervals for the Exponential distribution.

3.6.3 Likelihood Ratio Test

Section 3.6.3 discussed the asymptotic distribution of likelihood ratio of the likelihood function with MLE against likelihood function with true parameter. In data analysis, albeit not common in reliability data analysis, a situation is for two statistical models: one simple and one complex, such as in the regression data analysis, complex model contains all the factors, while simple model contains part of the factors. The likelihood ratio function can be used to compare the performance between the two models. The likelihood ratio function is defined as:

$$L_R(t_1, t_2, \ldots, t_n) = \frac{\prod_{i=1}^{n} f\left(t; \widehat{\boldsymbol{\theta}}\right)}{\prod_{i=1}^{n} f(t; \widehat{\boldsymbol{\varphi}})} \tag{3.78}$$

Suppose the $\widehat{\boldsymbol{\varphi}}$ is of dimension r, the $\widehat{\theta}$ is of dimension k and $r < k$. The $\prod_{i=1}^{n} f\left(t; \widehat{\boldsymbol{\theta}}\right)$ corresponds to complex model, while the $\prod_{i=1}^{n} f(t; \widehat{\boldsymbol{\varphi}})$ corresponding to simple model. To contain more general situation, let $\theta_j = g_j(\varphi_1, \varphi_2, \ldots, \varphi_r)$; $j = 1, 2, \ldots, k$. The function $g_i(.)$ has continuous first derivative. The derivation of ratio test here is adopted from [2] with slight modification as follows.

Let

$$b_{ij} = \frac{\partial g_j}{\partial \varphi_i}; B = (b_{ij}); i = 1, 2, .., r; j = 1, 2, \ldots, k$$

According to the result from Eq. (3.74) $\mathcal{L}_n(\boldsymbol{\theta_0}) - \mathcal{L}_n\left(\widehat{\boldsymbol{\theta}}\right) = \frac{1}{2}\left(\boldsymbol{\theta_0} - \widehat{\boldsymbol{\theta}}\right)' \mathcal{L}_n''(\boldsymbol{\theta_*})\left(\boldsymbol{\theta_0} - \widehat{\boldsymbol{\theta}}\right)$, we can obtain

$$\mathcal{L}_n\left(\widehat{\boldsymbol{\theta}}\right) - \mathcal{L}_n(\boldsymbol{\theta_0}) = \frac{1}{2}\sqrt{n}\left(\widehat{\boldsymbol{\theta}} - \boldsymbol{\theta_0}\right)' I(\boldsymbol{\theta_0})\sqrt{n}\left(\widehat{\boldsymbol{\theta}} - \boldsymbol{\theta_0}\right) \tag{3.79}$$

Let $M_n = \sqrt{n}\left(\widehat{\boldsymbol{\theta}} - \boldsymbol{\theta_0}\right)$. Equation (3.79) is rewritten as

$$\mathcal{L}_n\left(\widehat{\boldsymbol{\theta}}\right) - \mathcal{L}_n(\boldsymbol{\theta_0}) = \frac{1}{2} M_n' I(\boldsymbol{\theta_0}) M_n \tag{3.80}$$

Let $V_n = \frac{1}{\sqrt{n}}\left(\frac{\partial \ln f(t; \boldsymbol{\theta})}{\partial \theta_1}, \quad \ldots \quad \frac{\partial \ln f(t; \boldsymbol{\theta})}{\partial \theta_k}\right)_{k \times 1}\Bigg|_{\boldsymbol{\theta} = \boldsymbol{\theta_0}}$,

According to the central limit theorem (Recall the Sect. 3.4), the

$$V_n \overset{n \to +\infty}{\longrightarrow} N(0, I(\boldsymbol{\theta_0})) \tag{3.81}$$

Recall Eq. (3.76),

$$\sqrt{n}\left(\widehat{\boldsymbol{\theta}}_n - \boldsymbol{\theta}_0\right) \overset{n \to \infty}{\longrightarrow} N\left(0, I^{-1}(\boldsymbol{\theta_0})\right)$$

We can find the M_n has the same distribution as $I^{-1}(\boldsymbol{\theta_0})V_n$. Then, Eq. (3.80) can be rewritten as

$$\mathcal{L}_n\left(\widehat{\boldsymbol{\theta}}\right) - \mathcal{L}_n(\boldsymbol{\theta_0}) = \frac{1}{2} V_n' I^{-1}(\boldsymbol{\theta_0}) V_n \tag{3.82}$$

Similarly, for the $\prod_{i=1}^{n} f(\boldsymbol{t}; \widehat{\boldsymbol{\varphi}})$, the

$$\mathcal{L}_n(\widehat{\boldsymbol{\varphi}}) - \mathcal{L}_n(\boldsymbol{\varphi}) = \frac{1}{2} U_n' I^{-1}(\boldsymbol{\varphi_0}) U_n \tag{3.83}$$

where $U_n = \frac{1}{\sqrt{n}} \left(\frac{\partial \ln f(\boldsymbol{t}; \boldsymbol{\varphi})}{\partial \varphi_1}, \quad \dots \quad \frac{\partial \ln f(\boldsymbol{t}; \boldsymbol{\varphi})}{\partial \varphi_k} \right)_{r \times 1} \Bigg|_{\boldsymbol{\varphi} = \boldsymbol{\varphi_0}} = BV_n.$

Combining Eq. (3.82) and (3.83), let

$$\begin{aligned} Y_n &= 2\left(\mathcal{L}_n\left(\widehat{\boldsymbol{\theta}}\right) - \mathcal{L}_n(\boldsymbol{\theta_0})\right) - \left(\mathcal{L}_n(\widehat{\boldsymbol{\varphi}}) - \mathcal{L}_n(\boldsymbol{\varphi})\right) = V_n' I^{-1}(\boldsymbol{\theta_0}) V_n \\ &- U_n' I^{-1}(\boldsymbol{\varphi_0}) U_n = V_n' I^{-1}(\boldsymbol{\theta_0}) V_n - (BV_n)' I^{-1}(\boldsymbol{\varphi_0}) BV_n \\ &= V_n' \left\{ I^{-1}(\boldsymbol{\theta_0}) - B' I^{-1}(\boldsymbol{\varphi_0}) B \right\} V_n \end{aligned} \tag{3.84}$$

Since $V_n \ {}^{n \to +\infty} \to N(0, I(\boldsymbol{\theta_0}))$, and let $Z \sim N(0, \mathbf{1}_k)$,and let $I^{\frac{1}{2}}(\boldsymbol{\theta_0})$ written as $I^{\frac{1}{2}}$, Eq. (3.84) is rewritten as

$$\begin{aligned} V_n' \left\{ I^{-1}(\boldsymbol{\theta_0}) - B' I^{-1}(\boldsymbol{\varphi_0}) B \right\} V_n &= Z' I^{\frac{1}{2}} \left\{ I^{-1}(\boldsymbol{\theta_0}) - B' I^{-1}(\boldsymbol{\varphi_0}) B \right\} I^{\frac{1}{2}} Z \\ &= Z' \left\{ \mathbf{1}_k - I^{\frac{1}{2}} B' I^{-1}(\boldsymbol{\varphi_0}) B I^{\frac{1}{2}} \right\} Z \end{aligned} \tag{3.85}$$

Some matrix theory knowledge requires to find the structure of $\mathbf{1}_k - I^{\frac{1}{2}} B' I^{-1}(\boldsymbol{\varphi_0}) B I^{\frac{1}{2}}$. According to the definition of $g_j(.)$, it is easy to find $I(\boldsymbol{\varphi_0}) = BI(\boldsymbol{\theta_0})B'$, the matrix $I^{\frac{1}{2}} B' I^{-1}(\boldsymbol{\varphi_0}) B I^{\frac{1}{2}}$ is a idempotent matrix, which can be check by letting $A = I^{\frac{1}{2}} B' I^{-1}(\boldsymbol{\varphi_0}) B I^{\frac{1}{2}}$, It can be seen

$$A.A = I^{\frac{1}{2}} B' I^{-1}(\boldsymbol{\varphi_0}) B I^{\frac{1}{2}} . I^{\frac{1}{2}} B' I^{-1}(\boldsymbol{\varphi_0}) B I^{\frac{1}{2}} = I^{\frac{1}{2}} B' I^{-1}(\boldsymbol{\varphi_0}) B I^{\frac{1}{2}} = A.$$

And the trace

$$tr(A) = tr\left(I^{\frac{1}{2}}B'I^{-1}(\boldsymbol{\varphi_0})BI^{\frac{1}{2}}\right) = tr\left(I^{-1}(\boldsymbol{\varphi_0})BI^{\frac{1}{2}}I^{\frac{1}{2}}B'\right)$$
$$\times \text{ (by applying } tr(AB) = tr(BA)) = tr\left(I^{-1}(\boldsymbol{\varphi_0})I(\boldsymbol{\varphi_0})\right) = tr(\mathbf{1}_r) = r$$

The $tr\left(\mathbf{1}_k - I^{\frac{1}{2}}B'I^{-1}(\boldsymbol{\varphi_0})BI^{\frac{1}{2}}\right) = k - r$. The trace $k - r$ and the idempotent matrix of $I^{\frac{1}{2}}B'I^{-1}(\boldsymbol{\varphi_0})BI^{\frac{1}{2}}$ implies the $\mathbf{1}_k - I^{\frac{1}{2}}B'I^{-1}(\boldsymbol{\varphi_0})BI^{\frac{1}{2}}$ is of the form $\begin{pmatrix} 1 & \cdots & 0 \\ \cdots & \cdots & \cdots \\ 0 & \ldots 0 & 1 \end{pmatrix}$, only 1 and 0 in the diagonal.

According to Eq. (3.85), the $V_n{}'\{I^{-1}(\boldsymbol{\theta_0}) - B'I^{-1}(\boldsymbol{\varphi_0})B\}V_n$ thus follows $\chi^2(k - r)$. According to Eq. (3.84), the $2\left(\mathcal{L}_n\left(\widehat{\boldsymbol{\theta}}\right) - \mathcal{L}_n(\boldsymbol{\theta_0})\right) - \left(\mathcal{L}_n(\widehat{\boldsymbol{\varphi}}) - \mathcal{L}_n(\boldsymbol{\varphi})\right)$ then also follows $\chi^2(k - r)$. When the $\mathcal{L}_n(\boldsymbol{\theta_0}) = \mathcal{L}_n(\boldsymbol{\varphi})$ by assumption of the two are equal, the

$$2\left(\mathcal{L}_n\left(\widehat{\boldsymbol{\theta}}\right) - \mathcal{L}_n(\widehat{\boldsymbol{\varphi}})\right) \sim \chi^2(k - r) \tag{3.86}$$

Then recall Eq. (3.78), we obtain

$$2 \ln L_R \sim \chi^2(k - r) \tag{3.87}$$

Simulation Validation

The simulation is to validate if the 2 ln L_R~$\chi^2(k - r)$ holds. The simulation uses the two well known distribution in reliability engineering, Exponential and Weibull distribution with PDF as

$$f(t) = \lambda e^{-\lambda t}; \lambda > 0, t \geq 0$$
$$f(t) = \frac{\beta}{\alpha}\left(\frac{t}{\alpha}\right)^{\beta - 1} e^{-(t/\alpha)^\beta}; \alpha > 0, \beta > 0, t \geq 0$$

It is well-known the Exponential is a special case of Weibull distribution, where the $\alpha = \frac{1}{\lambda}; \beta = 1$. The data are drawn from an Exponential distribution with $= 0.1$, that is the real distribution. The empirical likelihood ratio values are obtained by $2\left(\mathcal{L}_n\left(\widehat{\boldsymbol{\theta}}\right) - \mathcal{L}_n(\widehat{\boldsymbol{\varphi}})\right)$, where the $\mathcal{L}_n(\widehat{\boldsymbol{\varphi}})$ is the log likelihood function value from Exponential by fitting the data to Exponential distribution, denoting simple model; and the $\mathcal{L}_n\left(\widehat{\boldsymbol{\theta}}\right)$ is the log likelihood function value from Weibull by fitting the data to Weibull distribution, denoting complex model. Total 5000 iterations are run. The sample size is selected $n = 5$ and $n = 1000$ respectively. The histograms are plotted by the 5000 of 2 ln L_R values, as shown in Fig. 3.15. It can be seen the empirical likelihood ratio values can fit the expected distribution $\chi^2(1)$ well, even for small data size $n = 5$. The fitting for the $n = 1000$ is even better.

If the data are drawn from a Weibull distribution ($\alpha = 10$; $\beta = 2$) instead of Exponential distribution. Sample size chooses 5. The empirical likelihood ratio

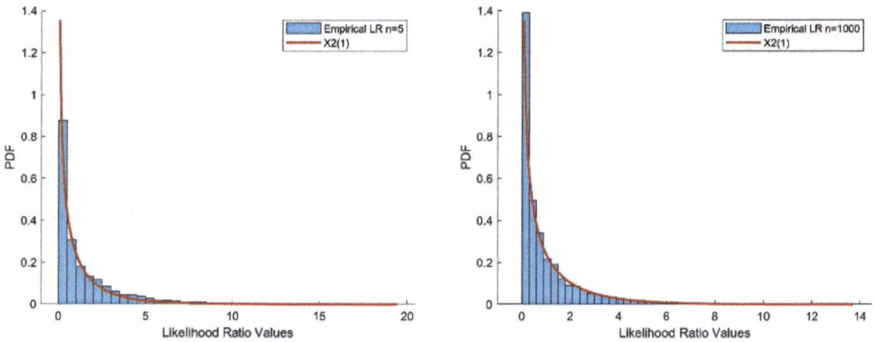

Fig. 3.15 Empirical likelihood ratio fitting with expected χ^2

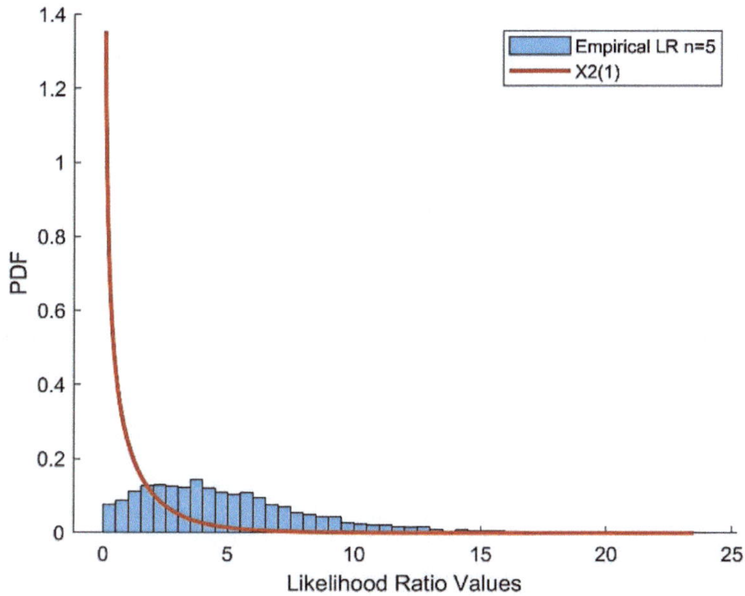

Fig. 3.16 Empirical likelihood ratio fitting χ^2 for Weibull data

values are obtained by $2\left(\mathcal{L}_n\left(\widehat{\boldsymbol{\theta}}\right) - \mathcal{L}_n(\widehat{\boldsymbol{\varphi}})\right)$. The asymptotic distribution $\chi^2(1)$ will then not hold. Figure 3.16 plots the empirical likelihood ratio values fail to fit the $\chi^2(1)$, which suggests the Exponential distribution does not fit the data. From the two simulations, it finds the likelihood ratio approaches to χ^2 when the simple model holds; and if not, the likelihood ratio will not approach to χ^2. It implies using Eq. (3.87) to compare the performance of models is plausible.

Likelihood Ratio for Test
The 2 ln $L_R \sim \chi^2(k - r)$ is upon the assumption of the $\mathcal{L}_n(\boldsymbol{\theta_0}) = \mathcal{L}_n(\boldsymbol{\varphi})$. If the complex model is much better than the simple model, the assumption will be violated and the $\mathcal{L}_n(\boldsymbol{\theta_0}) > \mathcal{L}_n(\boldsymbol{\varphi})$ resulting to a large likelihood ratio value. The rejection area is on the right side of the PDF of χ^2, as shown in Fig. 3.17. If the 2 ln $L_R > \chi^2_{1-\alpha}$, it indicates the complex model is significantly better than the simple model.

The likelihood ratio test gains popularity as the simple calculation since the likelihood function is easy to be obtained. And we only need to compare the likelihood function of the different models, which is highly practical for many cases.

Example 3.14
A group of mining data is as follows (in hours) (Data adapted from [7]):

149; 33; 8; 5; 31; 12; 188; 2; 36; 63; 53; 56; 6; 4; 147; 36; 4; 76; 36; 36; 61; 48; 58; 68; 3; 12; 33; 7; 71; 149; 6; 54; 9; 115; 58; 96; 18; 16; 25; 139; 18; 34; 15; 41

Which one, the Exponential or the Weibull distribution, is better?

Solution: The log likelihood function of exponential is

$$\ln L(\lambda) = n \ln \lambda - \lambda \sum_{i=1}^{i=n} t_i$$

The ln $L(\lambda)$ value for maximum likelihood estimator $\widehat{\lambda} = 0.0206$ is 214.8094.

The log likelihood function of Weibull is

$$\ln L = n \ln \beta - n\beta \ln \alpha + (\beta - 1) \sum \ln t_i - \sum (t_i/\alpha)^\beta$$

The ln $L(\alpha, \beta)$ value for maximum likelihood estimator $\widehat{\alpha} = 49.1079; \widehat{\beta} = 1.0293$ is −214.7800.

The Exponential distribution is a special case of Weibull distribution. The Weibull is the complex model, and the Exponential is the simple model. Therefore,

$$2\left(\mathcal{L}_n\left(\widehat{\theta}\right) - \mathcal{L}_n(\widehat{\varphi}) = 2(-214.7800 - (-214.8094)) = 0.0589$$

According to Eq. (3.86), it follows $\chi^2(k - r) = \chi^2(1)$. For a significant level 0.01, the $\chi^2_{0.99}(1) = 6.6349$ where the 0.0589 is not bigger than 6.6349. Thus the Weibull distribution model does not show significantly better than the simple Exponential model.

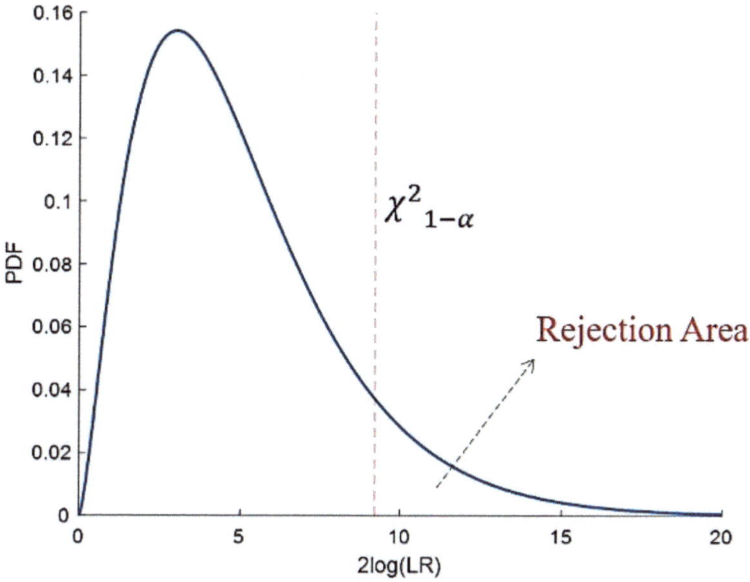

Fig. 3.17 Rejection area of likelihood ratio test

3.7 Statistical Test on Distribution

The distribution analysis assumes the observed data follows certain type of distribution. The distribution test in this section discusses if the data are from the specified distribution or if two population, i.e. two groups of observed data follow the same distribution.

3.7.1 Pearson χ^2 Test

In general, the Pearson test is upon a statistics as [8]

$$\chi^2(k-1) = \sum_{i=1}^{k} \frac{(O_i - E_i)^2}{E_i} \tag{3.88}$$

The hypothesis is

H_0: Data follow the distribution
H_1: Data do not follow the distribution

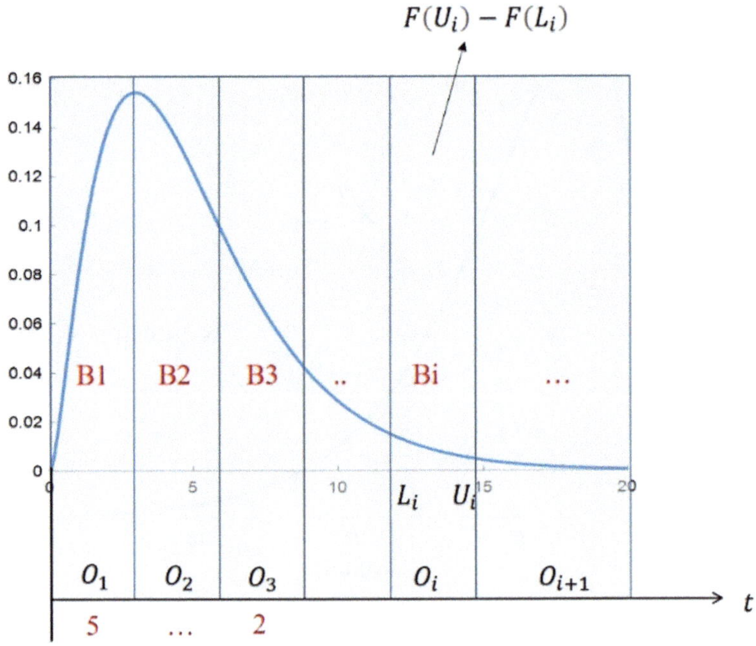

Fig. 3.18 Bins in Pearson test

At significant level α, if the $\sum_{i=1}^{k} \frac{(O_i - E_i)^2}{E_i} \geq \chi^2{}_{1-\alpha}(k-1)$, reject the H_0, suggesting the data does not follow the specified distribution; otherwise, accept the H_0, suggesting the data follows the distribution.

In Eq. (3.88), the O_i is the number of observed failures in group (or bin) i. The E_i is the number of expected failures in group i. The Pearson statistics has degree of freedom $k - 1$. For distribution, either discrete or continuous, when the distribution is known, the probability of each bin can be empirically computed. Suppose the edge of the bin i is L_i, U_i. The probability in this interval is then $F(U_i) - F(L_i)$, as demonstrated in Fig. 3.18. Let the total number of observations is n. The expected number of observations in this bin is then $n(F(U_i) - F(L_i))$. It is preferable the $n(F(U_i) - F(L_i)) \geq 5$ to ensure a close approximate of $\sum_{i=1}^{k} \frac{(O_i - E_i)^2}{E_i}$ to the $\chi^2(k-1)$. A critical question is to determine the number of bins. Matlab chooses the number of bins 10 by default in its Pearson test function chi2gof(.).

Example 3.15

Using the same data as in Example 3.14. The maximum likelihood estimates $\hat{\lambda}=0.0206$ for an Exponential distribution. Does the distribution fit well?

Solution: Choose the number of bins 4. The observed number of failures and the bin edges are shown in the following Fig. 3.19.

The E_1=15.22; E_2=9.16; E_3=6.25; E_4=13.37

$$\sum_{i=1}^{4} \frac{(O_i - E_i)^2}{E_i} = 0.3210$$

At significance level 0.05, the $\chi^2_{1-0.05}(k-1) = 7.8147$ where $k = 4$. The null hypothesis is accepted since $\sum_{i=1}^{4} \frac{(O_i - E_i)^2}{E_i} \leq \chi^2_{1-0.05}(k-1)$. The data is thus said to follow the Exponential distribution with $\hat{\lambda}=0.0206$ well.

When the parameters of the distribution are estimated from the same data set that is used in $\sum_{i=1}^{4} \frac{(O_i - E_i)^2}{E_i}$, the degree of freedom for the test statistics can be adjusted as [9]

$$\chi^2\left(k - 1 - n_{parameter}\right) = \sum_{i=1}^{k} \frac{(O_i - E_i)^2}{E_i} \tag{3.89}$$

The $n_{parameter}$ is the number of parameters in the distribution. For the Example 3.15, the degree of freedom is $k - 1 - n_{parameter} = 2$. The $\chi^2_{1-0.05}(2) = 5.9915$. The $\sum_{i=1}^{4} \frac{(O_i - E_i)^2}{E_i} \leq \chi^2_{1-0.05}(2)$. The null hypothesis remains accepted.

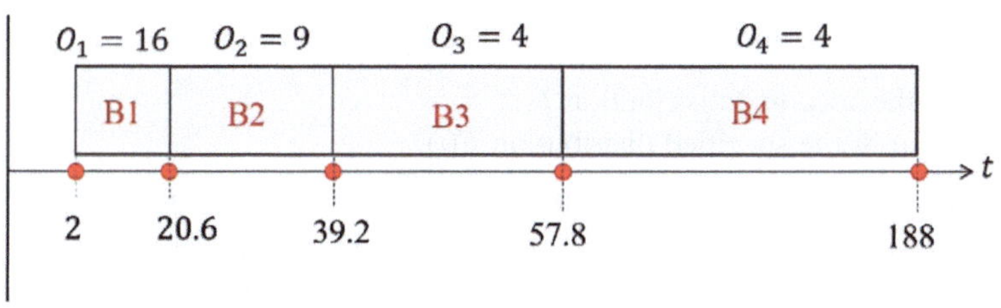

Fig. 3.19 Bins for Pearson test for Example 3.15

3.7.2 Kolmogorov–Smirnov Test

The Kolmogorov–Smirnov(K-S) test compares the difference between the CDF. A large difference suggests the two distributions differ. The statistics is written as follows [10]

$$D_n = \sup_t |F_n(t) - F(t)| \tag{3.90}$$

The $F_n(t)$ is empirical CDF obtained from observation data, the $F(t)$ is the target distribution to be tested. The $F(t)$ in reliability data analysis is normally a continuous function. For example, for the observed time to failure data as

$$t_1, t_2, t_3, \ldots, t_n$$

We assume the data follows Exponential distribution $f(t) = \lambda e^{-\lambda t}$ and the parameter is estimated as $\widehat{\lambda}$ for the distribution. The $F(t)$ is then $1 - e^{-\widehat{\lambda t}}$. For the $F_n(t)$, the empirical distribution, we can estimate it by

$$F_n(t) = \frac{\sum_{i=1}^{n} I(t_i \leq t)}{n} \tag{3.91}$$

For easier understanding, Eq. (3.91) can be obtained in another way. The $F_n(t)$ is the total number of failures till time t divided by the total number of observations. Ascendingly order the time to failure data as $t_{(1)}, t_{(2)}, t_{(3)}, \ldots, t_{(n)}$. The $F_n(t)$ is

$$F_n(t) = \begin{cases} 0, & t < t_{(1)} \\ \frac{r}{n}, & t_{(r)} \leq t < t_{(r+1)} \\ 1, & t \geq t_{(r+1)} \end{cases} \tag{3.92}$$

For hypothesis test, when the D_n is larger than a critical value, the data does not fit the distribution $F(t)$; otherwise, accept it. The hypothesis is:

H_0: Data fit the specified distribution $F(t)$
H_1: Data dont fit the specified distribution $F(t)$

The critical value (CV) is independent of the specified distribution and can be obtained by simulation or find it from standard table in some handbooks. If the H_0 holds, the $F(t)$ follows uniform distribution [0, 1]. One method to obtain the CV is from the empirical distribution of the D_n given the sample size n by using of simulation. Alternatively, for large data set $n > 40$, the critical value can be approximated by as follows Table 3.4.

Table 3.4 Critical value of K-S test [11]

α	0.20	0.10	0.05	0.02	0.01
CV	$1.07/\sqrt{n}$	$1.22/\sqrt{n}$	$1.36/\sqrt{n}$	$1.52/\sqrt{n}$	$1.63/\sqrt{n}$

Example 3.16

For the data from Example 3.14, using K-S test to decide if the maximum likelihood estimator $\widehat{\lambda} = 0.0206$ for Exponential distribution fitting these data.

Solution: Calculate the $|F_n(t_i) - F(t_i)|$ for all the t_i as shown in Table 3.5. The maximum $|F_n(t_i) - F(t_i)|$, corresponding to $D_n = 0.0630$.

According to Table 3.4, the CV for $\alpha = 0.05$ is $1.36/\sqrt{n} = 1.36/\sqrt{44} = 0.2050$. The hypothesis H_0 cannot be rejected since $D_n = 0.0630 < 0.2050$, i.e. the Exponential distribution $f(t) = \lambda e^{-\lambda t}$ for $\widehat{\lambda} = 0.0206$ can be accepted. Figure 3.20 shows the plots of $F_n(t_i)$ and $F(t_i)$. It shows the $F(t_i)$ can fit the $F_n(t_i)$ fit in general. But the visual check is not a good approach. In some plots, even ill-fitting cases do not show obvious deviation between the two. The plots in this example are just to demonstrate the plots of $F_n(t_i)$.

3.7.3 Q-Q Plot

Quantile-quantile (Q-Q) plot is a graphical approach to evaluate if the observed data can fit the specified distribution. The Q-Q plot is upon a simple principle: If the observed data can follow the specified distribution perfectly, the quantile estimated from the observed data will be same as the quantile calculated from the specified distribution. Therefore, graphically, if we let the quantile from observed data and the quantile from specified distribution be x and y direction respectively in a two-dimensional plot, the quantile point should be on the straight line $y = x$. If the quantile points deviate from the line $y = x$ significantly, the observations are then considered not fitting the specified distribution.

In the K-S test, we have used Eq. (3.92) to approximate the CDF. If the $F_n\left(t_{(r)}\right) = \frac{r}{n}$, the $t_{(r)}$ is quantile for the CDF of probability $\frac{r}{n}$. The theoretical CDF function is the assumed distribution, e.g. $F(t) = 1 - e^{-\lambda t}$ for Exponential distribution. The quantile for the same $\frac{r}{n}$ obtained from the specified CDF is then

$$t_{xr} = -\frac{\ln\left(1 - \frac{r}{n}\right)}{\lambda} \tag{3.93}$$

In general, the t_{xr} can be obtained from $F^{-1}\left(\frac{r}{n}\right)$. For a total n observations, we can then obtain n point $(t_{xr}, t_{(r)})$. Plotting the $(t_{xr}, t_{(r)})$ and a straight line $y = x$ together, we therefore have obtained the Q-Q plot. It is worth noting that the CDF $F_n(t_{(r)})$ can be approximated by other approaches, such as $\frac{r-0.5}{n}$ used in the Matlab, or other approximations such as $\frac{r-0.3}{n+0.4}$, or $\frac{r}{n+1}$. We will discuss this approximation later since it is used widely throughout this book.

Table 3.5 Calculation results for Example 3.16

i	**1**	**2**	**3**	**4**	**5**	**6**	**7**	**8**	**9**	**10**	**11**	**12**
t_i	2.00	3.00	4.00	4.00	5.00	6.00	6.00	7.00	8.00	9.00	12.00	12.00
$F_n(t_i)$	0.02	0.05	0.07	0.09	0.11	0.14	0.16	0.18	0.20	0.23	0.25	0.27
$F(t_i)$	0.04	0.06	0.08	0.08	0.10	0.12	0.12	0.13	0.15	0.17	0.22	0.22
$\lvert F_n(t_i) - F(t_i)\rvert$	0.02	0.01	0.01	0.01	0.02	0.02	0.04	0.05	0.05	0.06	0.03	0.05
i	**13**	**14**	**15**	**16**	**17**	**18**	**19**	**20**	**21**	**22**	**23**	**24**
t_i	15.00	16.00	18.00	18.00	25.00	**31.00**	33.00	33.00	34.00	36.00	36.00	36.00
$F_n(t_i)$	0.30	0.32	0.34	0.36	0.39	**0.41**	0.43	0.45	0.48	0.50	0.52	0.55
$F(t_i)$	0.27	0.28	0.31	0.31	0.40	**0.47**	0.49	0.49	0.50	0.52	0.52	0.52
$\lvert F_n(t_i) - F(t_i)\rvert$	0.03	0.04	0.03	0.05	0.02	**0.0630**	0.0616	0.04	0.03	0.02	0.00	0.02
i	**25**	**26**	**27**	**28**	**29**	**30**	**31**	**32**	**33**	**34**	**35**	**36**
t_i	36.00	41.00	48.00	53.00	54.00	56.00	58.00	58.00	61.00	63.00	68.00	71.00
$F_n(t_i)$	0.57	0.59	0.61	0.64	0.66	0.68	0.70	0.73	0.75	0.77	0.80	0.82
$F(t_i)$	0.52	0.57	0.63	0.66	0.67	0.68	0.70	0.70	0.72	0.73	0.75	0.77
$\lvert F_n(t_i) - F(t_i)\rvert$	0.04	0.02	0.01	0.03	0.01	0.00	0.01	0.03	0.03	0.05	0.04	0.05
i	**37**	**38**	**39**	**40**	**41**	**42**	**43**	**44**				
t_i	76.00	96.00	115.00	139.00	147.00	149.00	149.00	188.00				
$F_n(t_i)$	0.84	0.86	0.89	0.91	0.93	0.95	0.98	1.00				
$F(t_i)$	0.79	0.86	0.91	0.94	0.95	0.95	0.95	0.98				
$\lvert F_n(t_i) - F(t_i)\rvert$	0.05	0.00	0.02	0.03	0.02	0.00	0.02	0.02				

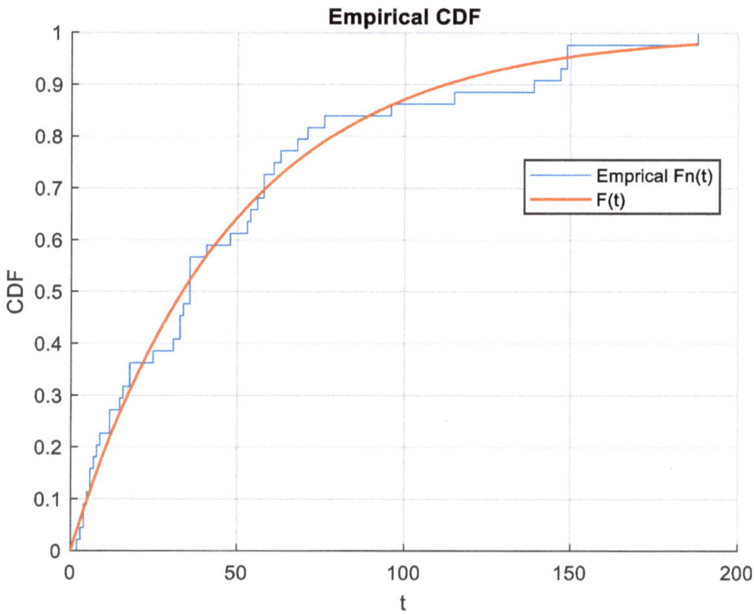

Fig. 3.20 Plot of empirical CDF

Example 3.17
A group of observed time to failure data as follows [12]

8 11 12 18 19 19 21 31 41 47 51 61 99

Does the data fit the Exponential distribution with parameter estimated from maximum likelihood estimation?

Solution: The maximum likelihood estimator from the data is $\widehat{\lambda} = \frac{n}{\sum_{i=1}^{i=n} t_i} = 0.0297$ (Refer to Chap. 3 for MLE on Exponential distribution). The specified distribution is then $f(t) = \widehat{\lambda} e^{-\widehat{\lambda} t}$. The CDF is $1 - e^{-\widehat{\lambda} t}$. Order the time ascendingly, shown in the 2nd row of Table 3.6. If we use the $\frac{r-0.5}{n}$ to approximate CDF, for $r = 1$, as shown in Table 3.6, the $F_n(t_{(r)}) = \frac{r-0.5}{n} = \frac{1-0.5}{13} = 0.0385 \approx 0.04$. The quantile $t_{xr} = -\frac{\ln\left(1 - \frac{r-0.5}{n}\right)}{\widehat{\lambda}} = -\frac{\ln(1-0.0385)}{0.0297} = 1.3214$. The first quantile point is then $(1.3214, 8)$. The other points can be calculated similarly.

Plot the point (x_r, y_r) and the line $y = x$, shown in Fig. 3.21. The empirical quantile point can fit the straight line in general. The Exponential distribution with $\widehat{\lambda} = 0.0297$ is then acceptable.

Table 3.6 Quantile calculation results

r	1	2	3	4	5	6	7	8	9	10	11	12	13
$y_r = t_r$	8.00	11.00	12.00	18.00	19.00	19.00	21.00	31.00	41.00	47.00	51.00	61.00	99.00
$\frac{r-0.5}{n}$	0.04	0.12	0.19	0.27	0.35	0.42	0.50	0.58	0.65	0.73	0.81	0.88	0.96
$x_r = F^{-1}\left(\frac{r-0.5}{n}\right)$	1.32	4.13	7.20	10.57	14.32	18.53	23.35	28.98	35.74	44.21	55.55	72.76	109.77

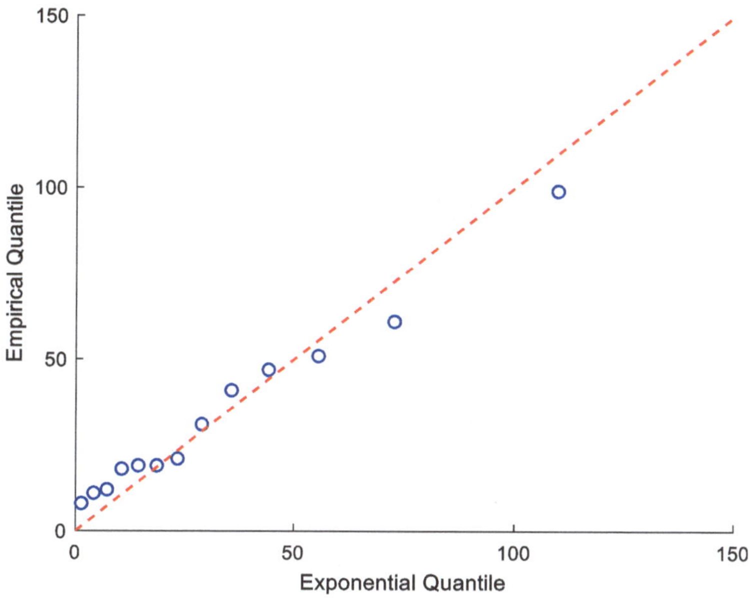

Fig. 3.21 QQ-plot for Example 3.17

The graphical Q-Q plot is simple in its principle. It only requires an explicit expression of CDF function $F(t)$, which is available for most continuous distribution in reliability data analysis.

3.7.4 CDF Approximation

The K-S test, Q-Q plot and some regression approaches for parameter estimation require the approximation of the CDF, as the approximated CDF is considered as the observed CDF to be fitted by the estimated CDF function. In state of art, there are several popular approximations as

- Mean (Herd-Johnson) approximation: $\frac{i}{n+1}$
- Mode (Kaplan-Meier approximation): $\frac{i}{n}$
- Median rank (Benard) approximation: $\frac{i-0.3}{n+0.4}$
- Modified Kaplan-Meier approximation: $\frac{i-0.5}{n}$

In this section, we present the derivation for the first three approximations.

Mean (Herd-Johnson) Approximation

Given the ith order at time t_i, the underlying probability of failure is $F(t_i)$. The probability of number of i failure at t_i is

$$\binom{n}{i} F(t_i)^i (1 - F(t_i))^{n-i} \tag{3.94}$$

The probability of $t_{(i)} \leq t$ is

$$F_i(t) = \sum_{j=i}^{n} \binom{n}{j} F(t)^j (1 - F(t))^{n-j} \tag{3.95}$$

Let $F(t) = y$. Rewritten Eq. (3.95) as for simplicity of denotation as [13]

$$G_i(y) = \sum_{j=i}^{n} \binom{n}{j} y^j (1 - y)^{n-j} \tag{3.96}$$

which is equal to [13]

$$G_i(y) = \frac{n!}{(i-1)!(n-i)!} \int_0^y u^{i-1} (1 - u)^{n-i} du \tag{3.97}$$

which is the Pearson's incomplete Beta distribution, and it is considered as the PDF of y.

$$Beta(i, n - i + 1)$$

The expectation for Beta distribution $Beta(\alpha, \beta)$ is $E(x) = \frac{\alpha}{\alpha+\beta}$, then for the ith order at time t_i, the

$$E(x) = \frac{\alpha}{\alpha + \beta} = \frac{i}{n+1}$$

Therefore, $F(t_i) \approx \frac{i}{n+1}$, which is the ***mean*** approximation. This approximation is also named Herd-Johnson approximation [14].

Mode (Kaplan-Meier) Approximation

From Eq. (3.97), we can easily get the PDF of the order i as $f(y) = \frac{n!}{(i-1)!(n-i)!} y^i (1 - y)^{n-i}$. The maximal point of $f(y)$, named mode in statistics, can be found from the $\ln f(y)$, similar to the maximum likelihood estimation, which is

$$\ln f(y) \sim i \ln y + (n - i) \ln(1 - y) \tag{3.98}$$

Take the first derivative of Eq. (3.98) and equal it to zero, it is obtained

$$\frac{i}{y} = \frac{n-i}{1-y} \tag{3.99}$$

The $\widehat{y}$, corresponding to the estimated CDF $F(t_i)$, is

$$\widehat{y} = \frac{i}{n}$$

Therefore, $F(t_i) \approx \frac{i}{n}$, which is the **mode** approximation. The model approximation is also named Kaplan-Meier approximation [14].

Median Rank (Benard) Approximation

The median rank is a popular approximation, but also a controversial approach. The derivation of the Benard rank here follows the paper published by Benard et al. [15]. The derivation is necessary as many publications use this approximation without explanation.

Since $G_i(y) = \sum_{j=i}^{n} \binom{n}{j} y^j (1-y)^{n-j}$, the $\sum_{j=i}^{n} \binom{n}{j} y^j (1-y)^{n-j} + \sum_{j=0}^{i-1} \binom{n}{j} y^j (1-y)^{n-j} = 1.$ The $G_i(y) = 1 - \sum_{j=0}^{i-1} \binom{n}{j} y^j (1-y)^{n-j}$. Then $\sum_{j=0}^{i-1} \binom{n}{j} y^j (1-y)^{n-j}$ is equivalent to $\sum_{j=n-i+1}^{n} \binom{n}{j} y^{n-j} (1-y)^j = G_{n-i+1}(1-y)$. Therefore,

$$G_i(y) = 1 - G_{n-i+1}(1-y) \tag{3.100}$$

The median rank is the point where $G_i(y^*) = 0.5$. Then

$$1 - G_{n-i+1}(1-y^*) = 0.5,$$

implying the $1 - y^*$ is the median of $G_{n-i+1}(.)$. Suppose the $y^* = \frac{i-a}{n+b}$. For the $G_{n-i+1}(.)$, the median is (substitute the i by $n - i + 1$)

$$\frac{n-i+1-a}{n+b} = 1 - y^* = 1 - \frac{i-a}{n+b} \tag{3.101}$$

It is easy to obtain

$$b = 1 - 2a \tag{3.102}$$

Therefore,

$$y^* = \frac{i-a}{n+b} = \frac{i-a}{n+1-2a} \tag{3.103}$$

Then

$$\sum_{j=i}^{n} \binom{n}{j} y^{*j}(1-y^*)^{n-j} = 0.5 \tag{3.104}$$

Or

$$\sum_{j=0}^{i-1} \binom{n}{j} y^{*j}(1-y^*)^{n-j} = 0.5 \tag{3.105}$$

For given n and j, we can obtain a from Eq. (3.104) or (3.105) by numerical method. Benard calculated the a is around 0.3. Let $n = \infty$, the limit of $n \to \infty$ above is

$$\lim_{n\to\infty} G_i(y) = \sum_{j=0}^{i-1} \binom{n}{j} y^{*j}(1-y^*)^{n-j} = \sum_{j=0}^{i-1} e^{-(i-a)} \frac{(i-a)^j}{j!} = 0.5 \tag{3.106}$$

From Eq. (3.106), we can find the a for various i from the equation as follows

$$\sum_{j=0}^{i-1} e^{-(i-a)} \frac{(i-a)^j}{j!} = 0.5 \tag{3.107}$$

Benard et al. [15] obtained the a for various i as in the Table 3.7.

The Benard roughly used the constant value $a = 0.3$ for simplicity. According to Eq. (3.102),

$$b = 1 - 2a = 1 - 0.6 = 0.4$$

Therefore,

Table 3.7 Values a for different i (Taken from [15])

i in Eq. (3.107)	a
1	0.307
2	0.321
5	0.329
10	0.331
100	0.333

$$F(t_i) \approx \frac{i-a}{n+b} = \frac{i-0.3}{n+0.4}.$$

which is the median rank approximation. The median rank approximation, seen from the above derivation procedure, is a rough approximation. The Benard approximation is the most common CDF approximation in reliability engineering.

Appendix 3A: Two Properties of Exponential Family

The Exponential family is a group of distributions with $E(T(t)) = \frac{\partial c}{\partial \theta}$ and $\frac{\partial^2 c}{\partial \theta^2} = var(T(t))$. This appendix proves the two properties. The PDF of Exponential family is of

$$f(t;\theta) = b(t)e^{\theta T(t) - c(\theta)} \tag{3.4}$$

Then the $E(T(t)) = \frac{\partial c}{\partial \theta}$ and $\frac{\partial^2 c}{\partial \theta^2} = var(T(t))$.

First derivative of (3.4) is $f'(t) = (T(t) - c')b(t)e^{\theta T(t) - c(\theta)}$, and $\int f(t;\theta)dt = 1$, then

$$\int f'(t)dt = \frac{\partial}{\partial \theta}\int f(t;\theta)dt = \frac{\partial}{\partial \theta}(1) = 0 \tag{3.A1}$$

Substitute the $f'(t) = (T(t) - c')b(t)e^{\theta T(t) - c(\theta)} = (T(t) - c')f(t;\theta)$ into the (3.A1).

$$\int (T(t) - c')b(t)e^{\theta T(t) - c(\theta)}dt = \int T(t)b(t)e^{\theta T(t) - c(\theta)}dt - \int c'b(t)e^{\theta T(t) - c(\theta)}dt = 0$$

Obviously, the $\int T(t)b(t)e^{\theta T(t) - c(\theta)}dt = E(T(t))$, and $\int c'b(t)e^{\theta T(t) - c(\theta)}dt = c'$ since $\int f(t;\theta)dt = 1$

Therefore,

$$c' = E(T(t)) \tag{3.A2}$$

The second derivative of (3.4) is

$$f''(t) = (-c'')b(t)e^{\theta T(t) - c(\theta)} + (T(t) - c')^2 b(t)e^{\theta T(t) - c(\theta)} \tag{3.A3}$$

Obviously, $\int f''(t)dt = 0$ since $\int f'(t)dt = 0$. Take the integral of both sides of (3.A3), we obtain

$$\int \left\{ (-c'')b(t)e^{\theta T(t) - c(\theta)} + (T(t) - c')^2 b(t)e^{\theta T(t) - c(\theta)} \right\} dt = 0 \tag{3.A4}$$

Rewrite the (3.A4) as

$$\int \left\{ (-c'')b(t)e^{\theta T(t) - c(\theta)} \right\} dt = c'' = \int \left\{ (T(t) - c')^2 b(t)e^{\theta T(t) - c(\theta)} \right\} dt = Var(T(t))$$

We obtain

$$c'' = Var(T(t)). \tag{3.A5}$$

Appendix 3B: Slutsky Theorem

Let X_n and Y_n be sequences of random elements. The X_n approaches to a random variable X when $n \to +\infty$. The Y_n approaches to a constant C. Then we have [16]

$$X_n + Y_n \xrightarrow{d} X + C$$

$$X_n Y_n \xrightarrow{d} CX$$

$$X_n / Y_n \xrightarrow{d} X/C; C \neq 0$$

The $\xrightarrow{d}$ denotes the sequences converge in probability.

Problems and Questions

1. Derive the Fisher information matrix for the two parameter Normal distribution.
2. Obtaining the critical value for Kolmogorov–Smirnov Test for sample size $n = 5$, 6, … using simulation method.
3. A group of mining data is as follows (in hours) [7], using Q-Q plot to check if the data follows Weibull distribution.

$149; 33; 8; 5; 31; 12; 188; 2; 36; 63; 53; 56; 6; 4; 147; 36; 4; 76; 36; 36; 61; 48; 58; 68; 3;$

$12; 33; 7; 71; 149; 6; 54; 9; 115; 58; 96; 18; 16; 25; 139; 18; 34; 15;$

41

References

1. Lehmann EL, Casella G (1998) Theory of point estimation. Springer, New York, pp xxvi–589
2. Xiru C (2004) Advanced mathematics statistics. Higher Education Press, Beijing
3. Shisong M, Jinglong W, Xiaolong P (2004) Advanced mathematics statistics. Higher Education Press, Beijing
4. Montgomery DC, Peck EA, Vining GG (2012) Introduction to linear regression analysis. Wiley

5. Myers RH (1990) Classical and modern regression with applications. Duxbury Press
6. Johnson RA, Wichern DW (2023) Applied multivariate statistical analysis. Pearson
7. Kumar U, Klefsjo B, Granholm S (1989) Reliability investigation for a fleet of load haul dump machines in a Swedish mine. Reliab Eng Syst Safe 26(4):341–361. https://doi.org/10.1016/0951-8320(89)90004-5
8. Pearson K (1900) X. On the criterion that a given system of deviations from the probable in the case of a correlated system of variables is such that it can be reasonably supposed to have arisen from random sampling. London Edinburgh Dublin Philos Magaz J Sci 50(302):157–175
9. Hogg RV, McKean JW, Craig AT (2012) Introduction to mathematical statistics. Pearson
10. Lawless JF (2003) Statistical models and methods for lifetime data, Wiley series in probability and statistics, 2nd edn. Wiley-Interscience, Hoboken, NJ
11. Rohatgi VK, Saleh AKME (2015) An introduction to probability and statistics, 3rd edn. Wiley, Hoboken, NJ
12. Negreiros ACSVD, Lins ID, Moura MJD, Droguett EL (2020) Reliability data analysis of systems in the wear-out phase using a (corrected) q-exponential likelihood. Reliab Eng Syst Saf 197
13. Arnold BC, Balakrishnan N, Nagaraja HN (2008) A first course in order statistics. SIAM
14. Bartlett HJG (1998) Reliability analysis with suspended items. School of Mechanical, Manufacturing and Medical Engineering, Queensland University of Technology
15. Benard A, Bos-Levenbach EC (1955) The plotting of observations on probability-paper, Stichting Mathematisch Centrum. Statistische Afdeling
16. Gut A (2013) Probability: a graduate course. Springer

Chapter 4
Statistical Exponential Distribution

Exponential distribution is the most used distribution in reliability engineering. It is widely applied to model life data for non-repairable systems or the time between failures for repairable systems. Two forms of Exponential distribution can have: the one-parameter Exponential and the two-parameter Exponential. In the one-parameter Exponential distribution, the single parameter represents the failure rate, which characterizes the system's reliability. The simplicity of having just one parameter makes parameter estimation straightforward. Unless otherwise specified, this chapter refers to Exponential distribution as the one-parameter one.

4.1 Life Data with Exponential Distribution

The Exponential distribution describes situations where the frequency of failure occurrences remains constant over time. For many complex systems—such as electronic systems composed of many components, or very complex systems like airplanes, the time between failures often follows Exponential distribution. The following examples show some cases where the failure time follows or does not follow Exponential distribution.

Life Data Example 1

The failure times from mining trucks were collected from a Swedish mining company. The time between failure is as follows [1].

```
225.15 255.35 109.40 52.10 44.40 280.11 39.20 180.55 254.10 22.45 45.40
252.10 33.50 09.20 56.25 22.15 31.35 44.20 169.41 20.45 149.05 326.30
61.35 33.55 17.55 37.35 203.20 97.55 110.25 45.25 86.45 85.10
```

The time unit is in hours. Data size is 32. The histogram and a cumulative number of failures against time is plotted. A curvature is shown in the cumulative failure plot,

F. Yuan, *Essentials of Reliability Engineering*, Springer Series in Reliability Engineering, https://doi.org/10.1007/978-3-032-15544-3_4

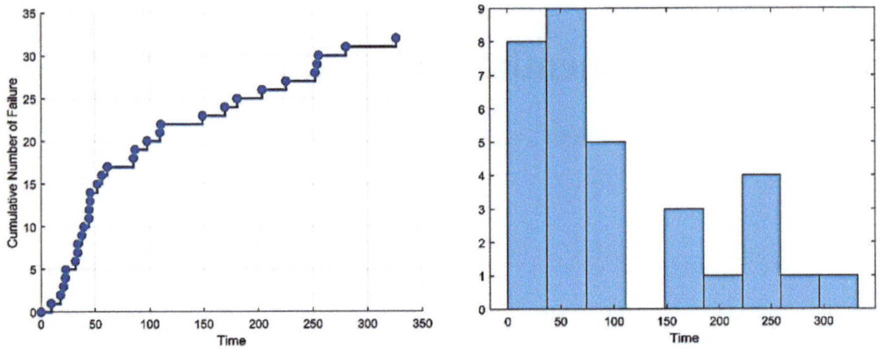

Fig. 4.1 Histogram and cumulative number of failure for life dataset 1

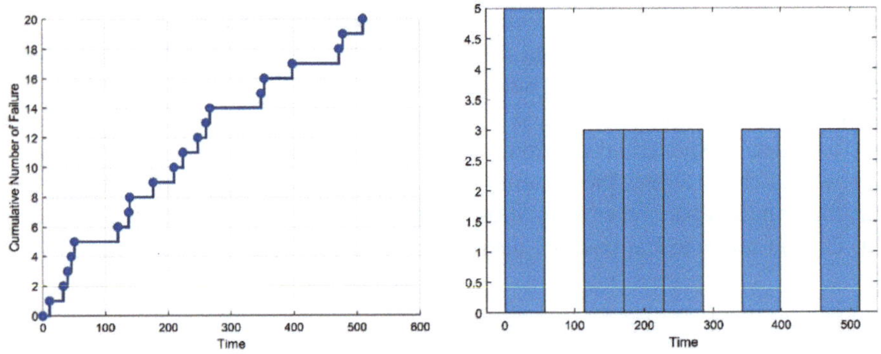

Fig. 4.2 Histogram and cumulative number of failure for life dataset 2

as shown in Fig. 4.1. It implies the frequency of failure occurrence is not constant and could be time-dependent, as otherwise if frequency of failure occurrence is not time-dependent, the curvature tends to be linear, i.e. the number of cumulative failures is proportional to the operating time.

Life Data Example 2

Mechanical machine failure times from a manufacturing center are as follows [2]. The time unit is in hours with data size 20. The cumulative number of failures shows no significant curvature, as shown in Fig. 4.2. The frequency of failure occurrence is approximately constant.

```
176.00 248.00 10.50 472.00 45.00 39.00 209.33 261.25 510.00 120.00
224.00 267.50 32.00 50.00 138.50 398.00 478.00 353.00 348.00 137.06
```

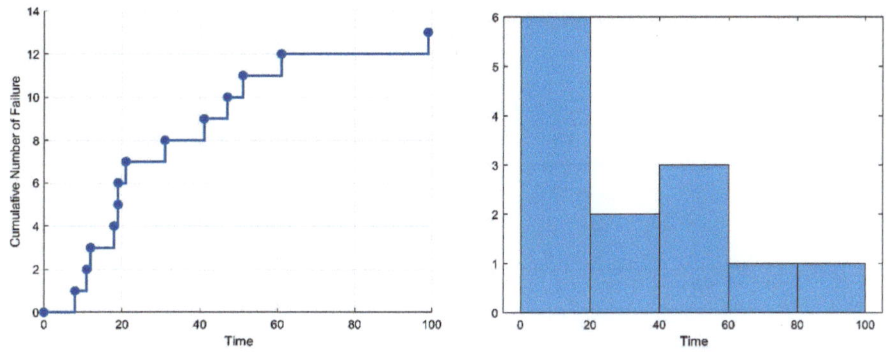

Fig. 4.3 Histogram and cumulative number of failure for dataset 3

Life Data Example 3

The last data sets are failure times from magnetic resonance imaging (MRI) scanner [3]. Time unit in in days with data size 13. This example is of a small data set. Curvature is also shown in the cumulative failure figure, as shown in Fig. 4.3. But the curvature is not significant if excluding the last one.

```
99 11 12 21 8 31 19 41 19 51 18 61 47
```

The visual analyses from plots are subjective and empirical analysis, not quantitative analysis. Here we use statistical methods to analyze these example data. A premise assumption assumes the data follows Exponential distribution. As presented in Chap. 2, the Probability Density Function (PDF) of Exponential is

$$f(t) = \lambda e^{-\lambda t} \tag{4.1}$$

The Cumulative Distribution Function (CDF) and reliability function can be easily derived from Eq. (4.1) as CDF $F(t) = 1 - e^{-\lambda t}$ and reliability $R(t) = e^{-\lambda t}$ respectively. Figure 4.4 shows plots of the PDF, CDF, reliability function and the failure rate of the Exponential. The PDF is monotonically decreasing with time. The monotonic decreasing of PDF is a remarkable characteristic of Exponential distribution. The CDF monotonically increases with time, while the reliability monotonically decreases with time. The monotonic increase of CDF is general for all kinds of statistical distributions. It tends to the maximum value 1 when time is infinite. This implies the failure of a system will be certain when the operating time approaches infinite. The reliability function is monotonically decreasing with time. It can be understood as: The longer a machine operates, the higher the likelihood of failure, and thus its reliability decreases over time.

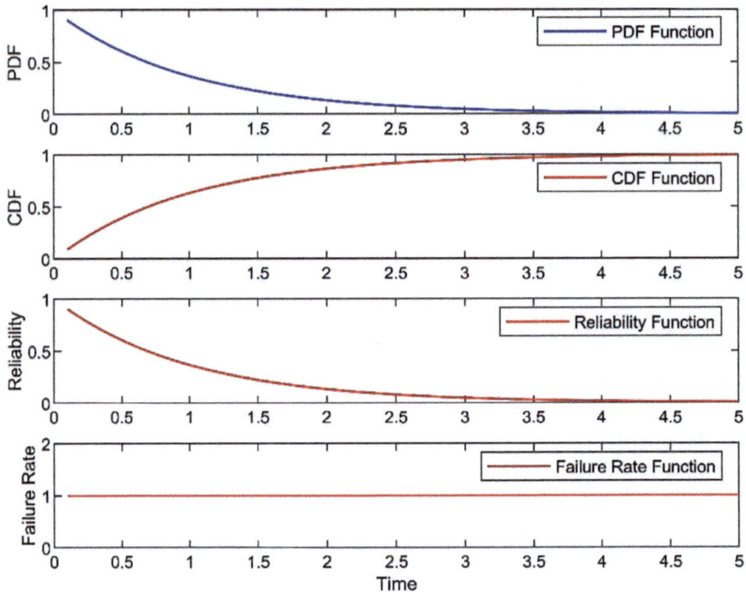

Fig. 4.4 The PDF/CDF and reliability plot

4.2 Maximum Likelihood Parameter Estimation

The classical Exponential distribution has only one parameter. This section discusses methods using maximum likelihood to find the best estimator for the sole parameter. Let the observed failure times be t_1, t_2, t_3, …, t_n. Assume the failures are statistically independent from each other. The likelihood function is then

$$L(\lambda) = f(t_1)\, f(t_2) f(t_3) \ldots f(t_n) \tag{4.2}$$

where $f(t_i) = \lambda e^{-\lambda t_i}$. Equation (4.2) is equal to

$$L(\lambda) = \lambda^n e^{-\lambda \sum_{i=1}^{i=n} t_i} \tag{4.3}$$

Take negative natural logarithm, we obtain

$$-\ln L(\lambda) = -n \ln \lambda + \lambda \sum_{i=1}^{i=n} t_i \tag{4.4}$$

To find the maximum point, we differentiate it with respect to the unknown parameter λ as follows:

$$\frac{\partial - \ln L(\lambda)}{\partial \lambda} = -n/\lambda + \sum_{i=1}^{i=n} t_i = 0 \tag{4.5}$$

It is not hard to obtain the solution to Eq. (4.5), as the maximum likelihood estimator is

$$\widehat{\lambda} = n/\sum_{i=1}^{i=n} t_i \tag{4.6}$$

Obviously, the estimator is the inverse of mean time to failure (MTTF)

$$\widehat{\lambda} = 1/MTTF \tag{4.7}$$

since the MTTF is defined as $MTTF = \sum_{i=1}^{i=n} t_i/n$.

4.2.1 *Bias of the Estimator*

It is necessary to check the bias of the estimator, as this ensures the estimator converges to the true parameter when data size increases. Instead of checking the estimator directly, we examine the expectation of MTTF first. If the t_i is assumed to be statistically independent from each other, then the expectation of MTTF is

$$E(MTTF) = \frac{\sum_{i=1}^{i=n} E(t_i)}{n}. \tag{4.8}$$

From Sect. 2.7.2 in Chap. 2, we know the mean of the Exponential distribution $E(t_i) = 1/\lambda$. Equation (4.8) is thus

$$E(MTTF) = \frac{\sum_{i=1}^{i=n} E(t_i)}{n} = 1/\lambda \tag{4.9}$$

It is well known $\lambda = \frac{1}{MTTF_{real}}$, where $MTTF_{real}$ means the MTTF when $n \to \infty$. Then from Eq. (4.9), we find

$$E(MTTF) = \frac{\sum_{i=1}^{i=n} E(t_i)}{n} = 1/\lambda = MTTF_{real}$$

Equation (4.8) for MTTF is then not biased as the empirical value tends to be the real one when $n \to \infty$. However, for the estimator on the failure rate λ, the situation is different. Since $E\left(\frac{1}{x}\right) \neq \frac{1}{E(x)}$, because

$$E\left(\widehat{\lambda}\right) = E\left(\frac{1}{MTTF}\right) \neq \left(\frac{1}{E(MTTF)} = \lambda\right)\left(\text{Use property } E\left(\frac{1}{x}\right) \neq \frac{1}{E(x)}\right).$$

it means $E\left(\widehat{\lambda}\right) \neq \lambda$, which suggests the maximum likelihood estimator on failure rate λ for Exponential distribution is biased [4]. The bias implies: for small data size, the MLE will not be able to center on the true parameter value λ. When biased, an estimator tends to have a higher risk to deviate from the true parameter. The bias should not be ignored when the data size is small. As for small data size, the bias is larger than a large size. A simulation was conducted to demonstrate how the biased estimator deviates from the true parameter. Practitioners can learn from this simulation to understand the risks associated with using a biased estimator, particularly when the sample size is small.

The simulation is carried out by using Matlab. The samples were drawn from Exponential distribution with $\lambda = 1$ and the sample size is as small as 5. In the simulation, for each 5 data sets simulated, we use Eq. (4.6) to obtain an estimate. Total 50,000 iterations have ran, suggesting 50,000 estimates evaluated. The distribution of these estimates is shown in Fig. 4.5. The red dashed line is the true parameter value 1.

The evaluated estimate differs at a large range spreading from 0.27 to 14.17, while the true estimator is 1. To check the bias, we compute out the mean of these 50,000 estimates is 1.2496, not close the true 1. If the estimator is unbiased, the mean of the 50,000 estimates should be very close to 1. In this case, the bias of using Eq. (4.6) is obvious.

The bias not only depends on the parameter estimating method but also on the data size. When the data size increases, the bias will decrease. For example, let the

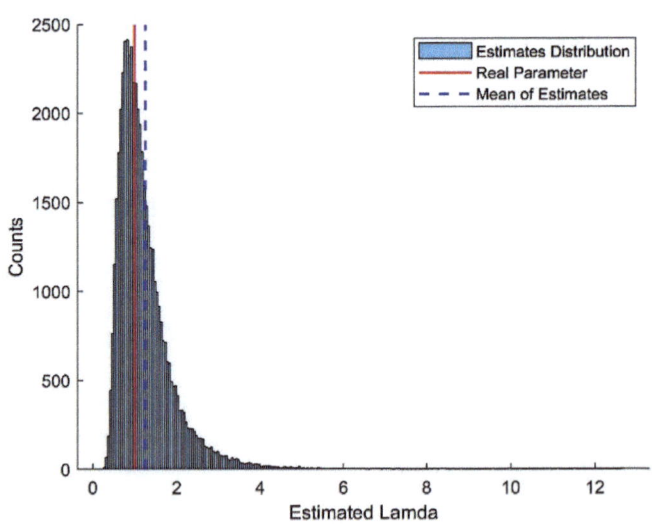

Fig. 4.5 Simulated estimates distribution

sample size in the previous case increase to 100, we run the same simulation. The mean of the estimators is then around 1.0101. Obviously, the bias has been reduced significantly.

For the 5 sample size case, since the mean as 1.2496 is significantly different with the true value 1, the bias of the estimator should then be a big concern. Despite the existence of many other methods that provide an unbiased estimator for the Exponential parameter, many state-of-the-art applications still use the biased MLE, which risks leading to incorrect decision-making. The following section discusses the real distribution of the MLE estimator of Eq. (4.6), from which a bias correction method is thereafter obtained.

4.3 Estimator Distribution Analysis

Since the estimator himself is a random variable. One thought we have is to obtain the explicit distribution of the estimator. Significant convenience would bring if the distribution of the estimator could be determined, as this would allow for precise control over the uncertainty of the estimator. For most MLE estimators, such as for Weibull distribution, the estimator doesn't have the explicit expression. Analyzing the performance from the distribution of the estimator is thus impossible. Fortunately, for the Exponential distribution, the distribution of the MLE can be explicitly obtained.

The MLE estimator distribution for Exponential can be obtained by using Gamma distribution. Some other publications obtained the same distribution by using the moment generating method [5–7]. Some of the derivation procedures are also shown in [8]. The probability density function of Gamma distribution is

$$G(x;k,\lambda) = \frac{x^{k-1}e^{-\lambda x}}{\lambda^{-k}\Gamma(k)}; \tag{4.10}$$

Let the samples from Gamma distribution denoted by x_1, x_2, x_3, ..., x_n. By the additive property of Gamma distribution, the sum of the Gamma random variable x_i also follows a Gamma distribution, written as

$$y = \sum x_i \sim G(x;nk,\lambda) \tag{4.11}$$

For simplicity of notion, we write the $G(x;nk,\lambda)$ as $G(nk,\lambda)$.

Compare the PDF of Gamma distribution shown in Eq. (4.10) with the PDF of Exponential distribution. It is obvious that the Exponential is a special case of Gamma distribution with $k = 1$. The Gamma function for Exponential distribution is $\Gamma(k = 1) = 1$. Therefore, the sum of Exponential distribution random variables is

$$y=\sum x_i \sim G(n,\lambda) \tag{4.12}$$

By the Gamma distribution Eq. (4.10), the corresponding probability density function is

$$f_y(y;n,\lambda)=\frac{y^{n-1}e^{-\lambda y}}{\lambda^{-n}\Gamma(n)}; \tag{4.13}$$

Let $z=\widehat{\lambda}=\frac{n}{\sum_{i=1}^{i=n}t_i}=\frac{n}{y}$, and rewrite it as $y=\frac{n}{z}$. According to the transformation method shown in Appendix 4A of this chapter, for which the first derivative in respect to z is $\frac{dy}{dz}=-\frac{n}{z^2}$, the PDF of $z=\widehat{\lambda}$ is then

$$f_z(z;n,\lambda)=\frac{(n/z)^{n-1}e^{-n\lambda/z}}{\lambda^{-n}\Gamma(n)}\cdot\left|-\frac{n}{z^2}\right|=\frac{n^n e^{-n\lambda/z}}{z^{n+1}\lambda^{-n}\Gamma(n)}$$

The distribution of $\widehat{\lambda}$ is then

$$f\left(\widehat{\lambda};n,\lambda\right)=\frac{n^n e^{-n\lambda/\widehat{\lambda}}}{\widehat{\lambda}^{n+1}\lambda^{-n}\Gamma(n)} \tag{4.14}$$

The transformation method presented in Appendix 4A of this chapter is a common approach for deriving the PDF of a function of a random variable. This method can be seen in many textbooks on probability and statistics [9].

4.3.1 Expectation of Estimator

Section 4.2 has demonstrated that the MLE estimator is biased. This section further analyzes it by using the exact distribution of the estimator shown in Eq. (4.14). it is advantageous to examine the bias when the exact distribution has been known. As we know, the MLE of Exponential distribution for observed samples t_1, t_2, t_3, ..., t_n is:

$$\widehat{\lambda}=\frac{n}{\sum t_i}=\frac{n}{Y} \tag{4.15}$$

According to Eq. (4.13), the expectation of $\widehat{\lambda}$ is then

$$E\left(\widehat{\lambda}\right)=\int_{-\infty}^{+\infty}\frac{n}{y}\frac{y^{n-1}e^{-\lambda y}}{\lambda^{-n}\Gamma(n)}dy \tag{4.16}$$

Then

$$\int_{-\infty}^{+\infty} \frac{n}{y} \frac{y^{n-1} e^{-\lambda y}}{\lambda^{-n}\Gamma(n)} dy = \frac{n\lambda}{n-1} \int_{-\infty}^{+\infty} \frac{y^{n-2} e^{-\lambda y}}{\lambda^{-(n-1)}\Gamma(n-1)} dy \tag{4.17}$$

It is easy to find the $\int_{-\infty}^{+\infty} \frac{y^{n-2}e^{-\lambda y}}{\lambda^{-(n-1)}\Gamma(n)} dy$ also follows Gamma distribution as $G(n-1, \lambda)$. The $\int_{-\infty}^{+\infty} \frac{y^{n-2}e^{-\lambda y}}{\lambda^{-(n-1)}\Gamma(n)} dy = 1$. Therefore, the expectation of estimator $\widehat{\lambda}$ has been analytically obtained as

$$E\left(\widehat{\lambda}\right) = \frac{n\lambda}{n-1} \tag{4.18}$$

Analytically, we have proved the MLE estimator of Eq. (4.15) is biased since $E\left(\widehat{\lambda}\right) \neq \lambda$.

We can readily correct the estimator by using Eq. (4.18). The estimator can be corrected by multiplying a correction factor $\frac{n-1}{n}$, as $\frac{n-1}{n} E\left(\widehat{\lambda}\right) = \frac{n-1}{n} \frac{n\lambda}{n-1} = \lambda$. The corrected estimator is then

$$\widehat{\lambda} = \frac{n-1}{n} \frac{n}{\sum t_i} = \frac{n-1}{\sum t_i} \tag{4.19}$$

We demonstrate the impact of bias by simulation. The left on Fig. 4.6 is empirical distribution of estimates for 50,000 iterations with a sample size 5 and the true failure rate for the Exponential distribution is 1. Two distributions are plotted: one is uncorrected original estimate, one is corrected (adjusted). It is clear the corrected is shifted to the left from the uncorrected. The *x*-axis in the figure is the estimator value. The left-shift tells the corrected becomes smaller, which agrees with the analytical analysis, since the uncorrected $\frac{n}{\sum t_i} > \frac{n-1}{\sum t_i}$, where the former is the uncorrected, the latter is the corrected.

Using the 50,000 iterations, by using the corrected estimator, the mean of the estimates is 1.0051, which is around the true value 1, while for the uncorrected, the

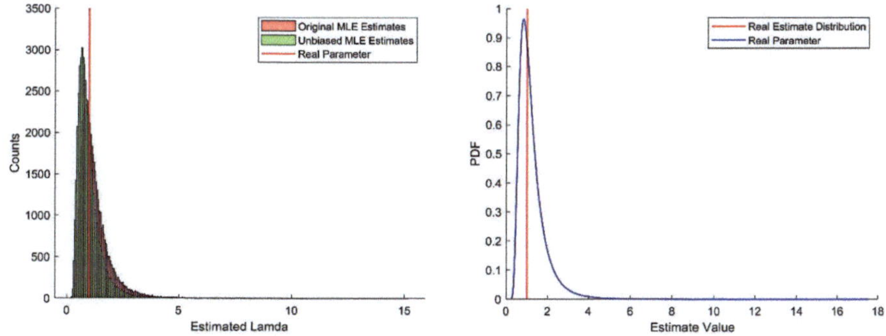

Fig. 4.6 Empirical and real estimate distribution

mean is 1.2496. The estimator has thus been considerably improved. The uncorrected estimates with mean 1.2496 agree well with the theoretical expected mean by using Eq. (4.18) as $E\left(\widehat{\lambda}\right) = \frac{n\lambda}{n-1} = \frac{5}{4} = 1.25$. The theoretical mean and the real empirical mean are very close. The small deviation results from the error of simulation software. The simulation of this example was using Matlab. In Matlab, the drawn random numbers are not the perfect Exponential distribution, which leads to some inevitable system errors. The right on Fig. 4.6 is the real MLE PDF using Eq. (4.14) for the estimator $\widehat{\lambda}$. It exhibits a similar pattern with the simulated left figure, implying the theoretical PDF of Eq. (4.14) agrees with the simulation results quite well.

We can further analyze the uncorrected estimator has a larger error for smaller data size. The error is reducing when the data size increases. According to Eq. (4.18), the $\frac{E\left(\widehat{\lambda}\right)}{\lambda} = \frac{n}{n-1}$. When the $n \longrightarrow +\infty$, the $\frac{E\left(\widehat{\lambda}\right)}{\lambda} = 1$. It implies the estimator (4.18) asymptotically reaches the true parameter when sample size becomes larger, which is a desirable property. It suggests for large sample size, the bias need not to be corrected, since the bias is small. We have also shown in the Sect. 4.2.1 when sample size increases from 5 to 100, the bias has been reduced significantly. The following Fig. 4.7 shows an asymptotical consistency, where we can learn how the bias is reduced with the increasing sample size. It approaches 1 when n is large, suggesting no notable bias. When the sample size reaches 100, the bias becomes very small, around only 1%.

Table 4.1 tabulates the numerical bias for various sample sizes. From the table it can find for data size less than 50, the bias level is larger than 2%. When n>100, the bias is less than 1%. In statistics, a small sample size problem is normally referring to

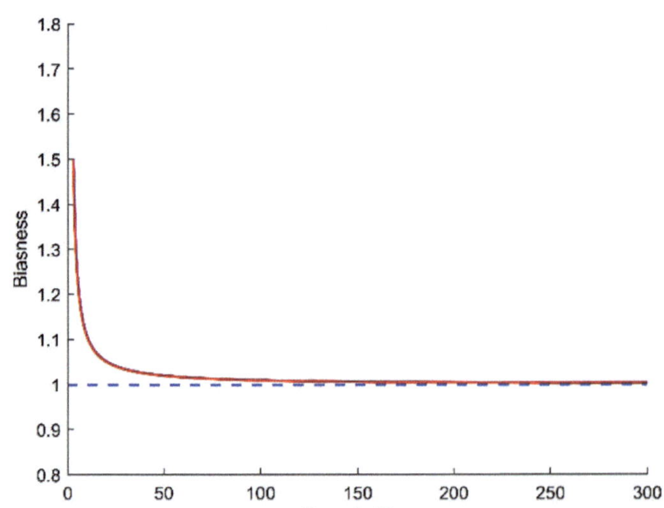

Fig. 4.7 Bias level against sample size

Table 4.1 Bias level with sample size. The bias level is measured as $\frac{E(\widehat{\lambda})}{\lambda}$. When no bias, it is 1

n	$\frac{E(\widehat{\lambda})}{\lambda}$	n	$\frac{E(\widehat{\lambda})}{\lambda}$	n	$\frac{E(\widehat{\lambda})}{\lambda}$
2	2.00	8	1.14	100	1.010
3	1.50	9	1.13	500	1.002
4	1.33	10	1.11	1000	1.001
5	1.25	15	1.07	5000	1.0002
6	1.20	20	1.05	10,000	1.0001
7	1.17	50	1.02	50,000	1

a data size less than 50. For sample size less than 10, it is then called extremely small data size. In reliability engineering, the data size is normally between 15 to 100. For such range data size, from the table, considering the bias is very necessary.

4.3.2 Exact Variance of Estimator

The variance can be used to control estimator error. In experiment design, the variance can be used to decide the test sample size. For Exponential distribution, taking advantage of the known estimator distribution, the estimator $\widehat{\lambda}$'s exact variance can also be analytically obtained. It is a clear advantage for using Exponential distribution, as for most other distributions, since the exact distribution is not available, the exact variance is not available as well.

The distribution of y is known as Eq. (4.13), then the

$$E\left(\widehat{\lambda}^2\right) = E\left(\left(\frac{n}{y}\right)^2\right) = \frac{n^2\lambda^2}{(n-1)(n-2)}\int_{-\infty}^{+\infty}\frac{y^{n-3}e^{-\lambda y}}{\lambda^{-(n-2)}\Gamma(n-2)}dy \tag{4.20}$$

Since $\int_{-\infty}^{+\infty}\frac{y^{n-3}e^{-\lambda y}}{\lambda^{-(n-2)}\Gamma(n-2)}dy = 1$, then

$$E\left(\widehat{\lambda}^2\right) = \frac{n^2\lambda^2}{(n-1)(n-2)} \tag{4.21}$$

As $Var\left(\widehat{\lambda}\right) = E\left(\widehat{\lambda}^2\right) - \left[E\left(\widehat{\lambda}\right)\right]^2$,the

$$Var\left(\widehat{\lambda}\right) = \frac{n^2\lambda^2}{(n-1)(n-2)} - \frac{n^2\lambda^2}{(n-1)^2} = \frac{n^2\lambda^2}{(n-1)^2(n-2)} \tag{4.22}$$

It is straightforward to find from Eq. (4.22): when the $n \to +\infty$, the variance approaches to 0. The variance of the estimator is also dependent on the true parameter λ. An entity independent from true parameter can be proposed as follows

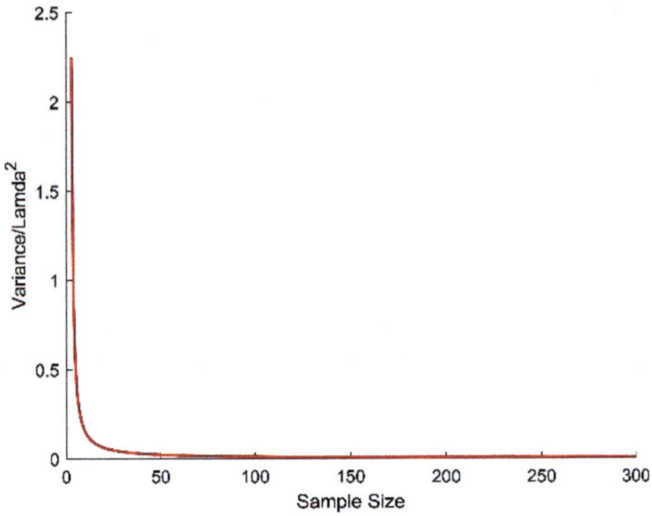

Fig. 4.8 Variance level against sample size

Table 4.2 Bias level against sample size

n	$\frac{Var(\hat{\lambda})}{\lambda^2}$	n	$\frac{Var(\hat{\lambda})}{\lambda^2}$	n	$\frac{Var(\hat{\lambda})}{\lambda^2}$
2	Not applicable	8	0.22	100	0.01
3	2.25	9	0.18	500	0.002
4	0.89	10	0.15	1000	0.001
5	0.52	15	0.09	5000	$2*10^{-4}$
6	0.36	20	0.06	10,000	10^{-4}
7	0.27	50	0.02	50,000	$2*10^{-5}$

$$\frac{Var(\hat{\lambda})}{\lambda^2} = \frac{n^2}{(n-1)^2(n-2)} \tag{4.23}$$

Equation (4.23) analytically shows the variance dependent on the sample size. Figure 4.8 plots Eq. (4.23). It shows the variance approaches to 0 when sample size increases.

Table 4.2 tabulates the numerical variance with various sample sizes. The values can be used to guide the sample size selection as it shows the degree of estimated one deviating from the real value.

Let the maximum allowed deviation from the true value be p, where $0 < p < 1$. Table 4.2 shows the $\frac{Var(\hat{\lambda})}{\lambda^2} < 1$ when $n > 3$. For sample size determination purpose, if only the $n > 3$ is considered, the problem to find optimal sample size is then converted to a problem to find the optimal n which satisfies

$$\frac{Var\left(\widehat{\lambda}\right)}{\lambda^2} \leq p^2 \tag{4.24}$$

Equation (4.24) is rewritten as

$$\frac{n^2}{(n-1)^2(n-2)} \leq p^2 \tag{4.25}$$

Solving the problem (4.25) is equivalent to find the solution of

$$n^3 - \left(4 + \frac{1}{p^2}\right)n^2 + 5n - 2 = 0 \tag{4.26}$$

It finds the solution to Eq. (4.26) is

$$n = \frac{4 + \frac{1}{p^2} \mp \sqrt{\left(4 + \frac{1}{p^2}\right)^2 - 15}}{3} \text{ or } n = \frac{4 + \frac{1}{p^2}}{3} \tag{4.27}$$

The integer n,corresponding to the optimal sample size, is then

$$n \geq \max\left(\frac{4 + \frac{1}{p^2} + \sqrt{\left(4 + \frac{1}{p^2}\right)^2 - 15}}{3}, \frac{4 + \frac{1}{p^2}}{3}\right) \tag{4.28}$$

Example 4.1

The lifetime of a product follows Exponential distribution. Experiments are planned to conduct to find out the failure rate. It requires the maximum allowed deviation from the true value be $p = 10\%$. What is the minimum sample size required for the experiment?

Solution: Using Eq. (4.28) by setting the p as $10\% = 0.1$, then

$$\frac{4 + \frac{1}{p^2} + \sqrt{\left(4 + \frac{1}{p^2}\right)^2 - 15}}{3} = 69.3$$

$$\frac{4 + \frac{1}{p^2}}{3} = 34.7$$

Take the max as 69.3. The sample size is an integer, and we should take the nearest higher integer, i.e. 70 for this example.

4.3.3 Variance Derived from Fisher Information Matrix

The variance shown in Eq. (4.22) is the exact variance of the MLE estimator. In state art, it is more popular to use an approximate variance derived from Fisher Information Matrix [10]. As we have discussed in Chap. 3, the Fisher method is general and not Exponential distribution specific. In this section, we compare the two variances. The likelihood function of Exponential distribution is:

$$-\ln L(\lambda) = -\ln\left\{\lambda e^{-\lambda t_1}\lambda e^{-\lambda t_2}\ldots\lambda e^{-\lambda t_n}\right\} = -n\ln\lambda + \lambda\sum_{i=1}^{i=n} t_i$$

The second derivative of the log likelihood function is not hard to obtain as.

$$I = \frac{\partial^2 - \ln L(\lambda)}{\partial\lambda\partial\lambda} = n/\lambda^2 \tag{4.29}$$

The variance is obtained as.

$$I^{-1} = \left(\frac{\partial^2 - \ln L(\lambda)}{\partial\lambda\partial\lambda}\right)^{-1} = \lambda^2/_n \tag{4.30}$$

The true variance of the MLE estimator for Exponential is $\frac{n^2\lambda^2}{(n-1)^2(n-2)}$ as shown Eq. (4.22). Then

$$\frac{n^2\lambda^2}{(n-1)^2(n-2)} = \left(\frac{n}{n-1}\cdot\frac{n}{n-1}\cdot\frac{n}{n-2}\right)\cdot\frac{\lambda^2}{n} > \frac{\lambda^2}{n}$$

The variance of MLE estimator for Exponential is higher than the one obtained from the Fisher. The Crow-Rao lower bound, which was discussed in Chap. 3, has thus not achieved. This is a drawback of the MLE for the Exponential distribution, although it is not serious.

4.4 Confidence Interval Derivation

The distribution of the estimator can be applied to find confidence interval (CI). Moreover, in general, the confidence interval in statistics can use the asymptotically normality of the estimator, i.e. according to central limit theorem, the distribution of estimator is normal distribution when sample size is infinite. The Fisher information matrix method and the likelihood ratio method are based on this. This section derives the CI by using the three methods and compares their performance.

4.4.1 Estimator Distribution Based Confidence Interval

The exact distribution of the estimator shown in Eq. (4.14) can be used to develop confidence intervals. For easy of derivation, we use Eq. (4.12) to derive the confidence interval instead of using Eq. (4.14). The $\widehat{\lambda}$ is the inverse of $y = \sum t_i$ in Eq. (4.12). If we find the confidence interval for y, the interval of estimator $\widehat{\lambda}$ can be derived easily. It has been known y follows Gamma distribution $G(n, \lambda)$, as shown in Fig. 4.9 for the PDF. For a confidence interval $1 - \alpha$, the lower bound of y, shown as y_L in Fig. 4.9, is the y value corresponds to CDF of $\alpha/2$; the upper bound of y, shown as y_H in Fig. 4.9, is the y value corresponds to CDF of $1 - \alpha/2$.

A confidence interval of $1 - \alpha$ of y is thus.

$$\begin{cases} y_L = G^{-1}(\alpha/2; n, \lambda) \\ y_H = G^{-1}(1-\alpha/2; n, \lambda) \end{cases} \tag{4.31}$$

Since $\widehat{\lambda} = n/\sum t_i$, the CI of the $\widehat{\lambda}$ is obtained as

$$\begin{cases} \hat{\lambda}_U = 1/y_L = \mathrm{n}/G^{-1}(\alpha/2; n, \lambda) \\ \hat{\lambda}_L = 1/y_U = n/G^{-1}(1-\alpha/2; n, \lambda) \end{cases} \tag{4.32}$$

The CI calculation here requires the value of true parameter λ, which is a theoretical value not known in practice. To use Eq. (4.32), we can use the MLE estimator $\widehat{\lambda} = n/\sum t_i$ as the true λ.

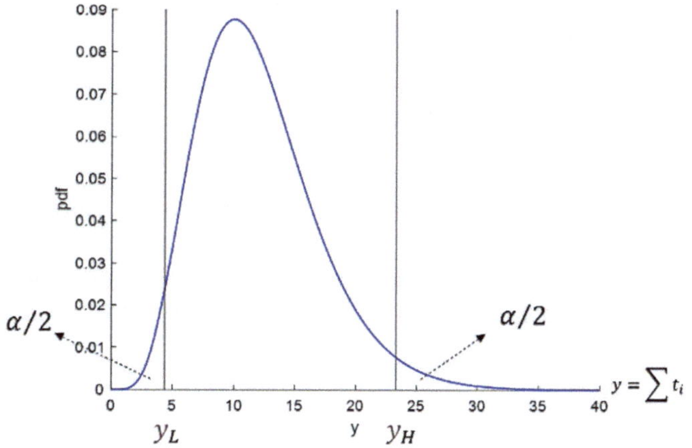

Fig. 4.9 Distribution of y

Example 4.2
For the data sets for Life Data Example 1 shown in Sect. 4.1, evaluate the failure rate of the system and evaluate its confidence interval of 95%.

Solution: The MLE of the failure rate is $\widehat{\lambda} = \frac{n}{\sum t_i} = 0.0094$ per hour. For a 0.95 confidence interval, the α is 1–0.95 = 0.05. The bound of y_L and y_H according to Eq. (4.31) is then

$$y_L = G^{-1}\left(\frac{\alpha}{2}; n, \lambda\right) = 2.33 \times 10^3$$

$$y_H = G^{-1}\left(1 - \frac{\alpha}{2}; n, \lambda\right) = 4.67 \times 10^3$$

The $G^{-1}\left(\frac{\alpha}{2}; n, \lambda\right)$ can be computed from software such as Matlab using function gaminv($\frac{\alpha}{2}$, n, $1/\lambda$), where λ is approximated by $\widehat{\lambda}$. The bound of failure rate according to Eq. (4.32) is as

$$\begin{cases} \hat{\lambda}_U = 1/y_L = 0.0068 \\ \hat{\lambda}_L = 1/y_U = 0.0138 \end{cases}$$

4.4.2 *Fisher Matrix Confidence Interval*

Many state-of-art applications and some commercial software use the Fisher matrix method to derive estimator's confidence interval. As we know, the Fisher Information provides a lower bound for the variance of estimator. And we let the true parameter as the mean of the estimator, which is held for unbiased estimator. According to the central limit theorem in statistics, when the sample size is infinite, the estimator $\widehat{\lambda}$ tends to be Normal distribution as

$$\widehat{\lambda} \sim Normal\left(\lambda, \widehat{\lambda}^2/n\right) \tag{4.33}$$

It is not hard to find the standardized $\frac{\widehat{\lambda} - \lambda}{\sqrt{\widehat{\lambda}^2/n}}$ follows standard Normal distribution.

Given the $Z_{\alpha/2}$ be the z value of the standard Normal distribution at probability $\alpha/2$, and let the two tails equal, the equal-tail confidence interval for $1 - \alpha$ is then

$$\begin{cases} \widehat{\lambda}_L = \widehat{\lambda} - Z_{\alpha/2}\dfrac{\widehat{\lambda}}{\sqrt{n}} \\ \widehat{\lambda}_U = \widehat{\lambda} + Z_{\alpha/2}\dfrac{\lambda}{\sqrt{n}} \end{cases} \tag{4.34}$$

A further transformation for Eq. (4.34) is necessary as the lower bound of Eq. (4.34) can be negative, which makes nonsense. To solve this limitation, one popular method is to use $\ln\widehat{\lambda}$ instead of $\widehat{\lambda}$ to estimate the interval [11]. Similar to Eq. (4.33), the distribution of $\ln\widehat{\lambda}$ is approximated as $\ln\widehat{\lambda} \sim Normal\left(\ln\lambda, \mathrm{Var}\left(\ln\widehat{\lambda}\right)\right)$. By the Delta method, the variance of $\ln\widehat{\lambda}$ can be approximated as

$$\mathrm{Var}\left(\ln\widehat{\lambda}\right) = \frac{Var\left(\widehat{\lambda}^2\right)}{\widehat{\lambda}^2} \tag{4.35}$$

Then the $\frac{\ln\widehat{\lambda} - \ln\lambda}{\mathrm{Var}\left(\ln\widehat{\lambda}\right)} \sim N(0, 1)$. A non-negative confidence interval of 1 − α is then written as

$$\frac{\ln\widehat{\lambda} - \ln\lambda}{\sqrt{\mathrm{Var}\left(\ln\widehat{\lambda}\right)}} = \frac{\ln\widehat{\lambda} - \ln\lambda}{\sqrt{\frac{Var\left(\widehat{\lambda}^2\right)}{\widehat{\lambda}^2}}} = \frac{\ln\widehat{\lambda} - \ln\lambda}{\sqrt{\frac{\widehat{\lambda}^2/n}{\widehat{\lambda}^2}}} = \frac{\ln\widehat{\lambda} - \ln\lambda}{\sqrt{\frac{1}{n}}}$$

And let it equal to $Z_{\alpha/2}$, then

$$\ln\widehat{\lambda}_L = \ln\widehat{\lambda} - \frac{Z_{\alpha/2}}{\sqrt{n}}$$

Finally, we obtain

$$\begin{cases} \widehat{\lambda}_L = \widehat{\lambda} e^{-\frac{Z_{\alpha/2}}{\sqrt{n}}} \\ \widehat{\lambda}_U = \widehat{\lambda} e^{\frac{Z_{\alpha/2}}{\sqrt{n}}} \end{cases} \tag{4.36}$$

Example 4.3
For the Data Sets 1 used in Example 4.2. Derive the confidence interval of 95% using the Fisher Matrix method.

Solution: The MLE of the failure rate is $\widehat{\lambda} = \frac{n}{\sum t_i} = 0.0094$ per hour. For a 0.95 confidence interval, the α is 1–0.95 = 0.05. The $Z_{\alpha/2} = 1.96$ for $\alpha = 0.05$. According to Eq. (4.36),

$$\begin{cases} \widehat{\lambda}_U = \widehat{\lambda} e^{-\frac{Z_{\alpha/2}}{\sqrt{n}}} = 0.0094 * e^{-\frac{1.96}{\sqrt{32}}} = 0.0133 \\ \widehat{\lambda}_L = \widehat{\lambda} e^{\frac{Z_{\alpha/2}}{\sqrt{n}}} = = 0.0094 * e^{\frac{1.96}{\sqrt{32}}} = 0.0066 \end{cases}$$

The CI is numerically close to the CI obtained from Example 4.2 which is [0.0068, 0.0138].

4.4.3 Likelihood Ratio Confidence Interval

As we have discussed in Chap. 3, the likelihood function using an MLE divided by the likelihood function value calculated with the true parameter can be approximated by $\chi^2_{\alpha;k}$ [12]. The α is the significance level which is the same as before. For example, for a 0.95 CI, the $\alpha = 0.05$. The k is the degree of freedom. In the case for Exponential distribution, the $k = 1$. The likelihood ratio function is as follows

$$-2.\ln\left(\frac{L(\lambda)}{L(\widehat{\lambda})}\right) \le \chi^2_{1-\alpha;1} \tag{4.37}$$

The likelihood function $L(\lambda) = \lambda e^{-\lambda t_1} \lambda e^{-\lambda t_2} \ldots \lambda e^{-\lambda t_n} = \lambda^n e^{-\lambda \sum t_i}$. Let Eq. (4.37) equal to 0 and therefore

$$-2.\ln\left(\frac{L(\lambda)}{L(\widehat{\lambda})}\right) = -2.\ln\left(\frac{\lambda^n e^{-\lambda \sum t_i}.}{\widehat{\lambda}^n e^{-\widehat{\lambda} \sum t_i}.}\right) = \chi^2_{1-\alpha;1}$$

Take logarithm of both sides,

$$\lambda^n e^{-\lambda \sum t_i} - \widehat{\lambda}^n e^{-\widehat{\lambda} \sum t_i} e^{-\frac{\chi^2_{1-\alpha;1}}{2}} = \lambda^n e^{-\lambda \sum t_i} - \widehat{\lambda}^n e^{-\widehat{\lambda} \sum t_i - \frac{\chi^2_{1-\alpha;1}}{2}} = 0 \tag{4.38}$$

which is a nonlinear equation. For the function

$$f(\lambda) = \lambda^n e^{-\lambda \sum t_i} - \widehat{\lambda}^n e^{-\widehat{\lambda} \sum t_i - \frac{\chi^2_{1-\alpha;1}}{2}} \tag{4.39}$$

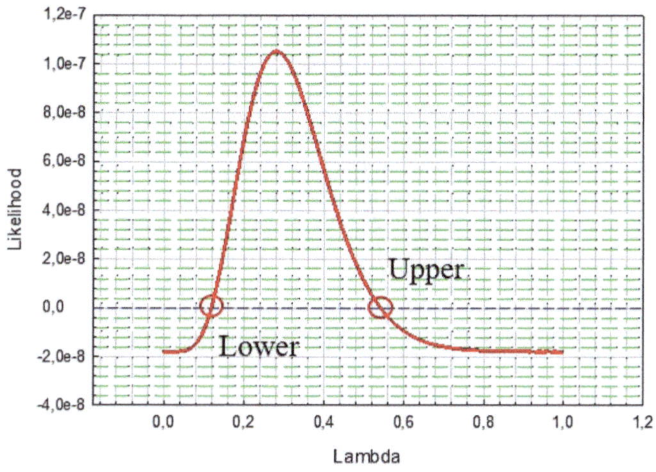

Fig. 4.10 Numerical plot of likelihood function

For Exponential distribution, the function (4.39) has only one maximum. The $\widehat{\lambda}$ corresponding to this maximum is the MLE estimator $\widehat{\lambda}$. The $f(\lambda) = 0$ is then the boundary of the confidence interval $1 - \alpha$. Fig. 4.10 shows the $f(\lambda) = 0$ has two solutions. The lower solution is the lower bound, while the higher solution is the upper bound. Therefore, the lower bound and upper bound are the two solutions of $\widehat{\lambda}$ obtained from the $-2.\ln\left(\frac{L(\lambda)}{L(\widehat{\lambda})}\right) = \chi^2_{1-\alpha;1}$.

A disadvantage of this method is that it requires a numerical procedure to find the roots of Eq. (4.38). For large sample sizes, this can be challenging because the magnitude of $\sum t_i$ could be very large.

Example 4.4
For the Data Sets 1 used in Example 4.2. Derive the confidence interval of 95% using the likelihood ratio method.

Solution: The 95% CI corresponds to $1 - \alpha = 0.95$. The $\chi^2_{1-\alpha;1} = \chi^2_{0.95;1} = 3.8415$. The failure rate is estimated as $\widehat{\lambda} = \frac{n}{\sum t_i} = 0.0094$ per hour. The $\sum t_i = 3399.8$. Substitute these values to Eq. (4.39) and then the remaining question is to find the roots for

$$f(\lambda) = \lambda^{32} e^{-3399.8\lambda} - 0.0094^{32} e^{-0.0094*3399.8 - \frac{3.8415}{2}} = 0$$

Plot the function $f(\lambda)$ shown in the Fig. 4.11. It is obvious there are two roots, one as lower bound, one as upper bound. From the plot, it is shown the bound is [0.00653, 0.01303]. The CI from likelihood ratio method is very similar to that from Fisher method.

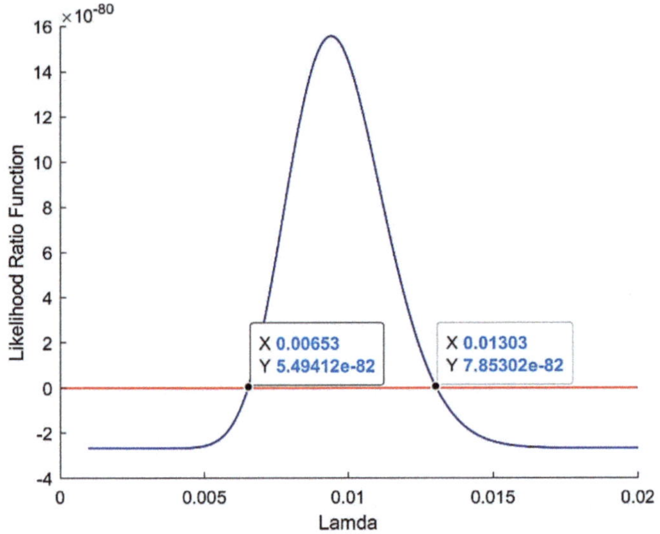

Fig. 4.11 Numerical plot of likelihood function

4.4.4 Simulation Comparison of CIs

The confidence interval serves as an indicator of the *reliability* of the estimator. Fisher-based and likelihood-based methods are the two most common approaches in state-of-the-art applications, while the CI from exact distribution could be as real CI. It is necessary to have a comparison among the CIs derived from these various methods discussed in Sect. 4.4.1 (Exact Distribution CI), Sect. 4.4.2 (Fisher Method CI) and Sect. 4.4.3 (Likelihood Ratio CI) to check their real performance.

The simulation chooses three Exponential distributions with failure rate bigger than 1, equal to 1 and less than 1 respectively. The sample size varies from 5 to 500. For each sample size, a total of 50,000 iterations has been run. CI is chosen as 0.95. The performance of the CI is defined as: The percentage of this CI covers the known true parameter out of the 50,000 iterations. For the exact distribution CI, the unbiased estimator shown as Eq. (4.19) is used as the true parameter. The results are shown in Table 4.3.

As shown in Table 4.3, all three methods work well for all ranges of data sizes, as for the 0.95 CI, the real CIs for all are around or above the 0.95. Especially, for the likelihood ratio method, the real CI can be even higher than the specified 0.95 for data size larger than 5. This means the real CI using likelihood ratio method will be significantly higher than the specified CI. The real estimator-distribution method

Table 4.3 CI performance comparison by simulation (using 0.95 confidence level)

Real Parameter	Sample Size in Simulation	Real Estimator Distribution	Likelihood Ratio CI	Fisher CI
2	5	0.9458	0.9458	0.9346
	10	0.9467	0.9802	0.9436
	20	0.9482	0.9790	0.9477
	50	0.9481	0.9770	0.9475
	100	0.9496	0.9760	0.9480
1	5	0.9463	0.9473	0.9356
	10	0.9472	0.9799	0.9421
	20	0.9500	0.9788	0.9477
	50	0.9484	0.9768	0.9476
	100	0.9499	0.9765	0.9493
0.1	5	0.9459	0.9465	0.9355
	10	0.9476	0.9807	0.9435
	20	0.9473	0.9783	0.9456
	50	0.9486	0.9767	0.9478
	100	0.9484	0.9759	0.9477
	500	0.9495	1	0.9489

Table 4.4 CI using bias correction estimator as failure rate at 95% confidence interval

Real λ	Sample Size	Biased Estimator	Unbiased Estimator
2	5	0.8981	0.9456
	20	0.9359	0.9482
1	5	0.8964	0.9445
	20	0.9369	0.9498
0.1	5	0.8992	0.9465
	20	0.9356	0.9479

shows very stable performance as it always shows around 0.95 for all failure rates 2, 1 and 0.1 and for all data sizes. The Fisher CI, however, depends more on data size, as it has lower accuracy for small data size 5.

Another simulation was done to test the effectiveness of the bias correction. In this case, only small data size is considered as for the large size, the biased estimator works well for all methods. The CI was using the estimation-distribution shown in Sect. 4.4.1. Two estimators are used as the real failure rate: one using the original MLE as Eq. (4.15), one using the corrected MLE of Eq. (4.19). The results are present in Table 4.4. It shows the unbiased one shows much better than the biased one for the small data sizes. As for the unbiased one, the performance of the estimation-distribution method can catch the performance of the other two CIs shown in Table 4.3.

The simulation conducted here demonstrates the importance of using the unbiased estimator for confidence intervals. It can also be concluded from this simulation: for reliability and risk data analysis, when the sample size is small, the maximum likelihood estimator should be corrected to be unbiased. Failing to do so poses a high risk of making incorrect decisions.

Example 4.5
A kind of mining machine with a population of 10. They ran to failure with the observed failure time as (in Days).

2.1 0.3 2.34 0.69 0.81 0.1 2.14 1.14 0.42 1.49.

Suppose the failure time follows Exponential distribution. Estimate the failure rate and provide a confidence interval.

Solution: The failure rate is $\widehat{\lambda}=n/\sum_{i=1}^{i=n}t_i$. In this example,

$$\widehat{\lambda}=\frac{10}{2.1+0.3+2.34+0.69+0.81+0.1+2.14+1.14+0.42+1.49}=0.867\,\text{per}\,\text{day}.$$

According to Eq. (4.36), the confidence interval using the Fisher method is

$$\begin{cases} \widehat{\lambda}_L=\widehat{\lambda}e^{-\frac{Z_\alpha\sqrt{Var\left(\widehat{\lambda}\right)}}{\widehat{\lambda}}}=\widehat{\lambda}e^{-\frac{Z_{\alpha/2}}{\sqrt{n}}} \\ \widehat{\lambda}_U=\widehat{\lambda}e^{\frac{Z_\alpha\sqrt{Var\left(\widehat{\lambda}\right)}}{\widehat{\lambda}}}=\widehat{\lambda}e^{\frac{Z_{\alpha/2}}{\sqrt{n}}} \end{cases}$$

If we choose the significance level at 0.05, then the $\alpha=\frac{1-0.95}{2}=0.4665$. The $Z_{\alpha/2}=1.96$. The $\widehat{\lambda}_L=0.4665$, $\widehat{\lambda}_U=1.6114$.

4.4.5 *Distribution of Total Work Time*

The analysis on the parameter estimation on the single parameter Exponential distribution is based on the total work time. In practice, the total work time is also of interest in experiment design, as it can represent the total experiment time. For a complete data set for Exponential distribution, the total work time is as follows:

$$\text{Total Work Time}=\sum_{i=1}^{i=n}t_i \tag{4.40}$$

As we discussed earlier, since the Exponential distribution is a special case of Gamma distribution, and the sum of Gamma distribution is still Gamma distribution, so does for the Exponential distribution. If the failure rate is λ, the distribution of total time follows Gamma distribution with $k=n$. The PDF function is

$$f(y;n,\lambda)=\frac{y^{n-1}e^{-\lambda y}}{\lambda^{-n}\Gamma(n)} \tag{4.41}$$

Let the total work time is $y = T_{Total}$. A variable written $z = 2\lambda T_{Total} = 2\lambda y$ then also follow Gamma distribution with PDF as follows by using the derivation in Example 4A.2 of Appendix 4A in this chapter.

$$f(z)=\frac{2^{-k}z^{k-1}e^{-\frac{z}{2}}}{\Gamma(k)} \tag{4.42}$$

According to Example 4A.2, Eq. (4.42) is identical to a χ^2 distribution with number of freedoms $2k$, i.e. $\chi^2(2k)$. We therefore can obtain the confidence interval for the total work time with the well-known χ^2 distribution, i.e. the $z = 2\lambda T_{Total}$ follows distribution $\chi^2(2k)$ where $k = n$. The n is the number of samples.

For a two-sided confidence interval, given the confidence length $1 - \alpha$, the z is

$$z_L=\chi^2_{\frac{\alpha}{2}}(2k);\quad z_U=\chi^2_{1-\alpha/2}(2k) \tag{4.43}$$

A confidence interval for the failure rate is then

$$\left[\lambda_L=\frac{\chi^2_{\alpha/2}(2k)}{2*T_{Total}},\lambda_U=\frac{\chi^2_{1-\alpha/2}(2k)}{2*T_{Total}}\right] \tag{4.44}$$

For one sided confidence interval, we normally pay attention on the upper bound, i.e. the worst case. The worst failure rate is

$$\lambda_U=\frac{\chi^2_{1-\alpha}(2k)}{2*T_{Total}} \tag{4.45}$$

Once we obtain the CI for the failure rate, other CI such as reliability is readily to be obtained. The bound for reliability is

$$\left[e^{-\lambda_U t},e^{-\lambda_L t}\right] \tag{4.46}$$

which follows the reliability function $R(t) = e^{-\lambda t}$.

Using Eq. (4.44) for CI on the failure rate is common in practical engineering. This method is adopted by multiple industrial standards.

Example 4.6
A kind of mining machine with a population of 10. They ran to failure with the observed failure time as (in Days).

$$2.1\ 0.3\ 2.34\ 0.69\ 0.81\ 0.1\ 2.14\ 1.14\ 0.42\ 1.49.$$

Suppose the failure time follows Exponential distribution. Estimate confidence interval of 95% for the failure rate with the approach in Eq. (4.44) and the lower bound of 95% CI for reliability of 5 days.

Solution: The ${\chi^2}_{\alpha/2}(2k)$ is $\alpha = 0.05$ and $k = 10$ in this example. Then ${\chi^2}_{\frac{\alpha}{2}}(2k) = 9.59$. ${\chi^2}_{1-\frac{\alpha}{2}}(2k) = 34.16$. The number of freedoms in the example is $2k = 20$. The $T_{Total} = 11.53$. The failure rate CI is then

$$\lambda_L = \frac{{\chi^2}_{\alpha/2}(2k)}{2 * T_{Total}} = 0.41, \lambda_U = \frac{{\chi^2}_{1-\alpha/2}(2k)}{2 * T_{Total}} = 1.48$$

The interval of failure rate is then [0.41, 1.48] for 95% CI.

The lower bound for reliability of 5 days chooses the highest failure rate is

$$R_{Lower}(t = 5) = \mathrm{e}^{-\lambda_U t} = \mathrm{e}^{-1.48*1} = 0.23$$

Examples 4.5 and 4.6 use the same data sets. The results show the confidence interval for the failure rate in the two examples show some difference. The ${\chi^2}_{\alpha/2}$ is narrower than the Fisher matrix one. Therefore, the ${\chi^2}_{\alpha/2}$ based one is recommended. The ${\chi^2}_{\alpha/2}$ based one uses the exact distribution of the total work time, which explains why it produces a narrower confidence interval in this example. The use of the χ^2 confidence interval shown in this section is a standard approach in the industry for estimating confidence intervals for failure rates or for MTTF/MTBF.

It is also necessary to emphasize that confidence intervals are not unique. For a given confidence level, such as 95%, many intervals can lead to it. Even for the same method, the confidence interval for a given confidence level varies, depending on how the tails are selected. In the case of the χ^2 method used in the above example, the tails on both sides are selected equally. But if unequal tails were selected, a different confidence interval will result.

4.5 Right Censored Context

In Accelerated Life Test (ALT) experiment, it is common that some test samples have not failed by the time the experiment is terminated. For example, desktop manufacturers typically conduct tests once some prototypes are ready. They run

24 samples until failure. After 6 months, only a few failed, while the majority did not. But the experiment must be stopped because the market cannot wait for all units to fail. This situation represents the right censoring, as we have discussed in Chap. 1. This section derives the parameter estimation method for the right censored situation.

4.5.1 Maximum Likelihood Estimation

Let a sample size of n products on test. Until time T, part of them have failed with failure times t_1, t_2, $\ldots t_i \ldots$, t_r. Assume the failures are statistically independent. The likelihood function is then

$$L(\lambda) = f(t_1)f(t_2)f(t_3)\ldots f(t_r)R(T)R(T)R(T) \tag{4.47}$$

where $f(t_i) = \lambda e^{-\lambda t_i}$. Equation (4.47) can be written to

$$L(\lambda) = \lambda^r e^{-\lambda \sum_{i=1}^{i=r} t_i} e^{-(n-r)\lambda T} \tag{4.48}$$

Let

$$\frac{\partial - \ln L(\lambda)}{\partial \lambda} = -r/\lambda + \sum_{i=1}^{i=r} t_i + (n-r)T = 0 \tag{4.49}$$

We obtain the maximum likelihood estimator as

$$\widehat{\lambda} = r / \left[\sum_{i=1}^{i=n} t_i + (n-r)T \right] \tag{4.50}$$

The r is the total number of failures. The denominator of Eq. (4.50) is

$$\text{Total Work Time} = \sum_{i=1}^{i=n} t_i + (n-r)T \tag{4.51}$$

The maximum likelihood estimator Eq. (4.50) implies the MTTF or MTBF can be simply estimated as

$$MTTF = \frac{\text{Total Work Time}}{\text{Total number of Failures}} \tag{4.52}$$

It implies regardless of the censored or complete context, Eq. (4.52) always holds.

Example 4.7
A PC company runs a life test experiment for their new PCs. The experiment runs 180 days. Total 24 PCs was collected to run the test. At the end of the test, only 3 failed with failure times (calendar time in days) as.

$$3, 150, 170.$$

What is MTBF or MTTF of the PCs?

Solution: PC is considered as a non-repairable system.

$$\begin{aligned}\text{MTTF} &= \text{Total Work Hours}/\text{Total Number of Failures}\\ &= (3 + 150 + 170 + 180 \times 21)/3 = 1367.7 \text{ days.}\end{aligned}$$

4.5.2 *Asymptotic Consistency*

As discussed in the previous section, for the right censored situation where time is right censored, the failure rate estimation can be achieved by maximum likelihood estimation. This section discusses the consistency of the estimator. For an individual sample, the work time is

$$y_i = \begin{cases} t_i; 0 \le t_i \le T; & \delta_i = 0 \\ T; & \delta_i = 1 \end{cases} \tag{4.53}$$

The probability density function of the right censored situation is.

$$f(t_i) = \begin{cases} \lambda e^{-\lambda t_i}, & 0 \le t_i \le T; \\ 0; & \text{Otherwise} \end{cases} \tag{4.54}$$

A Bernoulli distribution random δ_i is used to indicate it as censored or complete, which is defined as

$$\delta_i = \begin{cases} 1 & \text{Complete} \\ 0 & \text{Censored} \end{cases} \tag{4.55}$$

The expectation of y_i is

$$E(y_i) = \int_0^T t\lambda e^{-\lambda t} dt + T \int_T^{+\infty} \lambda e^{-\lambda t} dt = \frac{1}{\lambda}\left(1 - e^{-\lambda \mathrm{T}}\right) \tag{4.56}$$

The δ_i is Bernoulli distribution with probability $p = 1 - e^{-\lambda \mathrm{T}}$. Hence the

$$E(\delta_i) = 1 - e^{-\lambda T} \tag{4.57}$$

For the right censored situation, according to Eq. (4.50), the maximum likelihood estimator is

$$\widehat{\lambda} = \frac{r}{\left[\sum_{i=1}^{i=n} t_i + (n-r)T\right]} = \frac{\sum_1^n \delta_i}{\sum_1^n y_i} = \frac{\sum_1^n \delta_i/n}{\sum_1^n y_i/n} \tag{4.58}$$

According to large number theory in statistics, when the δ_i is independent from each other, then.

$$\frac{\sum_1^n \delta_i}{n} \xrightarrow{n\to\infty} E(\delta_i); \quad \frac{\sum_1^n y_i}{n} \xrightarrow{n\to\infty} E(y_i); \tag{4.59}$$

According to the Slutsky Theorem, Eq. (4.58) is [13].

$$\widehat{\lambda} = \frac{\sum_1^n \delta_i/n}{\sum_1^n y_i/n} \xrightarrow{n\to\infty} \frac{1 - e^{-\lambda T}}{\frac{1}{\lambda}(1 - e^{-\lambda T})} = \lambda \tag{4.60}$$

The estimator $\widehat{\lambda}$ is asymptotically approaches to the true failure rate λ when the $n \to \infty$. The asymptotic consistency implies for large sample size, the MLE for right censored situation is still able to center around its true parameter. However, the exact distribution of the total work time may not follow the Gamma distribution as it does in the complete-data case. For censored data, the exact distribution remains an open question.

4.6 Goodness of Fit Test

What we have discussed so far is to estimate the failure rate and its confidence interval, by priorly assuming the failure times follow Exponential distribution. The assumption of Exponential should be verified. Graphic method is one of them.

4.6.1 Graphic Method

The CDF of Exponential distribution can be approximated by empirical CDF. The CDF is the probability of failure. Suppose there are a total number of n products in the beginning. Until time t_i, the number of failures is n_i. The probability of failure can be approximated as

$$F(t_i) \approx \frac{n_i}{n} \tag{4.61}$$

As we have discussed in Chap. 1, the property of Eq. (4.61) can be proved by using the Binomial distribution. For a population of n products, suppose the probability of failure is p. The expectation of number of failures is np. The expected probability of failure is then

$$\frac{\text{Number of Failure}}{\text{Total Population}} = \frac{n\text{p}}{n} = \text{p} = \widehat{F}(t_i) \tag{4.62}$$

It means the expected probability of failure (4.61) can approximate the real probability $F(t_i)$.

On the other hand, given time t_i, the corresponding CDF of Exponential is

$$F(t_i) = 1 - e^{-\lambda t_i} \tag{4.63}$$

Rearrange it as

$$e^{-\lambda t_i} = 1 - F(t_i) \tag{4.64}$$

Take logarithm in both sides of Eq. (4.64), the

$$\lambda t_i = -\ln[1 - F(t_i)] \tag{4.65}$$

Equation (4.65) is an analytical expression of linear function in form of $y = ax$, where $y = -\ln[1 - F(t_i)]$, $x = t_i$, $a = \lambda$. If the λ is known as $\widehat{\lambda}$, e.g. already estimated from MLE or from other estimation method, we can plot a straight line from Eq. (4.65) for arbitrary values t_i.

On the other side, the $\widehat{F}(t_i)$ is a value approximated by Eq. (4.61), we can transform it to

$$-\ln\left[1 - \widehat{F}(t_i)\right] \tag{4.66}$$

Using Formula (4.66) as y value, the $x = t_i$. Then the data set $\left(t_i, -\ln\left[1 - \widehat{F}(t_i)\right]\right)$ is considered as the original data. If the straight line plotted from Eq. (4.65) fit the data set $\left(t_i, -\ln\left[1 - \widehat{F}(t_i)\right]\right)$ well, the $\widehat{\lambda}$ is then a correct estimator and the distribution follows Exponential.

Example 4.8

To continue the Example 4.6. A kind of mining machine with a population of 10. They ran to failure with the observed failure time as (in Days).

2.1 0.3 2.34 0.69 0.81 0.1 2.14 1.14 0.42 1.49

Check if this data follows Exponential distribution.

Solution: We first order the life data ascendingly as.

0.1 0.3 0.42 0.69 0.81 1.14 1.49 2.1 2.14 2.34

The total number of failures until t_i is then the i in the ordered life data. The probability of failure is $\frac{i}{n}$. The t_i and its corresponding approximate $F(t_i)$ is shown in the following Table 4.5.

We plot the data set, present as dot in the Fig. 4.12, t_i as the x-axis, the $-\ln\left[1-\widehat{F}(t_i)\right]$ as the y-axis, excluding the last data point with Inf value. Using the maximum likelihood estimator of $\widehat{\lambda} = 0.8673$, we obtain the estimated line is $y = 0.8673x$ by using Eq. (4.65), shown as the red line in Fig. 4.12.

It shows the data (presented as dots) can fit the line in general. It concludes the data follows Exponential distribution.

The Goodness of fit test using the graphic method is simple, but subjective. It requires software for computing. Another problem revealed in Example 4.8 is the last data point where the $-\ln\left[1-\widehat{F}(t_i)\right]$ value is infinite. One can use the following method as shown Table 4.6 to approximate $\widehat{F}(t_i)$ to avoid this problem.

Table 4.5 Empirical data sets

i	t_i	$\widehat{F}(t_i)$	$-\ln\left[1-\widehat{F}(t_i)\right]$
1	0.1	0.1	0.1054
2	0.3	0.2	0.2231
3	0.42	0.3	0.3567
4	0.69	0.4	0.5108
5	0.81	0.5	0.6931
6	1.14	0.6	0.9163
7	1.49	0.7	1.204
8	2.1	0.8	1.6094
9	2.14	0.9	2.3026
10	2.34	1	**Inf**

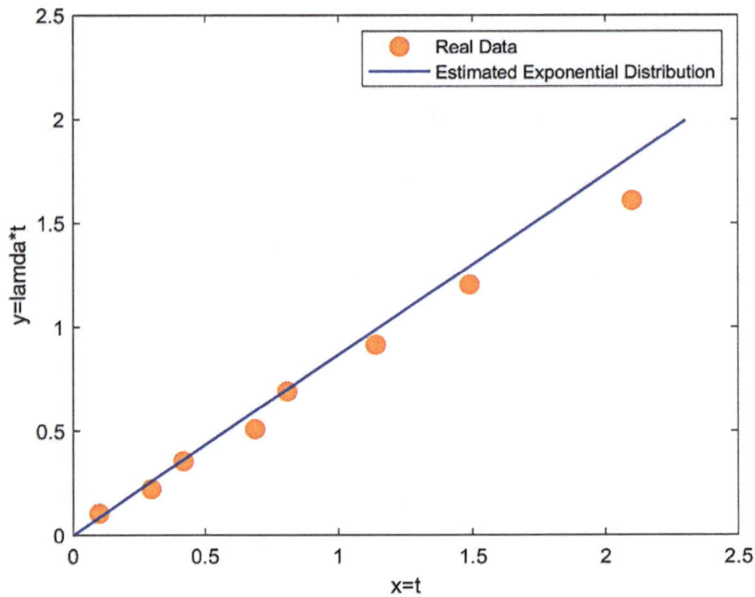

Fig. 4.12 GFT of Example 4.8

Table 4.6 Selected failure probability approximation

Method	$\widehat{F(t_i)}$
Mean rank	$\frac{i}{n+1}$
Median rank	$\frac{i-0.3}{n+0.4}$
Symmetrical CDF	$\frac{i-0.5}{n}$

The least squares method can also be used to estimate the failure rate. Using the least squares approach, for a linear function $y = ax$, given data sets $\langle x_i, y_i \rangle$, the slope is a, which is also the failure rate estimation

$$\widehat{\lambda} = a = \frac{\sum(x_i y_i)}{\sum x_i^2} \tag{4.67}$$

Nevertheless, using the least squares method to estimate the failure rate is not common. Since the maximum likelihood estimator on the λ has analytical solution, it is recommended to use maximum likelihood estimator as it has a more solid statistical foundation.

4.7 Other Issues on Exponential Distribution

4.7.1 Memoryless Property of Exponential Distribution

The Exponential is with constant failure rate. The constant failure rate attributes convenience for data analysis. For an operating product, if the previous working hours are unknown, for Exponential distribution, we only need to know the operating time after current time t_0, as shown in Fig. 4.13.

The probability that the product has survived until t_0 is

$$R(t_0) = e^{-\lambda t_0} \tag{4.68}$$

Conditioned on the survived t_0, the probability density function of the product will fail at time t_1 is

$$f(t_1) = \lambda e^{-\lambda t_1} / R(t_0) = \lambda e^{-\lambda(t_1 - t_0)} \tag{4.69}$$

Equation (4.69) implies the conditioned PDF remains Exponential distribution with the same failure rate. The probability of failure is a function only of time interval $t_1 - t_0$. The reliability is completely unrelated to starting time t_0, but only the length of $t_1 - t_0$. Therefore, in practice, if the question of interest is to investigate how long a system can work after t_0, we can simply assume starting time $t_0 = 0$. For example, if the $t_1 = 10$, the $t_0 = 9$, then $t_1 - t_0 = 1$. The $f(t_1) = \lambda e^{-\lambda(t_1 - t_0)}$ is equivalent to $f(t = 1) = \lambda e^{-\lambda t}$, where we implicitly specify the system starts from 0. Considering starting time from 0 can simplify problems, especially for the left censored situation.

4.7.2 Probability of Failure by MTBF

The MTBF/MTTF is the inverse of failure rate of Exponential distribution. In practice, it often misunderstands half of the product population (50%) would fail before MTBF/MTTF. This statement is equivalent as the probability of failure for a product is 50% before MTBF/MTTF. It seems true from our intuition. But we can verify it using mathematics. The probability of failure for Exponential distribution to MTBF/MTTF is

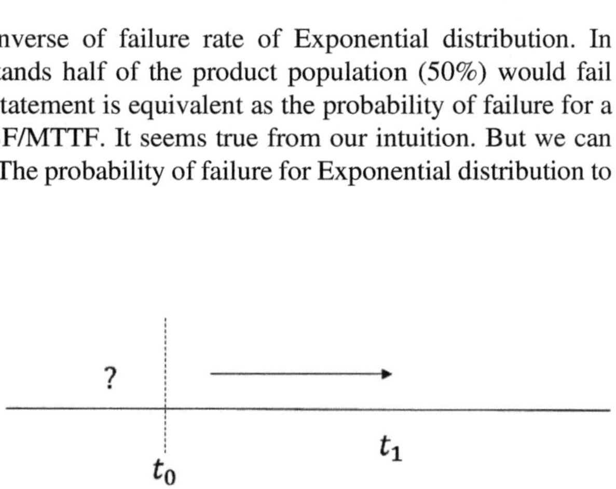

Fig. 4.13 Memoryless in Exponential Distribution

$$F(MTBF) = 1 - e^{-\lambda.MTBF} \tag{4.70}$$

Since the $\lambda = 1/MTBF$, Eq. (4.70) is rewritten as

$$F(MTBF) = 1 - e^{-1} = 0.6321 \tag{4.71}$$

It clearly proves the probability of failure before MTBF is not 50%, but 63.21%.

Appendix 4A: Probability Density Function Transformation

This appendix gives the transformed probability density function for a univariate random number. Suppose the probability density function $f(x)$ is continuous. Another random variable y is defined as

$$y = T(x) \tag{4.A1}$$

with inverse function written as

$$x = T^{-1}(y) \tag{4.A2}$$

The probability density function of y is then

$$g(y) = f\left(T^{-1}(y)\right).\left|\frac{dx}{dy}\right| \tag{4.A3}$$

Example 4A.1
The Exponential distribution is with PDF as $f(x) = \lambda e^{-\lambda x}$ where $x > 0$. A random variable y is with $y = 1/x$. The

$$x = T^{-1}(y) = 1/y$$

The

$$\frac{dx}{dy} - 1/y^2$$

According to (4.A3), the PDF of the random variable y

(continued)

$$g(y) = \frac{\lambda e^{-\lambda/y^2}}{y^2}; y > 0$$

Example 4A.2

The Gamma distribution is with PDF as $f(x; k, \lambda) = \frac{x^{k-1}e^{-\lambda x}}{\lambda^{-k}\Gamma(k)}$ where $x > 0$. A random variable y is with $y = ax$. The

$$x = T^{-1}(y) = y/a$$

The

$$\frac{dx}{dy} = -\frac{1}{a}$$

The PDF of the random variable y can be obtained from (4.A3)

$$g(y) = \frac{(y/a)^{k-1}e^{-\lambda y/a}}{\lambda^{-k}\Gamma(k)} \cdot \frac{1}{a} = \frac{(y/a)^{k-1}e^{-\lambda y/a}}{a\lambda^{-k}\Gamma(k)}$$

If we let the $a = 2\lambda$, the above equation is: then written as

$$g(y) = \frac{(y/2\lambda)^{k-1}e^{-\frac{\lambda y}{2\lambda}}}{2\lambda.\lambda^{-k}\Gamma(k)} = \frac{2^{-k}y^{k-1}e^{-\frac{y}{2}}}{\Gamma(k)} \tag{4.A4}$$

Refer to Eq. (2.48) in Chap. 2. The Chi-Squared distribution is with PDF as

$$f(x) = \frac{2^{-n/2}x^{n/2-1}e^{-x/2}}{\Gamma(n/2)} \tag{4.A5}$$

The (4.A5) is identical to (4.A4) when $k = n/2$. The n is the number of freedoms for the Chi-Squared distribution.

Problems and Questions

1. Europipe II has a pipeline from Kårstø to Dornum of length around 658 km. If we set a requirement as that, the reliability of this pipeline should be at least 99%. How many failures per km can the pipeline maximally tolerated? (Hint: support the failure follows Exponential distribution. In this case, PDF is a function of length instead of time)
2. A component has constant failure rate

$$r(t) = \lambda = 2.5 * 10^{-5}/\text{hours}$$

a. Determine the probability that the component survives for a period of 2 months without failure.
b. Find the MTTF of the component.
c. Find the probability that the components can survive by its MTTF.

3. One parameter Exponential distribution (PDF $f(t) = \lambda \exp(-\lambda t)$) is the simplest but an important distribution in reliability engineering. Suppose we have recorded lifetimes for a kind of product as follows

10.2 89.6 54.0 96.0 23.3 30.4 41.2 0.8 73.2 3.6 28.0 31.6

a. What is the failure rate λ?
b. Find the confidence interval of the estimated failure rate.
c. Estimate the reliability for a service time of 10 hours.

4. An experiment was conducted for 365 days. There are total 10 samples on test. And only 3 of them failed at the end of the test. The failure times are: 16, 48, 180 days. Suppose the life follows Exponential distribution, what is the failure rate?
5. If random variable x follows Exponential distribution with failure rate λ, what is the PDF function of random variable $1/x$? Is it still Exponential?
6. If random variable x follows Exponential distribution with failure rate λ, what is the pdf function of random variable ax, where a is a positive constant? Is it still Exponential?

References

1. Kumar U (1989) Statistical analysis on mining data. Lulea University of Technology, Lulea
2. Dai Y, Zhou YF, Jia YZ (2003) Distribution of time between failures of machining center based on type I censored data. Reliab Eng Syst Safe 79(3):377–379
3. de Negreiros ACSV, Lins ID, Moura MJD, Droguett EL (2020) Reliability data analysis of systems in the wear-out phase using a (corrected) q-exponential likelihood. Reliab Eng Syst Safe 197
4. Kendall MG, O'Hagan A, Forster J (2004) Kendall's advanced theory of statistics, Kendall's library of statistics, 2nd edn. Arnold, London
5. Childs A, Balakrishnan N, Chandrasekar B (2012) Exact distribution of the MLEs of the parameters and of the quantiles of two-parameter exponential distribution under hybrid censoring. Statistics 46(4):441–458
6. Cheng CH, Chen JY, Bai JM (2013) Exact inferences of the two-parameter exponential distribution and Pareto distribution with censored data. J Appl Stat 40(7):1464–1479
7. Gupta RD, Kundu D (2007) Generalized exponential distribution: existing results and some recent developments. J Stat Plan Infer 137(11):3537–3547

8. Yuan F, Lu J, Batalden BM (2016) Sample size determination and confidence interval derivation for exponential distribution, Presented at the the 26th international ocean and polar engineering conference
9. Rohatgi VK (1975) An introduction to probability theory and mathematical statistics, Wiley series in probability and mathematical statistics. Wiley, New York
10. Lehmann EL, Casella G (1998) Theory of point estimation, Springer texts in statistics, 2nd edn. Springer, New York
11. Meeker WQ, Escobar LA (1998) Statistical methods for reliability data, Wiley series in probability and statistics Applied probability and statistics section. Wiley, New York
12. Lehmann EL, Romano JP (2005) Testing statistical hypotheses, Springer texts in statistics, 3rd edn. Springer, New York
13. Allan G (2005) Probability: a graduate course. Springer, Berlin

Chapter 5
Weibull Distribution

The Exponential distribution, discussed in the previous chapter, assumes a constant failure rate that sometimes fails to represent field data. In practice, failure rates can vary—decreasing, increasing, or exhibiting complex patterns with multiple peaks. A non-constant failure rate is more realistic, as many products, especially in the wear-out phase, typically show an increasing failure rate. Consequently, a more flexible statistical distribution is needed to model such behavior. The Weibull distribution, widely used in reliability engineering, meets this need due to its versatility. This chapter examines the statistical properties of the Weibull distribution and explains how it can be effectively applied to reliability analysis.

5.1 Further on Weibull Distribution

The probability density function (PDF) of the two-parameter Weibull distribution, previously introduced in Chap. 2, is defined as:

$$f(t) = \frac{\beta}{\alpha}\left(\frac{t}{\alpha}\right)^{\beta-1} e^{-(t/\alpha)^{\beta}}; \alpha > 0, \beta > 0, t \geq 0 \tag{5.1}$$

In the two-parameter Weibull distribution, the α is the scale parameter and the β is the shape parameter. It is worth noting that the Weibull distribution has a three-parameter variant, which includes an additional location parameter beyond the two-parameter version shown in Eq. (5.1). However, the three-parameter Weibull is rarely used in reliability engineering and therefore will not be covered in this book. By default, any mention of the Weibull distribution in this book refers to the two-parameter form defined in Eq. (5.1). Figure 5.1 presents PDF plots for various Weibull parameters. At first glance, these plots may resemble those of other statistical distributions. However, the failure rate function of the Weibull distribution exhibits unique characteristics that distinguish it from other distributions.

F. Yuan, *Essentials of Reliability Engineering*, Springer Series in Reliability Engineering, https://doi.org/10.1007/978-3-032-15544-3_5

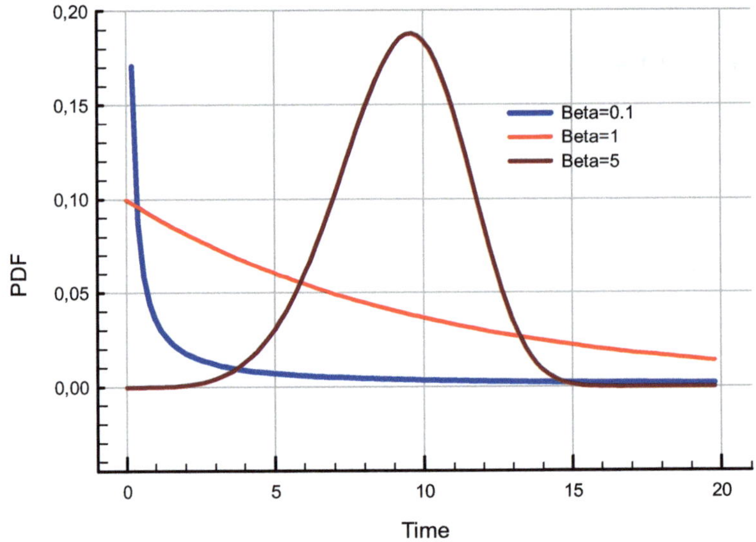

Fig. 5.1 PDF function of Weibull distribution

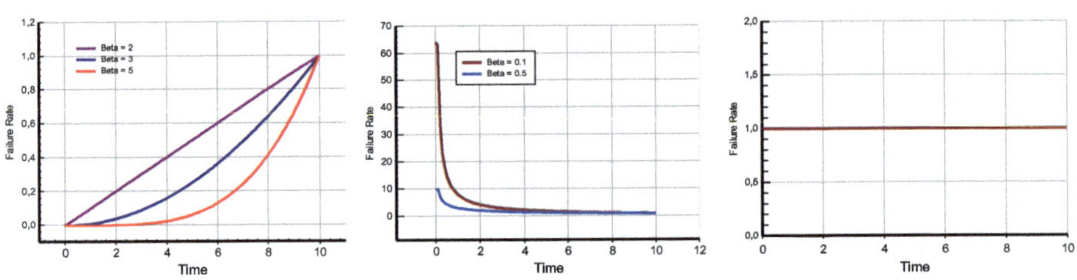

Fig. 5.2 Failure rate function of Weibull distribution

The failure rate function of Weibull distribution is as

$$\lambda(t) = \frac{\beta}{\alpha}\left(\frac{t}{\alpha}\right)^{\beta-1} \tag{5.2}$$

From mathematical perspective, it is not hard to find, the $\beta > 1$ leads to an increasing failure rate, the $\beta < 1$ leads to a decreasing failure rate, while the $\beta = 1$ leads to constant failure rate, as shown in Fig. 5.2. Constant failure rate is equivalent to Exponential distribution, i.e. Exponential is a special case of Weibull. One advantage of Eq. (5.2) is: with two parameters, Weibull can accommodate three types of failure rate. Therefore, when using the Weibull distribution for data analysis, we can avoid making assumptions about constant, decreasing, or increasing failure rates. The data, based on which to evaluate the β value, can determine the monotony of the failure rate function.

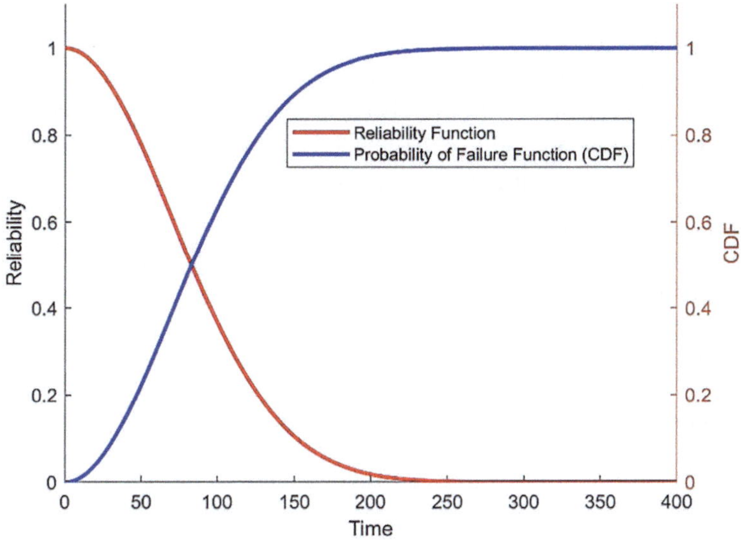

Fig. 5.3 Reliability and CDF function for Weibull distribution

Some argument state that the failure times of the mechanical system tend to follow Weibull distribution. It might be due to the Weibull model being initially developed for material and mechanical failures. As presented in the [1, 2], they are the earliest papers on the Weibull distribution. However, subsequent engineering experience has demonstrated that the applicability of the Weibull distribution extends well beyond mechanical failures; it can also describe the failure data of electronic components and complex systems.

Corresponding to the PDF of Eq. (5.1) and failure rate of Eq. (5.2), the reliability function of Weibull distribution is

$$R(t) = e^{-(t/\alpha)^{\beta}} \tag{5.3}$$

The probability of failure function, also known as the cumulative distribution function (CDF), is given by

$$F(t) = 1 - e^{-(t/\alpha)^{\beta}} \tag{5.4}$$

Figure 5.3 demonstrates the plots of reliability and CDF for Weibull distribution with $\alpha = 100$ and $\beta = 2$.

Example 5.1

LHD (Load, Haul, Dump) machines are a critical equipment in mining companies. Data was collected from a Swedish mining company on the time to failure as shown in Appendix 5A: Table 5.A1 in this chapter [3]. Assume the data follows Weibull distribution, the parameter $\widehat{\alpha} = 54; \widehat{\beta} = 1.07$ for LHD A.

(a) What is the reliability at $t = 100$ hours?
(b) At what time does the machine fail with a probability of 98%?
(c) Is the failure rate increasing or decreasing?

Solution:

(a) Reliability $\mathrm{R}(t = 100) = e^{-\left(\frac{100}{54}\right)^{1.07}} = 0.1446$
(b) As $\mathrm{F}(t) = 1 - e^{-\left(\frac{t}{\alpha}\right)^{\beta}}$, we obtain

$$\mathrm{t} = \alpha[-\ln(1 - \mathrm{F})]^{\left(\frac{1}{\beta}\right)}$$

The F(t) = 0.98. We obtain t = 193.2. With probability 98%, the machine will fail at t = 193.2 hours.

(c) Since $\widehat{\beta} = 1.07 > 1$, the failure rate is increasing. Figure 5.4 shows some plots of the reliability, CDF and the failure rate function for this example.

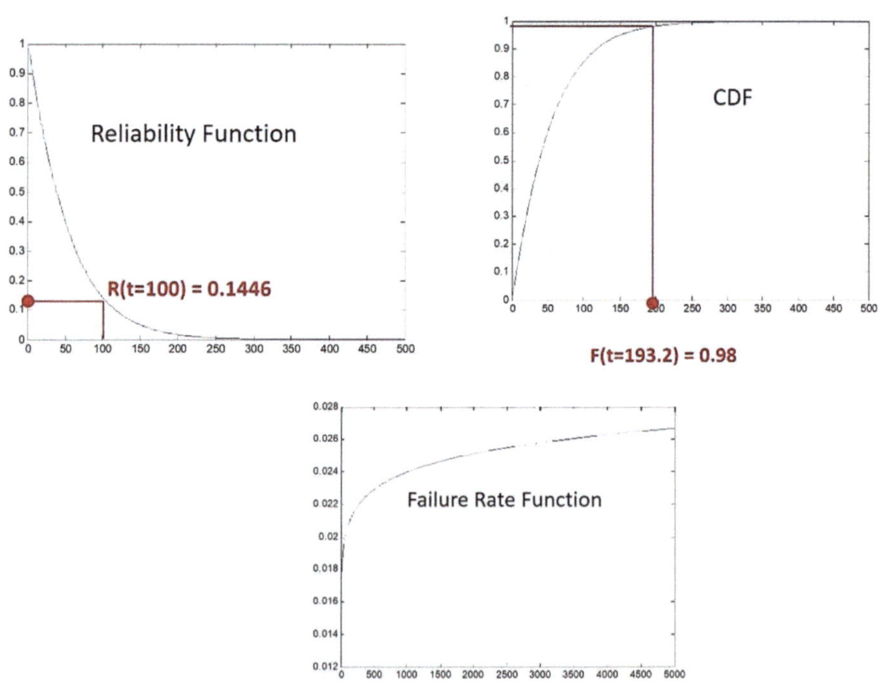

Fig. 5.4 Plots of Example 5.1

5.2 Parameter Estimation for Weibull

A key question in implementing the Weibull distribution for data analysis is how to evaluate the shape and scale parameters. Various methods can be employed to achieve this objective, while differing in performance. One of the most popular methods is the maximum likelihood estimation (MLE) method. In addition to MLE, the method of moments and least squares methods can also be used. This section discusses these methods.

5.2.1 *Maximal Likelihood Method*

The MLE assumes the most probably estimator is the best estimator. As we discussed in Chap. 3, the likelihood of the parameters can be derived from the joint PDF of observed data. Suppose the observed failure time are $t_1, t_2, \ldots, t_i, \ldots t_n$. The likelihood function is defined as

$$L(\alpha, \beta) = \prod_{i=1}^{n} f(t_i) \tag{5.5}$$

Substitute the Weibull pdf Eq. (5.1) into Eq. (5.5). The likelihood function becomes

$$L(\alpha, \beta) = \prod_{i=1}^{n} \frac{\beta}{\alpha} \left(\frac{t_i}{\alpha}\right)^{\beta-1} e^{-(t_i/\alpha)^{\beta}} \tag{5.6}$$

The $L(\alpha, \beta)$ is usually of very small magnitude, as demonstrated in Table 5.1 for various parameters for a given data set, where the likelihood function value is as small as a level at 10^{-130}. Calculator or computer handling such small easily run to overflow.

We can take logarithm of L to transform the small L into an easily tractable number. Finding the maximum of L is equivalent to find the minimum $-\ln L$, since $-\ln L$ is a monotonic function of L. Fig. 5.5 shows the plot of $-\ln L$ value for various estimators. It demonstrates that a minimum of $-\ln L$ exists and the minimum is unique, which is favorable. We can find the global optimal parameter by analyzing its first derivatives.

Table 5.1 L and $\ln L$ values

α	β	L	$-\ln L$
10	0.10	4.81E-130	297.7654
20	0.64	8.18E-102	232.7616
30	1.19	9,15E-103	234.9524
40	1.73	1.07E-110	253.214
...	...	...	...

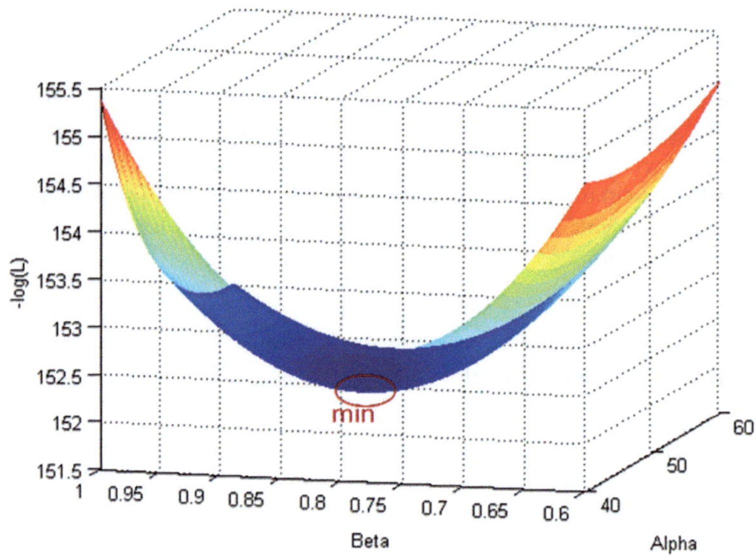

Fig. 5.5 Minimum value of $-\ln L$

The negative logarithm of Eq. (5.6) is

$$-\ln L = -n\ln\beta + n\beta\ln\alpha_i - (\beta-1)\sum \ln t_i + \sum (t_i/\alpha_i)^\beta \tag{5.7}$$

Take the first derivative in respect to parameter α and β, i.e.

$$\frac{\partial(-\ln L)}{\partial\alpha} = \frac{n\beta}{\alpha} - \beta\,\alpha^{-\beta-1}\sum t_i^\beta \tag{5.8}$$

$$\frac{\partial(-\ln L)}{\partial\beta} = -\frac{n}{\beta} + n\ln\alpha - \sum \ln t_i + \sum\left\{\left(\frac{t_i}{\alpha}\right)^\beta \ln\left(\frac{t_i}{\alpha}\right)\right\} \tag{5.9}$$

Setting both equations to zero and rearranging them, we obtain the ML estimators as

$$\widehat{\alpha} = \left(\sum t_i^{\widehat{\beta}}/n\right)^{1/\widehat{\beta}} \tag{5.10}$$

$$\frac{n}{\widehat{\beta}} - n\frac{\sum t_i^{\widehat{\beta}}\ln t_i}{\sum t_i^{\widehat{\beta}}} + \sum \ln t_i = 0 \tag{5.11}$$

where the $\widehat{\alpha}$ and $\widehat{\beta}$ is written instead of α, β to indicate they are estimates. The $\widehat{\alpha}$ depends on the $\widehat{\beta}$. The $\widehat{\beta}$ can be obtained from Eq. (5.11), while Eq. (5.11) has no closed-form expression. It has to use numerical method, e.g. the numerical Newton-Raphson method, to find the approximate $\widehat{\beta}$. The optimal solution of Eq. (5.11) is easy to find by numerical approach, as the optimum is globally unique as demonstrated in Fig. 5.5. Nevertheless, the non-analytical solution to Eq. (5.11) is a major weakness of MLE for shape parameter. It results in limitations, for example:

(a) The exact value of the shape parameter $\widehat{\beta}$ is hard to be obtained, as numerical solution is an approximated solution.
(b) The variance of the $\widehat{\beta}$ estimator is more difficult to evaluate.
(c) It is well known that the MLE estimator is biased. Correcting the bias is hard if no analytical solution is available.

Example 5.2
For the LHD (Load, Haul, Dump) machines data presented in Appendix 5A Table 5.A1 at this chapter [3]. Using maximum likelihood estimating method to find the Weibull parameters for LHD B.

Solution: The LHD B contains total $n = 44$ data. We firstly obtain the $\widehat{\beta}$ value from Eq. (5.11) and thereafter to obtain $\widehat{\alpha}$ from Eq. (5.10). We can plot the function to examine the solution.

$$f\left(\widehat{\beta}\right) = \frac{n}{\widehat{\beta}} - n\frac{\sum t_i^{\widehat{\beta}} \ln t_i}{\sum t_i^{\widehat{\beta}}} + \sum \ln t_i \tag{5.12}$$

We can find a unique solution to $f\left(\widehat{\beta}\right) = 0$ exists from Fig. 5.6. Numerical method can be used to find the exact unique solution. Here we use Matlab function wblfit() to obtain the solution and find out it is as follows

$$\widehat{\alpha} = 49.11$$

$$\widehat{\beta} = 1.0293$$

5.2.2 *Uniqueness of Weibull MLE*

In optimization theory, if a function has multiple optimal solutions—referred to as multiple local optima instead of global optimum—a solution from a numerical approach may be local rather than global. This section aims to verify whether Eq. (5.11) has a unique solution, i.e., a global optimum. Ensuring uniqueness is advantageous for numerical methods, as it guarantees that any solution obtained will

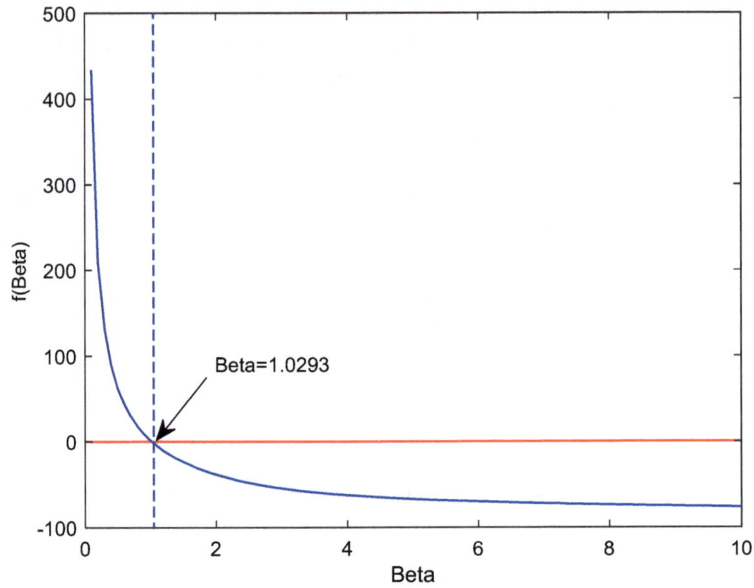

Fig. 5.6 Plot of shape function

be the best. Example 5.2 illustrates a rough graphical method, shown in Fig. 5.6, to examine whether a unique solution exists. Figure 5.6 suggests the solution is unique. It is not a special case for this specific data that it has unique solution. The solution to maximum likelihood estimation function of Eq. (5.11) is always unique. This section provides an analytical proof of this uniqueness, while a similar proof effort can be found in [4].

For convenience of analysis, we rewrite Eq. (5.11) as

$$\frac{n}{\widehat{\beta}} = n\frac{\sum t_i^{\widehat{\beta}} \ln t_i}{\sum t_i^{\widehat{\beta}}} - \sum \ln t_i \tag{5.13}$$

Let the right of Eq. (5.13) as a function $G\left(\widehat{\beta}\right)$

$$G\left(\widehat{\beta}\right) = n\frac{\sum t_i^{\widehat{\beta}} \ln t_i}{\sum t_i^{\widehat{\beta}}} - \sum \ln t_i \tag{5.14}$$

The monotonicity of the function $G\left(\widehat{\beta}\right)$ can be examined from its first derivative. The first derivative of $G\left(\widehat{\beta}\right)$ as

$$\frac{\partial G(\widehat{\beta})}{\partial \widehat{\beta}} = n \frac{\left(\sum t_i^{\widehat{\beta}} \ln^2 t_i\right) . \sum t_i^{\widehat{\beta}} - \left(\sum t_i^{\widehat{\beta}} \ln t_i\right)^2}{\left(\sum t_i^{\widehat{\beta}}\right)^2} \tag{5.15}$$

The lower part of Eq. (5.15) $\left(\sum t_i^{\widehat{\beta}}\right)^2$ is absolutely positive. Checking the monotonicity is equivalent to checking the positivity of the upper part of

$$\left(\sum t_i^{\widehat{\beta}} \ln^2 t_i\right) . \sum t_i^{\widehat{\beta}} - \left(\sum t_i^{\widehat{\beta}} \ln t_i\right)^2$$

Introducing A_i and B_i and let $A^2{}_i = t_i^{\widehat{\beta}} \ln^2 t_i$ and $B^2{}_i = t_i^{\widehat{\beta}}$, and $A_i B_i$ is therefore

$$A_i B_i = \sqrt{t_i^{\widehat{\beta}} \ln^2 t_i} . \sqrt{t_i^{\widehat{\beta}}} = t_i^{\widehat{\beta}} . \ln t_i$$

The upper of right of Eq. (5.15) is then

$$\left(\sum t_i^{\widehat{\beta}} \ln^2 t_i\right) \sum t_i^{\widehat{\beta}} - \left(\sum t_i^{\widehat{\beta}} \ln t_i\right)^2 = \sum A^2{}_i \sum B^2{}_i - \left(\sum A_i B_i\right)^2$$

Let the

$$C = \sum A^2{}_i \sum B^2{}_i - \left(\sum A_i B_i\right)^2 \tag{5.16}$$

According to Cauchy-Schwarz inequality, the $\sum A^2{}_i \sum B^2{}_i \geq (\sum A_i B_i)^2$.Thus

$$C \geq 0$$

We further check the $C = 0$. The equality in Eq. (5.17) holds only when:

(a) There exist constants μ and λ, and
(b) The μ and λ cannot both be 0 for all the $1 \leq i \leq n$ to let

$$\mu A_i = \lambda B_i \tag{5.17}$$

Write Eq. (5.17) explicitly as

$$\mu\sqrt{t_i^{\widehat{\beta}}\ln^2 t_i}=\lambda\sqrt{t_i^{\widehat{\beta}}} \tag{5.18}$$

The $t_i^{\widehat{\beta}}$ is not 0 since the t_i is positive as it is time unit. Equation (5.18) can then be simplified as

$$\mu|\ln t_i|=\lambda \tag{5.19}$$

Obviously, Eq. (5.19) holds for all t_i only when μ and λ are both 0, which contradicts to the condition: Not both of them can be 0. Therefore, the equality in Eq. (5.17) does not hold, i.e. the $C > 0$. Therefore, the

$$\sum A^2{}_i\sum B^2{}_i-\left(\sum A_iB_i\right)^2>0 \tag{5.20}$$

which suggests the $G\left(\widehat{\beta}\right)$ in Eq. (5.14) (the right side of Eq. (5.13)) is monotonically increasing. On the other hand, the left side of Eq. (5.13) as $\frac{n}{\widehat{\beta}}$ is obviously decreasing with $\widehat{\beta}$, as shown as the red line in Fig. 5.7. Figure 5.7 is plotted by using synthetic data: *1 4 6 8 10 14 19*. If these left and right side can cross, as demonstrated in Fig. 5.7, the solution to Eq. (5.11) then certainly exists and is unique.

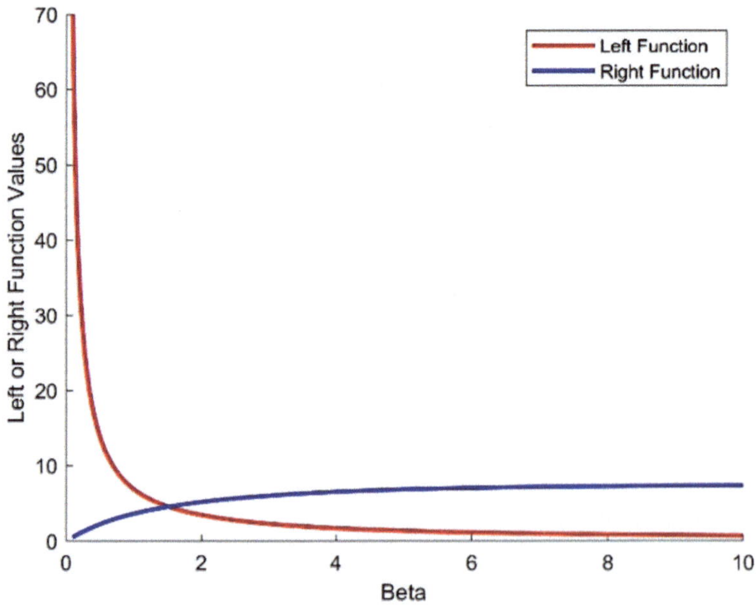

Fig. 5.7 Crossing of left and right sides of Eq. (5.13)

Appendix 5B proves the solution to Eq. (5.13) exists and is only one. It then further concludes that the MLE estimator for Weibull distribution exists, and the solution is unique. The proof of the uniqueness can simplify the numerical method of finding the solution for shape parameter $\widehat{\beta}$. Once an optimal solution is obtained, the solution is global and unique. The uniqueness proof, along with the moment method to be discussed in next section and the comparisons between them are also shown in reference [5].

5.2.3 *Moment Method of Weibull*

As another major parameter estimation method, moment method can be also applied to Weibull distribution, although it is not popular. This method can be an alternative estimator for MLE. The moment μ_k of Weibull distribution is defined as

$$\mu_k = \alpha^k \Gamma\left(1 + \frac{k}{\beta}\right) \tag{5.21}$$

Since Weibull distribution contains two parameters, two moments are thus required to determine the solution uniquely. It is usual to select the first two order moment ($k = 1, k = 2$). Let the moment of (5.21), which is theoretical moment value, equal to an empirical moment as [6].

$$\begin{cases} \mu_1 = \alpha\Gamma\left(1+\dfrac{1}{\beta}\right) = \bar{t} \\ \mu_2 = \alpha^2\Gamma\left(1+\dfrac{2}{\beta}\right) = \overline{t^2} \end{cases} \tag{5.22}$$

which means the $\mu_1 \approx \bar{t} = \sum t_i/n$; $\mu_2 \approx \overline{t^2} = \sum {t_i}^2/n$. From Eq. (5.22), we obtain

$$\mu_2/\Gamma\left(1 + \frac{2}{\widehat{\beta}}\right) = {\mu_1}^2/\Gamma^2\left(1 + \frac{1}{\widehat{\beta}}\right) \text{ or } \overline{t^2}/\Gamma\left(1 + \frac{2}{\widehat{\beta}}\right) = \bar{t}^2/\Gamma^2\left(1 + \frac{1}{\widehat{\beta}}\right) \tag{5.23}$$

$$\widehat{\alpha} = \mu_1/\Gamma\left(1 + \frac{1}{\widehat{\beta}}\right) \text{ or } \widehat{\alpha} = \bar{t}/\Gamma\left(1 + \frac{1}{\widehat{\beta}}\right) \tag{5.24}$$

Same as the MLE, the scale parameter depends on α, as shown in Eq. (5.24). The shape parameter β can be obtained from Eq. (5.23). Equation (5.23) has no analytical solution, and it has to be solved by numerical method, which is same as the MLE.

Example 5.3
Using the Moment method to evaluate the parameter for the LHD (Load, Haul, Dump) machines data used in **Example 5.2**.

Solution: Using $\mu_1 \approx \bar{t} = \sum t_i/n$; $\mu_2 \approx \overline{t^2} = \sum t_i^2/n$, from Eq. (5.23), we obtain an equation

$$\left(\sum \frac{t_i^2}{n}\right) / \Gamma\left(1 + \frac{2}{\beta}\right) = \left(\sum \frac{t_i}{n}\right)^2 / \Gamma^2\left(1 + \frac{1}{\beta}\right)$$

From numerical method, this example uses fsolve() function in Matlab to obtain the solution for above as

$$\widehat{\beta} = 1.0497$$

Then by Eq. (5.24)

$$\widehat{\alpha} = \frac{\mu_1}{\Gamma\left(1 + \frac{1}{\beta}\right)} = \frac{\sum \frac{t_i}{n}}{\Gamma\left(1 + \frac{1}{\beta}\right)} = 49.4674$$

The evaluated parameter estimators obtained from maximum likelihood method and moment method shows slight difference for this mining example. The MLE is with ($\widehat{\alpha} = 49.11$, $\widehat{\beta} = 1.0293$). Moment method evaluates the parameter as ($\widehat{\alpha} = 49.4674$, $\widehat{\beta} = 1.0497$). As the true parameters of are unknown, it is impossible to tell which method is more accurate. A simulation study is conducted in the next section to compare the performance.

5.2.4 Comparison of Moment and MLE

Different estimation methods will result in different estimates. This Section examines if the likelihood and moment method would have significantly different performance for Weibull distribution. The examination is not performed from theoretical perspective but performed by simulation. Matlab was used to implement it. The Weibull random numbers are generated from wblrnd() function, for three Weibull distributions: $\alpha = 10$, $\beta = 2$. The sample size, denoted by N, is from small to large as N = 5, N = 25, N = 50, N = 80, N = 160, N = 500, N = 5000. Each simulation runs 10,000 iterations. In this simulation study, the true parameters are known. We can then check the deviation of the estimate from the true parameter. The error is defined as the mean deviation over the 10,000 iterations. The criterion for performance is defined as the deviation between evaluated estimators and the true parameters, which is written as

$$Error\left(\widehat{\theta}\right) = \frac{\sqrt{\sum_{i=1}^{M}\left(\widehat{\theta}_i - \theta\right)^2}}{\theta} \tag{5.25}$$

The results are shown in Table 5.2.

For both two estimation methods, the error is decreasing with the sample size. Figure 5.8 plots the Error in % against sample size for scale parameter β, where the decreasing trend is obvious. For small sample size 5, the error is around 0.86% while for the large sample size, the error has been reduced to around 0.01%.

In terms of the criterion defined by Eq. (5.25), the moment and MLE do not show significant difference, while the MLE exhibits slightly better performance than the method of moments, as MLE errors tend to be lower in this simulation. Therefore, it is recommended to choose MLE for Weibull parameter estimation. MLE for the Weibull distribution is a standard method in many state-of-the-art applications.

5.2.5 *Equivariance of Estimators*

Equivariance of an estimator is not a primary concern when selecting an estimator. However, in life data analysis, equivariance is an important aspect to consider. The scale parameter of the Weibull distribution reflects the scale of the data, while the shape parameter should remain unaffected by changes in the selected scale. For instance, if the data unit changes from years to days, the shape parameter should remain the same. In statistics, this property of remaining unaffected by scale changes is known as equivariance. This section examines the equivariance of maximum likelihood and moment method for Weibull distribution.

Given the original failure time t_i, changing the scale is equivalent to multiply the t_i by a positive constant k, which is

$$y_i = kt_i; k > 0 \tag{5.26}$$

The k is a positive constant. According to Eq. (5.11), the formula used to estimate the shape parameter for the new scale is then

$$\frac{n}{\widehat{\beta}} = n\frac{\sum y_i^{\widehat{\beta}} \ln y_i}{\sum y_i^{\widehat{\beta}}} - \sum \ln y_i \tag{5.27}$$

Table 5.2 Increasing failure rate of Weibull distribution

	N = 5		N = 25		N = 50		N = 100		N = 150		N = 500		N = 5000	
(α=10, β=2)	*Moment*	*MLE*	*Moment*	*MLE*	*Moment*	*MLE*	*Moment*	*MLE*	*Moment*	*MLE*	*Moment*	*MLE*	*Moment*	*MLE*
Error α %	0.2327	0.2331	0.10560	0.1057	0.0740	0.074	0.0530	0.0529	0.0432	0.0432	0.0238	0.0238	0.0075	0.0075
Error β%	0.8650	0.8871	0.1810	0.1837	0.1200	0.1199	0.0818	0.0814	0.0663	0.0658	0.0353	0.0349	0.0111	0.0110

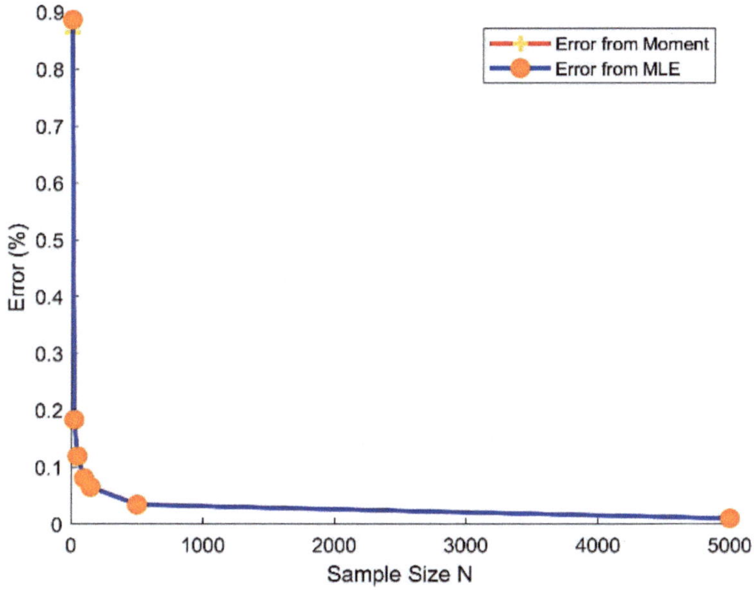

Fig. 5.8 Estimator error decreasing with sample size

We check if there is any difference between using t_i and scaled y_i. Substitute the (5.26) to (5.27), the (5.27) is rewritten as [5]

$$\frac{n}{\widehat{\beta}} = n\frac{\sum (kt_i)^{\widehat{\beta}} \ln(kt_i)}{\sum (kt_i)^{\widehat{\beta}}} - \sum \ln(kt_i) = n\frac{\sum t_i^{\widehat{\beta}}(\ln(k) + \ln(t_i))}{\sum t_i^{\widehat{\beta}}}$$
$$- \sum(\ln(k) + \ln(t_i)) = n\frac{\sum t_i^{\widehat{\beta}} \ln t_i}{\sum t_i^{\widehat{\beta}}} - \sum \ln t_i \tag{5.28}$$

We can find Eq. (5.28) is exact the same as Eq. (5.13) where the failure time was original. It means the solutions of Eqs. (5.28) and (5.13) will be the same. Scale parameter from MLE is thus of equivariance. The scale parameter will however change with the scale. According to Eq. (5.10),

$$\widehat{\alpha}_y = \left(\sum y_i^{\widehat{\beta}}/n\right)^{1/\widehat{\beta}} = k\widehat{\alpha}_t \tag{5.29}$$

i.e. the estimated scale parameter scales with failure time at a same scale factor of k.

For the moment method, according to Eq. (5.23), the shape parameter is estimated by

$$\sum t_i^2/\Gamma\left(1+\frac{2}{\widehat{\beta}}\right)=\left(\sum t_i\right)^2/\Gamma^2\left(1+\frac{1}{\widehat{\beta}}\right)$$

which can be rearranged as

$$\frac{\sum t_i^2}{\left(\sum t_i\right)^2}=\frac{\Gamma\left(1+\frac{2}{\widehat{\beta}}\right)}{\Gamma^2\left(1+\frac{1}{\widehat{\beta}}\right)} \tag{5.30}$$

Substitute the $y_i = kt_i$ to Eq. (5.30),

$$\frac{\sum y_i^2}{\left(\sum y_i\right)^2}=\frac{\sum (kt_i)^2}{\left(\sum kt_i\right)^2}=\frac{\sum t_i^2}{\left(\sum t_i\right)^2}$$

i.e. the right of Eq. (5.30) does not change with the scale factor k. The solution to β thus does not change with scale k. For parameter α for Weibull distribution, however it is different. According to Eq. (5.24), the parameter α is estimated as

$$\widehat{\alpha}_y=y\Gamma\left(1+\frac{1}{\beta}\right) \tag{5.31}$$

Substitute the $y_i = kt_i$ to Eq. (5.31), it is easy to find $\widehat{\alpha}_y=k\widehat{\alpha}_t$. The scale parameter adjusts in proportion to the scale of the data. For Weibull distribution, the likelihood method and the moment method exhibit equivariance for the shape parameter, which is desirable. However, the scale parameter varies proportionally with the data scale.

5.2.6 *Right Censor Situation for Weibull*

The Weibull analysis discussed so far considers a complete data set situation. For censored situation, the parameter estimation is like the complete data set situation. This chapter only considers the right censored situation. Suppose total number of products on test is n. The observed failure times are $t_1, t_2, \ldots, t_{r-1}, t_r$ at time T, where the $r \leq n$. The likelihood function of right censored situation is

$$L(\alpha,\beta)=\prod_{i=r+1}^{n} e^{-(T/\alpha)^\beta}\prod_{i=1}^{r}\frac{\beta}{\alpha}\left(\frac{t_i}{\alpha}\right)^{\beta-1}e^{-(t_i/\alpha)^\beta} \tag{5.32}$$

Take the negative logarithm of (5.32)

$$-\ln L = -r\ln\beta + r\beta\ln\alpha - (\beta-1)\sum \ln t_i + \sum\left(\frac{t_i}{\alpha}\right)^{\beta} + (n-r)(T/\alpha)^{\beta} \tag{5.33}$$

The first derivative in respect to parameter α and β is

$$\frac{\partial(-\ln L)}{\partial\alpha} = \frac{r\beta}{\alpha} - \beta\,\alpha^{-\beta-1}\sum t_i^{\beta} - (n-r)\frac{\beta}{T}\left(\frac{\alpha}{T}\right)^{-\beta-1} \tag{5.34}$$

$$\frac{\partial(-\ln L)}{\partial\beta} = -\frac{r}{\beta} + r\ln\alpha - \sum \ln t_i + \sum\left\{\left(\frac{t_i}{\alpha}\right)^{\beta}\ln\left(\frac{t_i}{\alpha}\right)\right\}$$
$$+ (n-r)(T/\alpha)^{\beta}\ln T/\alpha \tag{5.35}$$

Equal both of them to zero and rearrange them as

$$\widehat{\alpha} = \left(\frac{\sum t_i^{\widehat{\beta}} + (n-r)T^{\widehat{\beta}}}{r}\right)^{1/\widehat{\beta}} \tag{5.36}$$

$$\frac{\sum_{i=1}^{r} t_i^{\widehat{\beta}}\ln t_i + (n-r)T^{\widehat{\beta}}\ln \mathrm{T}}{\sum_{i=1}^{r} t_i^{\widehat{\beta}} + (n-r)T^{\widehat{\beta}}} - \frac{1}{\widehat{\beta}} = \frac{1}{r}\sum_{i=1}^{r}\ln t_i; \tag{5.37}$$

Again, same as the uncensored situation, Eq. (5.37) has no analytical solution. Numerical method has to be employed to obtain the solution.

Example 5.4
An experiment put total $n = 20$ products on test. Experiment is terminated at $T = 150$ hours. When it is terminated, total $r = 10$ product failed with time to failures as (in hours) [7]

$$23, 27, 38, 45, 58, 84, 90, 99, 109, 138$$

Estimate the parameter values by supposing the data follows Weibull distribution.

Solution: In the question, the $n = 20$; $r = 10$; $T = 150$. The β from Eq. (5.37) is written as

(continued)

Example 5.4 (continued)

$$f\left(\widehat{\beta}\right)=\frac{\sum_{i=1}^{r} t_i^{\widehat{\beta}}\ln t_i+(n-r)T^{\widehat{\beta}}\ln \mathrm{T}}{\sum_{i=1}^{r} t_i^{\widehat{\beta}}+(n-r)T^{\widehat{\beta}}}-\frac{1}{\widehat{\beta}}-\frac{1}{r}\sum_{i=1}^{r}\ln t_i$$

An empirical plot of the $f\left(\widehat{\beta}\right)$ is shown in Fig. 5.9. It shows there is one and only one point crossing $f\left(\widehat{\beta}\right)=0$, which is the solution. We use Matlab fsolve () to find the numerical solution as

$$\widehat{\beta}=1.3314.$$

The scale parameter is then

$$\widehat{\alpha}=\left(\frac{\sum t_i^{\widehat{\beta}}+(n-r)T^{\widehat{\beta}}}{r}\right)^{1/\widehat{\beta}}=192.3070$$

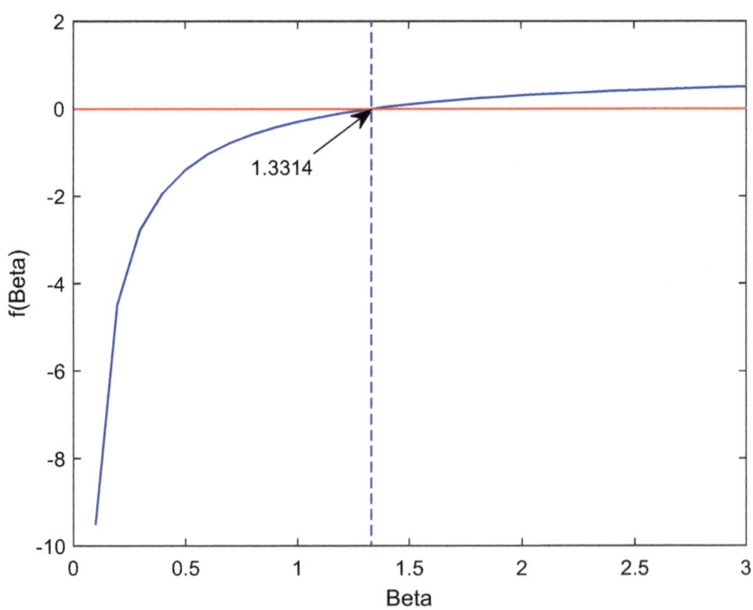

Fig. 5.9 Plot of $f\left(\widehat{\beta}\right)$ function

5.2.7 Bias Correction for Complete Data

Bias of an estimator is an issue for situation where sample size is small, as we have demonstrated in the Chap. 4 for Exponential distribution. Traditional method to correct bias is to find the out the expectation of the estimator if the analytical expression of the estimator and its expectation are available, then multiply the expectation by a constant factor, named bias correction factor, to let the expectation equal to the true parameter. For the Weibull distribution, unfortunately the shape parameter has no analytical expression. The bias correction based on the traditional method is thus hard. But there is a trick method that using simulation to find out the bias correction factor [8, 9].

The Weibull distribution is of a form of $f((t/\alpha)^{\beta})$. Let $v = (t/\alpha)^{\beta}$. The PDF of Weibull is then rewritten as

$$f(v) = e^{-v} \tag{5.38}$$

which is a typical Exponential distribution with $\lambda = 1$. For Weibull random variable t_i, we rewrite the $v_i = (t_i/\alpha)^{\beta}$ as $t_i = \alpha(v_i)^{1/\beta}$. For complete data, the MLE shape parameter function shown in Eq. (5.11) is then rewritten by substituting the $t_i = \alpha(v_i)^{1/\beta}$ into it, as follows:

$$\frac{\sum_{i=1}^{n} v_i^{\widehat{\beta/\beta}} \ln v_i}{\sum_{i=1}^{n} v_i^{\widehat{\beta/\beta}}} - \frac{1}{\widehat{\beta}/\beta} = \frac{1}{n} \sum_{i=1}^{n} \ln v_i \tag{5.39}$$

Equation (5.39) is an amazing function since it only depends on the v_i, which follows the Exponential distribution with parameter 1, and the $\widehat{\beta}/\beta$ is not dependent on the observed time to failure t_i. A distribution of the $\widehat{\beta}/\beta$ can then be able to be obtained empirically by using pseudo-random variables drawn from the standard Exponential distribution. The expectation $E\left(\frac{\widehat{\beta}}{\beta}\right)$ can be approximated from an empirical distribution.

Let the bias corrector factor as $B(n)$ and correct the bias by

$$E\left(\widehat{\beta} * B(n)\right) = \beta \tag{5.40}$$

Rearrange Eq. (5.40), we obtain

$$B(n) = 1/E\left(\frac{\widehat{\beta}}{\beta}\right) \tag{5.41}$$

The bias factor $B(n)$ depends on sample size n, as the $E\left(\frac{\widehat{\beta}}{\beta}\right)$ depends on sample size. An unbiased estimator is then

$$\widehat{\beta}_{ub} = \widehat{\beta} * B(n) \tag{5.42}$$

As we know from Eq. (5.39), the $\frac{\beta}{\beta}$ depends now only on the v_i from standard Exponential distribution. The random number v_i used in Eq. (5.38) can be generated from the standard Exponential distribution with PDF function e^{-v}. The random number is called Pseudo-random sample. The Pseudo-random number can be obtained by using Monte Carlo simulation. For sample size n, we can generate a number of n standard Exponential random numbers. And use them to obtain a $\frac{\widehat{\beta}}{\beta}$ by Eq. (5.39). Repeating it many times, we can get a group data of $\frac{\widehat{\beta}}{\beta}$, the mean of the group data can be approximated as $E\left(\frac{\widehat{\beta}}{\beta}\right)$.

An algorithm proposed to find the factor is as follows:

Algorithm 5.1

Step 1: Define the number of iterations M. It is recommended to be over 5000. Define a sample size as n.

Step 2: Generate Pseudo-random number with a size of n from the standard Exponential distribution e^{-v} using Monte Carlo method.

Step 3: For the number of n Pseudo-random number, using numerical method to obtain the solution to Eq. (5.39), The solution is donated as s_i, which is a sample of $\widehat{\beta}/\beta$. Repeat this procedure for all the number of M iterations.

Step 4: The bias corrector B(n) is evaluated as B(n) = $M/\sum s_i$, which approximates $1/E\left(\frac{\widehat{\beta}}{\beta}\right)$.

Table 5.3 presents the bias factor for sample size from 5 to 2000 by using the Algorithm 5.1 implementing by Matlab. The number of iterations chooses $M = 20, 000$. Result finds the factor is increasing with sample size and approaches to 1 when size becomes large as shown in Fig. 5.10. The approach to 1 is as expected, since the MLE is asymptotic unbiased. For larger data size, the bias is less and thus the factor

Table 5.3 Bias factor for complete data

	n is sample size. Below the sample size is the B(n).						
	n = 5	n = 6	n = 7	n = 8	n = 9	n = 10	
	0.6935	0.7495	0.7899	0.8168	0.8379	0.8545	
n = 11	n = 12	n = 13	n = 14	n = 15	n = 16	n = 18	n = 19
0.8685	0.8812	0.8905	0.8999	0.9057	0.9125	0.9233	0.9240
n = 20	n = 22	n = 26	n = 30	n = 40	n = 44	n = 50	n = 60
0.9292	0.9344	0.9450	0.9529	0.9655	0.9684	0.9724	0.9771
n = 70	n = 80	n = 90	n = 100	n = 200	n = 500	n = 1000	n = 2000
0.9798	0.9821	0.9844	0.9861	0.9937	0.9974	0.9986	0.9993

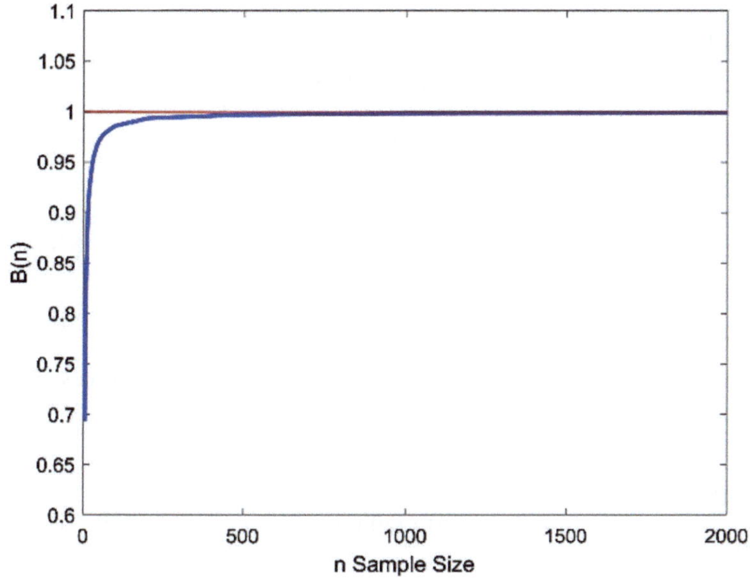

Fig. 5.10 Bias factor tends to unity when sample size increases

is near to 1. The last two digits of the bias factor could be random as it differs with simulation for each time. The random problem is due to the limit of Matlab. We should not be overly concerned with this negative impact of randomness, as even if the last digit is random, the precision of the bias factor remains sufficient for most applications.

Example 5.5

For LHD B data shown in Table 5A.1. Suppose the data follows Weibull distribution, estimate the Weibull parameter and correct its bias.

Solution: If the data is complete, using Eq. (5.10) and (5.11), we obtain the original biased MLE as

$$\widehat{\beta} = 1.0293$$

$$\widehat{\alpha} = 49.1079$$

For sample size 44 in this example, according to Table 5.3, which is from the Algorithm 5.1 using Matlab, the bias factor is

$$\mathrm{B(n)} = 0.9684$$

According to Eq. (5.42), the unbiased shape parameter is then $1.0293 * 0.9684 = 0.9968$, i.e.

(continued)

Example 5.5 (continued)

$$\widehat{\beta}_{ub} = 0.9968$$

According to Eq. (5.10), the new scale parameter is

$$\widehat{\alpha}_{UB} = \left(\sum t_i^{\widehat{\beta}}/n\right)^{1/\widehat{\beta}} = \left[\frac{149^{0.9968} + 33^{0.9968} + \ldots}{44}\right]^{1/0.9968} = 48.4586$$

Example 5.5 is with a medium data size, this Example 5.6 demonstrates an example with small data size.

Example 5.6
A product is tested with the time to failure (in hours) with sample size $n = 6$ as [7].

$$450, 600, 530, 120, 450, 650$$

Suppose the data follows Weibull distribution, estimate the Weibull parameter and correct its bias.

Solution: If the data is complete, using Eqs. (5.10) and (5.11), we obtain the original biased MLE is

$$\widehat{\beta} = 3.192$$

$$\widehat{\alpha} = 518.65$$

From Table 5.3, it shows the bias correction factor is 0.7495 for n =6, therefore

$$\widehat{\beta}_{ub} = 3.192 * 0.7495 = 2.3924$$

According to Eq. (5.10), the new scale parameter is

$$\widehat{\alpha}_{UB} = \left(\sum t_i^{\widehat{\beta}}/n\right)^{1/\widehat{\beta}} = \left[\frac{450^{2.3924} + 33^{2.3924} + \ldots}{6}\right]^{1/2.3924} = 505.4254$$

The scale parameter for biased and unbiased differ only a little. But the shape parameter differs notably. Figure 5.11 plots the reliability function for using both biased and unbiased, where it can see the reliability function having difference. The unbiased parameter in this case shows a lower reliability than that of the biased in general.

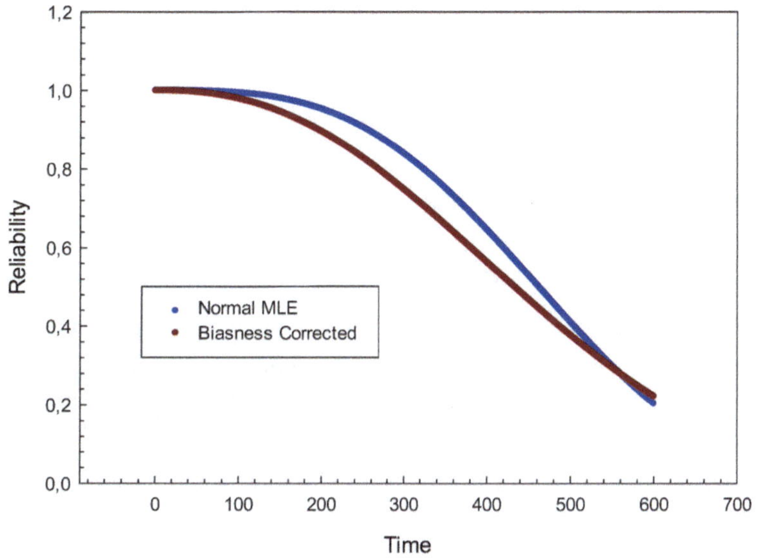

Fig. 5.11 Reliability with time for small sample size

5.2.8 *Bias Correction for Right Censored Data*

For right censored situation, the bias factor can be evaluated by a method same as the complete data set. But with an additional challenge, as Eq. (5.37) contains the censoring time T, which is not able to be converted to a v_i in the standard Exponential distribution $f(v) = e^{-v}$. Some tricks are needed to solve this problem [9]. Since the $\frac{\text{r}}{n}$ can approximate the probability of failure, the corresponding censoring time in standard Exponential can be derived from the $\frac{\text{r}}{n}$. The probability of failure of standard Exponential corresponds to $\frac{\text{r}}{n}$ by

$$\frac{\text{r}}{n} \approx 1 - e^{-c} \tag{5.43}$$

Rearrange the (5.43) as

$$c \approx -\ln\left(1 - \frac{\text{r}}{n}\right) \tag{5.44}$$

Based on Eq. (5.37), the formula independent of original failure time t_i is thus obtained as follows for right censored situation

$$\frac{\sum_{i=1}^{r} v_i^{\widehat{\beta/\beta}} \ln v_i + (n-r)c^{\widehat{\beta/\beta}} \ln c}{\sum_{i=1}^{r} v_i^{\widehat{\beta/\beta}} + (n-r)c^{\widehat{\beta/\beta}}} - \frac{1}{\widehat{\beta/\beta}} = \frac{1}{r}\sum_{i=1}^{r} \ln v_i \tag{5.45}$$

The expectation of $\widehat{\beta/\beta}$ can be evaluated from simulation.

Algorithm 5.2

Same as the Algorithm 5.1, this algorithm uses Monte Carlo simulation to obtain the $E\left(\frac{\widehat{\beta}}{\beta}\right)$ for right censored data. Pseudo-random samples are generated from the standard Exponential distribution with PDF function e^{-v}. Given sample size n and uncensored number of failures r. The detailed steps for the bias corrector factor procedure as follows:

Step 1: Define the number of iterations M. Calculate the truncation time c using Eq. (5.44).

Step 2: Generate Pseudo-random number with a size of n from the standard Exponential distribution e^{-v} using Monte Carlo method.

Step 3: Check if each Pseudo-random number in the total n is greater than c. If yes, label it as censored, otherwise label it as uncensored. If the number of uncensored is not exact r, go to *Step 2* to redo; otherwise go to *Step 4*.

Step 4: For the number of n Pseudo-random number, using numerical method to obtain the solution of Eq. (5.45). The solution is donated as s_i, which is a sample of $\widehat{\beta/\beta}$. Repeat this procedure for all the number of M iterations.

Step 5: The bias corrector B(n) is calculated by $\mathrm{B(n)} = M/\sum s_i$, which approximates $1/E\left(\frac{\widehat{\beta}}{\beta}\right)$.

Table 5.4 presents some factors obtained from Algorithm 5.2. It shows the bias correction factor increases with sample size and increases with the $\frac{r}{n}$. For large sample size and large $\frac{r}{n}$, the bias factor is near 1, which means estimator is not necessary to be corrected as the MLE estimator is asymptotic unbiased.

Table 5.4 Example of bias factor for censored data [9]

	n is sample size. r is the number of samples who are not censored.					
	B(n) is bias correction factor					
n = 5				r = 4		
				0.706		
n = 10			r = 5	r = 7	r = 8	r = 9
			0.8224	0.8819	0.9062	0.9135
n = 20	r = 4	r = 8	r = 10	r = 14	r = 16	r = 18
	0.7651	0.8911	0.9193	0.9401	0.9522	0.9507
n = 100	r = 10	r = 20	r = 40			
	0.8959	0.9477	0.9766			

To obtain the unbiased estimator, the $\widehat{\beta}_{ub} = \widehat{\beta} * B(n)$ which is Eq. (5.42) is still the same for this censored situation.

Example 5.7
Using the same data set of Example 5.4. The $n = 20$, $T = 150$ hours, $r = 10$. Estimate an unbiased parameter estimates for Weibull distribution.

Solution: From Example 5.4, the uncorrected MLE is

$$\widehat{\beta} = 1.3314.$$
$$\widehat{\alpha} = 192.3070$$

According to Table 5.4, the bias correction factor $B(n)$ is 0.9193 for $n = 20$, $r = 10$. The

$$\widehat{\beta}_{ub} = \widehat{\beta} * B(n) = 0.9193 * 1.3314 = 1.2240$$

$$\widehat{\alpha}_{ub} = \left(\frac{\left(\sum t_i^{\widehat{\beta}_{ub}}\right) + (n-r)T^{\widehat{\beta}_{ub}}}{r}\right)^{\frac{1}{\widehat{\beta}_{ub}}}$$

$$= \left(\frac{(23^{1.2240} + 27^{1.2240} + \ldots) + (20-10) * 150^{1.2240}}{10}\right)^{1/1.2240} = 199.29$$

5.2.9 *Asymptotic Unbiasedness of MLE for Weibull*

Asymptotic unbiasedness means that an estimator becomes unbiased when the sample size is large. For the MLE of the Weibull distribution with complete data, which has been discussed in previous sections, the bias factor approaches unity as the sample size approaches infinity, indicating that the estimator is asymptotically unbiased. This unbiasedness can be demonstrated theoretically. Referring to Sect. 5.2.7 and using the law of large numbers, when the $n \to \infty$, Eq. (5.39) can be rewritten as [9]

$$\frac{\mathrm{E}\left(v_i^{\widehat{\beta}/\beta} \ln v_i\right)}{\mathrm{E}\left(v_i^{\widehat{\beta}/\beta}\right)} - \frac{\beta}{\widehat{\beta}} = \mathrm{E}(\ln v_i) \tag{5.46}$$

Given a hypothesis as $\frac{\widehat{\beta}}{\beta} \to 1$, then on the left side of Eq. (5.46), the $\mathrm{E}\left(v_i^{\frac{\widehat{\beta}}{\beta}} \ln v_i\right) \to \mathrm{E}(v_i \ln v_i)$, $\mathrm{E}\left(v_i^{\frac{\widehat{\beta}}{\beta}}\right) \to \mathrm{E}(v_i)$. When if $\frac{\widehat{\beta}}{\beta} \to 1$, Eq. (5.46) is rewritten as

$$\frac{\mathrm{E}(v_i \ln v_i)}{\mathrm{E}(v_i)} - 1 = \mathrm{E}(\ln v_i)$$

The $\mathrm{E}(v_i) = 1$ is straightforward since the v_i is standard Exponential with failure rate 1. The above formula is rewritten as

$$\mathrm{E}(v_i \ln v_i) - 1 = \mathrm{E}(\ln v_i)$$

We can prove the $\mathrm{E}(v_i \ln v_i) - 1 = \mathrm{E}(\ln v_i)$ holds, and $\mathrm{E}(\ln v_i) = 0$, $\mathrm{E}(v_i \ln v_i) = 1$ when $\frac{\widehat{\beta}}{\beta} \to 1$ as follows:

Proof

For the expectation of lnt for t follows Exponential distribution with constant failure rate 1, the PDF function is e^{-t}. Then

$$E(\ln \mathrm{t}) = \int_{0^+}^{+\infty} \ln \mathrm{t}.e^{-t} dt$$

The $\int_{0^+}^{+\infty} \ln \mathrm{t}.e^{-t} dt = -\ln \mathrm{t}.e^{-t}\Big|_{0^+}^{+\infty} + \int_0^{+\infty} \frac{e^{-t}}{t} dt$. The

$$-\ln \mathrm{t}.e^{-t}\Big|_{0}^{+\infty} = -\lim_{t \to +\infty} (\ln \mathrm{t}.e^{-t}) + \lim_{t \to 0} (\ln \mathrm{t}.e^{-t}) = -\infty\ .$$

The $\int_0^{+\infty} \frac{e^{-t}}{t} dt$ is Gamma function for $\Gamma(0^+)$, which corresponds to $\int_0^{+\infty} \frac{e^{-t}}{t} dt = +\infty$. Therefore

$$E(\ln \mathrm{t}) = -\ln \mathrm{t}.e^{-t}\Big|_{0^+}^{+\infty} + \int_0^{+\infty} \frac{e^{-t}}{t} dt = -\infty\ + \infty\ = 0$$

As $E\left(\mathrm{t}^{\frac{\widehat{\beta}}{\beta}} \ln \mathrm{t}\right) = \int_0^{+\infty} \mathrm{t}^{\frac{\widehat{\beta}}{\beta}} \ln \mathrm{t}.e^{-t} dt$, since the $\frac{\widehat{\beta}}{\beta} \to 1$ by assumption, then

$$E\left(\mathrm{t}^{\frac{\widehat{\beta}}{\beta}} \ln \mathrm{t}\right) = \int_0^{+\infty} t. \ln \mathrm{t}.e^{-t} dt = -\int_0^{+\infty} t. \ln \mathrm{t} de^{-t}$$
$$= -t. \ln \mathrm{t}.e^{-t}\Big|_0^{+\infty} + \int_0^{+\infty} \ln \mathrm{t}.e^{-t} dt + \int_0^{+\infty} e^{-t} dt$$

We can find: $-t.\ln \mathrm{t}.e^{-t}\Big|_{0}^{+\infty} = -\lim_{t\to+\infty}(t.\ln \mathrm{t}.e^{-t}) + \lim_{t\to 0}(t.\ln \mathrm{t}.e^{-t}) = 0$. The $\int_0^{+\infty} e^{-t}dt = 1$ since e^{-t} is probability density function of Exponential with failure rate 1. Therefore *E*(*t*. ln t) = 1 + *E*(lnt), which is

$$E(t.\ln \mathrm{t}) - 1 = E(\ln \mathrm{t})$$

Therefore *E*(*t*. ln t)=1 since *E*(lnt) = 0.

For right-censored data, it a more complex scenario. Rewrite Eq. (5.45) as follows

$$\frac{\frac{\sum_{i=1}^{r} v_i^{\widehat{\beta}/\beta} \ln v_i}{n} + (1 - r/n)c^{\widehat{\beta/\beta}} \ln \mathrm{c}}{\frac{\sum_{i=1}^{r} v_i^{\widehat{\beta/\beta}}}{n} + (1 - r/n)c^{\widehat{\beta/\beta}}} - \frac{\beta}{\widehat{\beta}} = \frac{1}{r}\sum_{i=1}^{r} \ln v_i \tag{5.47}$$

When sample size $n \to \infty$, the $\frac{\sum_{i=1}^{r} .}{n} = E(.)$. Therefore, Eq. (5.47) becomes

$$\frac{\mathrm{E}\left(v_i^{\widehat{\beta}/\beta} \ln v_i\right) + (1 - r/n)c^{\widehat{\beta/\beta}} \ln \mathrm{c}}{\mathrm{E}\left(v_i^{\widehat{\beta/\beta}}\right) + (1 - r/n)c^{\widehat{\beta/\beta}}} - \frac{\beta}{\widehat{\beta}} = \mathrm{E}(\ln v_i) \tag{5.48}$$

Assume $\widehat{\beta/\beta} \to 1$, and E(ln$v_i$) = 0 and $E(v_i$ ln $v_i)$ = 1, Eq. (5.48) is rewritten as follows when if assumption $\widehat{\beta/\beta} \to 1$ holds.

$$\frac{1 + (1 - r/n)c \ln \mathrm{c}}{1 + (1 - r/n)c} - 1 = 0 \tag{5.49}$$

The 1 − *r*/*n* is the reliability at time c. The left side of Eq. (5.49) is thus

$$\frac{1 + \mathrm{e}^{-\mathrm{c}}c \ln \mathrm{c}}{1 + \mathrm{e}^{-\mathrm{c}}c} - 1 \tag{5.50}$$

When $\frac{r}{n} = 1$, corresponding to $1 - \frac{r}{n} = 0$, it implies the $c \to +\infty$. While $\lim_{c\to\infty} \frac{1+\mathrm{e}^{-\mathrm{c}}c \ln \mathrm{c}}{1+\mathrm{e}^{-\mathrm{c}}c} \to 1$, it means (5.49) holds when $c \to +\infty$. It means when $\frac{r}{n}$ is big, similar to complete data, the estimator is close to unbiased. When the $\frac{r}{n}$ is small, corresponding to smaller *c*, the estimator is biased. The level of bias for censored data is complex. From the plot of the left side of Eq. (5.50) shown in Fig. 5.12, the function is not monotonic, implying the level of bias is fluctuating with censoring situation. When *c* = 10, corresponding to the $\frac{r}{n} = 1 - e^{-10}$ equal to almost 1, the

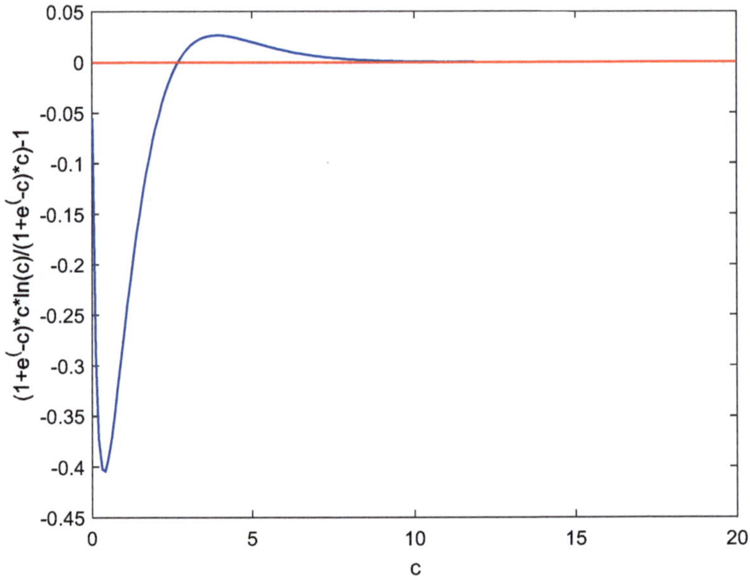

Fig. 5.12 Plot of Eq. (5.50)

estimator (for right censored data) is nearly unbiased as the left side of Eq. (5.49) is near 0 and it is equal to the right side. However, when the $\frac{r}{n}$ is less than $1 - e^{-10}$, Eq. (5.49) does not hold, implying estimator is biased.

5.3 Least Squares Estimation for Weibull

Both the maximum likelihood method and moment method discussed in the previous section requires numerical method to obtain optimal solution. Least squares estimation is a regression approach which requires no numerical iteration. This section discusses using regression to estimate the parameters for Weibull. As discussed in Chap. 4, the CDF can be approximated as media rank

$$\frac{i-0.3}{n+0.4} \tag{5.51}$$

The CDF approximated by median rank is empirical CDF. We can find parameters for Weibull that can let its CDF fit the empirical data well. The fitting is essentially regression. However, the original CDF against time is a nonlinear function as shown in Fig. 5.13.

To simplify the problem, we can convert the Weibull CDF function to a linear function, enabling the fitting to become linear. The theoretical CDF at time t_i is

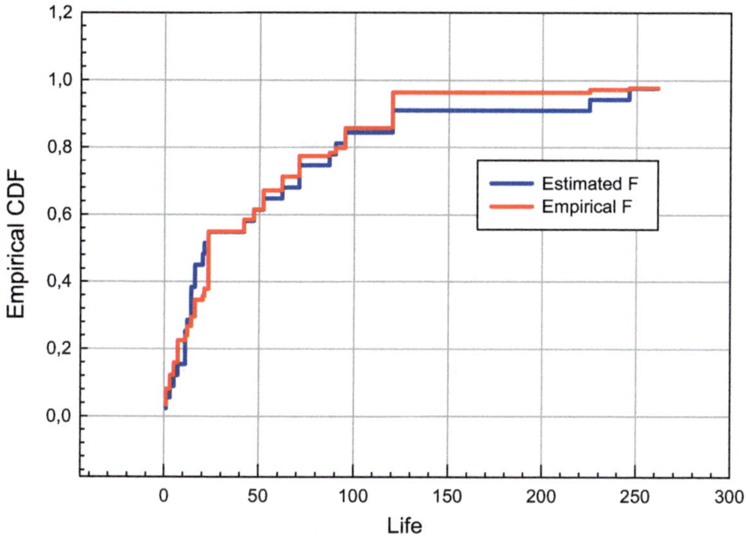

Fig. 5.13 Empirical CDF and estimated CDF

$$F(t) = 1 - e^{-(t/\alpha)^{\beta}} \tag{5.52}$$

It can be rewritten as

$$e^{-(t/\alpha)^{\beta}} = 1 - F(t) \tag{5.53}$$

Take logarithm for both sides of Eq. (5.53) and rearrange it. We obtain

$$\left(\frac{t}{\alpha}\right)^{\beta} = -\ln[1 - F(t)]$$

Take logarithm on the above equation and rearrange it again. An equation is obtained as

$$\ln\{-\ln[1 - F(t)]\} = \beta \ln t - \beta \ln \alpha \tag{5.54}$$

which is a linearized form of

$$y = Cx + D \tag{5.55}$$

where

Table 5.5 Ordered failure times

Index(i)	1	2	3	4	5	6	7	8	9	10	11	12	13	14	15
t_i	1	3	5	7	11	11	11	12	14	14	14	16	16	20	21
Index(i)	16	17	18	19	20	21	22	23	24	25	26	27	28	29	30
t_i	23	42	47	52	62	71	71	87	90	95	120	120	225	246	261

$$\begin{cases} y = \ln\{-\ln[1-F(t)]\} \\ x = \ln t \\ C = \beta \\ D = -\beta \ln \alpha \end{cases} \tag{5.56}$$

To demonstrate the regression method, this section uses the data from air condition as an example. The original failure times are [10]

23 261 87 7 120 14 62 47 225 71 246 21 42 20 5 12 120 11 3 14 71 11 14 11 16 90 1 16 52 95

Order the data ascendingly as shown in Table 5.5.

Using media rank defined in Eq. (5.51) for each failure time

$$F(t_1=1) \approx \frac{i-0.3}{n+0.4} = \frac{1-0.3}{30+0.4} = 0.02$$
$$F(t_2=3) \approx \frac{i-0.3}{n+0.4} = \frac{2-0.3}{30+0.4} = 0.06$$
$$F(t_3=5) \approx \frac{i-0.3}{n+0.4} = \frac{3-0.3}{30+0.4} = 0.09$$
$$F(t_4=7) \approx \frac{i-0.3}{n+0.4} = \frac{4-0.3}{30+0.4} = 0.12$$
$$F(t_5=11) \approx \frac{i-0.3}{n+0.4} = \frac{5-0.3}{30+0.4} = 0.15$$
$$F(t_6=11) \approx \frac{i-0.3}{n+0.4} = \frac{6-0.3}{30+0.4} = 0.19$$

....

Covert the empirical $F(t_i)$ value to linear line value y_i using Formula (5.56). The results are shown in Table 5.6.

The (t_i, $F(t_i)$) is converted to (x_i, y_i) values by assuming the t_i follows Weibull distribution. If the data really follows Weibull distribution, the plot of (x_i, y_i) will appear linearly; otherwise, it will appear in other nonlinear patterns. For parameter estimation purpose, the (x_i, y_i) can be considered as original data. The parameter estimation is performed by finding a linear line to fit it. The problem is then converted to a linear regression problem. In linear regression, as shown in Fig. 5.14, the error is classically defined as the deviation between the real value y_i and the regression value $\widehat{y_i} = Cx_i + D$, i.e.

Table 5.6 Linearized value

Index	t_i	$F(t_i)$	$x_i = \ln t_i$	$y_i = \ln \{-\ln [1 - F(t_i)]\}$
1	1	0.02	0.00	−3.76
2	3	0.06	1.10	−2.86
3	5	0.09	1.61	−2.38
4	7	0.12	1.95	−2.04
5	11	0.15	2.40	−1.78
6	11	0.19	2.40	−1.57
7	11	0.22	2.40	−1.39
8	12	0.25	2.48	−1.23
9	14	0.29	2.64	−1.09
10	14	0.32	2.64	−0.96
11	14	0.35	2.64	−0.84
12	16	0.38	2.77	−0.72
13	16	0.42	2.77	−0.61
14	20	0.45	3.00	−0.51
15	21	0.48	3.04	−0.41
16	23	0.52	3.14	−0.32
17	42	0.55	3.74	−0.23
18	47	0.58	3.85	−0.14
19	52	0.62	3.95	−0.05
20	62	0.65	4.13	0.04
21	71	0.68	4.26	0.13
22	71	0.71	4.26	0.22
23	87	0.75	4.47	0.32
24	90	0.78	4.50	0.41
25	95	0.81	4.55	0.52
26	120	0.85	4.79	0.62
27	120	0.88	4.79	0.74
28	225	0.91	5.42	0.88
29	246	0.94	5.51	1.06
30	261	0.98	5.56	1.33

$$\varepsilon_i = |\widehat{y_i} - y_i| \tag{5.57}$$

Considering mathematical convenience, instead of using (5.57), the error can be defined as

$$\varepsilon_i = (\widehat{y_i} - y_i)^2 = (Cx_i + D - y_i)^2 \tag{5.58}$$

For all the data sets, the total error is then in a squared form,

$$\text{Total Squared Error} = \sum\nolimits_{i=1}^{n} (Cx_i + D - y_i)^2 \tag{5.59}$$

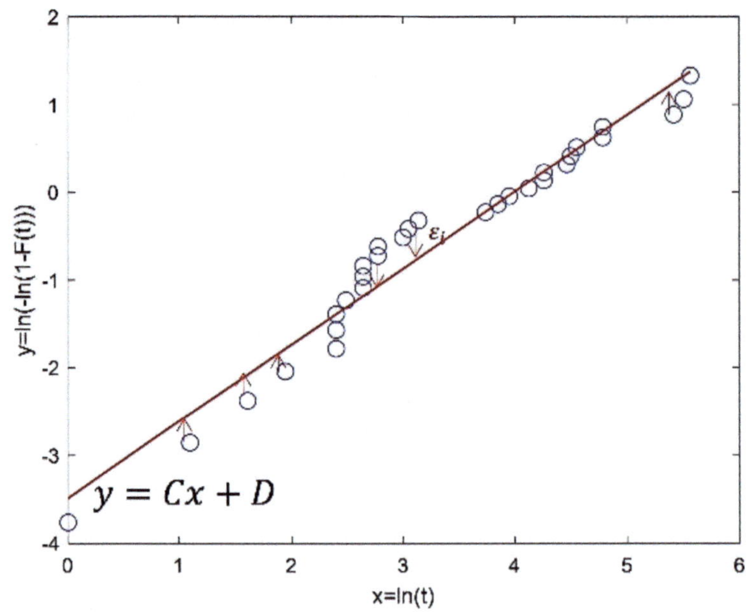

Fig. 5.14 Linear regression principle

The best line is supposed to have minimal total squared error. The task is then to find the C, D values with minimal total error, which is an unconstraint optimization problem. For such unconstraint problem, we can find the optimal solution from its first derivative. Expand Eq. (5.59) as

$$T=\sum_{i=1}^{n}(Cx_i+D-y_i)^2=\mathrm{n}D^2+2CD\sum x_i+C^2\sum x_i^2 \\ -2D\sum y_i-2C\sum x_iy_i+\sum y_i^2 \tag{5.60}$$

The first derivative of T is

$$\begin{cases}\dfrac{\partial T}{\partial D}=2nD+2C\sum x_i-2\sum y_i=0\\ \dfrac{\partial T}{\partial C}=2D\sum x_i+2C\sum x_i^2-2\sum x_iy_i=0\end{cases} \tag{5.61}$$

It is not hard to obtain the solution of Eq. (5.61) as

$$\begin{cases}C=\dfrac{n\sum x_iy_i-\sum x_i\sum y_i}{n\sum x_i^2-\left(\sum x_i\right)^2};\quad \overline{x}=\dfrac{\sum x_i}{n},\overline{y}=\dfrac{\sum y_i}{n}\\ D=\overline{y}-C\overline{x};\end{cases} \tag{5.62}$$

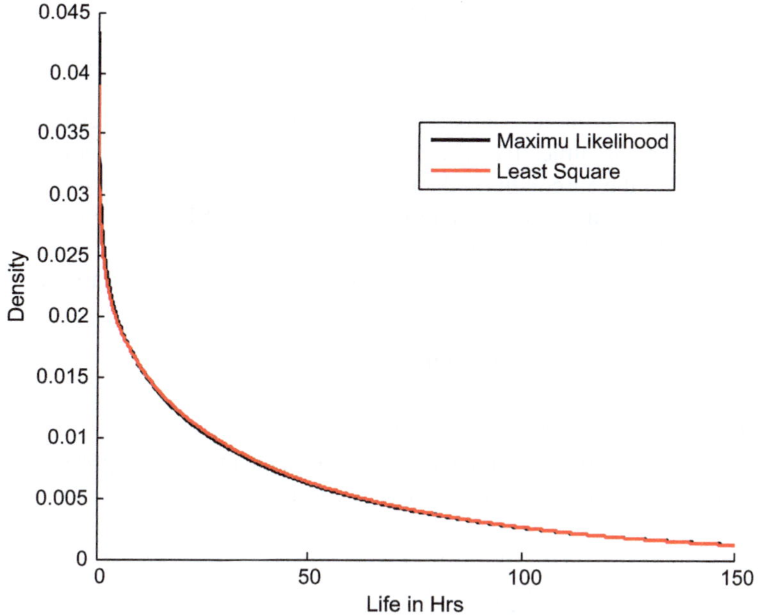

Fig. 5.15 Probability density function by using LSE and MLE

Recall the Formula (5.56) and rearrange the $D = -\beta \ln \alpha$ and $C = \beta$, we can obtain the least squares estimator of Weibull as

$$\begin{aligned} \widehat{\alpha} &= \exp(-D/C) \\ \widehat{\beta} &= C \end{aligned} \tag{5.63}$$

For the air condition data, the $C = 0.8713$, the $D = -3.4849$. Using Eq. (5.63), we obtain

$$\widehat{\alpha} = 54.1413; \widehat{\beta} = 0.8713$$

The maximal likelihood estimator of the same data is

$$\widehat{\alpha} = 54.6; \widehat{\beta} = 0.85$$

As shown in Fig. 5.15, the plot probability density function does not show significantly difference between the LSE and MSE for this case.

Regression R^2 Error

The fitting performance can be evaluated from R^2 Error which defines a perfect fitting with R^2 Error is unity [11].

$$R^2 = 1 - \frac{\sum_{i=1}^{n}(Cx_i + D - y_i)^2}{\sum_{i=1}^{n}(y_i - \bar{y})^2}; \bar{y} = \frac{\sum y_i}{n} \tag{5.63}$$

A good fitting is with the R^2 can be over 90%. Fitting performance can be checked from the plot, but it is subjective. The R^2 value provides a numerical indicator for the fitting. The R^2 for the air condition data is 0.9587, implying a good fitting. The good fitting also suggests the parameter obtained from the LSE is acceptable and the assumption of data following Weibull is verified.

The procedure of performing least squares estimation is summarised as **Algorithm 5.3:**

Step 1: Order the failure time ascendingly.
Step 2: Approximate CDF $F(t_i)$ by $F(t_i) = \frac{i-0.3}{n+0.4}$
Step 3: Computing $x_i = \ln(t_i)$ $y_i = \ln[-\ln(1 - F(t_i))]$
Step 4: Estimate regression parameters in $y = Cx + D$ using linear regression method.
The parameter values in linear function are obtained by

$$\begin{cases} C = \frac{n\sum x_i y_i - \sum x_i \sum y_i}{n\sum x_i^2 - (\sum x_i)^2}; \bar{x} = \frac{\sum x_i}{n}, \bar{y} = \frac{\sum y_i}{n} \\ D = \bar{y} - C\bar{x} \end{cases}$$

Step 5: Computing the Weibull parameter values using $\widehat{\alpha} = \exp(-D/C); \widehat{\beta} = C.$
Step 6: Computing the $R^2 = 1 - \frac{\sum_{i=1}^{n}(Cx_i + D - y_i)^2}{\sum_{i=1}^{n}(y_i - \bar{y})^2}.$

Adjustment for Right Censored Data

The previous example for the least squares estimating is limited for the uncensored data. It can be applied to censored data with small adjustment. Compared to a complete data set, the only issue to address for right-censored data lies in estimating $F(t_i) = \frac{i-0.3}{n+0.4}$. If the data are right censored at a fixed, rather than random, time, then the n for the median rank $F(t_i) = \frac{i-0.3}{n+0.4}$ is still the data size of all (not the total number of failures). The i is still the ith after ordering. The censored ones are just ignored. The other parts for Algorithm 5.3 are still the same.

Example 5.8

Using the same data set of Example 5.4. The $n = 20$, $T = 150$ hours, $r = 10$. Estimate the parameter values for Weibull distribution using least squares method.

Solution: Order the data and calculate the $F(t_i) = \frac{i-0.3}{20+0.4}$. Here the n uses 20 instead of total number of failures 10. Using the Algorithm 5.3 and implement the algorithm by using Matlab, the least squares estimator is obtained as:

(continued)

Example 5.8 (continued)

$$\widehat{\alpha} = 166.7978; \widehat{\beta} = 1.4380.$$

For comparison purpose, from **Example 5.4**, we know the MSE is

$$\widehat{\alpha} = 192.3070; \widehat{\beta} = 1.3314.$$

It is different for different parameter estimation method. The R^2 value is 0.9376, not an excellent fitting, but still acceptable. The plot of the fitting is shown in Fig. 5.16, which also shows the fitting is acceptable.

5.4 Goodness of Fit Test for Weibull

Both for uncensored and censored data analysis, the parameter estimation on Weibull distribution mainly uses maximal likelihood method. In this book, the purpose of introducing of least squares method is not for parameter estimation, but for the goodness of fit test. Section 5.3 discusses the linearization of Weibull CDF function. By some modifications, the method can be used for goodness of fit test. For goodness of fit test, the plot of the empirical data is the same as for the parameter estimation. But since for goodness of fit test, the parameter is already obtained, let us

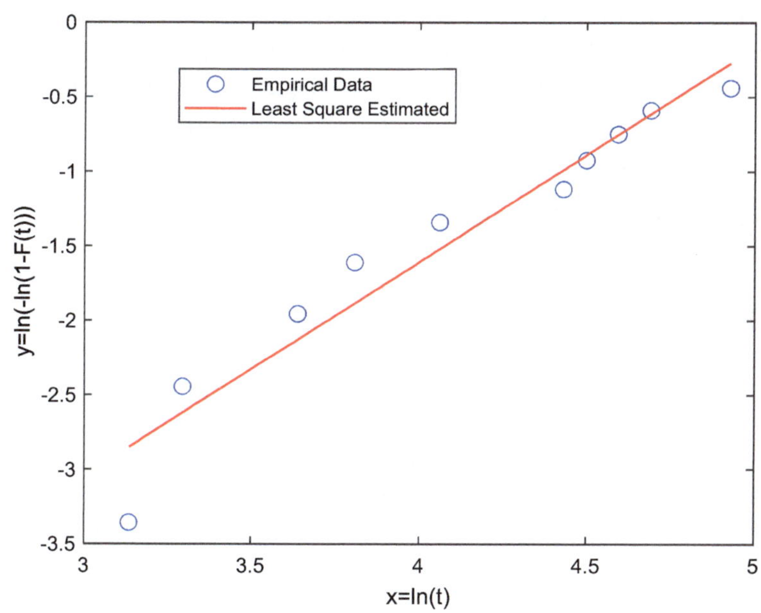

Fig. 5.16 Plot of fitting

say $\widehat{\alpha}$ for scale $\widehat{\beta}$ for shape, a straight-line, obtained from the estimated parameter values, is plot by using the following function

$$y = \ln\{-\ln[1 - F(t)]\} = \widehat{\beta}\ln t - \widehat{\beta}\ln\widehat{\alpha} = Cx + D \tag{5.64}$$

which is linearized from Weibull CDF $F(t) = 1 - e^{-(t/\widehat{\alpha})^{\widehat{\beta}}}$. The slope of the straight line is $C = \widehat{\beta}$, the threshold is $D = -\widehat{\beta}\ln\widehat{\alpha}$. If the estimated parameter is correct and the Weibull assumption holds, the linear line of Eq. (5.64) will fit the empirical data well.

The procedure of performing goodness of fit test using regression is presented in **Algorithm 5.4:**

Step 1: Order the failure time ascendingly.
Step 2: Approximate CDF $F(t_i)$ by $F(t_i) = \frac{i-0.3}{n+0.4}$
Step 3: Computing $x_i = \ln(t_i)$ $y_i = \ln[-\ln(1 - F(t_i))]$
Step 4: Using the given $\widehat{\alpha}$ for scale and $\widehat{\beta}$ for shape, let the slope of the line is $C = \widehat{\beta}$, threshold is $D = -\widehat{\beta}\ln\widehat{\alpha}$. plot the strait line $y = Cx + D$.
Step 5: Computing the $R^2 = 1 - \frac{\sum_{i=1}^{n}(Cx_i + D - y_i)^2}{\sum_{i=1}^{n}(y_i - \overline{y})^2}$.

Example 5.9
Using the same data set of Example 5.4. The $n = 20$, $T = 150$ hours, $r = 10$. Check if the maximal estimator $\widehat{\alpha} = 192.3070$; $\widehat{\beta} = 1.3314$ is acceptable?

Solution: Order the data and calculate the $F(t_i) = \frac{i-0.3}{20+0.4}$. Using the **Algorithm 5.4** and implement the algorithm by using Matlab. The slope of the line is $C = \widehat{\beta} = 1.3314$. Threshold $D = -\widehat{\beta}\ln\widehat{\alpha} = -7.00$. The plot is shown in Fig. 5.17, where we can say the line can fit the data in general.

The R^2 value is calculated as 0.9236, which is acceptable. The MLE is thus acceptable, and the data follows Weibull distribution.

5.5 Estimator Variance and CI for Weibull

Point estimation on the parameters is important for the Weibull distribution. The confidence interval and variance of the estimator contains the uncertainty of the estimator, which is also important for decision making. The variance for the Weibull can be obtained from maximal likelihood function. This section discusses the variance derivation and CI for uncensored and censored situation separately.

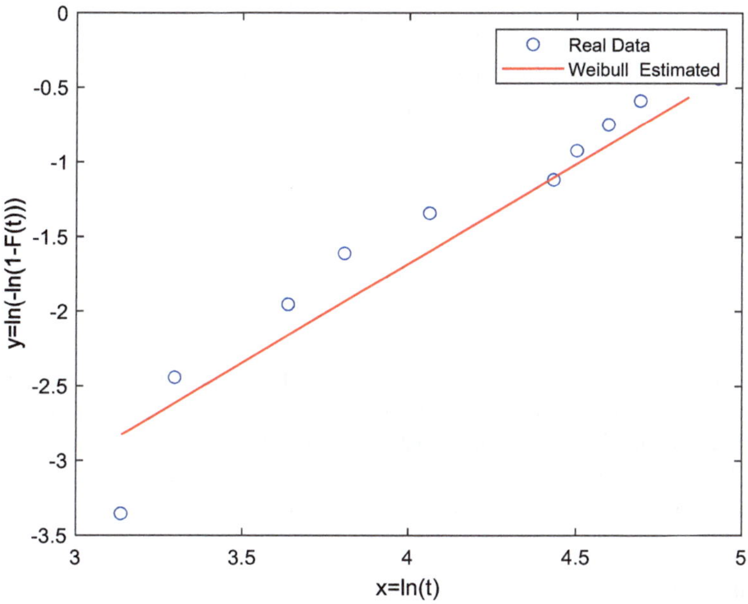

Fig. 5.17 Goodness of fit test for air condition data

5.5.1 *Variance for Uncensored Data*

A lower bound of variance can be obtained from the information matrix. From the likelihood function of Eq. (5.6) for Weibull distribution, the second derivative of the likelihood function is

$$\frac{\partial^2(-\ln L)}{\partial \alpha^2} = -\frac{n\beta}{\alpha^2} + \beta(\beta+1)\alpha^{-\beta-2}\sum t_i^\beta \tag{5.65}$$

$$\frac{\partial^2(-\ln L)}{\partial \alpha \partial \beta} = \frac{n}{\alpha} - \alpha^{-\beta-1}\sum t_i^\beta + \beta\alpha^{-\beta-1}\ln\alpha\sum t_i^\beta - \beta\alpha^{-\beta-1}\sum\left\{t_i^\beta \ln t_i\right\} \tag{5.66}$$

$$\frac{\partial^2(-\ln L)}{\partial \beta^2} = \frac{n}{\beta^2} + \sum\left\{\ln^2(t_i/\alpha)(t_i/\alpha)^\beta\right\} \tag{5.67}$$

The $\frac{\partial^2(-\ln L)}{\partial \alpha \partial \beta} = \frac{\partial^2(-\ln L)}{\partial \beta \partial \alpha}$ since the function ln$L$ is continuous. The Fisher information matrix [12] is constructed as

$$I(\alpha,\beta)=\begin{pmatrix}\frac{\partial^2(-\ln L)}{\partial\alpha^2} & \frac{\partial^2(-\ln L)}{\partial\alpha\partial\beta}\\ \frac{\partial^2(-\ln L)}{\partial\beta\partial\alpha} & \frac{\partial^2(-\ln L)}{\partial\beta^2}\end{pmatrix} \tag{5.68}$$

The variance of estimator $\widehat{\alpha},\widehat{\beta}$ can be obtained from the inverse of matrix $I(\alpha,\beta)$. Let us write

$$I^{-1}=\begin{vmatrix}V_{11} & V_{12}\\ V_{21} & V_{22}\end{vmatrix} \tag{5.69}$$

The variance of $\widehat{\alpha}$ is V_{11}, variance of $\widehat{\beta}$ is V_{22}. One important point needed to be considered is that the Fisher information matrix includes the true parameters in its expression. To obtain a numerical Fisher matrix, these parameter values must be known. In practice, we can approximate the true parameters using MLE. Following example is provided to demonstrate this approach.

Example 5.10

Using the LHD B data in Table 5A.1 from Appendix 5A. Using Fisher information matrix to evaluate the variance of the shape and scale parameters.

Solution: From **Example 5.2,** we have obtained MLE of this case is

$$\widehat{\alpha}=49.11$$

$$\widehat{\beta}=1.0293$$

According to Eq. (5.68), we obtain the Fisher Matrix as

$$I(\alpha,\beta)=\begin{pmatrix}\frac{\partial^2(-\ln L)}{\partial\alpha^2} & \frac{\partial^2(-\ln L)}{\partial\alpha\partial\beta}\\ \frac{\partial^2(-\ln L)}{\partial\beta\partial\alpha} & \frac{\partial^2(-\ln L)}{\partial\beta^2}\end{pmatrix}=\begin{pmatrix}0.0193 & -0.3849\\ -0.3849 & 75.0570\end{pmatrix}$$

The computation was carried by Matlab. The inverse matrix of $I(\alpha,\beta)$

$$I(\alpha,\beta)=\begin{pmatrix}0.0193 & -0.3849\\ -0.3849 & 75.0570\end{pmatrix}^{-1}=\begin{pmatrix}57.6108 & 0.2954\\ 0.2954 & 0.0148\end{pmatrix}$$

The procedure to obtain the inverse of a matrix is computationally intensive. This example use inv.() function in Matlab to perform it. The variance of α is then 57.6108. The variance of β is 0.0148.

5.5.2 Variance for Right Censored Data

The first derivative of right censored data of the likelihood function is

$$\frac{\partial(-\ln L)}{\partial\alpha}=\frac{r\beta}{\alpha}-\beta\,\alpha^{-\beta-1}\sum t_i^{\beta}-(n-r)\frac{\beta}{T}\left(\frac{\alpha}{T}\right)^{-\beta-1} \tag{5.34}$$

$$\frac{\partial(-\ln L)}{\partial\beta}=-\frac{r}{\beta}+r\ln\alpha-\sum\ln t_i+\sum\left\{\left(\frac{t_i}{\alpha}\right)^{\beta}\ln\left(\frac{t_i}{\alpha}\right)\right\}+(n-r)(T/\alpha)^{\beta}\ln T/\alpha \tag{5.35}$$

Just to simplify the second order derivative's derivation. Rewrite $\sum t_{i*}^{\beta}=\sum_{i=1}^{r}t_i^{\beta}+\sum_{i=r+1}^{n}T^{\beta}$, and $\sum\left\{\left(\frac{t_{i*}}{\alpha}\right)^{\beta}\ln\left(\frac{t_{i*}}{\alpha}\right)\right\}=\sum\left\{\left(\frac{t_i}{\alpha}\right)^{\beta}\ln\left(\frac{t_i}{\alpha}\right)\right\}+(n-r)(T/\alpha)^{\beta}\ln T/\alpha$. The t_{i*} is defined for $i\le r$, $t_{i*}=t_i$, which is same as the complete data; while for $i>r$, $t_{i*}=T$ for censored data. Equations (5.34) and (5.35) become as

$$\frac{\partial(-\ln L)}{\partial\alpha}=\frac{r\beta}{\alpha}-\beta\,\alpha^{-\beta-1}\sum t_{i*}^{\beta} \tag{5.70}$$

$$\frac{\partial(-\ln L)}{\partial\beta}=-\frac{r}{\beta}+r\ln\alpha-\sum\ln t_i+\sum\left\{\left(\frac{t_i*}{\alpha}\right)^{\beta}\ln\left(\frac{t_{i*}}{\alpha}\right)\right\} \tag{5.71}$$

which is of the same form as the first derivative for complete data. We therefore obtain the second derivative for censoring situation by just replacing the t_i by t_{i*} in Eq. (5.67), which is written as

$$\frac{\partial^2(-\ln L)}{\partial\alpha^2}=-\frac{r\beta}{\alpha^2}+\beta(\beta+1)\alpha^{-\beta-2}\sum t_{i*}^{\beta}$$

$$\frac{\partial^2(-\ln L)}{\partial\alpha\partial\beta}=\frac{r}{\alpha}-\alpha^{-\beta-1}\sum t_{i*}{}^{\beta}+\beta\alpha^{-\beta-1}\ln\alpha\sum t_{i*}{}^{\beta}-\beta\alpha^{-\beta-1}\sum\{t_{i*}{}^{\beta}\ln t_{i*}\}$$

$$\frac{\partial^2(-\ln L)}{\partial\beta^2}=\frac{r}{\beta^2}+\sum\left\{\ln^2(t_{i*}/\alpha)(t_{i*}/\alpha)^{\beta}\right\} \tag{5.72}$$

The remaining procedure for construction of the Fisher matrix is the same as for the uncensored data.

Example 5.11
Using the same data set of **Example 5.4**. The $n = 20$, $T = 150$ hours, $r = 10$. Estimate the variance for MLE of Weibull distribution.

Solution: From Example 5.4, the uncorrected MLE is

$$\widehat{\beta} = 1.3314.$$

$$\widehat{\alpha} = 192.3070$$

Substitute the (5.72) into the Fisher Matrix, the α, β is approximated by the $\widehat{\alpha} = 192.3070$, $\widehat{\beta} = 1.3314$.

$$I(\alpha, \beta) = \begin{pmatrix} \dfrac{\partial^2(-\ln L)}{\partial \alpha^2} & \dfrac{\partial^2(-\ln L)}{\partial \alpha \partial \beta} \\ \dfrac{\partial^2(-\ln L)}{\partial \beta \partial \alpha} & \dfrac{\partial^2(-\ln L)}{\partial \beta^2} \end{pmatrix} = \begin{pmatrix} 0.00047935 & 0.0277 \\ 0.0277 & 8.3752 \end{pmatrix}$$

$$I(\alpha, \beta) = \begin{pmatrix} 0.00047935 & 0.0277 \\ 0.0277 & 8.3752 \end{pmatrix}^{-1} = \begin{pmatrix} 2578.8 & -8.5269 \\ -8.5269 & 0.1476 \end{pmatrix}$$

The variance of α is 2578.8, variance of β is 0.1476.

5.5.3 *Confidence Interval*

Different with the Exponential distribution case, the exact distribution for estimators of Weibull is unavailable. The confidence interval derived from exact estimator distribution is infeasible. The maximum likelihood estimation method is well known asymptotically following Normal distribution. Let the estimator be $\widehat{\theta}$, the true parameter be θ. Then the maximum likelihood estimators converge to Normal distribution as

$$\left(\widehat{\theta} - \theta\right) \xrightarrow{n \to \infty} N\left(0, I^{-1}\right) \tag{5.73}$$

where I^{-1} is the inverse matrix of I as discussed before. A confidence interval can be obtained from Eq. (5.73).

As $\left(\widehat{\theta}_i - \theta\right) \to N(0, V_{ii})$, then

$$\frac{\widehat{\theta}_i - \theta}{\sqrt{V_{ii}}} \sim N(0, 1) \tag{5.74}$$

The V_{ii} is from inverse of the Fisher matrix I discussed in Sect. 5.5.2. At significance level α, the lower bound and upper bound are obtained at the sides of $\frac{\alpha}{2}$ and $1 - \frac{\alpha}{2}$, as shown in Fig. 5.18, i.e.

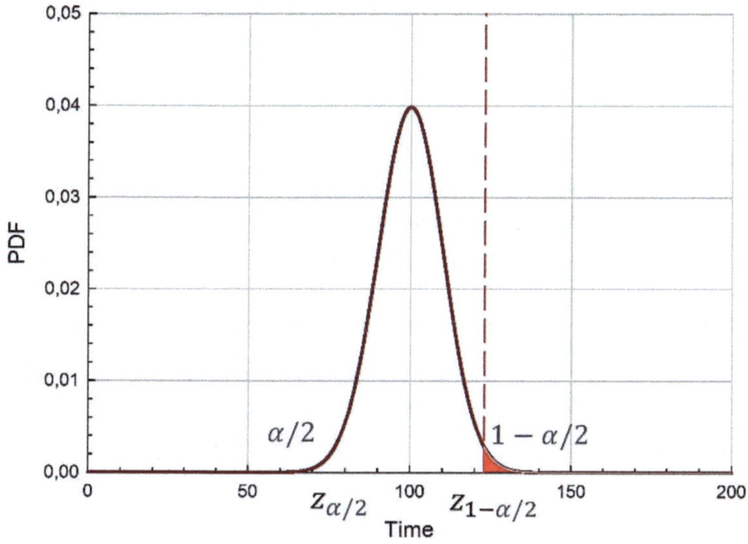

Fig. 5.18 Lower and upper bound of parameter

$$\frac{\widehat{\theta}_i - \theta}{\sqrt{V_{ii}}} = z_{\alpha/2}; \frac{\widehat{\theta}_i - \theta}{\sqrt{V_{ii}}} = z_{1-\alpha/2} \tag{5.75}$$

The confidence interval for parameter θ_i is then

$$\theta_i = \left[\widehat{\theta}_i - z_{1-\frac{\alpha}{2}}\sqrt{V_{ii}}, \widehat{\theta}_i + z_{1-\alpha/2}\sqrt{V_{ii}}\right] \tag{5.76}$$

However, for $\theta_i > 0$, Eq. (5.76) could cover negative value which does not make sense. One modification is to transform the θ_i into $\ln\theta_i$, and let the estimator of $\ln\theta_i$ follows Normal distribution with mean as the true one $\ln\theta_i$. By the Delta method shown in the Appendix 5B, the mean and variance of $\ln\theta_i$ are

$$E(\ln\theta_i) = \ln\widehat{\theta}_i; \tag{5.77}$$

$$Var(\ln\theta_i) = \frac{1}{\widehat{\theta}_i^2} Var(\theta_i) = \frac{1}{\widehat{\theta}_i^2} V_{ii} \tag{5.78}$$

Equation (5.76) is then rewritten as

$$\ln\theta_i = \left[\ln\widehat{\theta}_i - z_{1-\frac{\alpha}{2}}\sqrt{V_{ii}}/\widehat{\theta}_i, \ln\widehat{\theta}_i + z_{1-\frac{\alpha}{2}}\sqrt{V_{ii}}/\widehat{\theta}_i\right] \tag{5.79}$$

The corresponding confidence interval can be obtained from

$$[\theta_L, \theta_U] = \left[\widehat{\theta}_i/w, \widehat{\theta}_i \times w\right] \tag{5.80}$$

where $w = \exp\left[z_{\left(1-\frac{\alpha}{2}\right)}\frac{\sqrt{V_{ii}}}{\widehat{\theta_i}}\right]$. The similar derivation of Eq. (5.79) is also shown in [13].

Example 5.12
Using the LHD B data in Table 5A.1 from Appendix 5A. Evaluate the confidence interval of the shape and scale parameters of 95%.

Solution: According to Eq. (5.79), $[\theta_L, \theta_U] = \left[\widehat{\theta}_i/w, \widehat{\theta}_i \times w\right]$, $w = \exp\left[z_{\left(1-\frac{\alpha}{2}\right)}\frac{\sqrt{V_{ii}}}{\widehat{\theta_i}}\right]$. We have known from Example 5.10 the variance scale parameter is 57.6108. The variance shape is 0.0148. The 95% CI corresponds $\alpha = 0.05$ in $z_{\left(1-\frac{\alpha}{2}\right)}$. The scale and shape parameter from MLE are known as

$$\widehat{\alpha} = 49.11$$

$$\widehat{\beta} = 1.0293$$

The $z_{\left(1-\frac{\alpha}{2}\right)} = 1.96$ from the standard Normal distribution.

For scale parameter, the $\theta_L = \dfrac{\widehat{\theta_i}}{\exp\left[z_{\left(1-\frac{\alpha}{2}\right)}\frac{\sqrt{V_{ii}}}{\widehat{\theta_i}}\right]} = \dfrac{49.11}{\exp\left[1.96\frac{\sqrt{57.6108}}{49.11}\right]} = 36.2735$;

The $\theta_U = \widehat{\theta}_i \exp\left[z_{\left(1-\frac{\alpha}{2}\right)}\frac{\sqrt{V_{ii}}}{\widehat{\theta_i}}\right] = 49.11 * \exp\left[1.96\frac{\sqrt{57.6108}}{49.11}\right] = 66.48.$

For shape parameter, the $\theta_L = \dfrac{\widehat{\theta_i}}{\exp\left[z_{\left(1-\frac{\alpha}{2}\right)}\frac{\sqrt{V_{ii}}}{\widehat{\theta_i}}\right]} = \dfrac{1.0293}{\exp\left[1.96\frac{\sqrt{0.0148}}{1.0293}\right]} \approx 0.82$; The

$\theta_U = \widehat{\theta}_i \exp\left[z_{\left(1-\frac{\alpha}{2}\right)}\frac{\sqrt{V_{ii}}}{\widehat{\theta_i}}\right] = 1.0293 * \exp\left[1.96\frac{\sqrt{0.0148}}{1.0293}\right] \approx 1.30$

From the Matlab (2021a) wblfit() function, we obtain the same CI values.

5.6 Summary

Weibull analysis is an important analysis in reliability engineering. The general step for performing the life data analysis is summarized as follows:

Step 1: Assume the data following Weibull distribution.
Step 2: Parameter Estimation using maximal likelihood, moment or other methods.
Step 3: Perform goodness of fit test to verify the obtained parameters acceptable and verify the assumption of data following Weibull distribution.
Step 4: If necessary, obtain the Fisher matrix to compute the confidence interval.

In practice, goodness-of-fit test is a critical step, yet it is often overlooked in some state-of-the-art applications.

Appendix 5A: LHD Data (Table 5.A1)

Table 5.A1 Time between failures in hours

No.	TBFs (LHD A)	TBFs (LHD B)	TBFs (LHD C)
1	16	149	110
2	23	33	13
3	32	08	72
4	24	05	04
5	03	31	45
6	12	12	56
7	04	188	19
8	112	02	27
9	68	36	36
10	50	63	90
11	211	53	19
12	44	56	07
13	158	06	02
14	65	04	118
15	141	147	44
16	114	36	08
17	90	04	277
18	35	76	04
19	55	36	08
20	60	36	10
21	28	61	79
22	27	48	103
23	30	58	06
24	134	68	18
25	89	03	147
26	18	12	96
27	32	33	22
28	51	07	03
29	10	71	24
30	36	149	03
31	24	06	09
32	03	54	99
33	15	09	82
34	54	115	121
35	26	58	54
36	76	96	79
37	72	18	99
38	02	16	18
39	54	25	05

(continued)

Table 5.A1 (continued)

No.	TBFs (LHD A)	TBFs (LHD B)	TBFs (LHD C)
40	29	139	21
41	164	18	01
42	04	34	03
43	04	15	05
44	18	41	01
45			59
46			22
47			17
48			35
49			35
50			29

Data source: [3]

Appendix 5B

If the curve of left and right for Eq. (5.13) shown in Fig. 5.7 crosses, then the solution to Eq. (5.13) must exist. This appendix proves the two curves cross. Since both left and right of Eq. (5.13) [5]

$$\frac{n}{\widehat{\beta}} = n\frac{\sum t_i^{\widehat{\beta}} \ln t_i}{\sum t_i^{\widehat{\beta}}} - \sum \ln t_i$$

are monotone and continuous, to ensure the crossing, when the $\widehat{\beta}$ approaches to 0, if curve A is bigger or equal to Curve B, then for $\widehat{\beta}$ approaches to infinite, it will be opposite, the curve B should be bigger than A. Two conditions for the existence of the crossing are therefore

$$\lim_{\widehat{\beta}\to 0} \frac{n}{\widehat{\beta}} \geq \lim_{\widehat{\beta}\to 0} G\left(\widehat{\beta}\right)$$

$$\lim_{\widehat{\beta}\to +\infty} \frac{n}{\widehat{\beta}} \leq \lim_{\widehat{\beta}\to +\infty} G\left(\widehat{\beta}\right)$$

Or

$$\lim_{\widehat{\beta}\to 0} \frac{n}{\widehat{\beta}} \leq \lim_{\widehat{\beta}\to 0} G\left(\widehat{\beta}\right)$$

$$\lim_{\widehat{\beta}\to+\infty}\frac{n}{\widehat{\beta}}\ge\lim_{\widehat{\beta}\to+\infty}G\left(\widehat{\beta}\right)$$

Obviously, the $\lim_{\beta\to 0}\frac{n}{\beta}=+\infty$. For $\lim_{\widehat{\beta}\to 0}G\left(\widehat{\beta}\right)=n\frac{\sum t_i^{\widehat{\beta}}\ln t_i}{\sum t_i^{\widehat{\beta}}}-\sum\ln t_i=n\frac{\sum\ln t_i}{n}-\sum\ln t_i=0$. Therefore, the $\lim_{\widehat{\beta}\to 0}\frac{n}{\beta}>\lim_{\widehat{\beta}\to 0}G\left(\widehat{\beta}\right)$. The first condition meet.

For $\widehat{\beta}\to+\infty$,

$$\lim_{\widehat{\beta}\to+\infty}\frac{n}{\widehat{\beta}}=0.$$

To find out $\lim_{\widehat{\beta}\to+\infty}G\left(\widehat{\beta}\right)$, we let $t_{max}=$ max $(t_1,t_2,,\ldots t_i,\ldots.\ t_n)$. Then the

$$n\frac{\sum t_i^{\widehat{\beta}}\ln t_i}{\sum t_i^{\widehat{\beta}}}=n\frac{\sum\left(\frac{t_i}{t_{max}}\right)^{\widehat{\beta}}\ln t_i}{\left(\frac{t_i}{t_{max}}\right)^{\widehat{\beta}}}$$

When $\widehat{\beta}\to+\infty$, we have

$$\lim_{\widehat{\beta}\to+\infty}\left(\frac{t_i}{t_{max}}\right)^{\widehat{\beta}}=1;t_i=t_{max}$$

$$\lim_{\widehat{\beta}\to+\infty}\left(\frac{t_i}{t_{max}}\right)^{\widehat{\beta}}=0;t_i\ne t_{max}$$

Therefore, when $\widehat{\beta}\to+\infty$,

$$n\frac{\sum t_i^{\widehat{\beta}}\ln t_i}{\sum t_i^{\widehat{\beta}}}=n\frac{\sum\left(\frac{t_i}{t_{max}}\right)^{\widehat{\beta}}\ln t_i}{\left(\frac{t_i}{t_{max}}\right)^{\widehat{\beta}}}=n\ln t_{max}$$

Then the right side of Eq. (5.13)

$$n\frac{\sum t_i^{\widehat{\beta}} \ln t_i}{\sum t_i^{\widehat{\beta}}} - \sum \ln t_i = n \ln t_{max} - \sum \ln t_i$$

i.e.

$$\operatorname*{Lim}_{\widehat{\beta}\to+\infty} n\frac{\sum t_i^{\widehat{\beta}} \ln t_i}{\sum t_i^{\widehat{\beta}}} - \sum \ln t_i = n \ln t_{max} - \sum \ln t_i = \sum \ln(t_{max}/t_i)$$

Since $t_{max}/t_i \geq 1$, obviously the $\sum \ln (t_{max}/t_i) > 0$, which is equivalent to

$$\lim_{\widehat{\beta}\to+\infty} G\left(\widehat{\beta}\right) > 0$$

Therefore

$$\lim_{\widehat{\beta}\to+\infty} \frac{n}{\widehat{\beta}} < \lim_{\widehat{\beta}\to+\infty} G\left(\widehat{\beta}\right)$$

We have proved the condition ii also holds. Therefore, the two curves must cross.

Appendix 5C: Delta Method for Mean and Variance of Multivariate Function

The random number y is transformed from random number x_1, x_2, …, x_i… by a continuous function

$$y = g(x_1, x_2, \ldots, x_i \ldots, x_n) \tag{5.A.1}$$

Given random number x_i are independent from each other. The mean value of random number x_i is u_i, variance is σ_i^2. Using Taylor series of function y at a point $\boldsymbol{\mu} = (u_1, u_2, \ldots, u_i \ldots)$ by using only its first order.

$$y = g(u_1, u_2, \ldots, u_i, \ldots) + \sum_{i=1}^{n} \left.\frac{\partial g(u_1, u_2, \ldots, u_i, \ldots)}{\partial x_i}\right|_{u_1, u_2, \ldots, u_i, \ldots} (x_i - u_i) \tag{5.A.2}$$

The expansion can be rewritten as

$$y=g(u_1,u_2,\ldots,u_i,\ldots)-\sum_{i=1}^{n}\frac{\partial g(x_1,x_2,\ldots,x_i\ldots)}{\partial x_i}\bigg|_{\ldots}u_i+\sum_{i=1}^{n}\frac{\partial g(x_1,x_2,\ldots,x_i\ldots)}{\partial x_i}\bigg|_{\ldots}x_i \tag{5.A.3}$$

where $g(u_1,u_2,\ldots,u_i,\ldots)-\sum_{i=1}^{n}\frac{\partial g(x_1,x_2,\ldots,x_i\ldots)}{\partial x_i}\Big|_{\ldots}.u_i$ and $\frac{\partial g(x_1,x_2,\ldots,x_i\ldots)}{\partial x_i}$ are constant. The $\Big|_{\ldots}$ is the simplification of $\Big|_{u_1,u_2,\ldots,u_i,\ldots}$.

According to property of sum of two random variables, $Var(ax+by)=a^2Var(x)+b^2Var(y)$, the variance of random variable y by Eq. (5.A.3) is then

$$Var(y)=\sum_{i=1}^{n}\left[\frac{\partial g(.)}{\partial x_i}\right]^2 Var(x_i) \tag{5.A.4}$$

where we also using the result of the variance of a constant is 0.Given $E(x_i)=u_i$,

$$\begin{aligned}E(y)=E\Bigg[g(u_1,u_2,\ldots,u_i,\ldots)-\sum_{i=1}^{n}\frac{\partial g(x_1,x_2,\ldots,x_i\ldots)}{\partial x_i}\bigg|_{\ldots}u_i\Bigg]\\+\sum_{i=1}^{n}\frac{\partial g(x_1,x_2,\ldots,x_i\ldots)}{\partial x_i}\bigg|_{\ldots}E(x_i)=g(u_1,u_2,\ldots,u_i,\ldots)\\-\sum_{i=1}^{n}\frac{\partial g(x_1,x_2,\ldots,x_i\ldots)}{\partial x_i}\bigg|_{\ldots}u_i+\sum_{i=1}^{n}\frac{\partial g(x_1,x_2,\ldots,x_i\ldots)}{\partial x_i}\\\times\bigg|_{\ldots}E(x_i)=g(u_1,u_2,\ldots,u_i,\ldots)\end{aligned} \tag{5.A.5}$$

Problems and Questions

1. The CDF of Weibull distribution

$$F(t)=1-e^{-\left(\frac{t}{\alpha}\right)^{\beta}}$$

Please linearize the $F(t)$ into a linear function $y=ax+b$, and specify what is a and b in the linearized function.

2. If the CDF for times to failure is

$$F(t)=1-\frac{100}{(t+10)^2};t\geq 0$$

a. What is reliability function$R(t)$?
b. What is the failure rate as a function of time? Does the failure rate increase or decrease with time?

c. Suppose the system cannot run with reliability under 95%. What is the maximum service time for the system?

3. Derive the Fisher Information Matrix for parameters in Weibull distribution.
4. The life of a mining equipment has been known following Weibull distribution, the parameter estimated is known as (α is scale parameter, β is shape parameter)

$$\alpha = 1000\ hrs; \beta = 2$$

 a. What is the maximum operating hour, when it requires the reliability to be at least 95%?
 b. The mining company has 100 of such equipment. They are identical. How many equipment is expected to be survived after 500 hours?
 c. If the scale parameter is $\alpha = 1000\ hrs$, the shape parameter is unknown. What should the shape parameter be, in order to guarantee the reliability of 98% for 500 hours?

5. After life data analysis, we found the failure time of a product follows Weibull distribution. Supplier 1's product is with Weibull distribution of:

$$\alpha = 100\ days; \beta = 1.2$$

Supplier 2's product is with parameters

$$\alpha = 120\ days; \beta = 1.1$$

a. The product is planned to serve 200 days. From a reliability perspective, which supplier is better?
b. If we choose supplier 1, with the requirement on reliability at least 0.99, how long is its maximum service time?
c. If we choose supplier 1. How many products should be installed in parallel to ensure the reliability reaches 0.99 for the 200 days' service time?

6. A logic circuit is known to have a decreasing failure rate of the form

$$\lambda(t) = \frac{1}{20} t^{-1/2}, t > 0$$

where t is in years

a. What is the PDF function f(t), CDF function F(t)and reliability function R(t)?
b. If the design life is one year, what is the reliability?

7. For the data of LHD C in Appendix 5A. Suppose it follows Weibull distribution
   a. Estimate the parameters using maximal likelihood method.
   b. Obtain the variance using Fisher matrix method.
   c. Derive a confidence interval for the parameter estimators.

References

1. Weibull W (1951) A statistical distribution function of wide applicability. J Appl Mech-T Asme 18(3):293–297
2. Weibull W (1939) A statistical theory of the strength of materials. Ingeniorsvetenskapsakademiens handlingar 151
3. Kumar U, Klefsjo B, Granholm S (1989) Reliability investigation for a fleet of load haul dump machines in a Swedish mine. Reliab Eng Syst Saf 26(4):341–361
4. Balakrishnan N, Kateri M (2008) On the maximum likelihood estimation of parameters of Weibull distribution based on complete and censored data. Stat Probabil Lett 78(17):2971–2975
5. Yuan FQ, Barabadi A, Lu JM, Garmabaki AHS (2015) Performance evaluation for maximum likelihood and moment parameter estimation methods on classical two Weibull distribution. In: 2015 IEEE international conference on Industrial Engineering and Engineering Management (IEEM), 6–9 Dec. 2015
6. Yuan F (2016) Using a moment method to estimate parameters of Weibull distribution under right censored situation. Int J Perform Eng 12(6):589–595
7. M. Sisong, T. Yincai, W. Lingling, Reliability statistics. Beijing: Higher Education Press, 2008
8. Bain LJ (1972) Inferences based on censored sampling from Weibull or extreme-value distribution. Technometrics 14(3):693. https://doi.org/10.2307/1267297
9. Yuan F (2016) Biasness correction on maximum likelihood estimator using pivotal method for Weibull distribution for right censored situation. Int J Perform Eng 12(5):403–412
10. Barlow RE, Proschan F, Hunter LC (1996) Mathematical theory of reliability, Classics in applied mathematics, no. 17. SIAM, Philadelphia
11. Montgomery DC, Peck EA, Vining GG (2012) Introduction to linear regression analysis, Wiley series in probability and statistics, no. 821, 5th edn. Wiley, Hoboken, NJ, pp xvi–645
12. Kendall MG, Stuart A, Ord JK (1977) The advanced theory of statistics, 4th edn. Charles Griffin, London
13. Meeker WQ, Escobar LA (1998) Statistical methods for reliability data, Wiley series in probability and statistics Applied probability and statistics section. Wiley, New York, pp xxii–680

Chapter 6
System Reliability

System reliability considers the reliability of a system containing dependent or interdependent units, components, or subsystems. It concerns how the reliability of the whole system is related to its constituent parts. For instance, the reliability of an airplane can be assessed based on the reliability of its engines, avionics, navigation system, communication systems, and more. Similarly, the reliability of a laptop can be evaluated from that of its hard drive, screen, memory, power supply, battery, etc. The interactions among these subsystems, components, or units—referred to hereafter simply as "units" for simplicity—may not follow straightforward patterns but instead form a complex stochastic process. This chapter discusses methods for evaluating the reliability of an entire system based on the reliability of its constituent units.

6.1 Reliability Dependence

Within a system, the constituent unit can be independent from each other, i.e. the failure of a unit will not affect all the others. For example, the failure of wireless mouse used in a PC would not affect the function of the computer, the small crack of screen in mobile phone might not affect performance of the antenna. This is the scenario of independence from each other. Two other scenarios are partial dependence and total dependence. Total dependence implies unit A failure will cause unit B to fail. For example, the failure of generator in an autonomous ship will bring down the navigation unit as no power supply from the generator. Partial dependence is between total independence and total dependency. In this chapter, we confine the dependence of units as reliability dependence. If a system contains only two components, let us say unit A and B. The reliability of A and B is represented as $R(\mathrm{A})$, $R(\mathrm{B})$. For total independence, the system reliability is then

F. Yuan, *Essentials of Reliability Engineering*, Springer Series in Reliability Engineering, https://doi.org/10.1007/978-3-032-15544-3_6

$$R(S = A \cap B) = R(\text{A}) * R(B) \tag{6.1}$$

since A and B are independent, the event $S = A \cap B$. While when A and B are totally dependent, the

$$R(S = A \cap B) = R(\text{A}) = R(B) \tag{6.2}$$

Inequality holds as

$$R(\text{A}) * R(B) \leq R(\text{A}) \text{ or } R(\text{A}) * R(B) \leq R(B) \tag{6.3}$$

which implies that the reliability of the entire system tends to be lower in cases of total independence than in cases of total dependence. For partial dependency, no general equations as (6.1) and (6.2) exist as it is context related. In reliability engineering, for simplification, the units are often defined as totally independent from each other, unless explicitly specified. This total independence assumption would result in problems when the system is large. We will demonstrate some problems of this later. The reliability independence assumption in statistics is also named statistically independent. This independence is defined from the statistical perspective instead of physical perspective. Two statistical independence components do not mean the two components have no physical interaction or physical connection.

6.2 Series System

Failure from any one constituent unit in a series system can result in the whole system failure, which implies each constituent is indispensable and all units have to function to ensure the function of the whole system. For a PC, the CPU, memory, power supply all have to function to run the PC. The CPU, memory, power supply are then connected in series. The "series" is from reliability perspective, instead of the physical perspective. For aircraft with two engines, it is sometimes considered as connected in series for reliability, despite they are physically connected in parallel. In term of logic operator, the series is analogous to the "And" gate. We start with a series system with only two units for discussion. The True table for the series is shown in Table 6.1, where the "0" means False, "1" means True. According to definition of Series system, both two constituent units are functional, i.e. both are

Table 6.1 True table of series system

Scenario	Unit 1	Unit 2	System
1	0	0	0
2	0	1	0
3	1	0	0
4	1	1	1

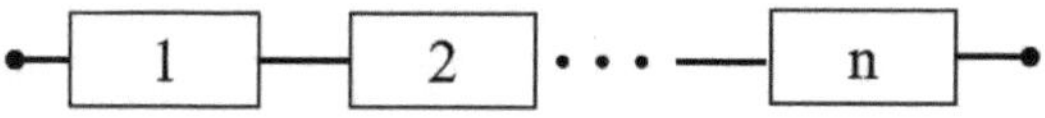

Fig. 6.1 Diagram of series system

true with "1", the system functions, denoted by 1; the other scenarios result to failure, denoted by 0.

Given the reliability of unit 1 and 2 as R_1 and R_2 respectively. The unreliability is then $1 - R_1$ and $1 - R_2$ respectively. The four scenarios in Table 6.1 are exclusive from each other. The unreliability of the system is then the sum of probability of scenario 1,2 and 3 that lead to failed system state.

$$\text{Probability of Failure} = (1 - R_1)(1 - R_2) + (1 - R_1)R_2 + R_1(1 - R_2) \quad (6.4)$$

Simplify Eq. (6.4), we obtain

$$\text{Probability of Failure} = 1 - R_1 R_2 \quad (6.5)$$

The reliability is sum of the probability of scenarios leading to system state "1", which is only the Scenario 4. The reliability is then

$$\text{Reliability} = R_1 R_2 \quad (6.6)$$

Obviously the Probability of Failure + Reliability = 1, which is certain. The sum of Eqs. (6.5) and (6.6) must be 1 as it is the sum of probability of all scenarios, which is sure to be 1. The reliability function Eq. (6.6) is same as the function that is derived from intersection event shown in Eq. (6.1). The reliability for the two constituent unit case can be generalized for any number of units. The reliability for n units of series system is

$$R_s = R_1 R_2 \ldots R_n \quad (6.7)$$

We can employ a diagram block to represent the series as in Fig. 6.1.

Example 6.1

A system consists of three subsystems connected in series. Reliability for each subsystem given a mission time of t is: $R_1(t) = 0.99$, $R_2(t) = 0.95$, $R_3(t) = 0.94$. What is the system reliability for mission time t?

Solution: According to Eq. (6.7), the system reliability is.

$$R_{system}(t) = R_1(t)R_2(t)R_3(t) = 0.99 * 0.95 * 0.94 = 0.8841$$

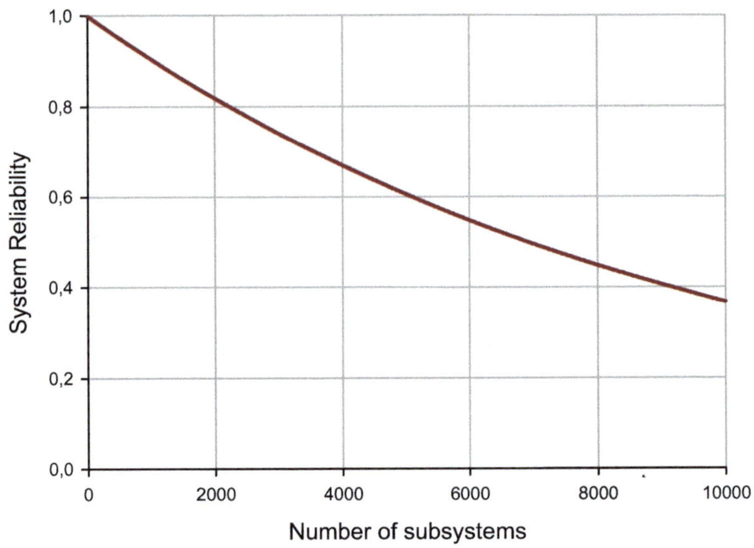

Fig. 6.2 System reliability decreases with number of units

System Reliability Reducing
One important finding from Eq. (6.7) is the whole system reliability is not higher than that of any of its constituent unit, i.e.

$$R_{system}(t) \leq R_i(t) \tag{6.8}$$

Example 6.1 can validate it. The system reliability in Example 6.1 is 0.8841, lower than that of any of its constituent units. This finding implies when the number of constituent units increases, the system reliability decreases. The proof of Eq. (6.8) is simple. By Eq. (6.7), $R_s = R_1 R_2 \ldots R_n$, and each $R_i \leq 1$. Multiplying the whole system reliability by a positive number less than 1, the system reliability will be reduced, i.e. the $R_s(t) \leq R_i(t)$. Figure 6.2 demonstrates that system reliability decreases with the number of units. The decreasing trend is not for any special case, but certain for all cases.

Large Number of Units Problems
The system reliability decreases with the number of units. An extreme scenario is the system reliability will approach 0 when the number of units approaches to infinite. Table 6.2 shows a system with identical reliability for its constituent unit. The system's reliability approaches 0 when the number of units is as large as around 5 million.

It seems the results from Table 6.2 for the large number of units are not consistent with our observations. For some large engineering systems such as aircraft, the number of components can reach 5 million. But the practice in the aviation industry tells us the reliability of an aircraft is not near 0 as described in Table 6.2. The

Table 6.2 System Reliability with number of units

	n is number of units. R_i is individual unit reliability.	
	$R_i = 0.9999$	$R_i = 0.99$
n = 10	System Reliability 0.9990	System Reliability 0.9044
n = 100	0.9900	0.3660
n = 1000	0.9048	4.32×10^{-5}
n = 10,000	0.3679	2.25×10^{-44}
n = 1 million	3.7×10^{-44}	0
n = 5 million	0	0

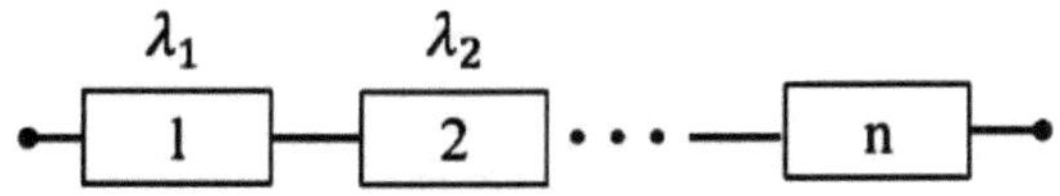

Fig. 6.3 Series system for exponential distribution

disagreement origins from the assumption of the series system: the component is assumed to be independent from others. But essentially most of the constituent units are dependent. Some are even fully dependent. For fully dependent two units, if one unit is with reliability 0.9, the system reliability for the whole series system is then same as the individual unit as 0.9. But if they are totally independent, the whole system's reliability is reduced to 0.9 × 0.9 = 0.81, lower than the dependent scenarios. It illustrates the dependence affects the system reliability.

Exponential Distribution in Series System

For each constituent unit, if the life data follows Exponential distribution with failure rate λ_i, and they are connected in series as shown in Fig. 6.3, the individual reliability for each unit is then

$$R_i(t) = e^{-\int_0^t \lambda_i d\tau} = e^{-\lambda_i t} \tag{6.9}$$

According to Eq. (6.7), the system reliability is then

$$R_{system}(t) = e^{-\lambda_1 t} * e^{-\lambda_2 t} * \ldots * e^{-\lambda_i t} * e^{-\lambda_n t} = e^{-(\lambda_1+\lambda_2+\ldots+\lambda_n)t} = e^{-(\sum \lambda_i)t} \tag{6.10}$$

It finds the system reliability is still of the form of Exponential distribution with constant failure rate. The failure rate for the whole system λ_s is as

$$\lambda_s = \sum \lambda_i \tag{6.11}$$

It suggests the system failure rate is the sum of its constituent unit's failure rate, implying the system failure rate is higher than that of any of its constituent unit. Equation (6.11) can simplify the calculation for system reliability as it can obtain reliability from its failure rate directly.

Example 6.2
Capacitors used in the helicopter are claimed with MTTF 5 million hours (around 571 years). If each helicopter has 10 capacitors installed in series. What is the system reliability of these capacitors? What is the system reliability for 1 year?

Solution: The failure rate for individual capacitor is

$$\lambda_i = \frac{1}{571} \approx 0.001751/\text{year}$$

According to Eq. (6.11), the system failure rate is

$$\lambda_s = \sum \lambda_i = 10 \times \frac{1}{571} \approx 0.01751 \text{ per year.}$$

System reliability for 1 year is then

$$R_s(t=1) = e^{-0.01751} \approx 0.9826$$

For context where MTTF is provided for each constituent unit, the MTTF for the whole system can be derived from that of the individual unit. Since for each individual unit,

$$\lambda_i = \frac{1}{MTTF_i} \tag{6.12}$$

The system MTTF is then

$$MTTF_s = \frac{1}{\lambda_s} = \frac{1}{\sum \lambda_i} = \frac{1}{\sum 1/MTTF_i} = \frac{1}{\frac{1}{MTTF_1} + \frac{1}{MTTF_2} + \cdots \frac{1}{MTTF_n}} \tag{6.13}$$

Example 6.3
Two components with MTTFs 100 hours and 200 hours respectively are connected in series. Calculate the MTTF of whole system?

Solution: According to Eq. (6.13),

$$MTTF_s = \frac{1}{\frac{1}{MTTF_1} + \frac{1}{MTTF_2} + \cdots \frac{1}{MTTF_n}} = \frac{1}{\frac{1}{100} + \frac{1}{200}} = \frac{100 * 200}{100 + 200} \approx 66.7 \text{ hours}$$

Weibull Distribution in Series System

Weibull distribution is common for the life data distribution, if the constituent unit follows Weibull distribution with reliability as

$$R_i(t) = e^{-\left(\frac{t}{\alpha_i}\right)^{\beta_i}} \tag{6.14}$$

If it is a series system, according to Eq. (6.7), the system reliability is

$$R_s(t) = \prod_{i=1}^{n} R_i(t) = e^{-\sum\left(\frac{t}{\alpha_i}\right)^{\beta_i}} \tag{6.15}$$

Obviously, the distribution for the system is not Weibull anymore, and the failure rate for the whole system is not constant but time dependent

$$\lambda_s(t) = \frac{f_s(t)}{R_s(t)} = \frac{[1 - R_s(t)]'}{R_s(t)} = \sum_{i=1}^{n} \frac{\beta_i}{\alpha_i}\left(\frac{t}{\alpha_i}\right)^{\beta_i - 1} \tag{6.16}$$

For example, if each subsystem follows Weibull distribution with

$$\alpha_1 = 5; \beta_1 = 3$$
$$\alpha_2 = 1; \beta_2 = 0.1$$

The failure rate of the system is demonstrated as the top curve shown in Fig. 6.4, which is not monotonous.

6.3 Parallel System

Series system is the minimal system to ensure the system to function as all constituent units must function to ensure the function of the whole system. Parallel system is a system with redundancy allowing the failure of some constituent units. A simple parallel system with two units is represented as in Fig. 6.5.

The whole system fails unless all the constituent units fail. The True table for the simple parallel system is shown in Table 6.3.

Only the first scenario leads to system failure, the corresponding probability, which is probability of failure for this scenario is

$$P = (1 - R_1)(1 - R_2) \tag{6.17}$$

The reliability is the complementary event of Scenario 1, corresponding to a probability of

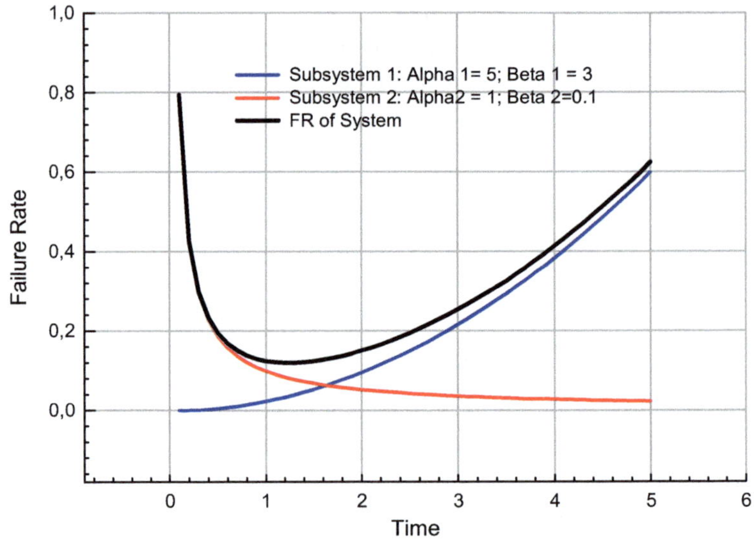

Fig. 6.4 Failure rate for Weibull distribution in series

Fig. 6.5 A simple parallel system

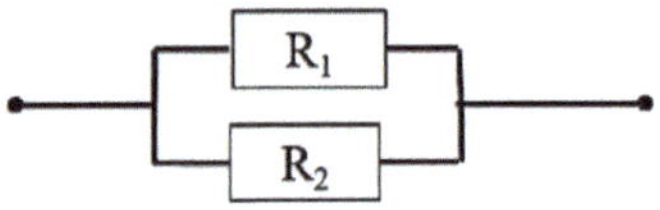

Table 6.3 True table of parallel system

Scenario	Unit 1	Unit 2	System
1	0	0	0
2	0	1	1
3	1	0	1
4	1	1	1

$$R = 1 - (1 - R_1)(1 - R_2) = R_1 + R_2 - R_1 R_2 \tag{6.18}$$

Extending it for a number of n units, the whole system reliability is then

$$R = 1 - (1 - R_1)(1 - R_2)\ldots(1 - R_n) \tag{6.19}$$

If the units are identical with reliability R_1, Eq. (6.19) can be simplified as

$$R = 1 - (1 - R_1)^n \tag{6.20}$$

Example 6.4
A system has two units connected in parallel. The first unit is with reliability $R_1 = 0.9$ for 1 year. The second unit is $R_2 = 0.8$ for 1 year. What is the system reliability?

Solution: According to Eq. (6.18),

$$R_s = 1 - (1 - R_1)(1 - R_2) = 1 - (1 - 0.9) \times (1 - 0.8) = 0.98$$

System Reliability Improving
The system reliability for parallel system is higher than that of its constituent unit, as demonstrated in the Example 6.4, where the system reliability is 0.98, higher than constituent unit's reliability 0.9 and 0.8. The conclusion can be proved by rewritten Eq. (6.19) as

$$1 - R = (1 - R_1)(1 - R_2)\ldots(1 - R_n)$$

The left hand of the above equation is not higher than any of $1 - R_i$, since the $1 - R_i$ is valued between 0 and 1. It is equivalent that the left hand is not lower than any of $1 - R_i$ on the right hand.

Parallel System for Exponential Distribution
The life distribution for parallel systems with constituent units following simple Exponential distribution, is not Exponential distribution. Suppose the failure rate of constituent units is $\lambda_1 = \lambda_2 = \lambda$. The system failure rate is then

$$\lambda_s(t) = \frac{f_s(t)}{R_s(t)} = \frac{2\lambda e^{-\lambda t} - 2\lambda e^{-2\lambda t}}{2e^{-\lambda t} - e^{-2\lambda t}} = \lambda \frac{e^{-\lambda t} - e^{-2\lambda t}}{e^{-\lambda t} - 0.5e^{-2\lambda t}} < \lambda \tag{6.21}$$

which is not constant, implying the lifetime is not Exponential distribution. From Eq. (6.21), we demonstrate the system failure rate is lower than that of constituent failure rate since $\lambda_s(t) < \lambda$. The Fig. 6.6 plots the failure rate curve for $\lambda_1 = \lambda_2 = 0.1$, where we can see a non-constant failure rate curve.

Example 6.5
For a system with two identical units in parallel, the failure rate is λ=0.1 per year. What is the system reliability for t = 10 years?

Solution: According to Eq. (6.18), the system reliability is

$$R_s = 1 - (1 - R_1)(1 - R_2)$$

$$R_s = 1 - \left(1 - e^{-\lambda t}\right)\left(1 - e^{-\lambda t}\right) = 1 - \left(1 - e^{-1}\right)\left(1 - e^{-1}\right) = 0.6004$$

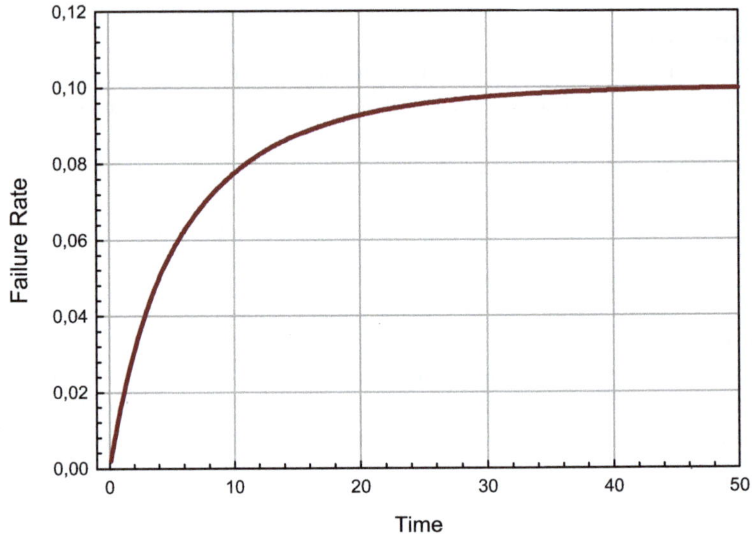

Fig. 6.6 Demonstration of non-constant system failure rate

Fig. 6.7 The m out n system

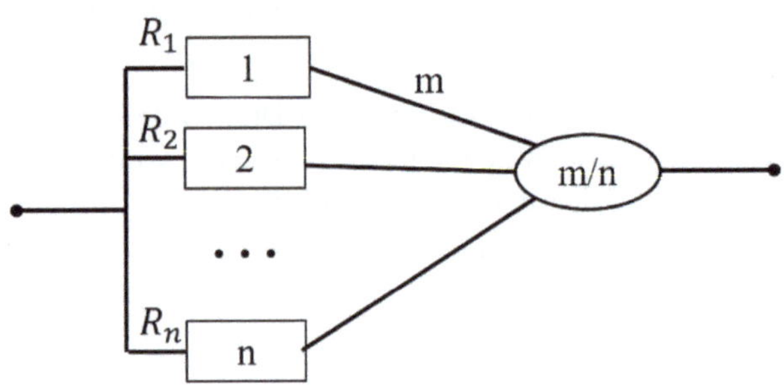

6.4 m Out n System

The parallel system is a typical reliability system with redundancy. The m out n system is another redundancy system that defines at least a number of m constituent units functional out of the total number of n to ensure the whole system functions. The parallel system can be considered as 1 out of n system, a special case of m out n system. Figure 6.7 illustrates the diagram of m out n system.

Analogous to the series and parallel analysis, we utilize the True table for reliability calculation. The True table for m = 2 out of n = 4 system is shown in Table 6.4, where 0 stands for failure, 1 stand for no-failure.

The probability of failure is the sum of scenarios corresponding to system failure "0", which is scenario 1, 2,3,5,9. In Scenario 1, all constituent units failed. In Scenario 2,3,5,9, only 1 out of 4 constituent units is functional. The combination of one unit failed is with number of $\begin{pmatrix} 4 \\ 1 \end{pmatrix} = 4$, The probability of scenarios 2,3,5,9 is identical as $(1 - R)^3 R^1$, the total probability of failure for only 1 functional is

Table 6.4 True table of m out n system

Scenario	Unit 1	Unit 2	Unit 3	Unit 4	System
1	0	0	0	0	**0**
2	0	0	0	1	**0**
3	0	0	1	0	**0**
4	0	0	1	1	1
5	0	1	0	0	**0**
6	0	1	0	1	1
7	0	1	1	0	1
8	1	1	1	1	1
9	1	0	0	0	**0**
10	1	0	0	1	1
11	1	0	1	0	1
12	1	0	1	1	1
13	1	1	0	0	1
14	1	1	0	1	1
15	1	1	1	0	1
16	1	1	1	1	1

$$\binom{4}{1}(1-R)^3R^1$$

which can be generalized to 2 out 4 functional is

$$\binom{4}{2}(1-R)^2R^2 \tag{6.22}$$

For i out n functional is then

$$\binom{n}{i}(1-R)^{n-i}R^i \tag{6.23}$$

The number of functional units less than m leads to failure, as the definition of m out n system, the probability of failure is then

$$\sum_{i=0}^{m-1}\binom{n}{i}(1-R)^{n-i}R^i \tag{6.24}$$

The reliability is straight forward as

$$R_s = 1 - \sum_{i=0}^{m-1}\binom{n}{i}(1-R)^{n-i}R^i \tag{6.25}$$

Especially, for the reliability of 1-out-of-2 Redundancy System for identical units, which is equivalent to the parallel system, according to Eq. (6.25), probability of all unit failed is

$$\binom{2}{0}(1-R)^2R^0 \tag{6.26}$$

The reliability is then

$$1-\binom{2}{0}(1-R)^2R^0 = 1-(1-R)^2 \tag{6.27}$$

which is the same as Eq. (6.18).

Example 6.6
Aircraft has multiple engines and half of the engines functional can ensure the safe return to airport. Airbus A340 is with 4 engines and at least 2 of them functional can ensure safe return. Boeing 777 has 2, and at least 1 of them functional ensure safe return. Suppose engines are all identical with reliability 0.95 and they are independent from each other. Do 4 engines and 2 engines have the same level of system reliability, regardless of the other considerations?

Solution: The 2 engines aircraft is a 1 out 2 redundancy system, the reliability is $1-(1-R)^2$, according to Eq. (6.27), the reliability is

$$1-(1-R)^2=0.9975.$$

The 4 engines aircraft is 2 out 4 system, the reliability is

$$R_s=1-\left[\binom{4}{0}(1-R)^4R^0+\binom{4}{1}(1-R)^3R^1\right]$$

$$\binom{4}{0}(1-R)^4R^0=6.25\times10^{-6}$$

$$\binom{4}{1}(1-R)^3R^1=4.75\times10^{-4}$$

$$R_s=0.9995$$

The aircraft does not have the same system reliability, the 4 engines aircraft has higher reliability. Figure 6.8 shows the difference between them for various reliability levels of a single engine, where the 4 engines aircraft always has higher reliability.

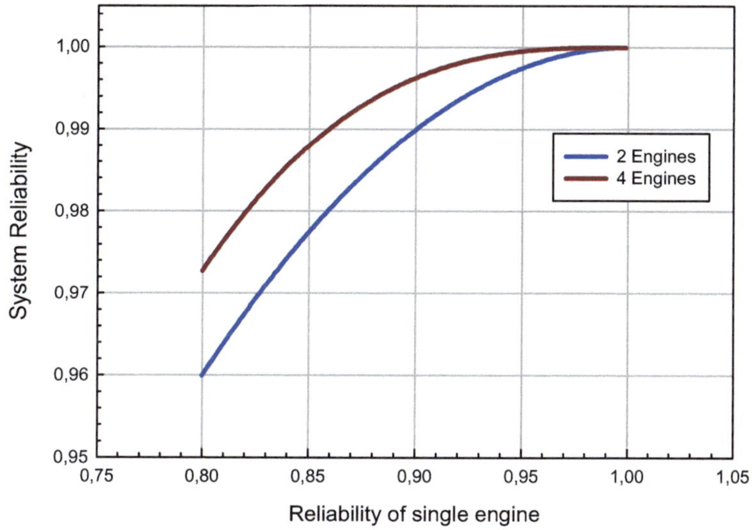

Fig. 6.8 The 2 engines and 4 engines reliability

6.5 Cold Standby System

In parallel reliability system, all parallel units are working simultaneously. One failed, the other is still functional without considering any switching time. This situation is called hot standby. In the engineering system, the redundancy unit probably does not work unless the primary unit fails. For example, in nuclear engineering, the equipment to switch off the reactor has several redundant units to ensure the safe switch. The backup switch will start to work unless the main switch fails. This type of system is named cold-standby. The UPS (Uninterruptible Power Supply) commonly used in IT systems to provide undisrupted power is another typical cold stand-by system. The IT system is supplied with power from normal AC power, the battery runs as standby. Once the main AC power is lost, a switch will turn on and the battery then starts to supply power, as Fig. 6.9 illustrated.

The reliability analysis of the previous reliability system is based on statistical methods. The cold standby is however more than statistic, but a stochastic process. The cold standby system can be illustrated as Fig. 6.10.

Suppose the primary and standby unit are with failure rate λ_1 and λ_2 respectively. To obtain the reliability for the Fig. 6.10, two scenarios could occur during time [0,t]:

Scenario **1**: Primary component never failed during [0,t]. The reliability is

$$R(t) = e^{-\lambda_1 t} \tag{6.28}$$

Scenario **2**: Primary component failed during [0,t]. Let us say it failed at time τ. Then the standby system is switched on work with probability R_{sw}. If it has been

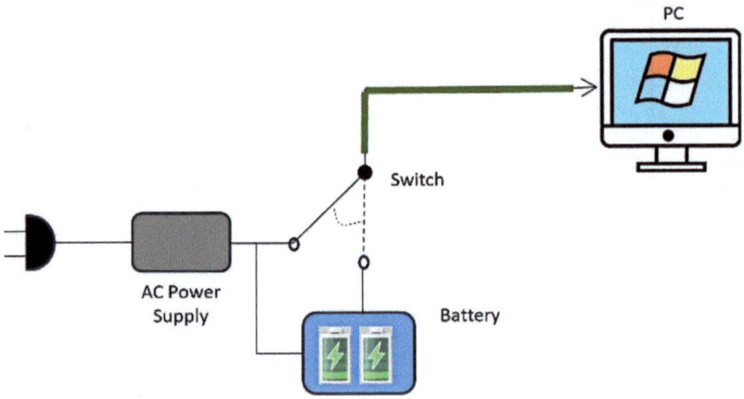

Fig. 6.9 Cold standby power supply system

Fig. 6.10 Cold standby diagram

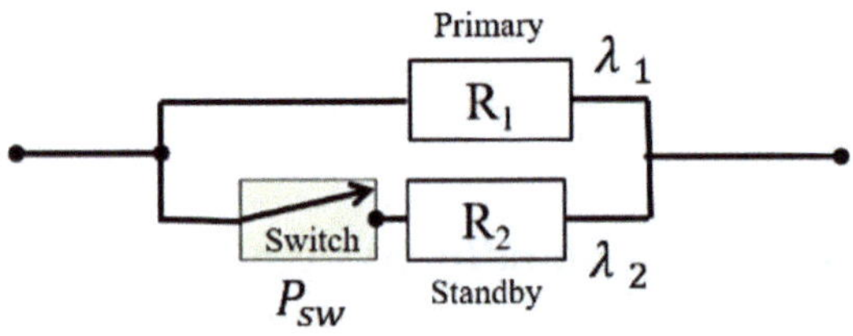

successfully switched on, and the standby unit has survived after time $[\tau, t]$, then the system is also functional for this time interval. The probability of this scenario is then

$$P(t) = R_{sw} f(\tau) e^{-\lambda_2 \tau} \Delta\tau = R_{sw} \lambda_1 e^{-\lambda_1 \tau} e^{-\lambda_2 (t-\tau)} \Delta\tau \tag{6.29}$$

The τ can be any between [0,t]. The system reliability of the scenario 2 is then

$$R(t) = \int_0^t R_{sw} \lambda_1 e^{-\lambda_1 \tau} e^{-\lambda_2 (t-\tau)} \mathrm{d}\tau = R_{sw} \lambda_1 e^{-\lambda_2 t} \int_0^t e^{-(\lambda_1 - \lambda_2)\tau} \mathrm{d}\tau \tag{6.30}$$

$$R(t) = \frac{R_{sw} \lambda_1}{\lambda_1 - \lambda_2} e^{-\lambda_2 t} - \frac{R_{sw} \lambda_1}{\lambda_1 - \lambda_2} e^{-\lambda_1 t} \tag{6.31}$$

All the other scenarios other than the above two will lead to failure. The whole system reliability is then from the two scenarios, and it is

$$R_s(t) = e^{-\lambda_1 t} + \frac{R_{sw} \lambda_1}{\lambda_1 - \lambda_2} e^{-\lambda_2 t} - \frac{R_{sw} \lambda_1}{\lambda_1 - \lambda_2} e^{-\lambda_1 t} \tag{6.32}$$

Equation (6.31) is only for the case of $\lambda_1 \neq \lambda_2$. When the $\lambda_1 = \lambda_2 = \lambda$, according to Eq. (6.30), the reliability for the second scenario is

$$R_{sw}\lambda_1 e^{-\lambda_2 t}\int_0^t e^{-(\lambda_1-\lambda_2)\tau}d\tau = R_{sw}\lambda t e^{-\lambda t} \quad (6.33)$$

The whole system reliability is

$$R_s(t) = e^{-\lambda t} + R_{sw}\lambda t e^{-\lambda t} \quad (6.34)$$

Example 6.7
What is the reliability of the system for 10 hours for the standby system shown on Fig. 6.10? The reliability of the switch is constant with 0.95 (The probability of the switch can successfully activate the standby system). And

$$\lambda_1 = 0.1/\text{hour}$$

$$\lambda_2 = 0.2/\text{hour}$$

Solution: According to Eq. (6.32), system reliability for this case is

$$R(t) = e^{-\lambda_1 t} + \frac{R_{sw}\lambda_1}{\lambda_1-\lambda_2}e^{-\lambda_2 t} - \frac{R_{sw}\lambda_1}{\lambda_1-\lambda_2}e^{-\lambda_1 t} =$$

$$R(t=10) = e^{-1} + \frac{0.95 * 0.1}{-0.1}e^{-2} - \frac{0.95 * 0.1}{-0.1}e^{-1}$$

$$R(t=10) = 0.58$$

Example 6.8
A system is composed of two identical pumps each with constant failure rate

$$\lambda = 10^{-3}\ \text{failures/hour}$$

The probability that the switch will fail to activate the standby pump has been estimated to be 1.5%. What is the reliability of the system at time t = 1000 hours?

Solution: The reliability of switch is R_{sw} = 1–0.015 = 0.985

$$R_s(t=1000) = e^{-\lambda t} + R_{sw}\lambda t e^{-\lambda t} = e^{-0.001\times 1000} + (1-0.015)\times 0.001$$

$$\times 1000 \times e^{-0.001\times 1000} = 0.7302$$

The reliability at 1000 hours is 0.7302.

6.6 Reliability Block Diagram (RBD)

Reliability Block Diagram (RBD) is a graphical representation of the reliability connection of a system, which shows the reliability connections among the components. RBD is not the same as the block schematic diagram of the system's functional layout. This section gives an example of RBD for a common PC system. Each component is a block for the RBD as shown in Fig. 6.11.

The failure from any of Memory, CPU, HDD, Mother Board or LCD will lead to failure of the PC, implying they are connected in series. Suppose the battery and AC power supply are both supplying the power to the PC at the same time. Then unless all of both fail, the system will fail. The AC power and battery are thus parallelly connected. The RBD is shown in Fig. 6.12.

Once the failure rate or reliability for each component is known, the whole system reliability can be calculated. The next Section addresses the calculation issue.

6.7 Complex System Reliability

The reliability system is typically complex containing series, parallel, cold standby or m out n system. For calculation, the complex system can be decomposed into simple series, parallel, etc. Figure 6.13 demonstrates an example of complex system, which is adapted from [1].

The missile system contains radar, launch and guidance and the missile system. Once the primary radar fails, standby radar will be switched on. The system is

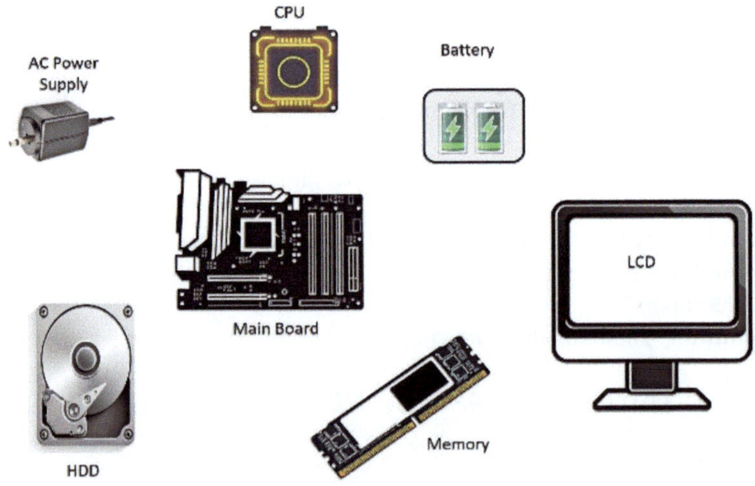

Fig. 6.11 Components of PC

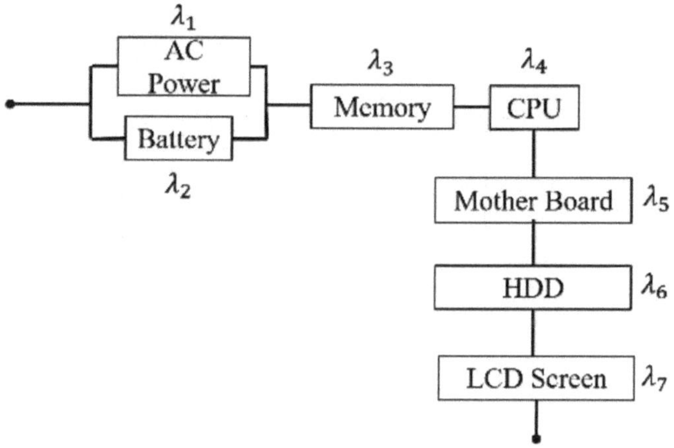

Fig. 6.12 RBD of PC

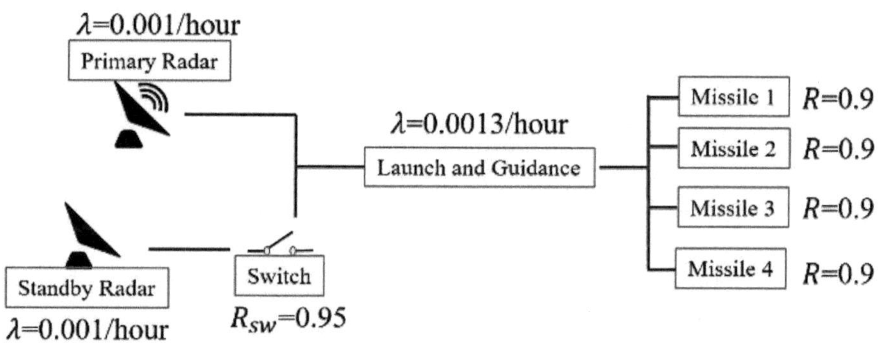

Fig. 6.13 Complex reliability system [1]

reliable if at least 3 out of 4 missiles can be fired. To calculate the reliability, the system is decomposed into 3 parts in series as shown in Fig. 6.14.

Assume all elements are independent. The radar system is a cold standby system. The missile system is 3 out 4 system. Considering for 24 hours, the reliability for radar system is

$$R_{cold}(t=24)=e^{-\lambda t}+R_{sw}\lambda te^{-\lambda t}=e^{-0.001*24}+0.95*0.001*24*e^{-0.001*24}=0.9985$$

The missile system is

$$R_{missile}=1-\sum_{i=0}^{m-1}\binom{n}{i}(1-R)^{n-i}R^{i};m=3,n=4$$

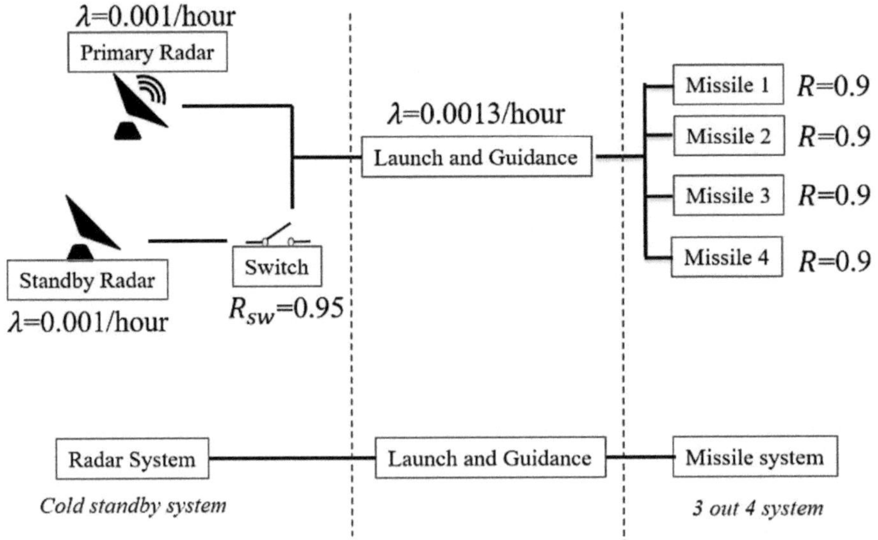

Fig. 6.14 Decomposition of complex system

$$R_{missile} = 0.9477$$

The Launch and Guidance is a simple Exponential system

$$R_{Launch\&Guidance}(t = 24) = \mathrm{e}^{-\lambda t} = e^{-0.0013*24} = 0.9693$$

The whole system is then

$$R_s = R_{cold} * R_{Launch\&Guidance} * R_{missile} = 0.9985 * 0.9693 * 0.9477 = 0.9172$$

Example 6.9

A reliability system is as Fig. 6.15.

Unit 1 and 2 are identical following Weibull distribution with $\alpha = 100$ days; $\beta = 1.3$. Unit 3 follows Exponential distribution with $\lambda = 0.003$ per day. What is the reliability for the whole system for 100 days?

Solution: $R_1 = R_2 = R(t = 100) = e^{-(t/\alpha)^\beta} = 0.3679;$

$$R_3(t = 100) = e^{-\lambda t} = 0.7408;$$

$$R_s = 1 - (1 - R_1R_2)(1 - R_3) = R_1R_2 + R_3 - R_1R_2R_3 = 0.7760$$

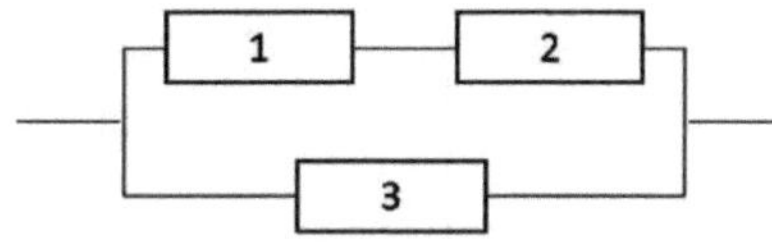

Fig. 6.15 A series-parallel RBD system

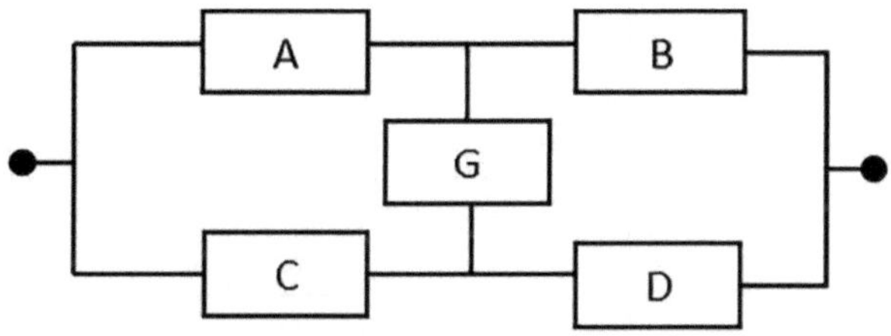

Fig. 6.16 Bridge RBD network

6.8 Network System Reliability

If reliability block diagram can be decomposed into basic series, parallel, m out n units etc., the system reliability can be computed from the basic units, as described in the previous section. However, in some contexts, for example, the communication system, power line network, internet network, the RBD could present in a form of network, where the bridge could be part of network, as shown in an example in Fig. 6.16.

For such network containing bridge, the decomposition method does not work. We can employ the minimal path set and minimal cut set. Minimal path set is set of units that can enable the system to be reliable, where anyone in the set failed, the set will result in failure. In Fig. 6.16, the path AB is a minimal path set, if A or B failed, the path failed. The units inside the path set are connected in series. We can identify all the path sets for Fig. 6.16 as

$$\{AB\}, \{CD\}, \{AGD\}, \{CGB\}$$

A diagram to represent the minimal path set is drawn in Fig. 6.17. The path sets are connected in parallel, while they are not independent from each other, since there are some common units in some path sets. The reliability calculation for parallel system does not apply. We can utilize some Boolean operation to obtain the numerical system reliability for the network, as expressed as

$$R_s = Pr(P1 \cup P2 \cup P3 \cup P4) \tag{6.35}$$

One point worth noting is that the Boolean operation is to obtain a disjoint event so the probability can be calculated from its constituent units. The joint events can be decomposed using Formula

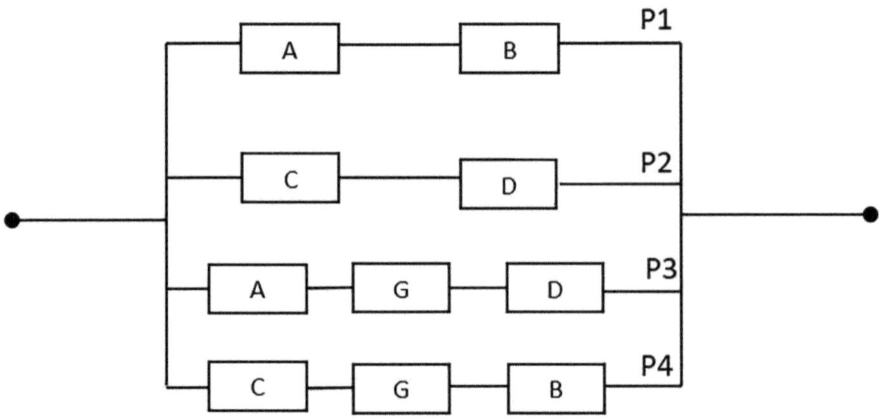

Fig. 6.17 Diagram of minimal path set

$$\Pr(P1 \cup P2 \cup \ldots \cup Pn) = \sum_{i=1}^{n} P(Pi) - \sum_{i<j} \Pr(Pi \cap Pj) + \sum_{i<j<k} \Pr(Pi \cap Pj \cap Pk) - \ldots + (-1)^{n-1} \Pr(P1 \cap \ldots \cap Pn) \tag{6.36}$$

It is easy to understand Eq. (6.36) by using an example. For Eq. (6.35), the

$$\begin{aligned} Pr(P1 \cup P2 \cup P3 \cup P4) = {} & \Pr(P1) + \Pr(P2) + \Pr(P3) + \Pr(P4) \\ & - \Pr(P1 \cap P2) - \Pr(P1 \cap P3) - \Pr(P1 \cap P4) - \Pr(P2 \cap P3) \\ & - \Pr(P2 \cap P4) - \Pr(P3 \cap P4) + \Pr(P1 \cap P2 \cap P3) \\ & + \Pr(P1 \cap P2 \cap P4) + \Pr(P1 \cap P3 \cap P4) + \Pr(P2 \cap P3 \cap P4) \\ & - \Pr(P1 \cap P2 \cap P3 \cap P4) \end{aligned} \tag{6.37}$$

$$\begin{aligned} Pr(P1 \cup P2 \cup P3 \cup P4) = {} & \Pr(AB) + \Pr(CD) + \Pr(AGD) + \Pr(CGB) \\ & - \Pr(ABCD) - \Pr(ABDG) - \Pr(ABCG) - \Pr(ACDG) - \Pr(BCDG) \\ & - \Pr(ABCDG) + \Pr(ABCDG) + \Pr(ABCDG) + \Pr(ABCDG) \\ & + \Pr(ABCDG) - \Pr(ABCDG) \end{aligned} \tag{6.38}$$

It can be further simplified to

$$\begin{aligned} Pr(P1 \cup P2 \cup P3 \cup P4) = {} & \Pr(AB) + \Pr(CD) + \Pr(ADG) + \Pr(BCG) \\ & - \Pr(ABCD) - \Pr(ABDG) - \Pr(ABCG) - \Pr(ACDG) - \Pr(BCDG) + 2 \\ & * \Pr(ABCDG) \end{aligned} \tag{6.39}$$

The $\Pr(Pi \cap Pj) = \Pr(Pi) \Pr(Pj)$. If the reliability for each unit is 0.9, according to Eq. (6.39), the system reliability is

$$R_s = Pr(P1 \cup P2 \cup P3 \cup P4) = 0.9 * 0.9 + 0.9 * 0.9 + 0.9 * 0.9 * 0.9 + 0.9 * 0.9 * 0.9 - 0.9 * 0.9 * 0.9 * 0.9 - 0.9 * 0.9 * 0.9 * 0.9 - 0.9 * 0.9 * 0.9 * 0.9 - 0.9 * 0.9 * 0.9 * 0.9 - 0.9 * 0.9 * 0.9 * 0.9 + 2 * 0.9 * 0.9 * 0.9 * 0.9 * 0.9 = 0.9785$$

We can obtain the reliability from the unreliability, which is a complementary event for reliability. The probability of failure as

$$PoF = 1 - Pr(P1 \cup P2 \cup P3 \cup P4) = Pr\left(\overline{P1 \cup P2 \cup P3 \cup P4}\right) \tag{6.40}$$

We use the "+" for $\cup$ for simplification of the expression, and according to Boolean law shown in Appendix 6A

$$\begin{aligned}
\overline{P1 \cup P2 \cup P3 \cup P4} &= \left(\overline{AB}\right)\left(\overline{CD}\right)\left(\overline{AGD}\right)\left(\overline{CGB}\right) \\
&= \left(\overline{A} + \overline{B}\right)\left(\overline{C} + \overline{D}\right)\left(\overline{A} + \overline{G} + \overline{D}\right)\left(\overline{C} + \overline{G} + \overline{B}\right) \\
&= \left(\overline{AC} + \overline{AD} + \overline{BC} + \overline{BD}\right)\left(\overline{A} + \overline{G} + \overline{D}\right)\left(\overline{C} + \overline{G} + \overline{B}\right) \\
&= \left(\overline{AC} + \overline{AD} + \overline{ABC} + \overline{ABD} + \overline{ACG} + \overline{ADG} + \overline{BCG}\right. \\
&\quad \left. + \overline{BGD} + \overline{ACD} + \overline{AD} + \overline{BCD} + \overline{BD}\right)\left(\overline{C} + \overline{G} + \overline{B}\right)
\end{aligned} \tag{6.41}$$

Equation (6.41) can be simplified from the law $A + (AB) = A$, also as shown in the Appendix 6A, then

$$\overline{AC} + \overline{AD} + \overline{ABC} + \overline{ABD} + \overline{ACG} + \overline{ADG} + \overline{BCG} + \overline{BGD} + \overline{ACD} + \overline{AD} + \overline{BCD} + \overline{BD} = \overline{AC} + \overline{AD} + \overline{BCG} + \overline{BD}$$

Equation (6.41) is rewritten as

$$\left(\overline{AC} + \overline{AD} + \overline{BCG} + \overline{BD}\right)\left(\overline{C} + \overline{G} + \overline{B}\right) = \overline{AC} + \overline{ACD} + \overline{BCG} + \overline{BCD} + \overline{ACG} + \overline{ADG} + \overline{BCG} + \overline{BDG} + \overline{ABC} + \overline{ABD} + \overline{BCG} + \overline{BD}$$

It further simplifies it as

$$\overline{AC} + \overline{BCG} + \overline{ADG} + \overline{BD} \tag{6.42}$$

The joint sets $\overline{AC}, \overline{BCG}, \overline{ADG}, \overline{BD}$ presented in Eq. (6.42) are named minimal cut sets. Using Eq. (6.36) to merge the minimal cut sets to disjoint sets. The union of cut sets is then

$$Pr(\overline{AC}\cup\overline{BCG}\cup\overline{ADG}\cup\overline{BD}) = \Pr(\overline{AC}) + \Pr(\overline{BCG}) + \Pr(\overline{ADG}) + \Pr(\overline{BD})$$
$$- \Pr(\overline{ABCG}) - \Pr(\overline{ACDG}) - \Pr(\overline{ABCD}) - \Pr(\overline{ABCDG}) - \Pr(\overline{BCDG})$$
$$- \Pr(\overline{ABDG}) + \Pr(\overline{ABCDG}) + \Pr(\overline{ABCDG}) + \Pr(\overline{ABCDG}) + \Pr(\overline{ABCDG})$$
$$- \Pr(\overline{ABCDG})$$

which is simplified into

$$\Pr(\overline{AC}) + \Pr(\overline{BCG}) + \Pr(\overline{ADG}) + \Pr(\overline{BD}) - \Pr(\overline{ABCG}) - \Pr(\overline{ACDG})$$
$$- \Pr(\overline{ABCD}) + 2 * \Pr(\overline{ABCDG}) - \Pr(\overline{BCDG}) - \Pr(\overline{ABDG}).$$

If the reliability for each unit is 0.9, then the unreliability for each unit is 1–0.9 = 0.1. The probability of failure of system is then

$$0.1 * 0.1 + 0.1 * 0.1 * 0.1 + 0.1 * 0.1 * 0.1 + 0.1 * 0.1 - 0.1 * 0.1 * 0.1 * 0.1$$
$$- 0.1 * 0.1 * 0.1 * 0.1 - 0.1 * 0.1 * 0.1 * 0.1 + 2 * 0.1 * 0.1 * 0.1 * 0.1 * 0.1$$
$$- 0.1 * 0.1 * 0.1 * 0.1 - 0.1 * 0.1 * 0.1 * 0.1 = 0.0215$$

The reliability is then 1 – 0.0215 = 0.9785, which is the same from Eq. (6.39) from the minimal path sets.

Example 6.10

A reliability system is as follows, same as Example 6.9, the reliability is 0.3679 for components 1 and 2. Reliability is 0.7408 for component 3. Using minimal path set method to calculate the system reliability

Solution: The minimal path sets are: {12}, {3} as shown in Fig. 6.18. According to Eq. (6.36),

$$R_s = Pr(12\cup 3) = \Pr(12) + Pr(3) - \Pr(12\cap 3) = \Pr(12) + \Pr(3)$$
$$- \Pr(123) = \Pr(1) * \Pr(2) + \Pr(3) - \Pr(1) * \Pr(2) * \Pr(3) = 0.3679$$
$$* 0.3679 + 0.7409 - 0.3679 * 0.3679 * 0.7409 = 0.7760$$

The system reliability is 0.7760, same as from the Example 6.9 by using the series and parallel approach.

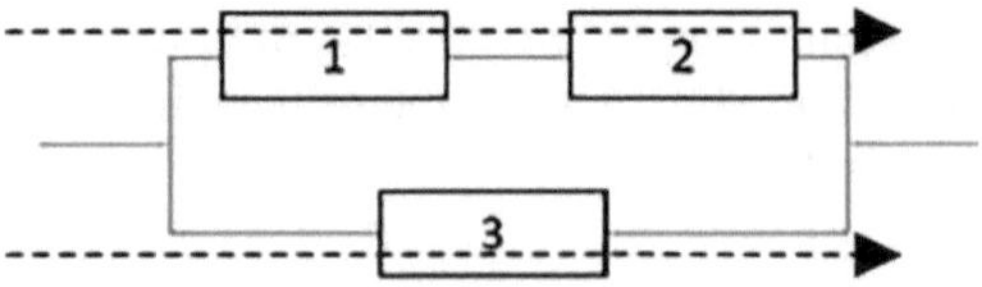

Fig. 6.18 Minimal path set for Example 6.9

In Boolean algebra, the path sets and cut sets are duality events. If the path set is {P1}, {P2},…,{Pn}, the cut sets C1, C2, … are then

$$\mathrm{C1, C2, \ldots Cn} = \overline{P1 \cup P2 \ldots \cup Pn} \quad (6.43)$$

Equation (6.43) holds for all path sets. For the Example 6.10, the cut sets is

$$\overline{12 \cup 3} = \overline{12} \cap \overline{3} = \left(\overline{1} \cup \overline{2}\right) \cap \overline{3} = \overline{13} + \overline{23}$$

The cut sets is $\{\overline{13}\}$, $\{\overline{23}\}$. The unreliability is

$$\Pr\left(\overline{13} + \overline{23}\right) = \Pr\left(\overline{13}\right) + \Pr\left(\overline{23}\right) - \Pr\left(\overline{13} \cap \overline{23}\right) = \Pr\left(\overline{13}\right) + \Pr\left(\overline{23}\right) - \Pr\left(\overline{123}\right)$$
$$= \Pr\left(\overline{1}\right)\Pr\left(\overline{3}\right) + \Pr\left(\overline{2}\right)\Pr\left(\overline{3}\right) - \Pr\left(\overline{1}\right)\Pr\left(\overline{2}\right)\Pr\left(\overline{3}\right)$$

For the Example 6.10, the $\Pr\left(\overline{1}\right) = \Pr\left(\overline{2}\right) = 1 - 0.3679 = 0.6321$, $\Pr\left(\overline{3}\right)$ = 1–0.7409 = 0.2591. The probability of failure is then

$$\Pr\left(\overline{1}\right)\Pr\left(\overline{3}\right) + \Pr\left(\overline{2}\right)\Pr\left(\overline{3}\right) - \Pr\left(\overline{1}\right)\Pr\left(\overline{2}\right)\Pr\left(\overline{3}\right) = 0.6321 * 0.2591 + 0.6321 * 0.2591 - 0.6321 * 0.6321 * 0.2591 = 0.2240$$

The reliability is 1 − 0.2240 = 0.7760, same as from the path sets.

6.9 Simulation for System Reliability

An approach to find out the system reliability is to employ simulation. The status of the constituent units is simulated using software. Firstly, we generate a random number r_i for each unit i between 0 and 1, and the probability of obtaining any random number between 0 and 1 is equal. The probability density function of the ideal random number thus follows uniform distribution with range 0 to 1, as shown in left of Fig. 6.19.

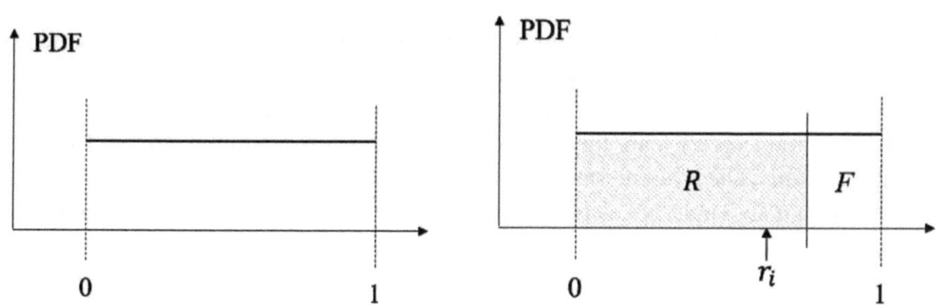

Fig. 6.19 Uniform distribution

Table 6.5 True rable for Fig. 6.16

Scenario	Unit A	Unit B	Unit C	Unit D	Unit G	System
1	0	0	0	0	**0**	**0**
2	0	0	0	1	**0**	**0**
3	0	0	1	0	**0**	**0**
4	0	0	1	1	0	1
5	0	1	0	0	**0**	**0**
6	0	1	0	1	0	**0**
7	0	1	1	0	**0**	**0**
8	1	1	1	1	**0**	1
9	1	0	0	0	**0**	0
10	1	0	0	1	0	0
11	1	0	1	0	**0**	0
12	1	0	1	1	0	1
13	1	1	0	0	0	1
14	1	1	0	1	0	1
15	1	1	1	0	0	1
16	1	1	1	1	0	1
17	0	0	0	0	**1**	**0**
18	0	0	0	1	**1**	**0**
19	0	0	1	0	**1**	**0**
20	0	0	1	1	1	1
21	0	1	0	0	**1**	**0**
22	0	1	0	1	1	**0**
23	0	1	1	0	**1**	1
24	1	1	1	1	**1**	1
25	1	0	0	0	**1**	**0**
26	1	0	0	1	1	1
27	1	0	1	0	**1**	**0**
28	1	0	1	1	1	1
29	1	1	0	0	1	1
30	1	1	0	1	1	1
31	1	1	1	0	1	1
32	1	1	1	1	1	1

Suppose the reliability for the unit is known as a numerical value R. The status of unit is simulated as functional if the generated random number $r_i \leq R$, i.e. if r_i locates in the shadowed area shown on the right of Fig. 6.19; otherwise, it is considered failed. A status vector can then be generated by simulating a status for each constituent unit. The system status can then be determined by checking the True table as shown in Table 6.5. The True table is drawn according to the RBD from Fig. 6.16.

Repeat the procedure for a number of iterations, let us say n. If the total number of functional states is n_f, the system reliability is simulated as

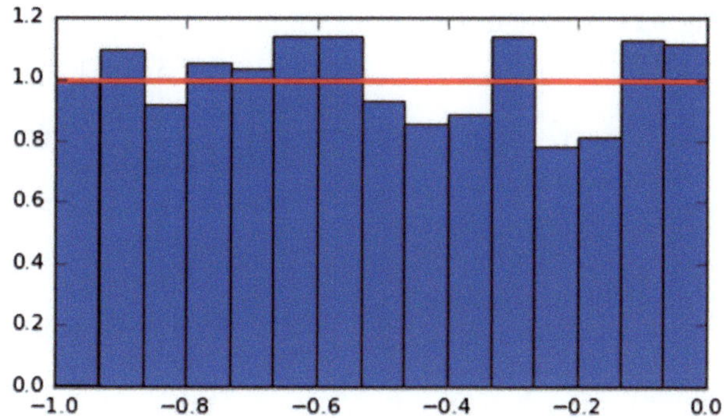

Fig. 6.20 Empirical distribution of random number

$$R = \frac{n_f}{n}$$

For the reliability simulation for a RBD as Fig. 6.16, the Matlab code is attached in Appendix 6B. For 50,000 iterations, the system reliability obtained is 0.97<u>90</u>, compared with the real reliability 0.9785. The last two digits are random, because of software system limitation. The random of the last two digits is due to the fundamental issue of the random number generating technique. The generated random number is not the ideal uniform distribution. Figure 6.20 shows the empirical PDF of the random number from the same code that is used for generating random numbers, which is obviously not ideal uniform distribution. The random number is thus not real random, which results in the randomness of the last two digits of 0.97<u>90</u>.

Simulation method can avoid the tedious mathematical derivation and formulas discussed in Sect. 6.8, but with the shortcoming of the randomness issue described above and time consuming. For practical applications, the randomness problem can be ignored as the precision requirement is normally not as precise as to 0.x%. As the known reliability value for each unit might not be accurate, the high precision requirement on the system's reliability is then not necessary.

6.10 Reliability Importance Measure

Importance measures is one method of identifying the relative importance of each component in a system from a reliability perspective. Take the case shown in Fig. 6.12 as example. In this case, the importance of each unit varies: The failure of AC power may not lead to failure, while the failure of CPU and Memory will surely result in failure. To demonstrate our importance measure method, we simplify Fig. 6.12 to include only 4 units as shown in Fig. 6.21.

The reliability for the power system in the RBD of Fig. 6.21 is

$$R_{power} = R_1 + R_2 - R_1 R_2 \tag{6.44}$$

For the whole system, the reliability is

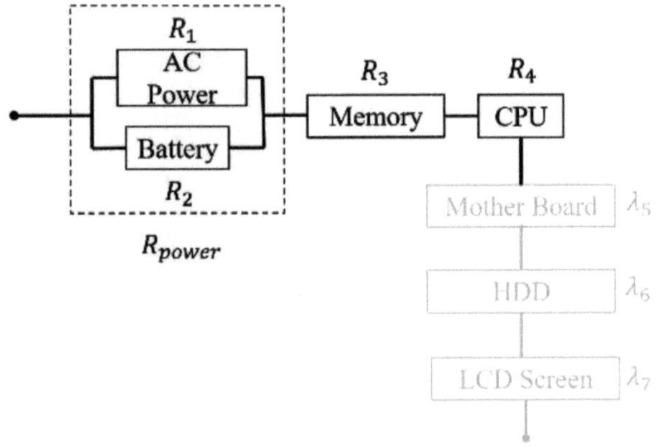

Fig. 6.21 RBD of four PC components

$$R_s = (R_1 + R_2 - R_1R_2)R_3R_4 \tag{6.45}$$

Reliability Importance Measure

The importance measure method introduced here can be explained from the concept of regression analysis. In the regression analysis, each variable corresponds to a coefficient, as the 1 for x_1, 0.2 for x_2 in the following example

$$y = x_1 + 0.2x_2 + 4x_3 \tag{6.46}$$

For each variable, the sensitivity varies according to their coefficients, when

x_1 increaes 1 − > y increases 1
x_2 increaes 1 − > y increases 0.2
x_3 increaes 1 − > y increases 4

i.e. when we increase a same quantity for each x_i, the gain of y varies. If y is the benefit of intervisitation, x is the cost. We know investing on the variable with the highest sensitivity will gain the highest benefit at a same cost. The sensitivity, i.e. the coefficient in Eq. (6.46) can be obtained in a general way by taking its first derivative

$$\frac{\partial y}{\partial x_1} = 1;\ \frac{\partial y}{\partial x_2} = 0.2;\ \frac{\partial y}{\partial x_3} = 4; \tag{6.47}$$

If the function as Eq. (6.46) is other general function, even non-linear, the first derivative method as shown in Eq. (6.47) still applies. For example, if the function is

$$f(x) = {x_1}^2 + 4{x_2}^4 \tag{6.48}$$

The sensitivity is then

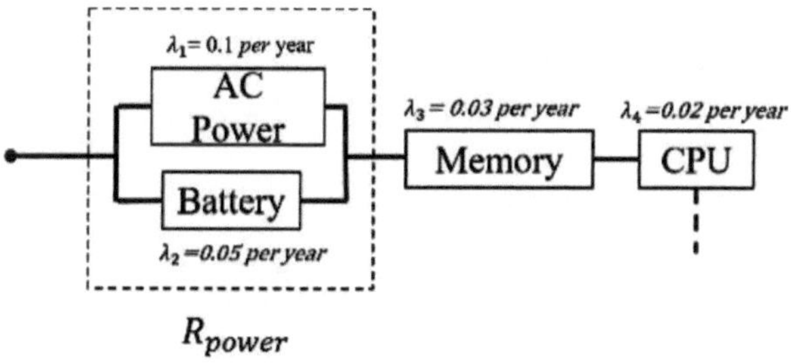

Fig. 6.22 Failure rate of constituent unit

$$\frac{\partial f}{\partial x_1} = 2x_1; \frac{\partial f}{\partial x_2} = 16{x_2}^3; \tag{6.49}$$

For reliability, if the system reliability is the $f(x)$, the reliability of constituent unit is the x_i, we can use the similar concept of sensitivity analysis: find the unit with the highest sensitivity to increase or decrease the system reliability. This sensitivity in reliability is a **reliability importance measure**.

Example 6.11

For example, for the case shown in Fig. 6.21, and the failure rate for each unit is shown in following Fig. 6.22.

The system reliability function as shown in Eq. (6.45) is $R_s = (R_1 + R_2 - R_1R_2)R_3R_4$. The reliability sensitivity of each unit is

$$\frac{\partial R_s}{\partial R_1} = (1 - R_2)R_3R_4;$$

$$\frac{\partial R_s}{\partial R_2} = (1 - R_1)R_3R_4R_5R_6R_7;$$

$$\frac{\partial R_s}{\partial R_3} = (R_1 + R_2 - R_1R_2)\ R_4;$$

$$\frac{\partial R_s}{\partial R_4} = (R_1 + R_2 - R_1R_2)\ R_3;$$

The reliability for each unit is

$$R_1(t) = e^{-0.1t}$$
$$R_2(t) = e^{-0.05t}$$
$$R_3(t) = e^{-0.03t}$$
$$R_4(t) = e^{-0.02t} \tag{6.50}$$

Then the system reliability is

(continued)

Example 6.11 (continued)

$$R_s(t) = \left(e^{-0.1t} + e^{-0.05t} - e^{-0.15t}\right)e^{-0.05t}$$

The sensitivity is then

$$\frac{\partial R_s}{\partial R_1} = (1 - R_2)R_3R_4 = \left(1 - e^{-0.05t}\right)e^{-0.05t};$$

$$\frac{\partial R_s}{\partial R_2} = (1 - R_1)R_3R_4 = \left(1 - e^{-0.1t}\right)e^{-0.05t};$$

$$\frac{\partial R_s}{\partial R_3} = (R_1 + R_2 - R_1R_2)\ R_4) = \left(e^{-0.1t} + e^{-0.05t} - e^{-0.15t}\right)e^{-0.03t};$$

$$\frac{\partial R_s}{\partial R_4} = (R_1 + R_2 - R_1R_2)R_3 = \left(e^{-0.1t} + e^{-0.05t} - e^{-0.15t}\right)e^{-0.02t}; \quad (6.51)$$

The reliability importance measure is a function of time when reliability is a function of time. The importance measure against time is plotted in Fig. 6.23.

For demonstrating the change of importance measure, the Fig. 6.24 shows the reliability importance of 1 and 9 years. For 1 year, the importance of Memory and CPU, which has no redundancy, show very high importance as 0.8869 and 0.8781. The corresponding system reliability for 1 year is $R_s = 0.9468$. If it is intended to improve system reliability to 0.98, we can achieve the goal by trying to improve the reliability of CPU, which is of the highest importance, to a new level. Rewrite Eq. (6.45) as

$$R_4 = R_s/(R_1 + R_2 - R_1R_2)R_3$$

Substitute the numerical value to the above formula letting $R_1 = 0.9048$, $R_2 = 0.9512$ $R_3 = 0.9704$ for $t = 1$ year from Eq. (6.50). The new R_4 required is 1.01, implying only improving the CPU cannot reach target. We can try to improve both R_3R_4. The $R_3R_4 = \frac{New\ R_s}{R_1+R_2-R_1R_2} = \frac{0.98}{0.9048+0.9512-0.9048*0.9512} = 0.9846$. If we improve the R_3 to 0.99, then the R_4 should be 0.9846/0.99 = 0.9945.

Importance measure is dependent on the reliability of constituent unit, but also depending on the time considered. For the case in Example 6.11, when the time considered is changed to 9 years, the reliability of constituent unit will change and the importance measure will change accordingly, as shown in the lower part of Fig. 6.24.

For the allocation of reliability during product design phase, the importance measure provides a smart mean to improve the system reliability with least cost. However, other constraints could play a role in the design, such as the economic

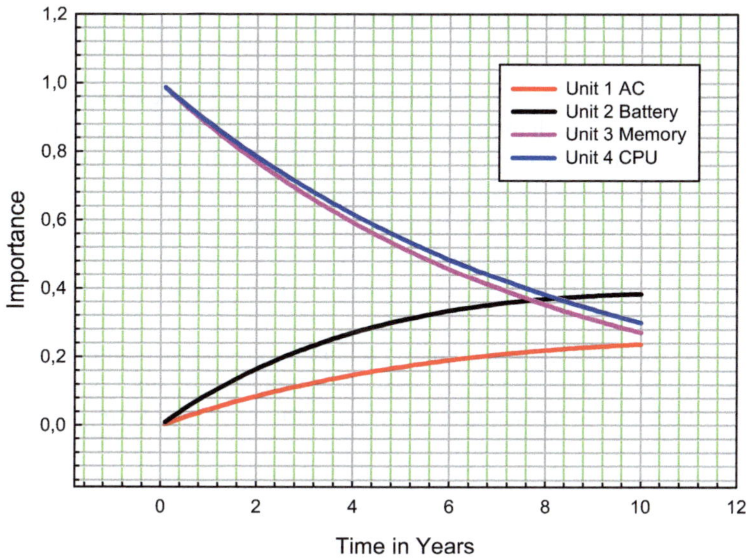

Fig. 6.23 Time dependence of importance measure

t = 1 year

Components	Importance Measures
Unit 1 AC	0.0464
Unit 2 Battery	0.0905
Unit 3 Memory	**0.8781**
Unit 4 CPU	**0.8869**

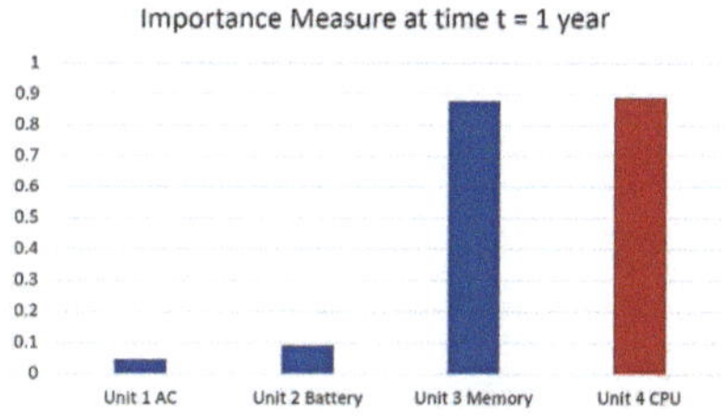

t = 9 year

Components	Importance Measures
Unit 1 AC	0.231
Unit 2 Battery	0.378
Unit 3 Memory	0.310
Unit 4 CPU	0.340

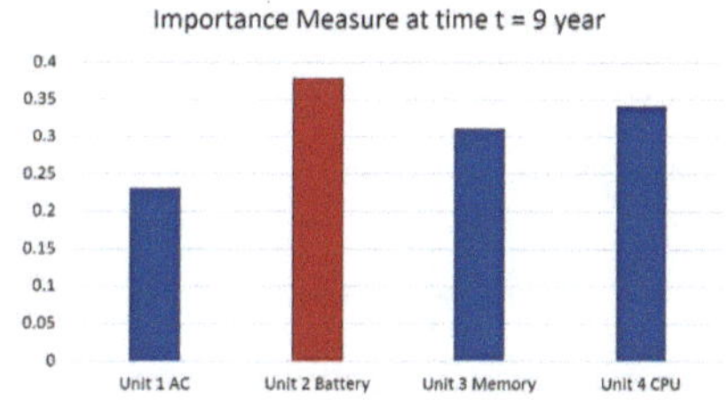

Fig. 6.24 Importance measure for 1 and 9 years

issue. In engineering, for example, to improve the reliability of a unit from 0.5 to 0.6 is easier than to improve from 0.95 to 0.99. When a unit is already with very high reliability, improving it further could require high economic cost and more tough technical issues needed to be solved. The importance measure just addresses the issue from reliability aspect.

6.11 System Reliability Prediction

During new product design, the reliability level must be predicted to ensure that the product can achieve its reliability goal. Reliability prediction is therefore an important activity in industry. Figure 6.25 illustrates the reliability activities involved in electronic device design. During the design phase, when the design is ready, the reliability level needs to be predicted. However, the available information at this stage is often very limited, as we may only know the reliability goal, design schemes, bill of materials, and similar data. Conducting reliability evaluation at this stage thus is a challenge.

The parts count method is a common reliability prediction technique adopted in several standards, including MIL-HDBK-217 (published by the US Department of Defense) which is widely using in electronic industries [2]. A simple parts count method is as

$$\lambda_s = \Sigma n_i \lambda_i \pi_i \tag{6.52}$$

The λ_i is the failure rate of the constituent unit i. The n_i is the number of identical units. The π_i is the quality factor which can be set as 1 for general circumstance. Implicit assumptions in Eq. (6.52) are the unit or part is with lifetime of Exponential distribution and the unit are connected in series. According to Eq. (6.11), the failure rate of whole system is with constant failure rate when the failure rate from constituent unit is constant and connected in series. If two constituent units are identical with failure rate λ_i, the system failure rate is $\lambda_i + \lambda_i = 2\lambda_i$, where the 2 is the n_i in Eq. (6.52). As the series system results in the lowest reliability compared with other system configurations, the parts count method is thus a lower bound for the system reliability.

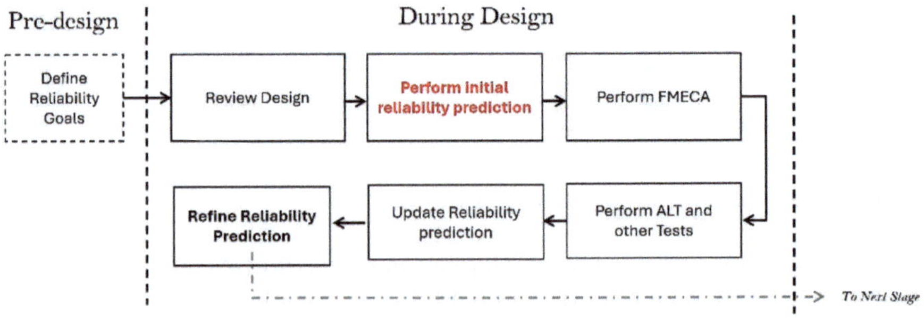

Fig. 6.25 Reliability issues in electronics development

Table 6.6 Bill of material for board (Note: The failure rate in this table is fabulated)

Item	Quantity	Ref	Value	Failure Rate
Capacitors	2	C1,C3	100 nF	$3.5/10^9$hours
Capacitor	1	C2	10uF	$3/10^9$hours
Jumpers	2	JP1, JP2	x	$2/10^9$hours
MEMS digital sensor	1	U1	x	0.02/year

Example 6.12

The bill of materials is tabulated in Table 6.6.

Using parts count according to Eq. (6.52), the system failure rate is

$\lambda_s = 2 * 3.5 * 10^{-9} + 1 * 3 * 10^{-9} + 2 * 3.5 * 10^{-9} + \frac{0.02}{365*24} = 2.3 * 10^{-6}$ per hour

The failure rate for the board is evaluated as $2.3 * 10^{-6}$ per hour.

The reliability prediction discussed in Example 6.12 is very rough. Some units such as PCB substrate and solder points are not considered. Nevertheless, the evaluated value can be used as guidance for part selection and give a rough prediction of system reliability which is useful during design phase.

6.12 A Case of Aviation System Reliability

This section discusses a case of the system reliability evaluation for a general aviation aircraft [3]. The airplane is decomposed into 6 subsystems: Airframe, Powerplant, Flight control, Ground control, Cockpit instrumentation, Electrical. The airframe is the mechanical structure of the aircraft. Powerplant system refers to the airplane engine and propeller to produce thrust. The flight control system controls the fly direction, fly attitude and the forces of flight. Flight control often contains redundancy, one is primary, and one is secondary since it is critical to aircraft. The ground control system is the system installed on the aircraft that controls the airplane's behaviors on the ground such as the landing gear, ground steering system [3]. Cockpit instrumentation is the instrument in the cockpit to provide the pilot with the flight condition, such as speed, altitude etc. Electrical system is the lighting system and other components involved in the source and distribution of electrical power [3]. In this example case, as no available data, the ground control and cockpit instrumentation are out of consideration.

The distributions of airframe components are shown in Table 6.7.

The reliability function for Weibull distribution is

Table 6.7 Airframe (Distribution data adapted from [3])

Airframe Subsystem	Distribution	Beta	Alpha	$R(t=10)$
Electrostatic devices	Weibull	2.53	5.89E+03	0.999999901913716
Empennage	Weibull	1.16	5.03E+03	0.999265449628180
Engine box and cabin fuselage	Weibull	1.42	6.28 + 03	0.999893616517257
Exterior coatings	Weibull	1.45	2.99 + 03	0.999742829631535
Seats	Weibull	2.66	6.77E+03	0.999999970444784
Upholstery	Weibull	1.79	4.29E+03	0.999980595733378
Wing	Weibull	1.79	4.25E+03	0.999980267617276

Table 6.8 Flight control system (Distribution data adapted from [3])

Flight Control	Distribution	Beta	Alpha	$R(t=10)$
Directional	Weibull	1.85	4729	0.999988736475107
Longitudinal	Weibull	1.57	4718	0.999936585836351
Lateral	Weibull	2.25	5843	0.999999404243038
Flaps	Weibull	0.95	3956	0.996596964172298
Trim	Weibull	0.73	2672	0.983221993367545
Hydraulic	Weibull	1.14	3977	0.998912897591354
LG	Weibull	0.92	2896	0.994580417727095
Steering	Weibull	1.65	3994	0.999948988790270

$$R(t) = exp^{-\left(\frac{t}{\alpha}\right)^{\beta}} \tag{6.53}$$

For 10 hours operation, the reliability for each component is presented in the last column of the Table. Consider the components are connected in series for reliability, according to Eq. (6.7),

$$\begin{aligned} R_s = R_1 R_2 \ldots R_n = & 0.999999901913716 * 0.999265449628180 \\ & * 0.999893616517257 * 0.999742829631535 * 0.999999970444784 \\ & * 0.999980595733378 * 0.999980267617276 = 0.9989 \end{aligned}$$

The system reliability for the airframe is then 0.9989.

The distributions of flight control system components are shown in Table 6.8. The system reliability for flight control system is

$$\begin{aligned} R_s = R_1 R_2 \ldots R_n = & 0.999988736475107 * 0.999936585836351 \\ & * 0.999999404243038 * 0.996596964172298 * 0.983221993367545 \\ & * 0.998912897591354 * 0.994580417727095 * 0.999948988790270 = 0.9734 \end{aligned}$$

The system reliability for the flight control is 0.9734.

The distributions of electrical system components are shown in Table 6.9.

Table 6.9 Electrical system (Distribution Data adapted from [3])

Electrical	Distribution	Beta	Alpha	$R(t = 10)$
Lighting	Weibull	1.66	5610	0.999972665211712
Source and distribution	Weibull	1.67	4950	0.999968377044941

Table 6.10 Powerplant system (Distribution Data adapted from [3])

Powerplant	Distribution	Beta	Alpha	$R(t = 6)$
Engine	Weibull	1.58	4830	0.999942541927066
Fuel	Weibull	1.44	5130	0.999874857790511
Heating and ventilation	Weibull	1.60	4190	0.999936255352934
Propeller	Weibull	1.63	3740	0.999935996254477

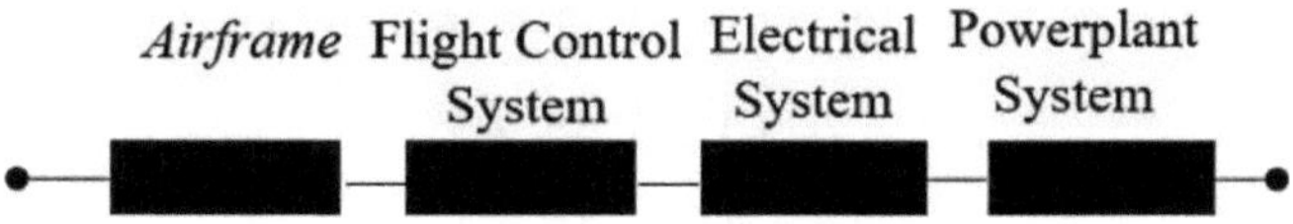

Fig. 6.26 RBD of aircraft system

$$R_s = R_1 R_2 \ldots R_n = 0.999972665211712 * 0.999968377044941 = 0.9719$$

The system reliability for the electrical system is 0.9999.

The distributions of powerplant system components are shown in Table 6.10.

$$R_s = R_1 R_2 \ldots R_n = 0.999942541927066 * 0.999874857790511 * 0.999936255352934 * 0.999935996254477 = 0.9997$$

The reliability for powerplant system is 0.9997.

For the whole system, they are connected in series as shown in Fig. 6.26.

The whole aircraft system reliability is then

$$R_s = R_1 R_2 R_3 R_4 = 0.9989 * 0.9734 * 0.9999 * 0.9997 = 0.9719$$

The analysis demonstrates an aircraft case study where we can see how system reliability decreases with the complex of the system. At component level, the reliability reaches as high as 0.999999901913716 for example for the Electrostatic Devices in the airframe. The subsystem for the airframe can still reach 0.9989 where 6 components are contained. While for the whole system, the reliability decreases to 0.9719.

Appendix 6A: Boolean Law

The Boolean law is shown in following. It presents the union as both ∪ or +, the intersection as both $A \cap B$ and AB

$$A \cup B = B \cup A; A + B = B + A$$

$$A \cap B = B \cap A; AB = BA$$

$$A \cap (B \cap C) = (A \cap B) \cup C; A(BC) = A(BC)$$

$$A \cup (B \cup C) = (A \cup B) \cup C; A + (B + C) = (A + B) + C$$

$$A \cap (B \cup C) = (A \cap B) \cup (A \cap C); A(B + C) = AB + AC$$

$$A \cup (B \cap C) = (A \cup B) \cap (A \cup C); A + (BC) = (A + B)(A + C)$$

$$A \cap A = A; AA = A$$

$$\overline{A \cap B} = \overline{A} \cup \overline{B}; \overline{AB} = \overline{A} + \overline{B}$$

$$\overline{A \cup B} = \overline{A} \cap \overline{B}; \overline{A + B} = \overline{AB}$$

$$A \cup (A \cap B) = A; A + (AB) = A$$

Appendix 6B: Matlab Code for System Reliability Simulation

```
%-----------------------------------------
nSim = 50000;
%-----------------------------------------
pCA = 0.9; %Component 1 Reliablity
pCB = 0.7; %Component 2 Reliablity
pCC = 0.9; %Component 3 Reliablity
pCD = 0.9; %Component 3 Reliablity
pCG = 0.9; %Component 3 Reliablity
nStateA = -1; %Component 1 Status 0 or 1
nStateB = -1; %Component 1 Status 0 or 1
nStateC = -1; %Component 1 Status 0 or 1
nStateD = -1; %Component 1 Status 0 or 1
nStateG = -1; %Component 1 Status 0 or 1
n = 1;
nFailure = 0; %Number of failure for the total number of iterations
while n < nSim
n = n + 1;
  rndCA = rand();
  rndCB = rand();
  rndCC = rand();
  rndCD = rand();
  rndCG = rand();

  %Check the state of each component
  if rndCA > pCA
    nStateA = 0; %Failed
  else
    nStateA = 1;
  end

  if rndCB > pCB
    nStateB = 0; %Failed
  else
    nStateB = 1;
  end

  if rndCC > pCC
    nStateC = 0; %Failed
  else
    nStateC = 1;
  end

  if rndCD > pCD
    nStateD = 0; %Failed
  else
    nStateD = 1;
  end

  if rndCG > pCG
    nStateG = 0; %Failed
  else
    nStateG = 1;
  end
```

```
if (nStateA == 0 && nStateB == 0 && nStateC == 0  && nStateD == 0  && nStateG == 0 ) %Fail 1
   nFailure = nFailure + 1;
elseif (nStateA == 0 && nStateB == 0 && nStateC == 0  && nStateD == 1  && nStateG == 0)%Fail 2
   nFailure = nFailure + 1;
elseif (nStateA == 0 && nStateB == 0 && nStateC == 1  && nStateD == 0  && nStateG == 0)%Fail 3
   nFailure = nFailure + 1;
elseif (nStateA == 0 && nStateB == 1 && nStateC == 0  && nStateD == 0  && nStateG == 0) %5
         nFailure = nFailure + 1;
elseif (nStateA == 0 && nStateB == 1 && nStateC == 0  && nStateD == 1  && nStateG == 0) %6
   nFailure = nFailure + 1;
elseif (nStateA == 0 && nStateB == 1 && nStateC == 1  && nStateD == 0  && nStateG == 0) %7
         nFailure = nFailure + 1;
elseif (nStateA == 1 && nStateB == 0 && nStateC == 0  && nStateD == 0  && nStateG == 0)%9
   nFailure = nFailure + 1;
elseif (nStateA == 1 && nStateB == 0 && nStateC == 0  && nStateD == 1  && nStateG == 0) %10
      nFailure = nFailure + 1;
elseif (nStateA == 1 && nStateB == 0 && nStateC == 1  && nStateD == 0  && nStateG == 0)%11
      nFailure = nFailure + 1;
elseif (nStateA == 0 && nStateB == 0 && nStateC == 0  && nStateD == 0  && nStateG == 1)%17
      nFailure = nFailure + 1;
elseif (nStateA == 0 && nStateB == 0 && nStateC == 0  && nStateD == 1  && nStateG == 1)%18
      nFailure = nFailure + 1;
elseif (nStateA == 0 && nStateB == 0 && nStateC == 1  && nStateD == 0  && nStateG == 1)%19
      nFailure = nFailure + 1;
elseif (nStateA == 0 && nStateB == 1 && nStateC == 0  && nStateD == 0  && nStateG == 1)%21
      nFailure = nFailure + 1;
elseif (nStateA == 0 && nStateB == 1 && nStateC == 0  && nStateD == 1  && nStateG == 1)%22
      nFailure = nFailure + 1;
elseif (nStateA == 1 && nStateB == 0 && nStateC == 0  && nStateD == 0  && nStateG == 1)%25
      nFailure = nFailure + 1;
elseif (nStateA == 1 && nStateB == 0 && nStateC == 1  && nStateD == 0  && nStateG == 1)%27
      nFailure = nFailure + 1;
  end
end
RSys = 1- (nFailure / nSim); %System Reliability
```

Problems and Questions

1. The series system model is

   a. Simplest of any system model
   b. Lowest reliability of any system model
   c. Lest total number of components of any system model
   d. All of the above

2. If all the subsystems in a series system have a constant failure rate, then

   a. The failure rate of the whole system is constant
   b. The failure rate of the whole system will increase if more subsystems will be added
   c. The failure rate of the system is the sum of the subsystem failure rates.
   d. All of the above

3. Regarding a complex system's reliability, which **one** is **correct**?
   a. System reliability will increase when a new component has been installed in parallel.
   b. System reliability will increase when a new component has been installed in series.
   c. Failure rate of a system's reliability will decrease when a new component has been installed in series.

4. Which one of the following approaches can improve the reliability?
   a. Remove the component with lowest reliability in the parallel system.
   b. Remove the component with highest reliability in the parallel system.
   c. Remove the component with highest reliability in the series system.
   d. Remove the cold standby system.

5. Regarding the system reliability for a series system, if each unit follows Exponential distribution, which **one is not** correct?
   a. The failure rate of the whole system reliability is also constant.
   b. The failure time of the whole system also follows Exponential distribution.
   c. The failure rate of the whole system will be lower than its constituent unit's failure rate.
   d. The failure rate of the whole system is a summation of its constituent unit's failure rate

6. Which one of the following approaches can improve the reliability?
   a. Add a component with very high reliability level in the series system.
   b. Add a component with very low reliability in the parallel system.
   c. Remove the cold standby system.
   d. None of above

7. Five subsystems have the following RBD Fig. 6.27. Please calculate the system reliability, when the reliability of A and B is 0.9, reliability of C and D is 0.99, reliability of the E is 0.95?
8. Engineers are designing a new system. There are two alternatives under consideration. The first alternative is with the two subsystems connected in parallel.

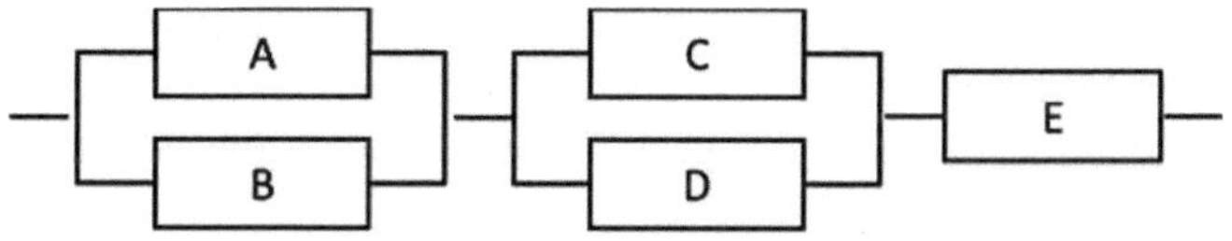

Fig. 6.27 RBD of Question 7

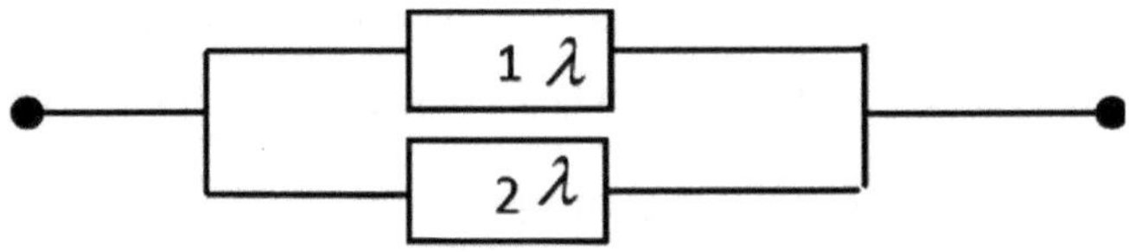

Fig. 6.28 RBD of Question 8 without switch

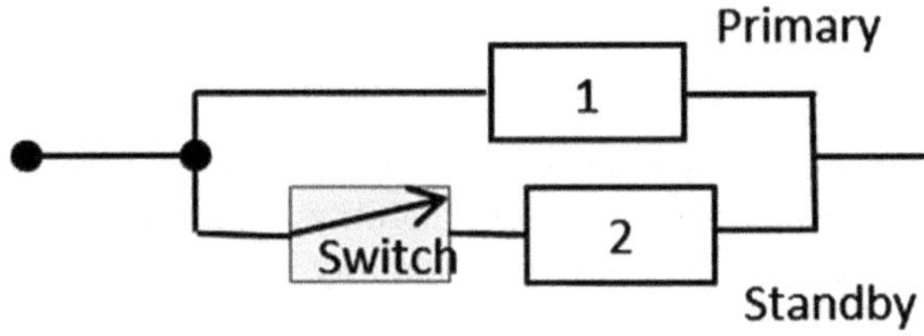

Fig. 6.29 RBD of Question 8 with switch

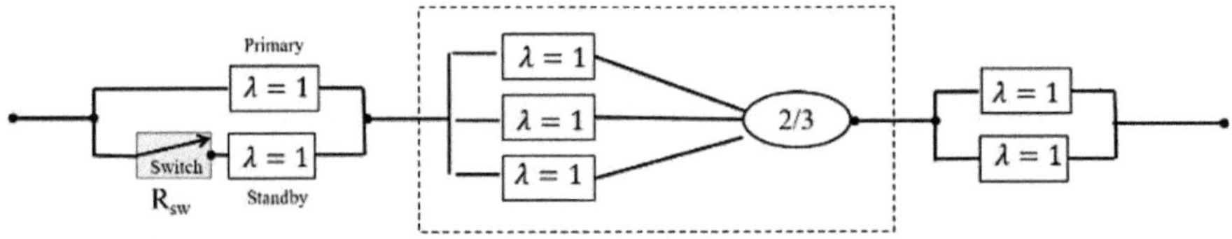

Fig. 6.30 RBD of Question 9

Fig. 6.31 RBD of Question 10

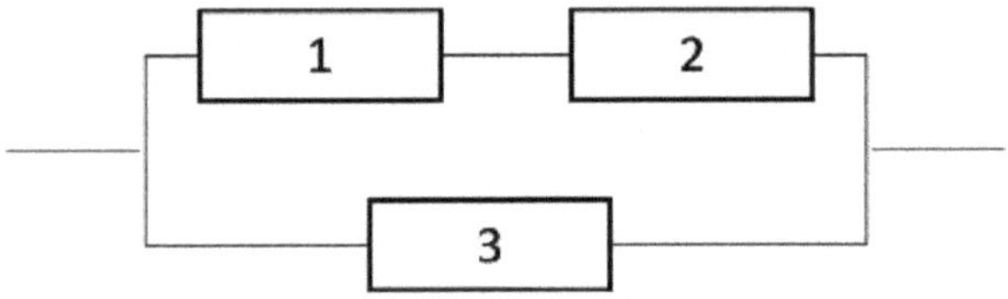

The failure rate of each subsystem is equal with $\lambda = 0.1$ per hour as shown in Fig. 6.28.

The second alternative connects the subsystems as a cold standby system as shown in Fig. 6.29. However, the subsystems are with higher failure rate $\lambda = 0.2$ per hour. The reliability of Switch is estimated as 0.95.

For a service time of 10 hours. Which alternative has higher reliability?

9. The reliability configuration of a system is as following Fig 6.30. The second diagram is a 2 out 3 system, i.e. at least 2 out of 3 should work so the whole system can work.

   Suppose the service time is 1 year. The reliability of the switch is $R_{sw} = 0.98$. What is the reliability?
10. A reliability diagram block is as following Fig 6.31.

Unit 1 and 2 are identical following Weibull distribution with

$$\alpha = 100\ days; \beta = 1.3$$

Unit 3 follows Exponential distribution with

$$\lambda = 0.003 \text{ per day}$$

a. What is the reliability for the whole system for 100 days.
b. Calculate the importance measure for each unit for the 100 days.

References

1. O'Connor PP, Kleyner A (2011) Practical reliability engineering, 5th edn. Wiley, Oxford
2. U. Department of Defense (1991) MIL-HDBK-217F reliability prediction of electronic equipment, Department of Defense, USA, pp 1–13
3. Pettit D, Turnbull A (2001) General aviation aircraft reliability study. NASA, Hampton

Chapter 7
Reliability Evaluation from Limit-State Function

Reliability evaluation we have discussed before has been focusing on methods of using event data analysis, mainly by using failure data to find out the reliability level. This method is essentially analyzing reliability from management perspective and statistic perspective, while less from the engineering perspective. Data from fields contain noise, outlier, inaccuracy. The resultant evaluation accuracy is thus compromised. With the growing concern of reliability issue for the modern equipment's such as cellphone, smart home, autonomous drone, the reliability evaluation has gone beyond the original domain. In the industries, reliability evaluation from field data might be too rough in its accuracy or infeasible as it often lacks data. There is another domain of reliability analysis which analyzes the probabilistic behavior of the system, where the reliability level is evaluated from a probabilistic analysis, for example, evaluating reliability from thermal models where some parameters and measurements are uncertain for electronic units; evaluating the probability of bridge collapsing by mechanical analysis where operating condition is uncertain in structural engineering. This kind of analysis is also named probabilistic design. This Chapter addresses this method. We start from the simplest stress-strength analysis then to the more general limit-state analysis.

7.1 Stress-Strength Analysis

It is well-known that the stress applied to an engineering system can lead to failure when it exceeds the system's strength limit. The term "stress" is general and can refer to any stress such as force per unit area, voltage, or current per unit area. System design should ensure that the strength is greater than the stress. In a deterministic context, where both stress and strength are represented by deterministic numerical values, the system is considered safe if

F. Yuan, *Essentials of Reliability Engineering*, Springer Series in Reliability Engineering, https://doi.org/10.1007/978-3-032-15544-3_7

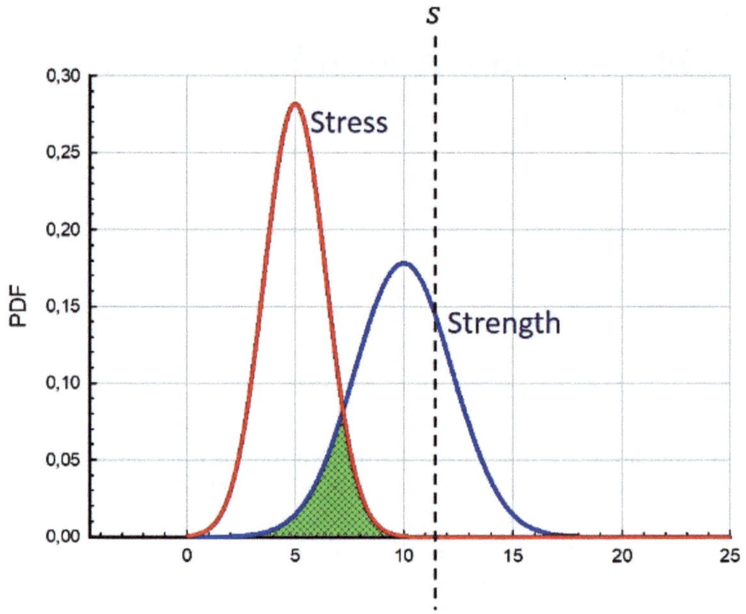

Fig. 7.1 Stress and strength are random

$$Stress \leq Strength$$

For example, if the strength of a support is $1000N/m^2$, the stress applied is $900N/m^2$. The support is then considered safe. The above formula can be generalized to be probabilistic to let both stress and strength be random variable, or anyone, either stress or strength be random variable. The probability of failure will become random accordingly. Figure 7.1 shows an example where both stress and strength follow Normal distribution. In random cases, evaluating failure is more complex than simply comparing the stress and strength value.

The overlap area shown as shadow in Fig. 7.1 are undesirable. Less size of this area can have higher reliability. We can examine it in an analytical way. Suppose the PDF function of stress is f_{stress}, strength is $f_{strength}$. The probability of failure is then

$$P(Stress \leq Strength) \tag{7.1}$$

Given any stress value s, the strength then should be $\geq s$, corresponding to a probability as

$$f_{stress}(s)\int_{s}^{+\infty} f_{strength}(\tau)d\tau \tag{7.2}$$

If the s is positive valued $0 \leq s \leq +\infty$, the overall probability, corresponding to reliability is then

$$R = \int_0^{+\infty} f_{stress}(s) \int_s^{+\infty} f_{strength}(\tau) d\tau ds \tag{7.3}$$

Therefore, if the PDF of f_{stress} and $f_{strength}$ is specified, a numerical value of reliability can be obtained. But the integration computation in Eq. (7.3) is a challenge, as it might have no closed form solution. But fortunately, if both stress and strength follow Normal distribution, there is closed-form solution. Rewrite Eq. (7.1) as

$$Diff = Strength - Stress \tag{7.4}$$

The $Diff \geq 0$ corresponds to safety, otherwise failure. If the strength and stress are Normal distribution, the *Diff* is also Normal distributed. It can be proved from a general case as

$$y = a_0 + a_1 x_1 + a_2 x_2 + a_3 x_3 + \ldots \tag{7.5}$$

The x_i is random number of Normal distributions, a_i are constant coefficients. The mean and variance of random x_i are μ_i and σ_i^2 respectively. The y is then a random number following Normal distribution with mean and variance as follows. The proof of it can refer to Appendix 7A.

$$\begin{aligned} \mu_y &= a_0 + a_1\mu_1 + a_2\mu_2 + a_3\mu_3 + \ldots \\ \sigma_y^2 &= a_1^2\mu_1^2 + a_2^2\mu_2^2 + a_3^2\mu_3^2 + \ldots \end{aligned} \tag{7.6}$$

Apply Eq. (7.6) to Eq. (7.4), where $a_0 = 0$, $a_1 = 1$, $a_2 = -1$, then

$$\mu_{Diff} = \mu_{Strength} - \mu_{Stress} \tag{7.7}$$

$$\sigma_{Diff}^2 = \mu_{Strength}^2 + \mu_{Stress}^2 \tag{7.8}$$

The probability of failure corresponds to the area of Diff value less than 0, reliability is the area larger than 0 in the PDF of Diff distribution, as follows

$$\begin{cases} P(z_{dif} < 0); Failure \\ P(z_{dif} \geq 0); Reliability \end{cases} \tag{7.9}$$

The failure probability and reliability are demonstrated in Fig. 7.2.

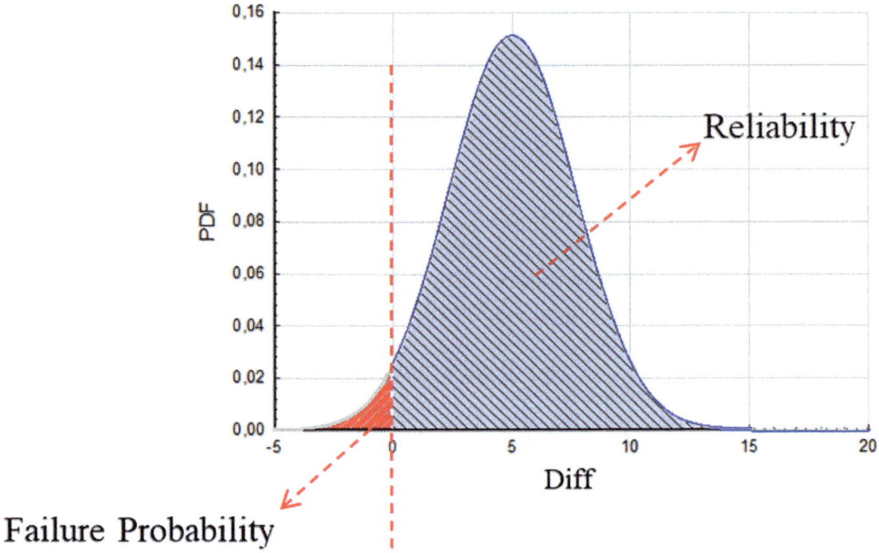

Fig. 7.2 PDF of Diff distribution

Example 7.1
A mechanical system has strength with Normal Distribution as (1 Pa = 1N/m^2).

$$\mu_{Strength} = 5000 \text{ Pa}; \sigma_{Strength} = 300 \text{ Pa}$$

Stress is also Normal Distribution as.

$$\mu_{Stress} = 4000 \text{ Pa}; \sigma_{Stress} = 400 \text{ Pa}$$

What is the failure probability and what is the reliability?

Solution: Since both strength and stress are normally distributed, then (Strength—Stress) is also Normal distributed with

$$\mu_{dif} = \mu_{Strength} - \mu_{Stress}$$

$$\sigma_{Dif} = \sqrt{\sigma^2_{Strength} + \sigma^2_{Stress}}$$

$$\mu_{Diff} = 5000 - 4000 = 1000 \text{ Pa}$$

(continued)

Example 7.1 (continued)

$$\sigma_{Diff} = \sqrt{300^2 + 400^2} = 500 \text{ Pa}$$

Convert the 0 to the z-score in a standard Normal distribution to obtain the Probability of Diff less than 0. We have.

$$z = \left(0 - \mu_{dif}\right)/\sigma_{Dif} = (-1000/500) = -2$$

$$\Phi(-2) = 0.023$$

Failure probability is then 0.023 and reliability = 1–0.023 = 0.977. The $\Phi(.)$ stands for the CDF of standard Normal distribution with mean 0 and variance 1.

In Example 7.1, the standardization on the Diff = 0 is to obtain the probability from standard Normal table. The standardization is not necessary when software such as Matlab is used to obtain the probability.

7.2 General Case of Stress-Strength Analysis

Example 7.1 shows above is the simplest case. We can write the strength and stress in a more general way as

$$\text{G(Strength, Stress)} = Strength - Stress \tag{7.10}$$

When $\text{G}(x_1, x_2) < 0$, system fails; otherwise, it is safe as $\text{G}(x_1, x_2) \geq 0$. The $\text{G}(x_1, x_2)$ is defined as safety function or limit-state function, also as safety margin. The G is a general function which can be linear or nonlinear. Figure 7.3 demonstrates an example of linear safety function where the strength and stress follow Normal distribution. The above discussed strength-stress analysis is a special case of the G (x_1, x_2).

The probability of failure is.

$$P_F = P(Strength - Stress < 0) = P(G < 0) \tag{7.11}$$

If the distribution of G can be known as f_M as Normal distribution with mean μ_M and standard deviation σ_M, the probability of failure is then

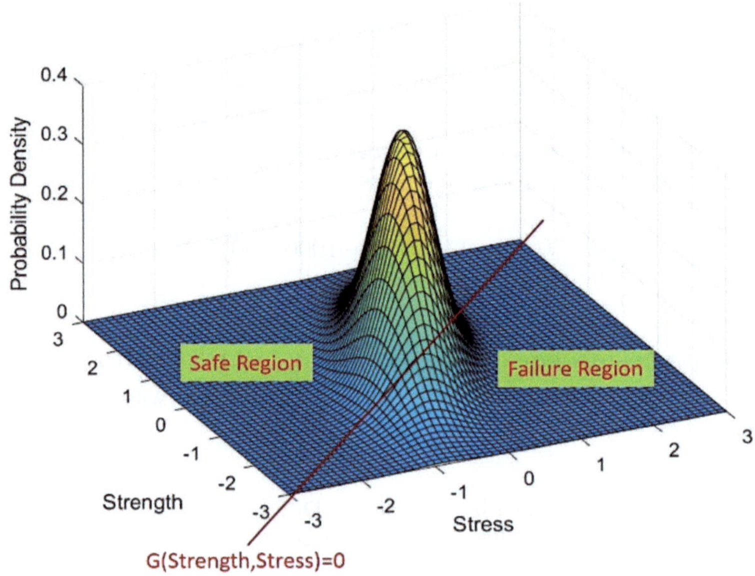

Fig. 7.3 Linear safety function

Fig. 7.4 Reliability index in standard normal distribution

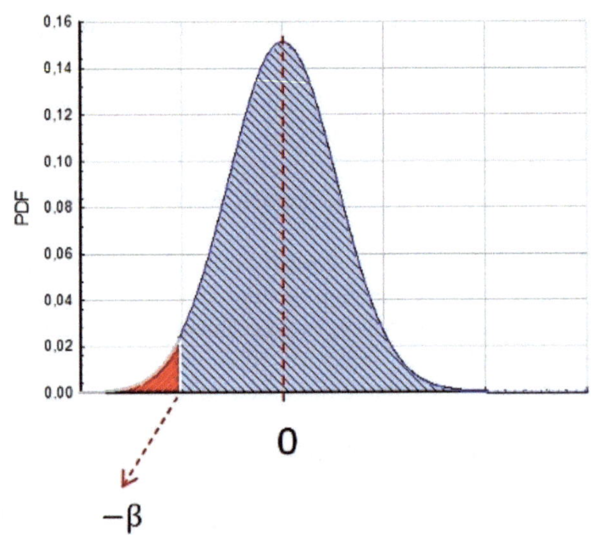

$$P_F = \int_{-\infty}^{0} f_G dG = \Phi\left(\frac{0-\mu_G}{\sigma_G}\right) = \Phi\left(\frac{-\mu_G}{\sigma_G}\right) = \Phi(-\beta) \tag{7.12}$$

The Φ(.) is the CDF of standard Normal distribution. This case is exact the stress-strength case. The β is defined as the reliability index. Fig. 7.4 shows the reliability

β	P_F
0	0.5
1	0.158655
2	0.02275
3	0.00135
4	3.17E-05
5	2.87E-07
6	9.87E-10
7	1.28E-12
8	6.22E-16

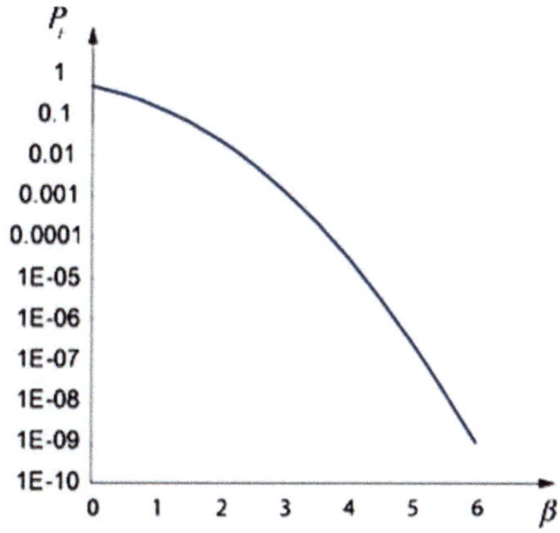

Fig. 7.5 Reliability index in standard normal distribution

index in the standard Normal distribution plot. The area less than the index corresponds to probability of failure. Larger β means to move the β left further, leading to smaller probability of failure. That means the larger β implies higher reliability level. This is why it is called reliability index.

Figure 7.5 plots a figure of probability of failure corresponding to β. It shows the probability of failure is decreasing with increasing β. Higher index implies the higher reliability. The table along with Fig. 7.5 shows when reliability index is 3, the probability of failure is 0.00135, the reliability is then 1–0.00135 = 0.99856, which is a very high reliability level.

We go further to a more general case. The distribution of margin G is not able to be specified but as an analytical joint PDF of the parameters x_i, where the stress and strength in the above example can be two of these parameters.

$$R = P(G(x_1, x_2, \ldots) > 0) = \int_{G(x_1, x_2, \ldots) > 0} f(x_1, x_2, \ldots) dx_1 dx_2 \ldots . \tag{7.13}$$

Each parameter is a random number following certain distribution, which might not be Normal distribution. To illustrate Eq. (7.13), Fig. 7.6 illustrates the contour of two joint Normal random variables x_1, x_2. The $G(x_1, x_2) = (x_2 + 3)^2 - 2 - x_1$. The shadowed area corresponds to $G(x_1, x_2) < 0$, i.e. the probability of failure, the rest area corresponds to safety.

Reliability is determined by obtaining the numerical values of Eq. (7.13) through integration over the region where $G(x_1, x_2) > 0$. In most cases, this integration has not closed-form solution. The following section will discuss methods for solving this problem.

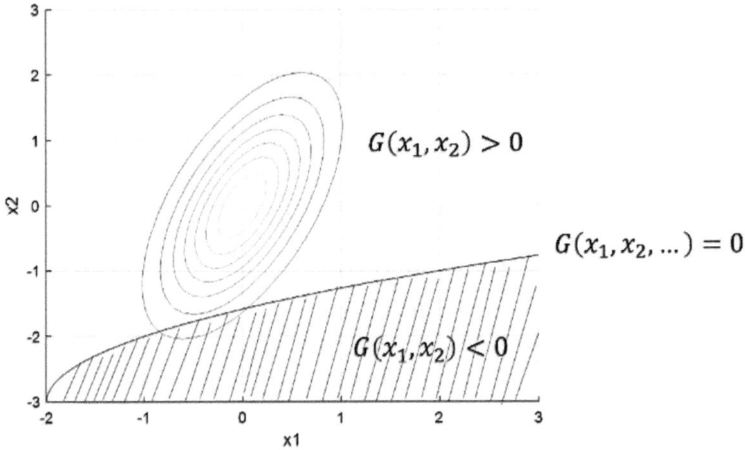

Fig. 7.6 Reliability region for nonlinear safety function

Example 7.2
Using safety function method to obtain the reliability for the question in Example 7.1.

Solution: The safety function for the strength-stress analysis is G(Strength, Stress) = *Strength* − *Stress*. According to Eq. (7.5), the linear function of *M* is Normal distribution with mean and standard deviation

$$\mu_G = \mu_{x_1} - \mu_{x_2}$$

$$\sigma_G = \sqrt{\sigma_{x_1}^2 + \sigma_{x_2}^2}$$

$$\mu_G = 5000 - 4000 = 1000pa$$

$$\sigma_G = \sqrt{400^2 + 300^2} = 500pa$$

According to Eq. (7.12), the reliability index is then $\beta = \frac{\mu_G}{\sigma_G} = \frac{1000}{500} = 2$.

$$P = \Phi(-\beta) = \Phi(-2) = 0.023$$

The reliability is then 1–0.023 = 0.977.

7.3 Reliability Approximating

Obtaining a solution from Eq. (7.13) is challenging and it is normally no closed form solution. If the parameters x_i is mutually independent and normal distributed, and the safety function can be rewritten in a linear function form, the safety margin G is then also normally distributed, according to Eq. (7.5). The strength-stress analysis is an example of it. However, the safety function is not linear in most applications. One solution is to approximate the safety function by a linear function, though this comes at the cost of some accuracy.

For a general safety function $G(x_1, x_2, \ldots)$, we approximate it to a linear function according to Eq. (A.7), supposing the expansion point is at $(x_{10}, x_{20}, \ldots, x_{n0})$. The approximation is as follows:

$$G(x_1, x_2, \ldots, x_n) \approx G(x_{10}, x_{20}, \ldots, x_{n0}) + \sum\nolimits_{j=1}^{n} \left.\frac{\partial G(x_{10}, x_{20}, \ldots, x_{n0})}{\partial x_j}\right|_{x_{j0}} \frac{(x_j - x_{j0})}{1!} \tag{7.14}$$

It can be written as

$$\tilde{G}(x_1, x_2, \ldots, x_n) = G(x_{10}, x_{20}, \ldots, x_{n0}) - \sum_{j=1}^{n} \left.\frac{\partial G}{\partial x_j}\right|_{x_{j0}} x_{j0} + \sum_{j=1}^{n} \left.\frac{\partial G}{\partial x_j}\right|_{x_{j0}} x_j \tag{7.15}$$

Let the mean of random variables x_1, x_2, …, x_n be μ_1, μ_2, …, μ_n, according to Eq. (7.6), the mean of $G(x_1, x_2, \ldots, x_n)$ is then

$$G(x_{10}, x_{20}, \ldots, x_{n0}) - \sum_{j=1}^{n} \left.\frac{\partial G}{\partial x_j}\right|_{x_{j0}} x_{j0} + \sum_{j=1}^{n} \left.\frac{\partial G}{\partial x_j}\right|_{x_{j0}} \mu_j \tag{7.16}$$

Variance of $G(x_1, x_2, \ldots, x_n)$ is

$$\sigma_G^2 = \sum\nolimits_{i=1}^{n} \sum\nolimits_{j=1}^{n} \frac{\partial G}{\partial x_i} \frac{\partial G}{\partial x_j} Cov(x_i, x_j) \tag{7.17}$$

where $Cov(x_i, x_j)$ is the covariance of x_i *and* x_j. If x_i, x_j are independent from each other, Eq. (7.17) is simplified as

$$\sigma_G^2 = \sum\nolimits_{i=1}^{n} \left(\frac{\partial G}{\partial x_i}\right)^2 \sigma_i^2 \tag{7.18}$$

where the σ_i^2 is the variance of random variable x_i.

Table 7.1 Parameter distribution for a mechanical system

Parameters	P	L	W	T
Mean	8kN	9 m	$0.0006 m^3$	$80,000 kN/m^2$
Standard deviation	1kN	0.5 m	$0.00009 m^3$	$20,000 kN/m^2$

In some cases, the mean of random variables $x_1, x_2, \ldots, x_n$, i.e. $\mu_1, \mu_2, \ldots, \mu_n$ can be used as the expansion point $(x_{10}, x_{20}, \ldots, x_{n0})$. According to Eq. (7.16), the mean of the safety function then becomes as

$$\mu_G \approx G(\mu_1, \mu_2, \ldots, \mu_n), \tag{7.19}$$

while the variance remains as Eq. (7.18). The reliability index is thus

$$\beta = \frac{\mu_G}{\sigma_G} = \frac{G(\mu_1, \mu_2, \ldots, \mu_n)}{\sqrt{\sum_{i=1}^{n} \left(\frac{\partial G}{\partial x_i}\right)^2 \sigma_i^2}} \tag{7.20}$$

The probability of failure can be obtained from CDF of standard Normal distribution $\Phi(-\beta)$. The reliability is then

$$1 - \Phi(-\beta) \tag{7.21}$$

Example 7.3

A mechanical system is subject to a central load P. The beam will fail if the load excesses the limit. The safety function is defined as follows with four parameters P, L, W, T.

$$G(P, L, W, T) = WT - PL/4$$

All the parameters P, L, W, T is random with Normal distribution whose distribution parameter values are shown in the following Table 7.1. We ignore the physical background of the parameters as it is not the concern of this book. One can refer [1, 2] for detail. Evaluate the reliability of the system.

Solution: According to Eq. (7.19), the mean of the safety function is

$$\mu_G \approx G(P, L, W, T) = WT - \frac{PL}{4} = 0.0006 \times 80{,}000 - 8 \times \frac{9}{4} = 48 - 18$$
$$= 30kNm$$

The first derivative of the parameter is

(continued)

Example 7.3 (continued)

$$\frac{\partial G}{\partial T} = W = 0.0006;$$

$$\frac{\partial G}{\partial W} = T = 80{,}000;$$

$$\frac{\partial G}{\partial P} = -\frac{L}{4} = -\frac{9}{4} = -2.25;$$

$$\frac{\partial G}{\partial L} = -\frac{P}{4} = -\frac{8}{4} = -2;$$

According to Eq. (7.18), the standard deviation of the safety function is

$$\sigma_G = \left[\sum_{i=1}^{n} \left(\frac{\partial G}{\partial x_i}\right)^2 \sigma_{x_i}^2\right]^{\frac{1}{2}}$$

$$= \sqrt{(0.0006 \times 20{,}000)^2 + (80{,}000 \times 0.00009)^2 + (-2.25 \times 1)^2 + (-2 \times 0.5)^2}$$

$$= 14.2092$$

Use Eq. (7.20) to obtain the reliability index as $\beta = \frac{\mu_G}{\sigma_G} = \frac{30}{14.2092} = 2.1113$.The probability of failure is $P_f = \Phi(-2.1113) = 0.0174$. Reliability is

$$1 - P_f = 1 - 0.0174 = 0.9826$$

7.4 First Order Reliability Method

The linear approximation discussed above using only the first order of the Taylor expansion. The reliability evaluation method using the first order of Taylor expansion is named the First Order Reliability Method (FORM). In the previous section, the mean of the random variables was used as the expansion point and uses only the first order, which means it is a FORM. While the method using the mean of the parameters for the Taylor expansion is simple and relatively rough. This Section describes an advanced method by using an optimal expansion point to evaluate the reliability. Before discussing the method, we would like to discuss a method to normalize distributions, for the purpose of simplifying the calculation.

7.4.1 Normalize Normal Distribution

For multivariate Normal distribution, for example two dimensions multivariate Normal distribution, the joint PDF function presents an ellipse shape contour, as shown in Fig. 7.7. Suppose the safety function is

$$G(x, y) = x^2 - (y + 4)^3 \tag{7.22}$$

The x and y are independent from each other with Normal distribution of mean and standard deviation $x \sim N\,(4{,}3)$ and $y \sim N\,(0,\,2)$, the joint PDF and its contour are shown on the right of Fig. 7.7.

A standard Normal distribution can be obtained by

$$z_x = (x - \mu_x)/\sigma_x \tag{7.23}$$

The z_x is the normalized Normal distribution for variable x. Rewrite Eq. (7.23) as

$$x = \mu_x + z_x\sigma_x \tag{7.24}$$

Similarly, the normalized random variable y is

$$y = \mu_y + z_y\sigma_y \tag{7.25}$$

Substitute Eqs. (7.24) and (7.25) into Eq. (7.22). The safety function Eq. (7.22) is then rewritten as

$$G(z_x, z_y) = (\mu_x + z_x\sigma_x)^2 - (\mu_y + z_y\sigma_y + 4)^3 \tag{7.26}$$

In original non-Normalized distribution, the $G(x, y) = 0$ corresponds to as follows by using Eq. (7.22)

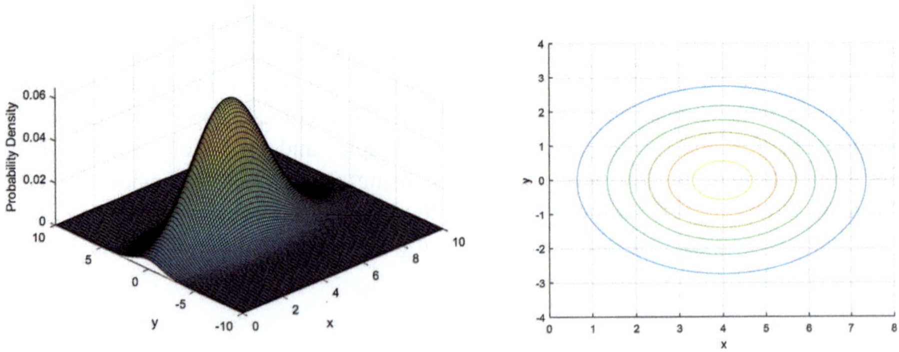

Fig. 7.7 Normal 2D distribution

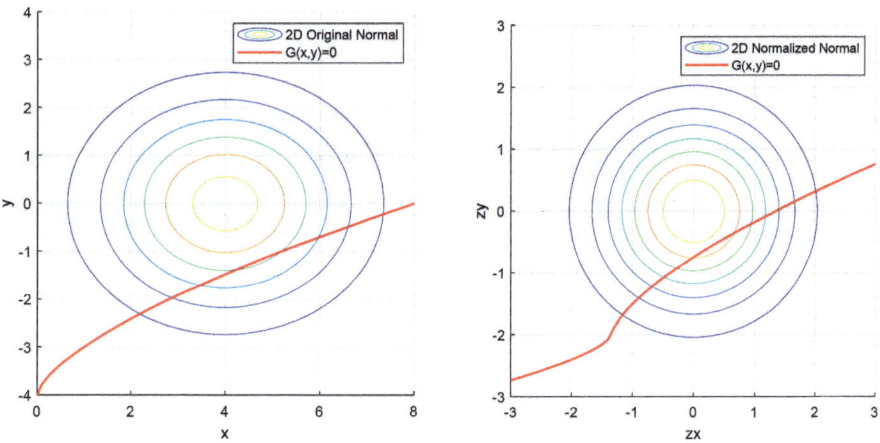

Fig. 7.8 Contour of original and normalized 2D normal distribution

$$y = x^{2/3} - 4 \tag{7.27}$$

For normalized Normal distribution, from Eq. (7.26), the $G(x, y) = 0$ corresponds to

$$z_y = \left((\mu_x + z_x \sigma_x)^{2/3} - \mu_y - 4 \right) / \sigma_y \tag{7.28}$$

The contour and the $G(x, y) = 0$ is plotted in Fig. 7.8 for both the original 2D Normal and the Normalized Normal distribution. After Normalization, the contour exhibits round circles instead of ellipses. We will discuss later to use the advantage of normalization for the reliability evaluation.

Converting the parameter x_i to standard Normal distribution is simple by just replacing the variable x_i by the $\mu_i + z_i\sigma_i$ in the safety function. The replacement and normalization will simplify the method to find the optimal solution for the calculation of reliability.

7.4.2 Select Optimal Point

The reliability calculation considers in this chapter only the normalized case. The Taylor expansion discussed in the Appendix B mentions the linear approximation is based on a point on which the linear line expands at. The expanding point for Traylor expansion can be any point but it is not hard to understand this point is better on the line $G(x, y) = 0$, as it is the boundary of safety and failure. However, the $G(x, y) = 0$ is a curve and which one point on the curve is the best? Let us firstly go to statistics and will come back to the best point later. After normalization, the joint PDF for the independent random variables z_1, z_2, … in the safety function is

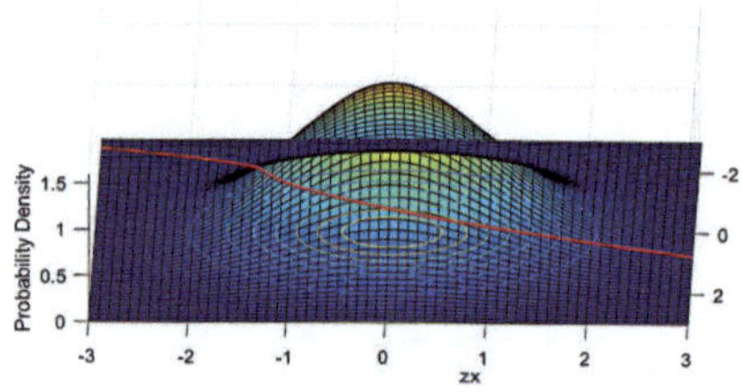

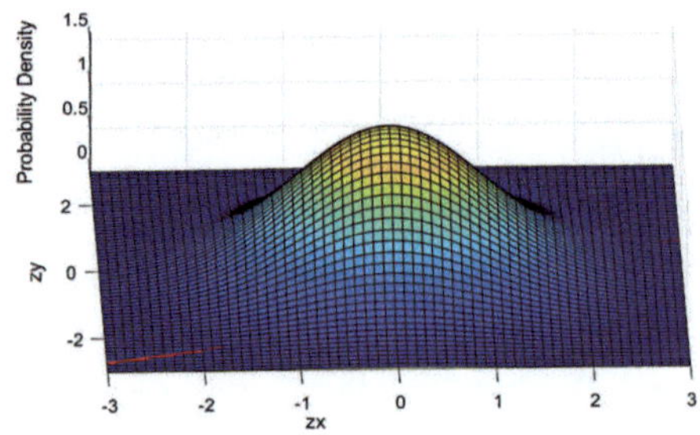

Fig. 7.9 Joint standard normal distribution

$$f(z_1, z_2, \ldots) = \prod_{i=1}^{n} \frac{1}{\sqrt{2\pi}} \exp\left(-\frac{1}{2}{z_i}^2\right) \tag{7.29}$$

Eq. (7.29) is the joint PDF of the n independent Normal distributions with PDF $\frac{1}{\sqrt{2\pi}} \exp\left(-\frac{1}{2}{z_i}^2\right)$. For a safety function $G(z_1, z_2, \ldots)$, the probability of failure is then

$$P_f = \iiint\limits_{G(z_1, z_2, \ldots) < 0} \prod_{i=1}^{n} \frac{1}{\sqrt{2\pi}} \exp\left(-\frac{1}{2}{z_i}^2\right) dz_1 dz_2 \ldots dz_n \tag{7.30}$$

Figure 7.9 plots the joint two-dimensional standard Normal distribution, where in the middle of the cone, corresponding to point (0,0) is the peak of PDF.

The PDF is decreasing away from the origin O(0,0), as shown in Fig. 7.10.

The point near the O is with higher probability than that farer from O. It is demonstrated in Fig. 7.10, where the point P0 has higher probability than P1 and P2 shown from the contour line. The safety function $G(zx, zy) = 0$ is the boundary between safety and failure. If the objective is to find a linear line to approximate the $G(zx, zy) = 0$, the P0 can be considered as optimal point as the expanding point, as it is the point with highest probability on the curve $G(zx, zy) = 0$. As shown in Fig. 7.11, the approximate function to the boundary line $G(zx, zy)$ is then the linear line L. This line L is the tangent line lying on the point P0, corresponding to the first order approximation using Taylor expansion at the point P0. The approximation will result in error as shown in the shadow area on Fig. 7.11. The shadow area is in safety region while it is accounted as failed region by the linear approximation. It demonstrates this method achieves calculation convenience at the cost of accuracy.

Therefore, the nearest point to the O is considered as the design point for linear approximation. Given any point $P(z_x, z_y)$ on the safety function, the distance from the O to the P is

$$OP = \sqrt{(z_x - 0)^2 + (z_y - 0)^2} = \sqrt{{z_x}^2 + {z_y}^2} \tag{7.31}$$

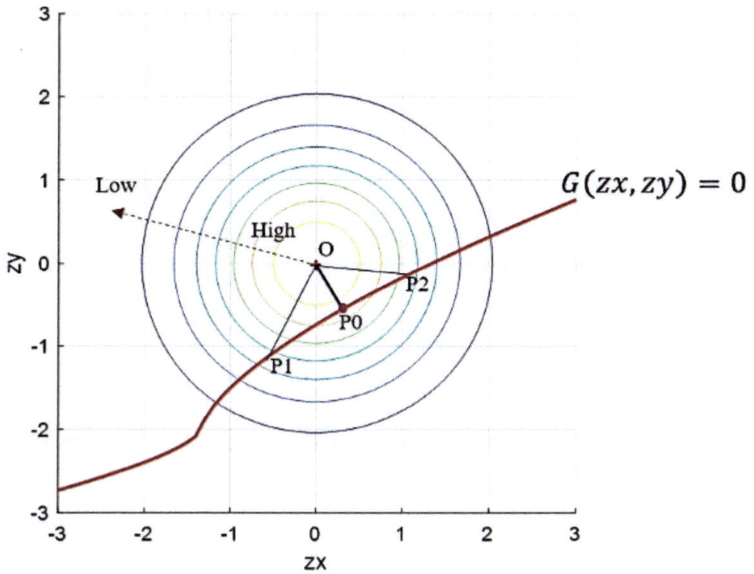

Fig. 7.10 Optimal point

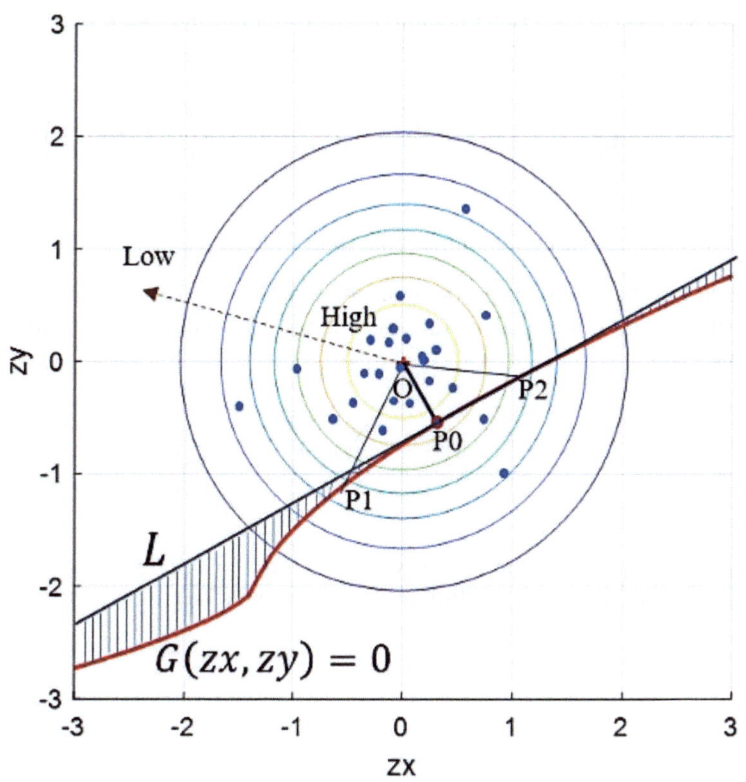

Fig. 7.11 Linear approximation from optimal point

The nearest point to the O is the minimal OP discussed in Eq. (7.31). We can write it in an optimization form as follows to find this nearest point.

$$\begin{aligned} \min OP &= \sqrt{z_x^2 + z_y^2} \\ s.t. G(z_x, z_y) &= 0 \end{aligned} \tag{7.32}$$

Since the minimal OP will have the same solution as minimal $z_x^2 + z_y^2$, Eq. (7.32) is equivalent to another problem

$$\begin{aligned} \min z_x^2 + z_y^2 \\ s.t. G(z_x, z_y) = 0 \end{aligned} \tag{7.33}$$

Equation (7.33) considers only two-dimensional cases. For a general situation where there are n-dimensional, Eq. (7.33) is rewritten as

$$\begin{aligned} \min \sum_{i=1}^{n} z_i^2 \\ s.t. G(z_1, z_2, \ldots, z_n) = 0 \end{aligned} \tag{7.34}$$

Recalling Eq. (7.29), Eq. (7.34) is equivalent to finding the maximum probability point in the joint PDF function, since

$$f(z_1, z_2, \ldots) = \prod_{i=1}^{n} \frac{1}{\sqrt{2\pi}} \exp\left(-\frac{1}{2} z_i^2\right) = \left(\frac{1}{\sqrt{2\pi}}\right)^n \exp\left(-\frac{1}{2} \sum_{i=1}^{n} z_i^2\right) \tag{7.35}$$

where $\min \sum_{i=1}^{n} z_i^2$ corresponds to maximal joint PDF $f(z_1, z_2, \ldots)$.

We therefore have found the optimal point, which is also called design point. Suppose the design point is $Z_* = (z_{1*}, z_{2*}, \ldots, z_{n*})$. According to first order Taylor expansion,

$$G(z_1, z_2, \ldots, z_n) \approx G(z_{1*}, z_{2*}, \ldots, z_{n*}) + \sum_{i=1}^{n} \frac{\partial G(z_{1*}, z_{2*}, \ldots, z_{n*})}{\partial z_i}\bigg|_{z_{i*}} \cdot \frac{(z_i - z_{i*})}{1!} \tag{7.36}$$

Since the Z_* is a point on the line of $G(z_1, z_2, \ldots, z_n) = 0$, $G(z_{1*}, z_{2*}, \ldots, z_{n*}) = 0$. Rewrite the linear Eq. (7.36) as (Referring to Appendix 7B)

$$G(z_1, z_2, \ldots, z_n) \approx \sum\nolimits_{i=1}^{n} \frac{\partial G(z_{1*}, z_{2*}, \ldots, z_{n*})}{\partial z_i}\bigg|_{z_{i*}} z_i - \sum\nolimits_{i=1}^{n} \frac{\partial G(z_{1*}, z_{2*}, \ldots, z_{n*})}{\partial z_i}\bigg|_{z_{i*}} z_{i*} \tag{7.37}$$

Let

$$a_0 = -\sum\nolimits_{i=1}^{n} \frac{\partial G(z_{1*}, z_{2*}, \ldots, z_{n*})}{\partial z_i}\bigg|_{z_{i*}} z_{i*} \tag{7.38}$$

$$a_i = \sum\nolimits_{i=1}^{n} \frac{\partial G(z_{1*}, z_{2*}, \ldots, z_{n*})}{\partial z_i}\bigg|_{z_{i*}} \tag{7.39}$$

Note that the a_0 is a constant value. Equation (7.37) is rewritten into a simple equation as

$$G(z_1, z_2, \ldots, z_n) \approx a_0 + \sum\nolimits_{i=1}^{n} a_i z_i \tag{7.40}$$

According to the Appendix 7.A and the z_i are standard Normal distribution with mean 0 and variance 1, the

$$\mu(G) = a_0 \tag{7.41}$$

$$Var(G) = \sum\nolimits_{i=1}^{n} a_i^2 \tag{7.42}$$

According to Eq. (7.20), the probability of failure is

$$P_f = \Phi\left(-\frac{\mu(G)}{\sqrt{Var(G)}}\right) = \Phi\left(-\frac{a_0}{\sqrt{\sum_{i=1}^{n} a_i^2}}\right) \tag{7.43}$$

The reliability index is then

$$\beta = \frac{a_0}{\sqrt{\sum_{i=1}^{n} a_i^2}} \tag{7.44}$$

7.4.3 Algorithm to Search Design Point

The algorithm to search design point requires some supports from optimization theory. Actually, this issue is a pure optimization problem. One can avoid this section since it is less relevant to reliability theory. To develop the algorithm to search the design point proposed in the previous section, we would like to rewrite Eq. (7.44) in a form of vector. Let

$$\boldsymbol{a} = (a_1, a_2, \ldots, a_n) = \left(\left.\frac{\partial G(z_{1*}, z_{2*}, \ldots, z_{n*})}{\partial z_1}\right|_{z_{1*}}, \left.\frac{\partial G(z_{1*}, z_{2*}, \ldots, z_{n*})}{\partial z_2}\right|_{z_{2*}}, \ldots, \left.\frac{\partial G(z_{1*}, z_{2*}, \ldots, z_{n*})}{\partial z_n}\right|_{z_{n*}} \right) \tag{7.45}$$

The bold letter $\boldsymbol{a}$ represents a vector. Let the bold $\boldsymbol{z}_*$ represents the design point vector

$$\boldsymbol{z}_* = (z_{1*}, z_{2*}, \ldots, z_{n*}) \tag{7.46}$$

Then

$$a_0 = -\sum_{i=1}^{n} \left.\frac{\partial G(z_{1*}, z_{2*}, \ldots, z_{n*})}{\partial x_i}\right|_{x_{i*}} z_{i*} = \boldsymbol{a}\boldsymbol{z}_*^{\,T} \tag{7.47}$$

$$\sum_{i=1}^{n} a_i^{\,2} = \boldsymbol{a}\boldsymbol{a}^T \tag{7.48}$$

The T in the $\boldsymbol{z}_*^{\,T}$ means the transform of row vector $\boldsymbol{z}_*$ to a column vector. The $\boldsymbol{a}\boldsymbol{z}_*^{\,T} = \boldsymbol{a}.\boldsymbol{z}_*^{\,T}$ means the **inner product** of the two vectors $\boldsymbol{a}$ and $\boldsymbol{z}_*^{\,T}$. The reliability index of Eq. (7.44) is then

$$\beta = \frac{\boldsymbol{a}\boldsymbol{z}_*^{\,T}}{\sqrt{\boldsymbol{a}\boldsymbol{a}^T}} \tag{7.49}$$

More generally, the vector can write as $\nabla G = \boldsymbol{a}$**,** where the ∇ means the vector of fist derivative of function G. Equation (7.49) is further rewritten as

$$\beta = \frac{\nabla G.\boldsymbol{z}_*^{\,T}}{\sqrt{\nabla G.\nabla G^T}} \tag{7.50}$$

From Eq. (7.50), without providing proof as it is only the basic of geometry, we can find out the absolute value of β is the distance between original O and the design point P, as shown in Fig. 7.12.

Numerical iterative method is employed to find out the design point from the optimization Formula (7.34). The updated formula for the numerical method is used as follows

$$\boldsymbol{z}_{k+1} = \frac{1}{\nabla G(\boldsymbol{z}_k)\nabla G(\boldsymbol{z}_k)^T} \left\{ \nabla G(\boldsymbol{z}_k)\boldsymbol{z}_k^{\,T} - G(\boldsymbol{z}_k) \right\} \nabla G(\boldsymbol{z}_k) \tag{7.51}$$

The $\boldsymbol{z}_{k+1}$ is the solution of the design point at iteration $k + 1$. The iteration is stopped when

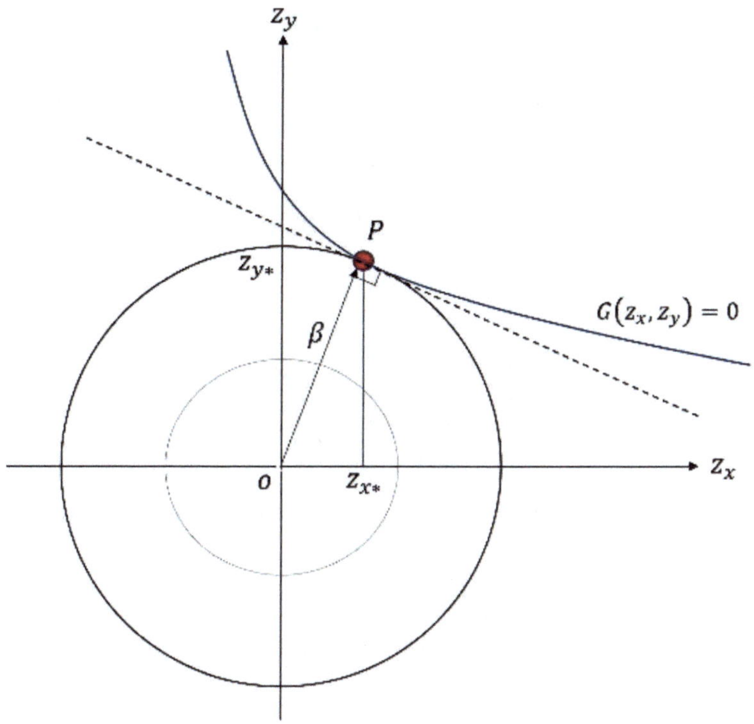

Fig. 7.12 The β value and the design point

$$|z_{k+1} - z_k| \le \varepsilon_1 \tag{7.52}$$

or

$$|\nabla G(z_{k+1}) - \nabla G(z_k)| \le \varepsilon_2 \tag{7.53}$$

One can arbitrarily choose a start point for $k = 1$, let us say z_1 as the tentative design point in the first iteration. The algorithm is named **Rackwitz Fiessler (RF) algorithm** which is widely used to solve the Eq. (7.34) [3].

Here, we present the proof of the key function (7.51) by using Lagrangian method. Firstly, we linearize the nonlinear function $G(z_k) = 0$ from the Taylor expansion, according to Eq. (7.36) as

$$G(z) \approx G(z_k) + \sum_{i=1}^{n} \frac{\partial G(z_k)}{\partial z_i}\bigg|_{z_{ik}} (z_i - z_{ik}) \tag{7.54}$$

Rewritten it as vector as

$$G(z) \approx G(z_k) + \nabla G(z_k)(z - z_k)^T \tag{7.55}$$

And $\sum_{i=1}^{n} z_i^2 = z_k z_k^T$. Lagrange function (with linearized constraint of (7.55)) of Eq. (7.34) is then

$$L(z, \lambda) = \frac{1}{2} z z^T + \lambda G(z) \tag{7.56}$$

Then

$$\begin{cases} \dfrac{\partial L(z, \lambda)}{\partial z} = z + \lambda \nabla G(z) \\ \dfrac{\partial L(z, \lambda)}{\partial \lambda} = G(z) \end{cases} \tag{7.57}$$

Substitute Eq. (7.55) into Eq. (7.57). We obtain the Kuhn-Tucker conditions

$$\begin{cases} \dfrac{\partial L(z, \lambda)}{\partial z} = z + \lambda \nabla G(z_k) = 0 \\ \dfrac{\partial L(z, \lambda)}{\partial \lambda} = G(z_k) + \nabla G(z_k)(z - z_k)^T = 0 \end{cases} \tag{7.58}$$

Let them equal to 0, we obtain

$$G(z_k) + \nabla G(z_k) z^T - \nabla G(z_k) z_k^T = 0 \tag{7.59}$$

Rearrange it as

$$\nabla G(z_k) z^T = \nabla G(z_k) z_k^T - G(z_k) \tag{7.60}$$

Transpose both side of Eq. (7.60),

$$z \nabla^T G(z_k) = z_k \nabla^T G(z_k) - G(z_k) \tag{7.61}$$

Since $z + \lambda \nabla G(z_k) = 0$, then $z = -\lambda \nabla G(z_k)$. Substitute the z into Eq. (7.61). We obtain

$$-\lambda \nabla G(z_k) \nabla^T G(z_k) = z_k \nabla^T G(z_k) - G(z_k) \tag{7.62}$$

According to Eq. (7.58),

$$z + \lambda \nabla G(z_k) = 0 \tag{7.63}$$

And rearrange Eq. (7.62), we obtain

$$\lambda = \frac{G(z_k) - z_k \nabla^T G(z_k)}{\nabla G(z_k) \nabla^T G(z_k)} \tag{7.64}$$

We obtain

$$z = \frac{\{z_k \nabla^T G(z_k) - G(z_k)\}}{\nabla G(z_k) \nabla^T G(z_k)} \nabla G(z_k) \tag{7.65}$$

The new z in Eq. (7.65) is the z_{k+1}. Equation (7.51) have thus obtained.

The geometrical explanation of the RF algorithm is illustrated in an example in Fig. 7.13. The point (x_1, y_1) are selected arbitrarily. The linear approximation of $G(x, y) = 0$ is L_1 which tangent to the point (x_1, y_1). The line perpendicular to L_1 originating from the origin O crosses the $G(x, y) = 0$ is (x_2, y_2), which is the updated point. Then plot a tangent line of (x_2, y_2), and whose perpendicular line originating from O crosses the $G(x, y) = 0$ is point (x_3, y_3). If the difference between $\beta 2$ and $\beta 3$ is less than a user-defined threshold, i.e. the ε_1 in Eq. (7.52) and ε_2 in (7.53), which means no significant difference, then stop the iteration. The point (x_3, y_3) is the Most Probable Point (MPP) or design point. From Fig. 7.13., the (x_3, y_3) is not the true MPP point P, but it already closes to the true, which means the (x_3, y_3) is an approximate MPP with acceptable error. In the RF algorithm, it is rare in the real case it will have the exact MPP, unless the $G(x, y) = 0$ is exactly a linear function.

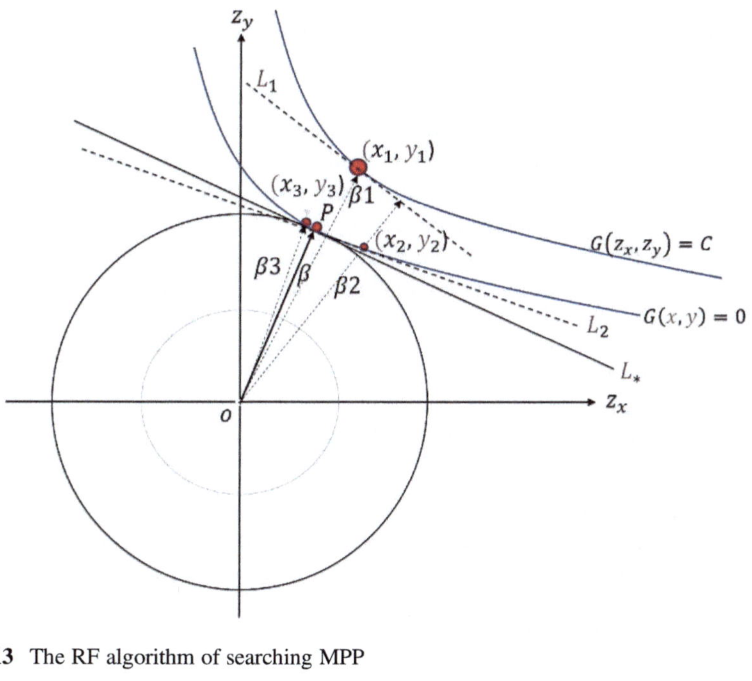

Fig. 7.13 The RF algorithm of searching MPP

From the demonstration of Fig. 7.13, we can find the solution approaches to the MPP point progressively after each iteration.

Algorithm 7.1 In summary, the RF algorithm to search the design point or the most probable point (MPP) is:

Step 1: Choose an arbitrary start point. It can be $\mathbf{z_1} = (0, 0, 0, ..0)$ or $\mathbf{z_1} = (1, 1, 1, ..1)$ or any point else. Let choose $\mathbf{z_1} = (1, 1, 1, ..1)$ for example. Set the error toleration in Eq. (7.52) and (7.53). For example, $\varepsilon_1 = \varepsilon_2 = 0.001$. Set the maximal iteration number, for example 50.

Step 2: Compute value vector $\nabla G(z_k) = \left(\frac{\partial G(z)}{\partial z_1}\Big|_{z_1=1}, \frac{\partial G(z)}{\partial z_2}\Big|_{z_2=1}, \ldots, \frac{\partial G(z)}{\partial z_n}\Big|_{z_3=1} \right)$ and value $G(z_k)$. Here the $\boldsymbol{k} = \mathbf{1}$.

Step 3: Compute the update z_{k+1} by using Eq. (7.51)

$$z_{k+1} = \frac{1}{\nabla G(z_k) \nabla G(z_k)^T} \left\{ \nabla G(z_k) {z_k}^T - G(z_k) \right\} \nabla G(z_k)$$

Step 4: Check if the stop criteria met from Eq. (7.52) or (7.53)

$$|z_{k+1} - z_k| \le \varepsilon_1$$

$$|\nabla G(z_{k+1}) - \nabla G(z_k)| \le \varepsilon_2$$

If it meets, stop the iterative procedure. The MPP or design point z_* is the point z_{k+1}. If not, let $k = k + 1$. Jump to the Step 2. Check if the maximum iteration number meets, if yes, terminate the iteration and it means the searching of MPP using this method failed. Some may wonder why to use the maximal iteration number. The RF algorithm assumes the iteration will converge, which means after each iteration, the new z_{k+1} will step toward the true MPP further. However, some safety function $G(z)$, in structure engineering also named limit-state function, does not converge, or the MPP does not actually exit. It leads a situation where the stop criteria will never meet. Setting a maximal iteration number can stop the algorithm to run indefinitely.

7.4.4 Procedure of Estimating Reliability

Suppose the safety function, or limit-state function $G(x_1, x_2, x_3, \ldots) = 0$ is known. The variables x_i in the $G(.)$ follows Normal distribution and are independent from each other. Variable x_i corresponds to a known mean μ_i and standard deviation σ_i. The overall procedure to obtain the reliability index and reliability in terms of probability is (**Algorithm 7.2**):

Step 1: Propose the safety function or limit-state function into a form of $G(x_1, x_2, x_3, \ldots) = 0$. Normalize the Normal distribution to a standard Normal distribution with mean 0 and standard deviation 1 by Formula (7.24)

$$x_i = \mu_i + z_i \sigma_i$$

Substitute the x_i into $G(x_1, x_2, x_3, \ldots) = 0$ to obtain an equivalent function $G(z_1, z_2, z_3, \ldots) = 0$, where all variable z_i are standard Normal distribution.

Step 2: Derive the gradient function vector $\nabla G(z_1, z_2, z_3, \ldots) = \left(\frac{\partial G(z_1, z_2, z_3, \ldots)}{\partial z_1}, \frac{\partial G(z_1, z_2, z_3, \ldots)}{\partial z_2}, \frac{\partial G(z_1, z_2, z_3, \ldots)}{\partial z_3} \ldots \right)$

Step 3: Employ the **Algorithm 7.1** to find the design point or MPP z_*.

Step 4: Compute reliability index by

$$\beta = \|z_*\| = \sqrt{z_{1*}{}^2 + z_{2*}{}^2 + \ldots}$$

Compute the reliability from $1 - \Phi(-\beta)$ or $\Phi(\beta)$, where $\Phi(.)$ is the CDF function of standard Normal distribution.

Step 5: If it is demanded, get back the design point $(x_1, x_2, x_3, \ldots)$ or MPP by

$$x_i = \mu_i + z_{i*} \sigma_i$$

We demonstrate the algorithm by a numerical example.

Example 7.4

The limit-state function is $G(x, y) = x^2 - (y + 4)^3$, which is Eq. (7.22). The $x \sim N$ (4,3) and $y \sim N$ (0, 2). Find out the reliability index and the reliability.

Solution: According to Algorithm 7.2,

Step 1: Normalize the Normal distribution.

$$x = \mu_x + z_x \sigma_x = 4 + 3z_x;$$

$$y = \mu_y + z_y \sigma_y = 0 + 2z_y;$$

Substitute the x and y into G

$$G(z_x, z_y) = (4 + 3z_x)^2 - (2z_y + 4)^3$$

Step 2: The gradient vector of $G(z_x, z_y)$

(continued)

Example 7.4 (continued)

$$\frac{\partial G(z_x, z_y)}{\partial z_x} = 2 * 3 * (4 + 3z_x) = 24 + 18z_x$$

$$\frac{\partial G(z_x, z_y)}{\partial z_y} = -3 * 2 * (2z_y + 4)^2 = -6 * (2z_y + 4)^2$$

$$\nabla G(z_x, z_y) = \left(24 + 18z_x, -6 * (2z_y + 4)^2\right)$$

Step 3: Find the MPP using Algorithm 7.1. The resulting MPP found is [0.32; −0.55] in the Normalized space.

Step 4: The

$$\beta = \|z_*\| = \sqrt{z_{1*}{}^2 + z_{2*}{}^2 + \ldots} = \sqrt{0.32^2 + (-0.55)^2} = 0.6363$$

Reliability is $1 - \Phi(-\beta) = \Phi(\beta) = 0.7377$

Step 5: The MPP is

$$x = \mu_x + z_x\sigma_x = 4 + 3z_x = 4 + 3 * 0.32 = 4.96;$$

$$y = \mu_y + z_y\sigma_y = 0 + 2z_y = 2 * (-0.55) = -1.1;$$

The design point or MPP is (4.96, −1.1).

The detailed procedure of algorithm 7.1 is shown as follows:

Step 1: The start point chosen is ($z_x = 1, z_y = 1$) The error tolerance threshold is $\varepsilon_1 = \varepsilon_2 = 0.001$. Set the maximal iteration number 50.

Iteration 1:

Step 2: For $\boldsymbol{k} = \mathbf{1}$, compute gradient vector value

$$z_1 = (z_x, z_y) = (z_x = 1, z_y = 1)$$

$$\nabla G(z_x, z_y) = \left(24 + 18z_x, -6 * (2z_y + 4)^2\right) = (42, -216)$$

$$G(z_x, z_y) = (4 + 3z_x)^2 - (2z_y + 4)^3 = (4 + 3)^2 - (2 + 4)^3 = -167$$

We get $\nabla G(z_1) = (42, -216)$, $G(z_1) = -167$.

(continued)

Example 7.4 (continued)

Step 3: Compute the update z_{k+1} by using Eq. (7.51)

$$z_{k+1} = z_2 = \frac{1}{\nabla G(z_k)\nabla G(z_k)^T}\left\{\nabla G(z_k){z_k}^T - G(z_k)\right\}\nabla G(z_k)$$

$$= \frac{1}{\nabla G(z_1)\nabla G(z_1)^T}\left\{\nabla G(z_1){z_1}^T - G(z_1)\right\}\nabla G(z_1)$$

$$= \frac{1}{(42,\ -216)(42,\ -216)^T}\left\{(42,\ -216)(1,1)^T + 167\right\}(42,\ -216) =$$

$$-\frac{7}{48420}(42,\ -216) = (-0.0061{,}0.0312)$$

Step 4: Check if the stop criteria met by Eq. (7.52) and (7.53)

$$\|z_{k+1} - z_k\| = \|z_2 - z_1\| = \|(-0.0061{,}0.0312) - (1,1)\| = 1.3967 > 0.001$$

Since $z_{k+1} = z_2 = (z_x, z_y) = (z_x = -0.0061, z_y = 0.0312)$

$$\nabla G(z_{k+1}) = \nabla G(z_2) = \left(24 + 18z_x,\ -6 * (2z_y + 4)^2\right)$$

$$= \left(24 + 18 * (-0.0061),\ -6 * (2 * (0.0312) + 4)^2\right) = (23.89,\ -99)$$

$$\|\nabla G(z_{k+1}) - \nabla G(z_k)\| = \|\nabla G(z_2) - \nabla G(z_1)\| = 118.37 > 0.001$$

Stop criteria is not met. The iteration continues.

Iteration 2: $k = 2$. We ignore the detailed procedure by only display key results for $k = 2$

$$z_2 = (-0.0061{,}0.0312)$$

$$z_{k+1} = z_3 = \frac{1}{\nabla G(z_k)\nabla G(z_k)^T}\left\{\nabla G(z_k){z_k}^T - G(z_k)\right\}\nabla G(z_k)$$

$$= (0.11,\ -0.46)$$

$$\|z_{k+1} - z_k\| = 0.50 > 0.001$$

$$\|\nabla G(z_{k+1}) - \nabla G(z_k)\| = 41.98 > 0.001$$

Criteria is not met. The iteration continues for $k = 3$.

(continued)

Example 7.4 (continued)

Iteration 3: $k = 3$.

$$z_3 = (0.11,\ -0.46)$$

$$z_{k+1} = z_4 = \frac{1}{\nabla G(z_k)\nabla G(z_k)^T}\left\{\nabla G(z_k)z_k^{\,T} - G(z_k)\right\}\nabla G(z_k)$$

$$= (0.2615,\ -0.5744)$$

$$\|z_{k+1} - z_k\| = 0.19 > 0.001$$

$$\|\nabla G(z_{k+1}) - \nabla G(z_k)\| = 8.75 > 0.001$$

Criteria is not met. The iteration continues for $k = 4$.

Iteration 4: $k = 4$.

$$z_4 = (0.2615,\ -0.5744)$$

$$z_{k+1} = z_5 = \frac{1}{\nabla G(z_k)\nabla G(z_k)^T}\left\{\nabla G(z_k)z_k^{\,T} - G(z_k)\right\}\nabla G(z_k)$$

$$= (0.3209,\ -0.5453)$$

$$\|z_{k+1} - z_k\| = 0.0662 > 0.001$$

$$\|\nabla G(z_{k+1}) - \nabla G(z_k)\| = 2.28 > 0.001$$

Criteria is not met. The iteration continues for $k = 5$.

Iteration 4: $k = 5$.

$$z_5 = (0.3209,\ -0.5453)$$

$$z_{k+1} = z_6 = \frac{1}{\nabla G(z_k)\nabla G(z_k)^T}\left\{\nabla G(z_k)z_k^{\,T} - G(z_k)\right\}\nabla G(z_k)$$

$$= (0.32,\ -0.545797601)$$

$$\|z_{k+1} - z_k\| = 0.0011 > 0.001$$

$$\|\nabla G(z_{k+1}) - \nabla G(z_k)\| = 0.0386 > 0.001$$

Criteria is not met. The iteration continues for $k = 6$.

Iteration 4: $k = 6$.

$$z_6 = (0.32,\ -0.545797601)$$

(continued)

Example 7.4 (continued)

$$z_{k+1} = z_7 = \frac{1}{\nabla G(z_k) \nabla G(z_k)^T} \left\{ \nabla G(z_k) {z_k}^T - G(z_k) \right\} \nabla G(z_k)$$
$$= (0.32, \; -0.54578117)$$

$$\|z_{k+1} - z_k\| = 0.000033 < 0.001$$

$$\|\nabla G(z_{k+1}) - \nabla G(z_k)\| = 0.0013 > 0.001$$

Criteria met. The iteration has stopped by successfully obtaining the MPP. Figure 7.14 plots the iteration procedure for the Example 7.4, where we can see the solution approach to real MPP with the iteration increasing.

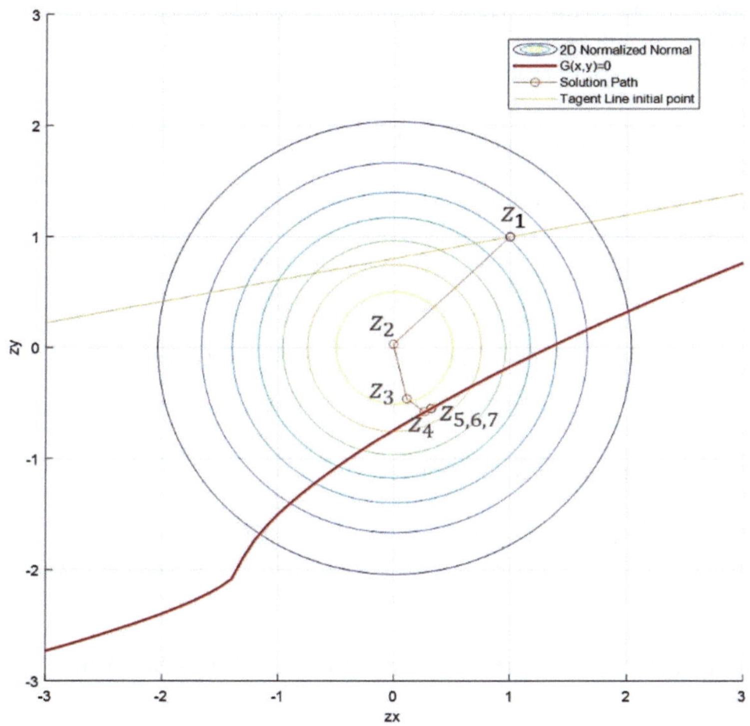

Fig. 7.14 Solution converges at true MPP

Example 7.5
Using the Algorithm 7.1 and 7.2 to find the MPP and calculate the reliability for the question of **Example 7.3**.

Solution: Normalize the variables in the limit-state function to standard Normal distribution variable. According to

$$x = \mu_x + z_x \sigma_x$$

The P is with mean 8 and standard deviation 1. The normalized random variable P can be written as 8 + z_P, the L, W, T can be converted similarly. The new limit-state function is

$$G(z_P, z_L, z_W, z_T) = (0.0006 + 0.00009 z_W)(80000 + 20000 z_T) - (8 + z_P) \times (9 + 0.5 z_L)/4$$

Applying the **Algorithm 7.2**, we obtain the reliability index $\beta = 2.3696$. The probability of failure is 0.0089. Reliability is then 0.9911. The MPP in the normalized limit-state function is.

$$(z_P, z_L, z_W, z_T) = (0.4669, 0.2170, -0.6580, -2.2174).$$

In the original limit-state function, the MPP is

$$\mathrm{P} = 8 + 0.4669 = 8.4669;$$

$$\mathrm{L} = 9 + 0.5^* \ 0.2170 = 9.1085;$$

$$\mathrm{W} = 0.0006 + 0.00009^* \ (-0.6580) = 0.00054;$$

$$\mathrm{T} = 80000 + 20000^* \ (-2.2174) = 35652.$$

MPP in the original safety function is

$$(P, L, W, T) = (8.4669, 9.1085, 0.00054, 35652)$$

7.5 Reliability Evaluation Using Monte Carlo Method

The above section discusses the method to approximate the safety function to a linear form when safety function is nonlinear. This method gains computing convenience at the cost of accuracy. This section discusses the method of Monte Carlo simulation. The simulation method is simpler in principle, but with extensive computational cost. The principle is: for each variable in the safety function, we generate a number according to the randomness of the variable. Then substitute all the generated values into the safety function to find out if it will fail. Suppose the safety function is $G(x_1, x_2, x_3, \ldots) = 0$. The algorithm of the Monte Carlo simulation is:

Step 1: Set number of iterations of the simulation as M. The number of failed states $n_f = 0$.

Step 2: Draw a number for each random variable x_i in the safety function, denoting it as t_i. Substitute them into the safety function $G(t_1, t_2, t_3, \ldots)$. If $G(t_1, t_2, t_3, \ldots) \geq 0$, it is safe; otherwise, it fails, and we increase the n_f by 1. Repeat this step until the number of iterations reaches the predefined M.

Step 3: The probability of failure is $P_f = \frac{n_f}{M}$;

The Monte Carlo simulation is simple and requires no approximation. The random number of the parameter can be any distribution but only requires an explicit expression of CDF function. But the Monte Carlo method relies on computer software such as Matlab for computation.

Example 7.6

Using Monte Carlo simulation method to evaluate the reliability in the Example 7.5. The number of iterations is set as 50,000. Apply the procedure of above-mentioned steps using Matlab. The random number is generated from normrnd() in Matlab.

```
%----------------------------------------
nSim = 50000; %Number of Iteration
%----------------------------------------
pmu=8;
psigma=1; %P distribution, mean and standard deviation
Lmu=9;
Lsigma=0.5;%L distribution, mean and standard deviation
wmu=0.0006;
wsigma=0.00009;%W distribution, mean and standard deviation
tmu=80000;
tsigma=20000;%T distribution, mean and standard deviation

n = 1;
nFailure = 0; %Count the number of failures
while n < nSim
  rP = normrnd(pmu,psigma);
  rL = normrnd(Lmu,Lsigma);
  rW = normrnd(wmu,wsigma);
  rT = normrnd(tmu,tsigma);
  stateNumber=rW*rT-rP*rL/4;
  if (stateNumber < 0)
    nFailure = nFailure + 1;
  end
  n = n +1;
end
RSys = 1- (nFailure / nSim);
```

(continued)

Example 7.6 (continued)

The evaluated reliability from the Monte Carlo ranges between 0.9903 and 0.9901. The evaluated reliability from Monte Carlo is accurate as it uses no approximation method. When the number of iterations is big, the estimated reliability from Monte Carlo can be considered as the real reliability. We can compare the accuracy of MPP method, and the Mean method discussed about. From Example 7.3 where Mean method is used, the reliability is 0.9826. The MPP method from Example 7.5 is 0.9911. The MPP method is closer to the Monte Carlo result which means it has higher accuracy than mean method for this example.

However, in general for Monte Carlo method, the result will exhibit certain uncertainty. The evaluated reliability as simulation result will have a certain degree of fluctuation when the number of iteration changes. Figure 7.15 shows the reliability evaluated (Y axis) against the number of iterations (X Axis) for the Example 7.6. At larger iteration number, the reliability with less level of fluctuation. As shown in Fig. 7.15, for large iteration number, the highest reliability is 0.99017, lowest is 0.990082. The true reliability should be around this range. The degree of fluctuation can be reduced by increasing the number of iterations. But we should be aware that the fluctuation is not able to be completely eliminated even though the number of iterations is very large.

In this book, the methods of evaluating reliability from safety function or limit-state function only cover the mean value for first order method, MPP FORM method and Monte Carlo method. When using the second order in the Taylor expansion to approximate the safety function, the method is named as Second Order Reliability

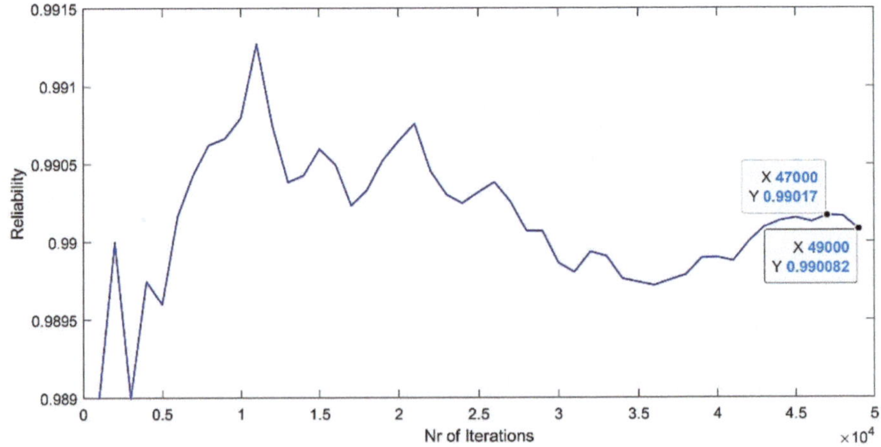

Fig. 7.15 Simulated reliability stabilizes with number of iterations

Method (SORM). The SORM is expected to have higher accuracy since the second order approximation is expected with less approximation error than that of FORM, but with much higher computational cost. Moreover, as the ultimate reliability accuracy is application specific, the SORM does not always have higher accuracy than that FORM. Many other methods such as Response Surface Method (RSM) are also popular for solving similar problems. This book does not cover them.

7.6 Ship Uncertainty Example

This section discusses a case example for ship design. The ship, as a structure, has complex formula in its analytical expression of stability or hull strength. In this case, the materials properties such as yield strength, Young's modulus exhibit certain degree of variation. Moreover, the operating condition exhibits as random, for example the wind speed, wave direction and power, as demonstrated in Fig. 7.16 [4]. The analysis on the stability and hull strength is then probabilistic.

For a failure mode of ultimate collapsing of the ship hull, the failure of ship hull is expected when the load exceeds its ultimate longitudinal strength. Taking account of uncertainties of load effects and material properties, the limit state function is defined as [5, 6].

$$G(\eta_u, M_u, \eta_s, M_s, \eta_w, M_w) = \eta_u M_u - \eta_s M_s - \phi \eta_w M_w \tag{7.66}$$

The M_u is the ultimate strength of ship, which represent the strength in the stress-strength analysis. The $M_u$is a function of materials properties and the ship geometry. M_s is the still-water moment, and M_w is the wave-induced moment. Both M_s and M_w acts as stress in the stress-strength analysis. The still-water moment on ship hulls comes from the action of the ship's lightweight, cargo, personnel and buoyancy [5]. The random distribution and weight of the personnel and cargo is the major contributors to variability in M_s. We know the M_s follows Normal distribution from the result of a statistical analysis of the still-water moment from about 2000 voyages of 100 steel ships. The wave-induced M_w relates to ocean condition, ship speed, ship travel routes and the specific type of ship. The η_u, η_s, η_w and ϕ are coefficients adjusting the hull strength and moment. All the η_u, M_u, η_s, M_s, η_w, M_w are with certain level of variation. The Table 7.2 presents the mean and standard deviation of

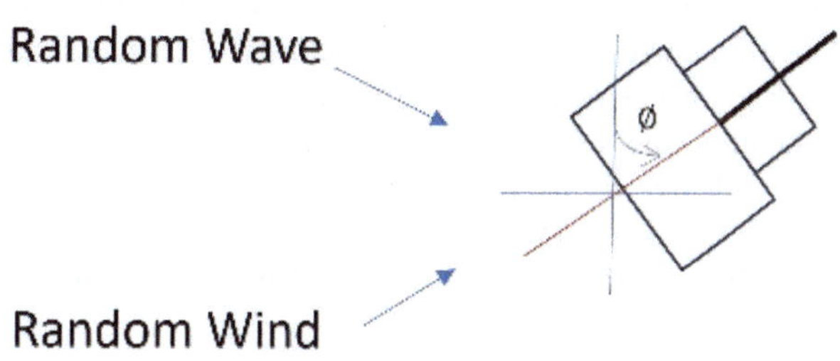

Fig. 7.16 Ship stability from wind and wave [4]

Table 7.2 Parameters distribution in the limit-state function (Data adapted from [5])

	η_u	M_u	η_s	M_s	ϕ (*not random*)	η_w	M_w
Mean	1	2.74×10^{10}nm	1	0	0.9	0.9	1.84×10^{7}nm
Standard deviation	0.15	1.24×10^{9}nm	0.05	1.65×10^{6}NM	–	0.15	1.84×10^{6}nm

those parameters. The M_u can be considered as following Normal distribution. The M_u in the Table 7.2 is adopted from a hull of a container ship.

Apply the FORM algorithm described in Algorithm 7.2. The start point for the MPP search is (1,1,1,1,1,1). After 6 iterations, the algorithm terminates as it has successful finding the solution. The solution converges to the true value but the converge cannot be visualized since the number of parameters is over 3 which means over 3-dimension. The Table 7.3 presents the updated solutions for each iteration.

The difference between the β_i and β_{i-1} becomes smaller when i increases, which implies the algorithm converges.

The design point or MPP in the Normalized space is.

$$\begin{pmatrix} \eta_{u*} = -6.66, \\ M_{u*} = -0.0011, \\ \eta_{s*} = 3.5744, \\ M_{s*} = 0.002675, \\ \eta_{w*} = 0.00402824, \\ M_{w*} = 0.00241798 \end{pmatrix}$$

Convert the design point MPP into original point, which is.

$$\begin{pmatrix} \eta_{u*} = 5.44 \times 10^{-4}, \\ M_{u*} = 2.74 \times 10^{10}, \\ \eta_{s*} = 1, \\ M_{s*} = 4.41 \times 10^{3}, \\ \eta_{w*} = 0.9006, \\ M_{w*} = 1.84 \times 10^{7} \end{pmatrix}$$

The reliability index by using $\beta = \|z_*\|$ showing in Step 4 in Algorithm 7.2 is 6.66304, corresponding to a probability of failure as low as $\Phi(-\beta) = 1.3411 \times 10^{-11}$. For very small probability of failure, the use of reliability index β is more indicative than use probability. This reliability index is a popular index used in the ship industry as its probability of failure is normally very small. This example considers six parameters, for which FORM proves to be efficient in evaluating reliability. In the shipbuilding industry, the number of parameters in a limit-state

Table 7.3 Iteration results

Iteration	i = 1	i = 2	i = 3	i = 4	i = 5	i = 6
η_u	1	−5.70297	−6.73766	−6.66186	−6.66303	−6.66304
M_u	1	−1.89302	−0.32138	0.02172	−0.00145	−0.0011
η_s	1	0.00011	3.40E-07	4.01E-07	3.63E-07	3.57E-07
M_s	1	0.0023	0.002958	0.002714	0.002672	0.002675
η_w	1	0.003627	0.004455	0.004087	0.004024	0.004028
M_w	1	0.002308	0.002674	0.002453	0.002415	0.002418
β_i	2.449489743	6.008945348	6.745324156	6.661901411	6.663036889	6.663038107

function can range from 40 to 50, or even reach hundreds. Despite this complexity, FORM can still serve as an efficient approach.

Appendix 7A: Linear Normal Distribution

In statistics, moment generating function for a random variable x is defined as [7].

$$M_x(t) = E(e^{xt}), t \in \mathbb{R} \tag{7A.1}$$

where $\mathbb{R}$ denotes for real value. The PDF of Normal distribution with mean μ and variance σ^2 is

$$f(x) = \frac{1}{\sigma\sqrt{2\pi}} e^{-\frac{1}{2}\left(\frac{x-\mu}{\sigma}\right)^2} \tag{7A.2}$$

The moment generating function of Normal distribution is then

$$M_x(t) = \int_{-\infty}^{+\infty} \frac{1}{\sigma\sqrt{2\pi}} e^{-\frac{1}{2}\left(\frac{x-\mu}{\sigma}\right)^2} e^{xt} dx \tag{7A.3}$$

Obtain the analytical solution to (7A.3) requires a little trick. Let the $z = \frac{x-\mu}{\sigma}$, and $x = \mu + z\sigma$. Rewrite the (7A.3)

$$\begin{aligned} M_x(t) &= \int_{-\infty}^{+\infty} \frac{1}{\sqrt{2\pi}} e^{-\frac{1}{2}z^2} e^{(\mu+z\sigma)t} dz = e^{\mu t} \int_{-\infty}^{+\infty} \frac{1}{\sqrt{2\pi}} e^{-\frac{1}{2}z^2} e^{z\sigma t} dz \\ &= e^{\mu t} \int_{-\infty}^{+\infty} \frac{1}{\sqrt{2\pi}} e^{-\frac{1}{2}\left(z^2 - 2z\sigma t + \sigma^2 t^2\right) + \frac{1}{2}\sigma^2 t^2} dz = e^{\mu t} e^{\frac{1}{2}\sigma^2 t^2} \int_{-\infty}^{+\infty} \frac{1}{\sqrt{2\pi}} e^{-\frac{1}{2}(z-\sigma t)^2} dz = e^{\mu t + \frac{1}{2}\sigma^2 t^2} \end{aligned} \tag{7A.4}$$

The (7A.4) uses $\int_{-\infty}^{+\infty} \frac{1}{\sqrt{2\pi}} e^{-\frac{1}{2}(z-\sigma t)^2} dz = 1$, since the $\frac{1}{\sqrt{2\pi}} e^{-\frac{1}{2}(z-\sigma t)^2}$ is the PDF function of Normal distribution random variable $z - \sigma t$. Additionally, the $a_i x_i$, where a_i is a constant real value, is also Normal distribution with mean $a_i \mu_i$ and variance ${a_i}^2 {\sigma_i}^2$.

Suppose Normal distribution random variable $x_1, x_2, x_3, \ldots$ are independent from each other. The mean and variance are $\mu_1, \mu_2, \mu_3, \ldots$. and ${\sigma_1}^2, {\sigma_2}^2, {\sigma_2}^2, \ldots$ .The linear combination of x_i is

$$Y = a_0 + a_1x_1 + a_2x_2 + \ldots \tag{7A.5}$$

The moment generating function of Y is

$$\begin{aligned} M_Y(t) &= E(e^{a_0 + a_1x_1t + a_2x_2t + \ldots}) = E(e^{a_0t})E(e^{a_1x_1t})E(e^{a_2x_2t})E(e^{a_3x_3t})\ldots \\ &= e^{a_0t}e^{a_1\mu_1t + \frac{1}{2}a_1{}^2\sigma_1{}^2t^2}e^{a_2\mu_2t + \frac{1}{2}a_2{}^2\sigma_2{}^2t^2}\ldots = e^{\left(a_0 + \sum a_i\mu_i\right)t + \frac{1}{2}\left(\sum a_i{}^2\sigma_i{}^2\right)t^2} \end{aligned} \tag{7A.6}$$

Equation (7A.6) is still of the form of (7A.4) by $\mu = a_0 \sum a_i\mu_i$; $\sigma^2 = \sum a_i{}^2\sigma_i{}^2$, i.e. the random variable Y is still Normal distribution with $\mu_Y = a_0 + \sum a_i\mu_i$; $\sigma_Y{}^2 = \sum a_i^2\sigma_i^2$.

Appendix 7B: Linear Approximation Using Taylor Series

The Taylor expansion is an expansion of series for a continuous function. For a general one variable continuous function $f(x)$, the Taylor expansion is

$$f(x) = f(x_0) + f'(x_0)\frac{(x - x_0)}{1!} + f''(x_0)\frac{(x - x_0)^2}{2!} + f'''(x_0)\frac{(x - x_0)^3}{3!} + \ldots \tag{7B.1}$$

The x_0 is the expansion point, the $f^n(x_0)$ is numerical value of the nth order of differentiation at point x_0. The nth order derivative is written as

$$f^n(x_0) = \left.\frac{\partial^n f(x)}{\partial x}\right|_{x_0} \tag{7B.2}$$

The expansion is abstract but if it is more understandable if it is interpreted from geometry perspective using an example. The function $f(x) = x^3$ is nonlinear function as plotted in Fig. 7B.1. It is obvious that

$$\begin{aligned} f'(x) &= 3x^2; \\ f''(x) &= 6x; \\ f'''(x) &= 6; \\ f^n(x) &= 0 \text{ for } n > 3 \end{aligned} \tag{7B.3}$$

The numerical value at an arbitrary point, let us say $x_0 = 4$ is

$$\begin{aligned} f(x_0) &= x^3 = 4^3 = 64; \\ f'(x_0) &= 3 * 4^2 = 48; \end{aligned}$$

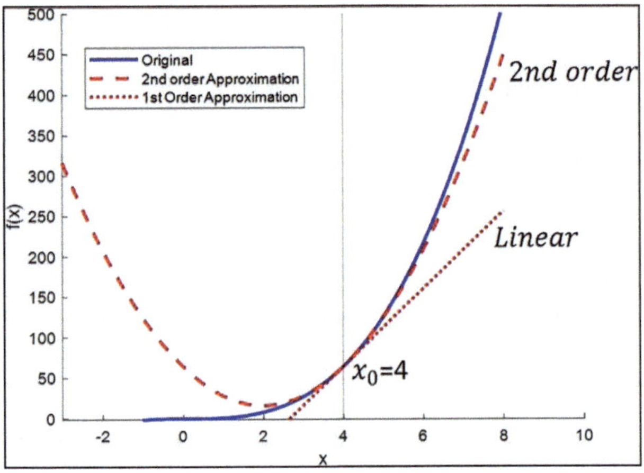

Fig. 7B.1 Approximate nonlinear function

$$
\begin{aligned}
f''(x_0) &= 6\text{x} = 6 * 4 = 24; \\
f'''(x_0) &= 6; \\
f^n(x_0) &= 0 \text{ for } n > 3
\end{aligned} \tag{7B.4}
$$

The Taylor expansion of Eq. (7B.1) is then

$$
\begin{aligned}
f(x) = & f(x_0) + f'(x_0)\frac{(x-x_0)}{1!} + f''(x_0)\frac{(x-x_0)^2}{2!} + f'''(x_0)\frac{(x-x_0)^3}{3!} \\
& = 64 + 48 * (x-4) + 24 * \frac{(x-4)^2}{2} + 6 * \frac{(x-4)^3}{3!} = (x-4)^3 \\
& + 12(x-4)^2 + 48 * (x-4) + 64
\end{aligned}
$$

The 2nd order Taylor expansion is

$$
\begin{aligned}
f(x) = & f(x_0) + f'(x_0)\frac{(x-x_0)}{1!} + f''(x_0)\frac{(x-x_0)^2}{2!} = 64 + 48 * (x-4) \\
& + 24 * \frac{(x-4)^2}{2} = 12(x-4)^2 + 48 * (x-4) + 64 = 12x^2 - 48x + 64
\end{aligned} \tag{7B.5}
$$

The 1st order Taylor expansion is

$$
f(x) = f(x_0) + f'(x_0)\frac{(x-x_0)}{1!} = 64 + 48 * (x-4) = 48x - 128 \tag{7B.6}
$$

From Fig. 7B.1 where the approximation of 2nd and 1st order are plotted, it shows the accuracy of the 2nd order is better than the first order, while the 1st order processes a desirable linear function. The linear Taylor approximation is equivalent to the tangent line on the nonlinear function at the expansion point.

For a general function with multiple variables $f(x_1, x_2, x_3, \ldots)$, the Taylor Expansion is

$$f(x_1, x_2, \ldots, x_n) = f(x_{10}, x_{20}, \ldots, x_{n0}) + \sum_{j=1}^{n} \frac{\partial f(x_{10}, x_{20}, \ldots, x_{n0})}{\partial x_j} \frac{(x_j - x_{j0})}{1!} + \sum_{j=1}^{n} \sum_{k=1}^{n} \frac{\partial^2 f(x_{10}, x_{20}, \ldots, x_{n0})}{\partial x_j \partial x_k} \frac{(x_j - x_{j0})(x_k - x_{k0})}{2!} + \ldots \tag{7B.7}$$

Equation (7B.7) only presents to the 2nd order as this is the interest of this Chapter. For an example function

$$f(x, y) = x^3 + y^3 \tag{7B.8}$$

The $\frac{\partial f(x,y)}{\partial x} = 3x^2$; $\frac{\partial f(x,y)}{\partial y} = 3y^2$. The linear approximation at point x_0, y_0 is

$$\begin{aligned} f(x, y) &= f(x_0, y_0) + \frac{\partial f(x, y)}{\partial x}(x - x_0) + \frac{\partial f(x, y)}{\partial y}(y - y_0) \\ &= x_0^3 + y_0^3 + 3x_0^2(x - x_0) + 3y_0^2(y - y_0) = 3x_0^2 x + 3y_0^2 y - 2x_0^3 - 2y_0^3 \end{aligned} \tag{7B.9}$$

Figure 7B.2 plots the linear approximation at expansion point (1,1). The linear function in three dimensions present as a plane.

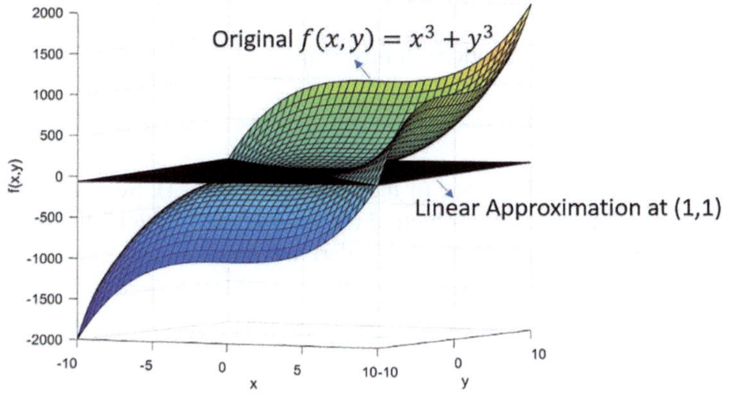

Fig. 7B.2 Approximate nonlinear function for multiple variables

Appendix 7C: Lagrange Optimization

The Lagrangian optimization method is a popular approach to solve optimization with equal constraints. The optimization problem is written as

$$\begin{aligned} &Max\, f(x_1, x_2, \ldots) \\ &s.t.\, g(x_1, x_2, \ldots) = 0 \end{aligned} \tag{7C.1}$$

We can take an example to interpret the method. Given the optimization problem is

$$\begin{aligned} &Max\, f(x, y) = x^2 + y^2 \\ &s.t.\, g(x, y) = x^2 + y + 2 = 0 \end{aligned} \tag{7C.2}$$

The contour of $x^2 + y^2$ and the constraint function $x^2 + y + 2 = 0$ are plotted in Fig. C.1. Since the contour is high in the center and lower in the outside, the point in the $g(x, y) = 0$ nearest to the higher contour region of $f(x, y)$ is the point with the maximal $f(x, y)$, i.e. the point is optimal solution. From Fig. C.1, it shows the point on $g(x, y) = 0$ tangent to the highest contour line. As a result, the point on the contour line of $f(x, y)$ will share a tangent line with a point on $g(x, y) = 0$. Surely the gradient of the point both on $g(x, y) = 0$, written as $\nabla g(x, y)$, and gradient of the point on $f(x, y)$, written as $\nabla f(x, y)$,will be the same direction. We can write it in an equation

$$\nabla f(x, y) = \lambda \nabla g(x, y) \tag{7C.3}$$

The λ is a constant named Lagrange multiplier [8].

The (7C.3) is with two equations, one for one partial derivative in respect to one variable. Another equation is the constraint $g(x, y) = 0$. The three equations are same as the partial derivatives from a function defined by

$$\mathcal{L}(x, y, \lambda) = f(x, y) + \lambda g(x, y) \tag{7C.4}$$

which is named Lagrange function. In this example, it is

$$\mathcal{L}(x, y, \lambda) = x^2 + y^2 + \lambda\left(x^2 + y + 2\right) = (\lambda + 1)x^2 + y^2 + \lambda y + 2\lambda \tag{7C.5}$$

The solution of (7C.1) will be one of the solutions of (7C.4). Solving the (7C.4) is easy since it is an unconstraint problem. For unconstraint problem, the maximum or minimum can be found by taking the first derivatives.

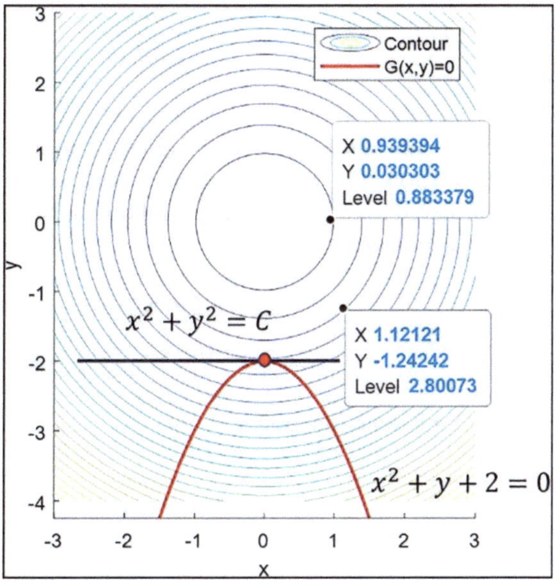

Fig. 7C.1 Demonstrating the Lagrange method

$$
\begin{aligned}
\frac{\partial \mathcal{L}}{\partial x} &= 2x(\lambda + 1) \\
\frac{\partial \mathcal{L}}{\partial y} &= 2y + \lambda \\
\frac{\partial \mathcal{L}}{\partial \lambda} &= x^2 + y + 2
\end{aligned}
\tag{7C.6}
$$

Let these partial derivatives equal to 0. The solution to (7C.6) is

$$x_* = 0; y_* = -2; \lambda_* = 4 \quad f(x_*, y_*) = 4$$

The solution is exactly the cross point (red point) shown in Fig. 7C.1, as expected.

Problems and Questions

1. What is the Taylor series of function e^{at}, where a is a constant value, t is variable?
2. Using Taylor expansion method to find out the linear approximation of nonlinear function $y = x^2$ at point $x_0 = 1$.
3. The random variable x follows Normal distribution with mean μ and variance σ^2. Using moment generating method to prove the ax is also Normal distribution with mean $a\mu$ and variance $a^2\sigma^2$.
4. Two Normal distribution random variables with parameters $x \sim N(2, 4)$, $y \sim N(1, 5)$. Find out the statistical distribution of $z = 2x + 3y$.
5. Solve the optimization problem using Lagrange method.

$$Max\ x + y$$
$$s.t.x^2 + y^2 = 1$$

6. Let us consider a steel beam subjected to a deterministic bending moment of Ma = 130 kNm. The yield stress F_y and the plastic modulus Z_p of the beam are considered as random variables following Normal distributions with the following parameters:

$$\mu Fy = 250\ \text{MPa}, \sigma Fy = 25\ \text{MPa}$$

$$\mu Zp = 9 \times 10^{-4}\ \text{m}^3, \sigma Zp = 4.5 \times 10^{-5}\ \text{m}^3$$

The limit-state function is

$$g(F_y, Z_p) = F_y - \frac{M_a}{Z_p}$$

Question: Estimate the reliability index and reliability using mean value first order method [9].

7. Let us suppose that the performance function of a problem is defined by

$$g(x_1, x_2, x_3) = 6.2\ x_1 - x_2 x_3$$

where the random variables x_1, x_2, x_3 follow Normal distributions with means $\mu_{x_1} = 20$, $\mu_{x_2} = 5$, and $\mu_{x_3} = 4$; and standard deviations $\sigma_{x_1} = 3.5$, $\sigma_{x_2} = 0.8$, and $\sigma_{x_3} = 0.4$.

Question: Estimate the reliability index by using FORM [9].

8. A brittle concrete wall will fracture when the external stress is over the critical stress. This critical stress of the concrete can be calculated from $\sigma_c = 0.2\sqrt{E.\gamma}$. The external stress σ_m applied on the wall is calculated from $\sigma_m = 0.1F$.

E—Young's modulus
γ—Surface energy.
F—External force

The E and the γ are with variation following Normal distribution. The external force applied on the wall is also random with Normal distribution. Their parameters are as shown in Table 7.4

The safety function for this case ($G(x_1, x_2, , \ldots) = 0$) is then $(E, \gamma, F) = 0.2\sqrt{E.\gamma} - 0.1F$.

Table 7.4 Parameter distribution for Question 8

	E	γ	F
Mean	12	0.2	2
Std	3	0.01	0.1

Table 7.5 Parameter distribution for Question 9

	V (in Volts)	$R(\Omega)$	Noise (In A)
Mean	22	2	0
Std	1	0.2	0.1

a. Calculate the reliability index? (Hint: you can use the mean of the random variables as the Taylor expansion point, or using RF algorithm to find out MPP)
b. What is the probability of failure?
c. What is the reliability?

9. A sensor won't work if the provided current is under 20 A. The formula used to calculate the current is

$$I = \frac{V}{R} + 10 + \mu$$

V: Voltage (V)
R: Resistance (Ω)
I: Current
μ: White Noise

The parameters I, R and Noise follow Normal distribution with parameters as shown in Table 7.5

a. Define is the safety function or limit-state function for this case? ($G(x_1, x_2, , \ldots) = 0$)
b. Calculate the reliability index?
c. What is the probability of failure?
d. What is the reliability?

10. Using Monte Carlo method to evaluate the reliability for Question 7. You can use Matlab to code the simulation method. Can you find out a stable reliability in your simulation?

References

1. Budynas RG, Nisbett JK (2015) Mechanical engineering design. Mc Graw Hill Education
2. Larsson O Lecture note "Reliability analysis". Lund University
3. Keshtegar B, Meng Z (2017) A hybrid relaxed first-order reliability method for efficient structural reliability analysis. Struct Saf 66:84–93
4. Kogiso N, Murotsu Y (2008) Application of first order reliability method to ship stability–final report of scape committee (part 5). In: Seakeeping and stability of ships, pp 266–273

5. Chen NZ, Soares CG (2007) Reliability assessment for ultimate longitudinal strength of ship hulls in composite materials. Probabilistic Eng Mech 22:330–342
6. Shi GJ, Gao DW (2021) Reliability analysis of hull girder ultimate strength for large container ships under whipping loads. Struct Infrastruct Eng 17(3):319–330
7. Kendall MG, Stuart A, Ord JK, O'Hagan A (1987) Kendall's advanced theory of statistics. Oxford University Press, New York
8. Chong EKP, Zak SH (2013) An introduction to optimization. Wiley, p 640
9. Arteaga EB, Soubra AH (2014) Reliability analysis methods, ALERT Doctoral School 2014 - Stochastic Analysis and Inverse Modellin, 2: 53–77

Chapter 8
Multivariate Distribution for Reliability Dependency Analysis

Classical life data analysis typically focuses on a single variable, usually failure time, using statistical distributions such as the Exponential and Weibull distributions, which have been discussed earlier in this book. However, in many applications, life data often involves multiple variables. For example, the lifespan of a car is determined by both age and mileage, the lifespan of an aircraft is influenced by both calendar years and total flying hours. This necessitates the use of multivariate distributions to address such cases. In statistics, a multivariate distribution refers to a method that incorporates more than one variable, making it suitable for analyzing complex relationships between multiple factors. This chapter explores the application of multivariate distributions in reliability analysis. Since multivariate distributions are less familiar in reliability engineering, the chapter begins with an introduction to the basic concepts of Normal multivariate distributions. Later, we will discuss using the Copula function to develop some special multivariate distributions for reliability dependency analysis.

8.1 Multivariate Normal Distribution

Multivariate Normal distribution is the simplest multivariate distribution with favorite statistical properties. The importance of multivariate Normal distribution in multivariate analysis is the same as the univariate Normal distribution in univariate distribution. Some non-Normal multivariate distribution can be transformed to or can be approximated as Normal multivariate distribution. As we know, the simple univariate Normal distribution is with a PDF as

$$f(t) = \frac{1}{\sigma\sqrt{2\pi}} e^{-\frac{1}{2}\left(\frac{t-\mu}{\sigma}\right)^2} \tag{8.1}$$

Rewrite it as

F. Yuan, *Essentials of Reliability Engineering*, Springer Series in Reliability Engineering, https://doi.org/10.1007/978-3-032-15544-3_8

$$f(t) = \frac{1}{\sigma\sqrt{2\pi}} e^{-\frac{1}{2}(t-\mu)\left(\sigma^2\right)^{-1}(t-\mu)} \tag{8.2}$$

Let $\sigma^2 = \Sigma$. Equation (8.2) is rewritten as [1]

$$f(t) = \frac{1}{(2\pi)^{1/2}|\Sigma|^{1/2}} e^{-\frac{1}{2}(t-\mu)\Sigma^{-1}(t-\mu)} \tag{8.3}$$

This is a form of multivariate distribution. For multivariate Normal distribution, the t and mean μ are row vector where each element corresponds to a value for the variable. Equation (8.3) is thus rewritten as

$$f(\boldsymbol{t}) = \frac{1}{(2\pi)^{p/2}|\Sigma|^{1/2}} e^{-\frac{1}{2}(\boldsymbol{t}-\boldsymbol{\mu})\Sigma^{-1}(\boldsymbol{t}-\boldsymbol{\mu})^T} \tag{8.4}$$

Where row vector $\boldsymbol{t} = (t_1, t_2, .., t_p)$, $\boldsymbol{\mu} = (\mu_1, \mu_2, .., \mu_p)$. The

$$\Sigma = \begin{bmatrix} {\sigma_{11}}^2 & {\sigma_{12}}^2 & \ldots {\sigma_{1p}}^2 \\ \cdots & \cdots & \cdots \\ {\sigma_{p1}}^2 & {\sigma_{p2}}^2 & \ldots {\sigma_{pp}}^2 \end{bmatrix} \tag{8.5}$$

The ${\sigma_{ij}}^2$ denotes the covariate of variable i and j. For $i = j$, ${\sigma_{ii}}^2$ is the variance of variable i. When the ${\sigma_{ij}}^2 = 0$ for all $i \neq j$, Eq. (8.5) is written as

$$\Sigma = \begin{bmatrix} {\sigma_{11}}^2 & 0 & \ldots 0 \\ \cdots & \cdots & \cdots \\ 0 & 0 & \ldots {\sigma_{pp}}^2 \end{bmatrix} \tag{8.6}$$

Equation (8.6) implies the variable t_i are independent from each other, for which reliability analysis commonly assumes. For independent case, using matrix analysis, it is not hard to find

$$|\Sigma| = {\sigma_{11}}^2 {\sigma_{22}}^2 {\sigma_{33}}^2 \ldots {\sigma_{pp}}^2 \tag{8.7}$$

and

$$(\Sigma)^{-1} = \begin{bmatrix} \frac{1}{{\sigma_{11}}^2} & 0 & \ldots 0 \\ \cdots & \cdots & \cdots \\ 0 & 0 & \ldots \frac{1}{{\sigma_{pp}}^2} \end{bmatrix} \tag{8.8}$$

Substitute Eqs. (8.6), (8.7), (8.8) to Eq. (8.4). We obtain

$$f(\boldsymbol{t})=\frac{1}{(2\pi)^{\frac{p}{2}}|\Sigma|^{\frac{1}{2}}}e^{-\frac{1}{2}(\boldsymbol{t}-\boldsymbol{\mu})(\Sigma)^{-1}(\boldsymbol{t}-\boldsymbol{\mu})^{T}}=\frac{1}{(2\pi)^{\frac{p}{2}}\left(\sigma_{11}{}^{2}\sigma_{22}{}^{2}\sigma_{33}{}^{2}\ldots\sigma_{pp}{}^{2}\right)^{\frac{1}{2}}}e^{-\frac{1}{2}(\boldsymbol{t}-\boldsymbol{\mu})\begin{bmatrix}\frac{1}{\sigma_{11}{}^{2}} & 0 & \ldots 0\\ \cdots & \cdots & \cdots\\ 0 & 0 & \ldots\frac{1}{\sigma_{pp}{}^{2}}\end{bmatrix}(\boldsymbol{t}-\boldsymbol{\mu})^{T}}$$
$$=\frac{1}{(2\pi)^{\frac{1}{2}}\sigma_{11}}e^{-\frac{1}{2}(t_1-\mu_1)\frac{1}{\sigma_{11}{}^{2}}(t_1-\mu_1)}\ldots\frac{1}{(2\pi)^{\frac{1}{2}}\sigma_{pp}}e^{-\frac{1}{2}\left(t_p-\mu_p\right)\frac{1}{\sigma_{pp}{}^{2}}\left(t_p-\mu_p\right)}=f(t_1)f(t_1)\ldots f\left(t_p\right) \tag{8.9}$$

which further proves the PDF of the multivariate Normal distribution is the multiplication of the marginal PDFs, implying all the factor t_i are independent.

8.1.1 CDF of Multivariate Normal Distribution

In reliability analysis, as a common situation, given a failure time t, we would like to compute the probability of failure, which is the CDF value. For multivariate Normal distribution, the CDF is more complex than simple Normal distribution.

Given $\boldsymbol{t} \leq \boldsymbol{t_0}$, which is specified as

$$P\left(t_1 \leq t_{10}, t_2 \leq t_{20}, .., t_p \leq t_{p0}\right) \tag{8.10}$$

From Eq. (8.4), the CDF of multivariate Normal distribution is the integral of the PDF as follows

$$F(\boldsymbol{t})=\iiint_{-\infty}^{\boldsymbol{t_0}}\frac{1}{(2\pi)^{p/2}|\Sigma|^{1/2}}e^{-\frac{1}{2}(\boldsymbol{t}-\boldsymbol{\mu})\Sigma^{-1}(\boldsymbol{t}-\boldsymbol{\mu})^{T}}d\boldsymbol{t} \tag{8.11}$$

The left of Fig. 8.1 shows the PDF of a Bivariate Normal which is of a symmetrical bell shape, the right figure shows its integral region shown as the shadow area which is of two-dimension. Unfortunately, closed-form solution to the CDF of Bivariate Normal distribution is hard to obtain. Numerical or simulation approach needs be employed, requiring a heavy computational cost [2].

The $F(\boldsymbol{t})$ is dependent on only the exponential part of Eq. (8.11). Let

$$r^2=(\boldsymbol{t}-\boldsymbol{\mu})\Sigma^{-1}(\boldsymbol{t}-\boldsymbol{\mu})^{T} \tag{8.12}$$

Without providing proof, for Bivariate Normal distribution, the CDF can be approximated as [3]

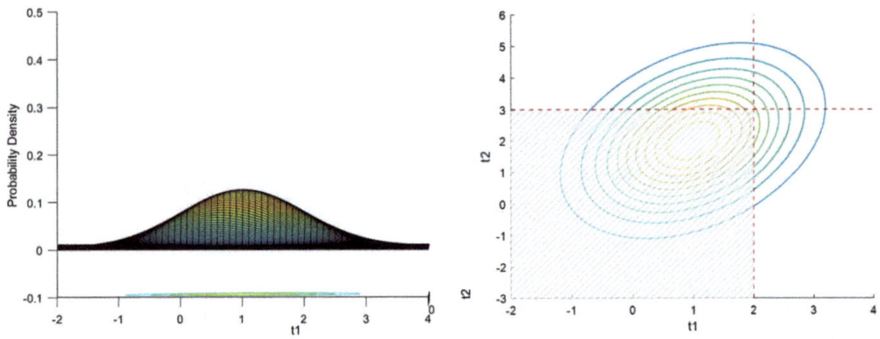

Fig. 8.1 PDF of Bivariate normal and its contour

$$F(\boldsymbol{t}) = 1 - e^{-r^2/2} \tag{8.13}$$

Example 8.1

The multivariate Normal distribution with $\boldsymbol{\mu} = (1\ 2)$, $\Sigma = \begin{bmatrix} 1 & 0.5 \\ 0.5 & 2 \end{bmatrix}$, compute the probability of

$$P(t_1 \leq 2, t_2 \leq 4)$$

Solution: Method 1: Approximate method

$\boldsymbol{t} - \boldsymbol{\mu} = (2\ 4) - (1\ 2) = (1\ 2)$; $\Sigma^{-1} = \begin{bmatrix} 1.1429 & -0.2857 \\ -0.2857 & 0.5714 \end{bmatrix}$

Since $r^2 = (\boldsymbol{t} - \boldsymbol{\mu})\Sigma^{-1}(\boldsymbol{t} - \boldsymbol{\mu})^T$

$$r^2 = (1\ 2)\begin{bmatrix} 1.1429 & -0.2857 \\ -0.2857 & 0.5714 \end{bmatrix}\begin{pmatrix} 1 \\ 2 \end{pmatrix} = 2.2857$$

$$F(\boldsymbol{t}) = 1 - e^{-\frac{r^2}{2}} = 0.6811$$

Method 2: Numerical method

The numerical method uses the Matlab mvncdf() function. The CDF is obtained as $F(\boldsymbol{t}) = 0.7910$. We can see there are some approximation errors by using Eq. (8.12).

8.1.2 Parameter Estimation

Parameter estimation for multivariate distribution is much more complicated than the univariate distribution. However, for data analysis, we shall estimate the parameters in matrix $\boldsymbol{\mu}$ and Σ in order to apply it. Given the observed data are

$$t_1, t_2, t_3, \ldots t_n$$

The likelihood function, same as the univariate distribution, is the joint PDF of the observed data

$$\begin{aligned} L &= \frac{1}{(2\pi)^{\frac{p}{2}}|\Sigma|^{\frac{1}{2}}} e^{-\frac{1}{2}(t_1 - \boldsymbol{\mu})\Sigma^{-1}(t_1 - \boldsymbol{\mu})^T} \frac{1}{(2\pi)^{\frac{p}{2}}|\Sigma|^{\frac{1}{2}}} e^{-\frac{1}{2}(t_2 - \boldsymbol{\mu})\Sigma^{-1}(t_2 - \boldsymbol{\mu})^T} \ldots \\ &= \frac{1}{(2\pi)^{\frac{np}{2}}|\Sigma|^{\frac{n}{2}}} e^{-\frac{1}{2}(\Sigma t_i - n\boldsymbol{\mu})\Sigma^{-1}(\Sigma t_i - n\boldsymbol{\mu})^T} \end{aligned} \tag{8.14}$$

Take logarithm of Eq. (8.14) as

$$\ln L = -\ln (2\pi)^{\frac{np}{2}} - \frac{n}{2} \ln|\Sigma| - \frac{1}{2} \sum_{i=1}^{n} (t_i - \boldsymbol{\mu})\Sigma^{-1} \sum_{i=1}^{n} (t_i - \boldsymbol{\mu})^T \tag{8.15}$$

Take the first derivative in respect of $\boldsymbol{\mu}$

$$\frac{\partial \ln L}{\partial \boldsymbol{\mu}} = \sum_{i=1}^{n} (t_i - \boldsymbol{\mu})\Sigma^{-1} \tag{8.16}$$

Set Eq. (8.16) equal to 0. We obtain

$$\widehat{\boldsymbol{\mu}} = \frac{\mathbf{1}}{\boldsymbol{n}} \sum_{i=1}^{n} \boldsymbol{t_i} \tag{8.17}$$

Without providing the detail due to the complex matrix operation, the maximum likelihood estimator for Σ is [1]

$$\widehat{\Sigma} = \frac{1}{n} \sum_{i} (t_i - \widehat{\boldsymbol{\mu}})^T (t_i - \widehat{\boldsymbol{\mu}}) \tag{8.18}$$

Example 8.2
The failure of a machine depends on two conditions which are measured by sensors. Their values are tabulated in Table 8.1. Suppose the two parameters follows multivariate Normal distribution and the system will fail when the values over P1 = 5 and P2 = 6. Compute the reliability of the system.

Solution: The measurements follow two-dimension multivariate Normal distribution. The parameters of the multivariate Normal can be evaluated by Formulas (8.17) and (8.18).

The $\boldsymbol{t}$ is a 10 × 2 matrix. The t_i is row vector 1 × 2.

$$t = \begin{pmatrix} 2.54\ 3.83 - 0.26\ 2.86\ 2.32\ 0.69\ 1.57\ 2.34\ 5.58\ 4.77 \\ 2.64\ 8.38\ 0.24\ 4.24\ 4.10\ 0.86\ 2.24\ 4.80\ 9.59\ 8.38 \end{pmatrix}^T$$

According to Formula (8.17),

$$\hat{\boldsymbol{\mu}} = \frac{1}{n}\sum_{i=1}^{n} t_i = \begin{pmatrix} \hat{\mu}_{P1} \\ \mu_{P1} \end{pmatrix}^T$$

$$\hat{\mu}_{P1} = \frac{2.54+3.83-0.26+2.86+2.32+0.69+1.57+2.34+5.58+4.77}{10} = 2.6240;$$

$$\hat{\mu}_{P2} = \frac{2.64+8.38+0.24+4.24+4.10+0.86+2.24+4.80+9.59+8.38}{10} = 4.5470;$$

According to Formula (8.18),

$$\hat{\Sigma} = \frac{1}{n}\sum_{i}(t_i - \hat{\mu})^T(t_i - \hat{\mu}) = \frac{1}{10}\sum_{i}\left(\begin{pmatrix} t_i - \hat{\mu}_{P1} \\ t_i - \hat{\mu}_{P2} \end{pmatrix}(t_i - \hat{\mu}_{P1}\ t_i - \hat{\mu}_{P2}\right) = \begin{bmatrix} 2.82 & 4.99 \\ 4.99 & 9.63 \end{bmatrix}$$

The reliability is $P(t_{p1} \leq 5, t_{p1} \leq 6)$. Since the parameters are known, the reliability can be computed from Eq. (8.11), using the numerical method. This example uses Matlab mvncdf() for calculation. The reliability is 0.6802.

8.1.3 A Simulated Study

The multivariate Normal distribution has significant higher number of parameters than that of univariate distribution. A univariate Normal contains 2 parameters. For p-dimensional multivariate Normal distribution contains a number of $3p/2 + p^2/2$

Table 8.1 Two dimensions measurement data

P1	2.54	3.83	−0.26	2.86	2.32	0.69	1.57	2.34	5.58	4.77
P2	2.64	8.38	0.24	4.24	4.10	0.86	2.24	4.80	9.59	8.38

parameters. Accordingly, the data size required to achieve accurate parameter estimation is much higher than univariate distribution. The simulation study exhibits how the estimators converge to the true parameter with increasing number of data size.

The 2 dimensional multivariate contains total 5 parameters. The real parameters are

$$\boldsymbol{\mu} = (2\ 3);\quad \Sigma = \begin{bmatrix} 1 & 1.5 \\ 1.5 & 3 \end{bmatrix}$$

The simulation uses Monte Carlo simulation to generate pseudo-random numbers and then uses the maximum likelihood method discussed in the previous Section to estimate parameters. Figure 8.2 demonstrates the parameter converges with the data size. The converging is expected, while the estimators will close the true parameter when the data size reaches around 1000, which is a large size. In reliability engineering, data size reaching to 1000 is not common. The example tells when we choose the multivariate distribution for data analysis, we should have sufficiently large data size for reliable parameter estimation.

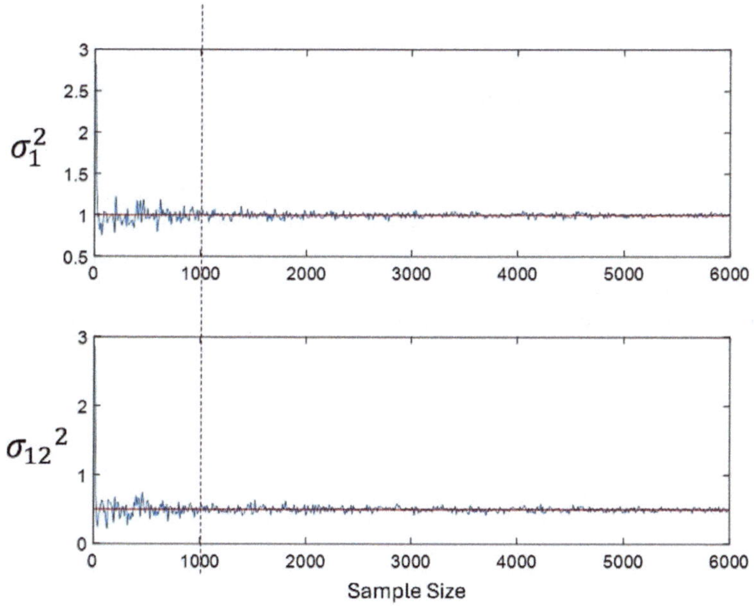

Fig. 8.2 Estimator converges with data size

8.2 Bivariate Normal Distribution

This section discusses the use of bivariate Normal distribution for dependency analysis. According to Eq. (8.4), the PDF for bivariate Normal distribution is

$$f(t) = \frac{1}{(2\pi)^{p/2}|\Sigma|^{1/2}} e^{-\frac{1}{2}(t-\mu)\Sigma^{-1}(t-\mu)^T} \tag{8.19}$$

$$\Sigma = \begin{bmatrix} {\sigma_1}^2 & {\sigma_{12}}^2 \\ {\sigma_{12}}^2 & {\sigma_2}^2 \end{bmatrix}$$

where ${\sigma_1}^2$ is variance of variable 1, written as ${\sigma_{11}}^2$ in Eq. (8.4), the ${\sigma_{12}}^2$ is the covariate of the two random variables defined as

$${\sigma_{12}}^2 = \rho\sigma_1\sigma_2 \tag{8.20}$$

The ρ is a coefficient defined between -1 and 1. Then

$$|\Sigma| = {\sigma_1}^2{\sigma_2}^2\left(1-\rho^2\right), \quad \Sigma^{-1} = \frac{1}{{\sigma_1}^2{\sigma_2}^2(1-\rho^2)}\begin{bmatrix} {\sigma_2}^2 & -\rho\sigma_1\sigma_2 \\ -\rho\sigma_1\sigma_2 & {\sigma_1}^2 \end{bmatrix} \tag{8.21}$$

The ρ is a parameter that can model the dependency of the two variables. If one variable corresponds to one component, the ρ can then be employed to model the dependency of the two components. In the Chap. 6 system reliability, when number of substituent systems increases to certain level, the system reliability approach to 0. One reason was the substituent components are independent from other. If the components are dependent from other, we can use the multivariate Normal distribution to model their dependency. At fixing values for both variables, we can examine how the CDF of the bivariate Normal changing with the ρ. As shown in Fig. 8.3, the probability CDF changes with the ρ efficient. At the $\rho = 0$, meaning no dependency, the probability is low. When it increases the ρ, the probability CDF is also increasing. Higher ρ leads to higher probability CDF. If the probability CDF is the reliability, then it means higher dependency leads to higher reliability for whole system. The coefficient ρ is a parameter representing the dependency of the variables.

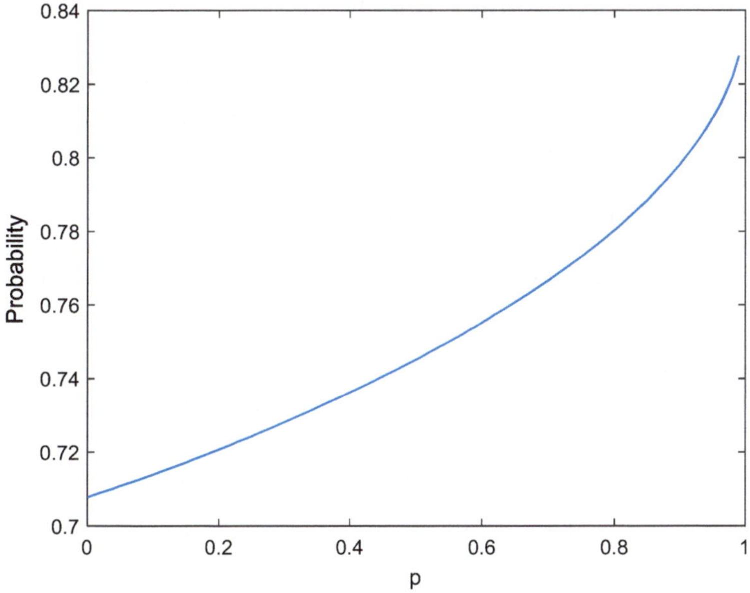

Fig. 8.3 Probability CDF changing with coefficient ρ

Example 8.3

Use the same data in Example 8.2. Suppose the two variables are independent from each other and each variable follows Normal distribution. The same as in Example 8.2, the system will fail when the values over P1 = 5 and P2 = 6. Compute the reliability of the system.

Solution: Table 8.1 contains two independent Normal distribution *P*1 and *P*2.

$$P1 = (2.54\ 3.83 - 0.26\ 2.86\ 2.32\ 0.69\ 1.57\ 2.34\ 5.58\ 4.77)$$

$$P2 = (2.64\ 8.38\ 0.24\ 4.24\ 4.10\ 0.86\ 2.24\ 4.80\ 9.59\ 8.38)$$

Using Matlab normfit() function, we obtain the mean and variance for *P*1 is 2.6240 and 1.7702, the reliability is the CDF function integrated from $-\infty$ to 5, as 0.9102.

For *P*2, the mean and variance are 4.5470 and 3.2707. The reliability is the CDF function integrated from $-\infty$ to 6, as 0.6716.

Since *P*1 and *P*2 are independent from each other, the system reliability is $0.9102 \times 0.6716 = 0.6113$, which is lower than that of 0.6802 in Example 8.2 where the dependency of the two variables is considered.

8.3 Bivariate Distribution Models

The above section discusses the use of multivariate Normal distribution, where for two-variate distribution, the coefficient ρ plays a role representing the dependency of the two components. This section discusses other methods to develop bivariate distribution for dependency modelling. Bivariate distribution is a common distribution used for reliability analysis, of which the abovementioned Bivariate Normal distribution is just one of them. In practise, the dependency of two failure modes in the mechanics system, the two-dimensional life data of automobile, are all cases of bivariate distribution.

There is a model named Marshall-Olkin model that is popular for bivariate distribution model [4]. In this model, the two variates are assumed to be exponentially distributed, and are connected by a term Max (t, s), which is written as:

$$R(t,s) = exp[-\lambda_1 \mathrm{t} - \lambda_2 \mathrm{s} - \lambda_0 \mathrm{Max\ (t, s)}]. \tag{8.22}$$

Applying the same thinking, the Exponential distribution can be generalized into a Weibull distribution case by replacing the Exponential distribution by a Weibull distribution [5] as

$$R(t,s) = exp[-\alpha_1 t^{\mathrm{c}} - \alpha_2 s^{\mathrm{c}} - \alpha_3 Max(t,s)^{\mathrm{c}}] \tag{8.23}$$

Or a simpler form of bivariate Weibull distribution is [6]:

$$R(t,s) = exp\left[-\alpha_1 t^{\beta_1} - \alpha_2 s^{\beta_2} - \alpha_3 t^{\beta_1} s^{\beta_2}\right]. \tag{8.24}$$

The dependence of the two variates is represented by the term $\alpha_3 t^{\beta_1} s^{\beta_2}$. Another way to combine the two variates is using a common exponent [7–11] as follows, which has some physical background.

$$R(t,s) = exp\left[-\left(\alpha_1 t^{\beta_1/\gamma} + \alpha_2 s^{\beta_2/\gamma}\right)^{\gamma}\right]. \tag{8.25}$$

In Eq. (8.26), the parameter γ plays a role of modelling the dependency of variables t, s. The Model Eq. (8.25) can be further simplified as

$$R(x,y) = exp\left\{-\left[(x/\alpha_1)^{\beta_1} + (y/\alpha_2)^{\beta_2}\right]^{\gamma}\right\}. \tag{8.26}$$

or

$$R(x,y) = exp\left[-\left(\alpha_1 x^{\beta_1} + \alpha_2 y^{\beta_2}\right)^{\gamma}\right]. \tag{8.27}$$

Equations (8.26) and (8.27) are of much simple form that can facilitate the parameter estimation [7, 11].

8.3.1 Maximum Likelihood Estimation Method

According to Larry [7], the joint probability density function (PDF) corresponding to Eq. (8.26) is as follows:

$$f(z_1,z_2) = \left[\gamma(1-\gamma)(z_1+z_2)^{\gamma-2} + \gamma^2(z_1+z_2)^{2\gamma-2}\right] exp[-(z_1+z_2)^{\gamma}] \quad (8.28)$$

If we use Eq. (8.26), the

$$z_1 = (x/\alpha_1)^{\beta_1}; z_2 = (y/\alpha_2)^{\beta_2} \quad (8.29)$$

And

$$\frac{\partial z_1}{\partial x} = \frac{\beta_1}{\alpha_1}\left(\frac{x}{\alpha_1}\right)^{\beta_1-1}; \frac{\partial z_2}{\partial y} = \frac{\beta_2}{\alpha_2}\left(\frac{y}{\alpha_2}\right)^{\beta_2-1} \quad (8.30)$$

Since

$$f(x,y) = f(z_1,z_2)\frac{\partial z_1}{\partial x}\frac{\partial z_2}{\partial y} \quad (8.31)$$

Rewrite Eq. (8.28) as

$$\begin{aligned} f(x,y) &= \left[\gamma(1-\gamma)(z_1+z_2)^{\gamma-2} + \gamma^2(z_1+z_2)^{2\gamma-2}\right] exp[-(z_1+z_2)^{\gamma}] \\ &= \left[\gamma(1-\gamma) + \gamma^2(z_1+z_2)^{\gamma}\right](z_1+z_2)^{\gamma-2} exp[-(z_1+z_2)^{\gamma}] \\ &= \gamma^2[1/\gamma - 1 + (z_1+z_2)^{\gamma}](z_1+z_2)^{\gamma-2} exp[-(z_1+z_2)^{\gamma}] \end{aligned} \quad (8.32)$$

According to Eq. (8.31), the

$$f(x,y) = \gamma^2\left\{\frac{1}{\gamma} - 1 + \left[\left(\frac{x}{\alpha_1}\right)^{\beta_1} + \left(\frac{y}{\alpha_2}\right)^{\beta_2}\right]^{\gamma}\right\}\left[\left(\frac{x}{\alpha_1}\right)^{\beta_1} + \left(\frac{y}{\alpha_2}\right)^{\beta_2}\right]^{\gamma-2}\cdot$$
$$\exp\left(-\left[\left(\frac{x}{\alpha_1}\right)^{\beta_1} + \left(\frac{y}{\alpha_2}\right)^{\beta_2}\right]^{\gamma}\right)\cdot\frac{\beta_1}{\alpha_1}\left(\frac{x}{\alpha_1}\right)^{\beta_1-1}\cdot \quad (8.33)$$
$$\frac{\beta_2}{\alpha_2}\left(\frac{y}{\alpha_2}\right)^{\beta_2-1}$$

The PDF is of complex expression. Suppose the observed failure data are given as pairwise as (x_i, y_i) $i = 1, 2, \ldots$. The likelihood function corresponding to Eq. (8.33) is then

$$L = \prod\nolimits_{i=1}^{n} f(x_i, y_i) \quad (8.34)$$

Take logarithm of both sides, the likelihood function is then

$$\begin{aligned}\ln L = n(-\ln\alpha_1 - \ln\alpha_2 + 2\ln\gamma + \ln\beta_1 + \ln\beta_2) + (\beta_1 - 1)\sum\frac{y_i}{\alpha_1}\\ + (\beta_2 - 1)\sum x_i/\alpha_2 + (\gamma - 2)\sum \ln \mathrm{B}_i + \sum \ln(1/\gamma - 1 + \mathrm{B}_i^{\gamma}) - \sum \mathrm{B}_i^{\gamma}\end{aligned} \quad (8.35)$$

where $\mathrm{B} = (x/\alpha_1)^{\beta_1} + (y/\alpha_2)^{\beta_2}$. No closed-form solution is available for α_1, β_1, α_2, β_2, γ from Eq. (8.35) for maximum likelihood estimation for ln$L$ = 0. Numerical methods should be engaged to find the solution.

8.3.2 Bayesian Estimation Method

Even engaging numerical method, solving the ln$L$ = 0 for Eq. (8.35) is challenging as it involves 5 parameters. For the numerical method with 5 variables, computing is time consuming. More critically, it is hard to verify the obtained solution is optimal and global as the shape of the ln$L$ is not clear. The Bayesian method may perform better when the optimal points of the likelihood function are not clear. The Bayesian method considers the parameters as random variables. The discussion of the Bayesian approach can refer to Chap. 10 of this book. The distribution for each parameter can be assigned subjectively, which is the prior distribution. For the classical Weibull distribution, the Gamma distribution is said to be a good prior distribution for the scale parameter [12]. According to Eq. (2.39) in Chap. 2, the Gamma distribution is with PDF as

$$f(x) = \frac{x^{k-1} e^{-\frac{x}{\theta}}}{\theta^k \Gamma(k)}; x > 0, k > 0, \theta > 0 \tag{8.36}$$

For scale parameter α_1, and let $k = a_1$ and $\theta = b_1$, the a_1, b_1, are numerical parameter values which can be specified arbitrarily. The a_1, b_1 are called hyper-parameters. Then the distribution for scale parameter α_1 is

$$f(\alpha_1) = \frac{{\alpha_1}^{k-1} e^{-\frac{\alpha_1}{\theta}}}{\theta^k \Gamma(k)} \tag{8.37}$$

Similar to parameter α_2, β_1, β_2, we can have a Gamma distribution for each. The prior distributions are not limited to any specific distribution; all distributions can also be used. For the γ, if we confine the parameter γ between 0 and 1. The Gamma distribution is then not suitable. In statistics, the Beta function is defined within (0,1) as

$$Beta(\gamma) = \frac{\Gamma(a_5 + b_5)}{\Gamma(a_5)\Gamma(b_5)} \gamma^{a_5 - 1} (1 - \gamma)^{b_5 - 1} \tag{8.38}$$

where the a_5, b_5 are hyper-parameters. By combining the likelihood function Eq. (8.35) and the prior distribution presented in Eqs. (8.36) and (8.37), the posterior Bayesian expression is as follows:

$$\begin{aligned} \pi(\alpha_1, \alpha_2, \beta_1, \beta_2, \gamma | Data) \propto L(Data) &\times \frac{1}{\Gamma(a_1) b_1^{a_1}} {\alpha_1}^{a_1 - 1} e^{-\frac{\alpha_1}{b_1}} \times \frac{1}{\Gamma(a_2) b_2^{a_2}} {\alpha_2}^{a_2 - 1} e^{-\frac{\alpha_2}{b_2}} \\ &\times \frac{1}{\Gamma(a_3) b_3^{a_3}} {\beta_1}^{a_3 - 1} e^{-\beta_1 / b_3} \times \frac{1}{\Gamma(a_4) b_4^{a_4}} {\beta_2}^{a_4 - 1} e^{-\beta_2 / b_4} \\ &\times \frac{\Gamma(a_5 + b_5)}{\Gamma(a_5)\Gamma(b_5)} \gamma^{a_5 - 1} (1 - \gamma)^{b_5 - 1} \end{aligned} \tag{8.39}$$

The a_1, b_1, a_2, b_2, a_3, b_3, a_4, b_4, a_5, b_5 are hyper-parameters which are assigned arbitrarily. For available data, the posterior $\pi(\alpha_1, \alpha_2, \beta_1, \beta_2, \gamma | Data)$, as usual, has no analytical solution. To solve the computing problem, in Bayesian analysis, Eq. (8.39) is usually performed by using MCMC (Markov chain Monte Carlo) simulation, for which Gibbs sampling is a popular method [13]. Refer to Chap. 10 in this book for detail for this computation method.

By simulation, we can obtain the empirical posterior distribution where parameter values are samples from the posterior distribution. The best estimate can be taken as the *mean* of the posterior empirical distributions. The obtained posterior distribution is an advantage of the Bayesian approach. From posterior, the confidence intervals of the estimates can be easily obtained. A drawback of Bayesian approach is however it engages very heavy computational cost, and it requires computer language coding. Checking the convergence of the simulation is another challenging.

8.3.3 Moment Method

As shown before, both maximum likelihood and Bayesian approach involves heavy computational cost. The numerical method based on maximum likelihood function may obtain unpredictable and unverifiable results. The Bayesian approach has stable performance as it requires no information of optimal points in the function, while the simulation depends on prior distribution and estimated estimator has certain uncertainty due to simulation. It is beneficial if other method can be engaged to solve the parameter estimation problem. Specially for model Eq. (8.27), Yuan proposed a moment method with much lower computational cost [14, 15]. The moment method follows an approach as: firstly, analytically obtain the theoretical moment of random variable by using some tricks. Then equal this theoretical moment to empirical moment to find out the estimates.

For model Eq. (8.27), a moment of x and y [14]

$$E\left(x^{i}\right)=\frac{\Gamma\left(1+\frac{i}{\gamma\beta_{1}}\right)}{\alpha_{1}^{i}} \tag{8.40}$$

$$E\left(y^{j}\right)=\Gamma\left(1+\frac{j}{\gamma\beta_{2}}\right)/\alpha_{2}^{j} \tag{8.41}$$

And

$$E\left(x^{i}y^{j}\right)=\frac{\Gamma\left(1+\frac{i}{\gamma\beta_{1}}+\frac{j}{\gamma\beta_{2}}\right)\Gamma\left(\frac{i}{\beta_{1}}+1\right)\Gamma\left(\frac{j}{\beta_{2}}+1\right)}{\Gamma\left(\frac{i}{\beta_{1}}+\frac{j}{\beta_{2}}+1\right)\alpha_{1}^{i}\alpha_{2}^{j}} \tag{8.42}$$

The i and j is order of the moment. Suppose data are available in pair as

$$(x_1, y_1);\quad (x_2, y_2); \ldots (x_k, y_k); \ldots (x_n, y_n) \tag{8.43}$$

According to the large sample theory, when data size n is large, the empirical mean of (x_i, y_i) will approach the expectation of this variable. It tells

$$E\left(x^{i}\right)\rightarrow\overline{x^{i}}=\sum_{k}{x_k}^{i}/n.$$

$$E(y^j) \to \overline{y^j} = \sum_k {y_k}^j / n.$$

$$E(x^i y^j) \to \overline{x^i y^j} = \sum_k {x_k}^i {y_k}^j / n \tag{8.44}$$

In order to determine the 5 unknown parameters, at least five equations are required to uniquely determine the parameters. We can choose the simplest first and second order for both variable x and y, and the first order for $x.\ y$. A total number of 5 moments is used for parameter estimation as follows [14]:

$$\begin{cases} E(x) = \dfrac{\Gamma\left(1+\dfrac{1}{\gamma\beta_1}\right)}{\alpha_1} = \overline{x} \\ E(x^2) = \dfrac{\Gamma\left(1+\dfrac{2}{\gamma\beta_1}\right)}{\alpha_1^2} = \overline{x^2} \\ E(y) = \dfrac{\Gamma\left(1+\dfrac{1}{\gamma\beta_2}\right)}{\alpha_2} = \overline{y} \\ E(y^2) = \dfrac{\Gamma\left(1+\dfrac{2}{\gamma\beta_2}\right)}{\alpha_2^2} = \overline{y^2} \\ E(xy) = \dfrac{\Gamma\left(1+\dfrac{1}{\gamma\beta_1}+\dfrac{1}{\gamma\beta_2}\right)\Gamma\left(\dfrac{1}{\beta_1}+1\right)\Gamma\left(\dfrac{1}{\beta_2}+1\right)}{\Gamma\left(\dfrac{1}{\beta_1}+\dfrac{1}{\beta_2}+1\right)\alpha_1\alpha_2} = \overline{xy} \end{cases} \tag{8.45}$$

Again, there is still no analytical solution to Eq. (8.45). Numerical approach is needed to find the solution. But the numerical approach for moment approach is much easier than that in the maximum likelihood method. Tricks can be played to simplify Eq. (8.45). Merging the first two equations of (8.45) to a one-variable equation as follows

$$\frac{\Gamma^2\left(1+\frac{1}{\gamma\beta_1}\right)}{\Gamma\left(1+\frac{2}{\gamma\beta_1}\right)} = \overline{x}^2 / \overline{x^2} \tag{8.46}$$

As the $\gamma\beta_1$ can be considered as one variable. For one-variable Eq. (8.46), the numerical method is easy and stable. After the $\gamma\beta_1$ has been obtained by Eq. (8.46). The parameter α_1 can then be obtained as

$$\alpha_1 = \frac{\Gamma\left(1 + \frac{1}{\gamma\beta_1}\right)}{\overline{x}} \tag{8.47}$$

Applying the same trick, the value of α_2 and $\gamma\beta_2$ can also be obtained. Special concerns needed on the γ and β_1, β_2. Let $\frac{\gamma\beta_2}{\gamma\beta_1} = Q$ and the Q can be obtained since the $\gamma\beta_1$ and $\gamma\beta_2$ are known by Eq. (8.46). The last equation in Eq. (8.45) is then written into a one variable equation as follows

$$\frac{\Gamma\left(\frac{1}{\beta_1} + 1\right)\Gamma\left(\frac{1}{Q\beta_1} + 1\right)}{\Gamma\left(\frac{1}{\beta_1} + \frac{1}{Q\beta_1} + 1\right)} = \widehat{\alpha_1}\widehat{\alpha_2}\overline{xy}/\Gamma\left(1 + \frac{1}{\widehat{\gamma\beta_1}} + \frac{1}{\widehat{\gamma\beta_2}}\right) \tag{8.48}$$

All the Q, $\widehat{\alpha_1}$, $\widehat{\alpha_2}$, $\widehat{\gamma\beta_1}$, $\widehat{\gamma\beta_2}$ are known already. A edge advantage of the moment method is the solutions to Eqs. (8.46) and (8.47) are unique as proved in [14]. Equation (8.48) is also unique when

$$\frac{\overline{xy}}{\overline{x}\,\overline{y}} \leq 1 \text{ or } \frac{\Gamma\left(1 + \frac{1}{\widehat{\gamma\beta_1}}\right)\Gamma\left(1 + \frac{1}{\widehat{\gamma\beta_2}}\right)}{\Gamma\left(1 + \frac{1}{\widehat{\gamma\beta_1}} + \frac{1}{\widehat{\gamma\beta_2}}\right)} \times \frac{\overline{xy}}{\overline{x}\,\overline{y}} \leq 1 \tag{8.49}$$

The proof can be referred to in Yuan [14].

Example 8.4
For the truck 2 data presented in Table 8A.1, use the moment method to estimate the parameter for the model $R(x, y) = exp\left[-\left(\alpha_1 x^{\beta_1} + \alpha_2 y^{\beta_2}\right)^{\gamma}\right]$ in Model (8.27).

Solution: Matlab fsolve() was used to solve the numerical solution for Eq. (8.46) and Eq. (8.48).

$$\overline{x} = 103.0234; \overline{x^2} = 1.902 \times 10^4$$

$$\overline{y} = 1.4042 \times 10^4; \overline{y^2} = 3.3090 \times 10^8$$

$$\overline{xy} = 2.4351 \times 10^6$$

By using fsolve() according to Eq. (8.46), the

$$\gamma\beta_1 = 0.8884$$

For x, according to Eq. (8.47),

(continued)

Example 8.4 (continued)

$$\alpha_1 = \frac{\Gamma\left(1+\frac{1}{\gamma\beta_1}\right)}{\overline{x}} = \frac{\Gamma\left(1+\frac{1}{0.8884}\right)}{103.0234} = 0.0093$$

For y,

$$\gamma\beta_2 = 0.8194$$

$$\alpha_2 = \frac{\Gamma\left(1+\frac{1}{\gamma\beta_2}\right)}{\overline{x}} = \frac{\Gamma\left(1+\frac{1}{0.8194}\right)}{1.4042\times 10^4} = 6.6708\times 10^{-5}$$

According to Eq. (8.48) by using fsolve() in Matlab, we obtain the

$$\beta_1 = 6.5964$$

Since

$$\frac{\gamma\beta_1}{\gamma\beta_2} = \frac{\beta_1}{\beta_2}$$

Then

$$\beta_2 = \beta_1 . \frac{\gamma\beta_2}{\gamma\beta_1} = 6.5964 * \frac{0.8194}{0.8884} = 6.0841$$

Therefore,

$$\gamma\beta_1 = 0.8884 = \gamma * 6.5964$$

$$\gamma = \frac{\gamma\beta_1}{\beta_1} = \frac{0.8884}{6.5964} = 0.1347$$

The uniqueness condition of Eq. (8.49) was checked. The $\frac{\overline{xy}}{\overline{\overline{x}}\,\overline{y}} = 0.9716 < 1$, confirming the uniqueness condition $\frac{\overline{xy}}{\overline{\overline{x}}\,\overline{y}} \leq 1$ is satisfied. The solution to for moment method shown in Eq. (8.45) thus exists and they are unique. The obtained estimates are as following Table 8.2.

Using the estimated parameter, the $R(x,y) = exp\left[-\left(\alpha_1 x^{\beta_1} + \alpha_2 y^{\beta_2}\right)^{\gamma}\right]$ is plotted for $x = 10$ and y is ranged from 0 to 500,000, shown as the left of Fig. 8.4, and for $y = 100{,}000$ and x is 1 to 5000. Both are seen as a decreasing trend, and the reliability tends to zero when the x or y tends to infinite, which agrees with the reality.

Table 8.2 Estimated parameter for Example 8.4

Parameter	Estimates
$\widehat{\alpha_1}$	0.0093
$\widehat{\beta_1}$	6.5964
$\widehat{\alpha_2}$	6.6708×10^{-5}
$\widehat{\beta_2}$	6.0841
$\widehat{\gamma}$	0.1347

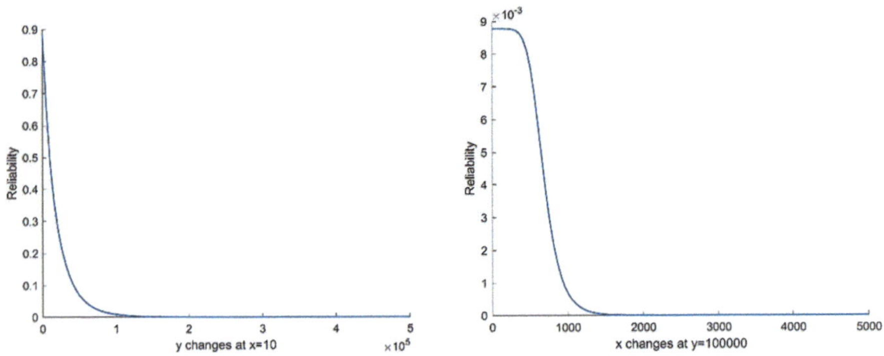

Fig. 8.4 Reliability for Bivariate Weibull distribution for a mining data

8.3.4 Goodness-of-Fit Test

The adequacy of model (8.26) or (8.27) should be checked by using goodness of fit test. A poorly fitting model for the available data has no practical value. Goodness of fit test is another challenge for use of bivariate distribution model. Let

$$B_i = \alpha_1 x_i^{\beta_1} + \alpha_2 y_i^{\beta_2} \tag{8.50}$$

We can write the reliability function for Model (8.26) and (8.27) as follows

$$R_i = exp[-B_i^{\gamma}] \tag{8.51}$$

Since parameter γ is known prior to the test, the quantity B_i^{γ} as a variable can be calculated. Equation (8.51) is degenerated into an Exponential distribution with failure rate 1 as PDF as follows

$$f(C_i) = e^{-C_i} \tag{8.52}$$

The $C_i = B_i^{\gamma}$. For fitting test, we linearize Eq. (8.52) by taking logarithm on both sides, then

$$\ln R_i = -C_i \tag{8.53}$$

which is of linear form $y = -x$.

Same as the graphic goodness of fit test method introduced in Chap. 5 for Weibull distribution, we then use Eq. (8.53) for test. The empirical R_i can be approximated, for example by using median rank as follows

$$R(i) = 1 - \frac{i - 0.3}{n + 0.4} \tag{8.54}$$

The

$$C_i = \left[\left(\frac{x_i}{\widehat{\alpha_1}}\right)^{\widehat{\beta_1}} + \left(\frac{y_i}{\widehat{\alpha_2}}\right)^{\widehat{\beta_2}}\right]^{\widehat{\gamma}} \tag{8.55}$$

Or

$$C_i = \left(\widehat{\alpha_1} x^{\widehat{\beta_1}} + \widehat{\alpha_2} y^{\widehat{\beta_2}}\right)^{\widehat{\gamma}} \tag{8.56}$$

If the data can fit the line Eq. (8.53), the parameter as well as the model are then both acceptable; If not, then either the estimated parameter values are incorrect or the Model (8.26) or (8.27) is not suitable for the data.

Example 8.5

For the Truck 1 data presented in Table 8A.1, use the Bayesian MCMC approach to estimate the parameter for reliability model $R(x, y) = exp\left[-\left[\left(\frac{x}{\alpha_1}\right)^{\beta_1} + \left(\frac{y}{\alpha_2}\right)^{\beta_2}\right]^{\gamma}\right]$. The $= \left[\left(\frac{x}{\alpha_1}\right)^{\beta_1} + \left(\frac{y}{\alpha_2}\right)^{\beta_2}\right]^{\gamma}$. Following estimates are obtained as shown in Table 8.3.

Use goodness of fit test to evaluate if the estimates are acceptable?

Solution: For truck 1, part of the values for B_i^{γ} and $\ln R_i$, which are empirical data, are shown in the following Table 8.4.

If the empirical data fit the straight line $y = -x$, the data is then considered fitting the model; otherwise not. For Truck 1, the data in general fit the data as shown in Fig. 8.5. The estimates in Table 8.3 are then acceptable.

Table 8.3 Bayesian estimator for model (8.27) [15]

Parameter	Bayesian estimates
α_1	11,210
β_1	4.526
α_2	79.33
β_2	4.224
γ	0.2237

Table 8.4 Examples of data sets for goodness of fit test

i	1	2	3	4	5	6	7	8	9	10	11	…
B_i^{γ}	0.0573	0.0677	0.0844	0.1013	0.1460	0.1525	0.1608	0.1632	0.1934	0.2197	0.2636	…
$\ln R_i$	−0.0143	−0.0350	−0.0562	−0.0799	−0.1000	−0.1226	−0.1458	−0.1694	−0.1937	−0.2186	−0.2441	…

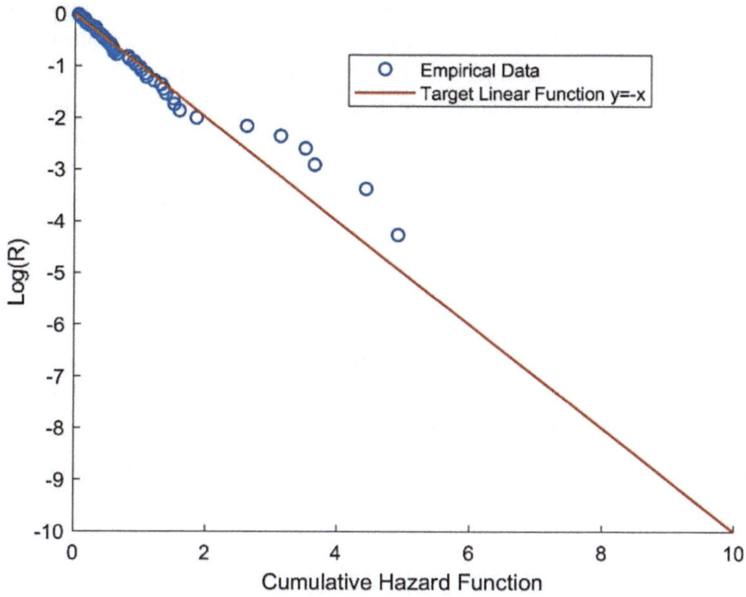

Fig. 8.5 Goodness of fit test for Truck 1 data

8.4 Copula Function for Reliability

In reliability data analysis, we often collect and analyze the life data limited to individual components and ignore the dependency of it on other components. In the chapter of system reliability, some methods assume the reliability of each component is independent. However, as we have discussed so far in this Chapter, the independence assumption does not hold in some context, or even we can say for most situations it does not hold. The dependency among the components should be considered in some situations. Other than the multivariate method and bivariate distribution discussed above, Copula function is another method that can be used for the dependency analysis.

8.4.1 Definition of Copula

The Copula is referred to as a function to join multivariate distribution to one distribution. For each variate, its distribution is in a form of marginal distribution, i.e. the CDF of the variable, or sometimes a reliability function. The Copula is to connect the CDF (by default it is CDF but can be reliability function as well) of several marginal distribution. For reliability data analysis, the CDF corresponds to the probability of failure. The Copula function for this CDF is thus for the coupling of probability of failure function. Our discussion of Copula starts with the simple bivariate Copulas. As above-mentioned, the Copula integrates the multivariate CDF to a CDF [16]. The CDF is valued in range $I \in [0, 1]$. For bivariate Copula, the Copula function is a map from $[0, 1] \times [0, 1] \rightarrow [0, 1]$, which is written as a function as $C(u, v)$. The $u = F(x)$, $v = G(y)$ which are uniformly distributed between 0 and 1, since they are the marginal distributions. The $C(u, v)$ should be incremental to satisfy the condition as CDF of a distribution. Other conditions are: for any value $u, v \in [0, 1]$, there are $C(u, 1) = u$; $C(1, v) = v$. Rewrite the Copula function in a more understandable form as

$$H(x, y) = C(u = F(x), v = G(y)) \tag{8.57}$$

The x, $y \in [-\infty, +\infty]$. The Sklar's Theorem, which is the fundamental theorem for Copula function, describes the conversion from $[0, 1] \times [0, 1] \rightarrow [0, 1]$ is unique when the marginal functions $F(x)$, $G(y)$ are continuous.

Sklar's Theorem [16] Let the $H(x, y)$ be the joint CDF with margins $F(x)$ and $G(y)$. Then there exists a Copula function

$$C(u, v) = C(F(x), G(y)) = P(X1 \leq x, Y1 \leq y) = H(x, y)$$

In other words, the Sklar's theorem states that the joint multivariate distribution can be rewritten in term of its constituent margin univariate distributions. One attention is needed to pay is the variables in $C(u, v)$ are u, v which is CDF of margin univariate distributions, while in the joint CDF $H(x, y)$ the variables are x and y. An example is presented as follows to illustrate the Theorem and show how to connect the Copula function with multivariate distribution.

Example 8.6
Given a bivariate Exponential distribution is with joint CDF as

$$H(x, y) = \begin{cases} 1 - e^{-x} - e^{-y} + e^{-(x+y+\theta xy)}; x \geq 0, y \geq 0 \\ 0, \mathit{Otherwise} \end{cases}$$

The $\theta \in [0, 1]$. Find out the Copula function for the $H(x, y)$.

Solution: The marginal function for variable x is

$$u = F(x) = H(x, y = \infty) = 1 - e^{-x}$$
$$v = G(y) = H(\infty, y) = 1 - e^{-y}$$

Taking the inverse function of the above two equations, we obtain

(continued)

Example 8.6 (continued)

$$x = -\ln(1-u);$$

$$y = -\ln(1-v)$$

Substitute the x, y to function $H(x, y)$, we obtain the Copula function

$$C(u, v) = u + v - 1 + (1-u)(1-v)e^{-\theta \ln(1-u)\ln(1-v)}$$

The **Sklar's** theorem assures the conversion from $H(x, y)$ to $C(u, v)$ is obtainable and unique.

The Frechet-Hoeffding Bounds for Copula function is described as

$$\max(u + v - 1, 0) \leq C(u, v) \leq \min(u, v) \tag{8.58}$$

The bound (8.58) is important for reliability analysis to find the bound. We will discuss it later.

8.4.2 *Series System Dependency*

In Chap. 6, the reliability of a series system is multiplication of that of its substituent components by assuming components independent from each other. We can obtain the system reliability in terms of probability of failure, i.e. the CDF of the component, by using Copula function. The CDF of substituent component is $F(t) = P(T \leq t)$. The probability of system failure is

$$F_s(t) = F_1(t) + F_2(t) - F_1(t) \cap F_2(t) \tag{8.59}$$

The $F_1(t) \cap F_2(t)$ might contains dependency of the two components, so we use the Copula function. Let $u = F_1(t)$, $v = F_2(t)$, Eq. (8.59) is rewritten into a reliability function as

$$R_s = 1 - C(u, 1) - C(1, v) + C(u, v) \tag{8.60}$$

where $C(u, 1) = u = F_1(t)$, $C(1, v) = v = F_2(t)$.

For a system with number of n substituent components, the substituent components are dependent. Formula (8.59) extends to

$$R_s(t)=1-F_s(t)=1-\left\{\sum F_i(t)-\sum_{1\le i\le j\le n} F_i(t)F_j(t)\ldots+(-1)^{n-1}F_1(t)\ldots F_n(t)\right\} \tag{8.61}$$

Rewrite the (8.61) in terms of Copula function as

$$R_s= C(1,1,\ldots)-\left\{\sum C(1,1,\ldots.F_i(t),\ldots)-\sum_{1\le i\le j\le n} C(1,1,\ldots F_i(t),\ldots.,F_j(t)\ldots)\ldots + (-1)^{n-1}C(F_1(t)\ldots F_n(t))\right\} \tag{8.62}$$

Example 8.7
A two-unit system are connected in series, while the two units are dependent from each other. The reliability of the two units is 0.99, 0.98 respectively. Suppose the dependency of the two units follows a Copula function as $C(u,v;\theta)=\exp\left\{-\left[(-\ln u)^{\frac{1}{\theta}}+(-\ln v)^{\frac{1}{\theta}}\right]^{\theta}\right\}$. The $\theta = 0.01$. What is the reliability of the system?

Solution: The CDF for the two units are $u = 1 - 0.99 = 0.01$; $v = 1 - 0.98 = 0.02$. The

$$C(u,v;\theta)=\exp\left\{-\left[(-\ln u)^{\frac{1}{\theta}}+(-\ln v)^{\frac{1}{\theta}}\right]^{\theta}\right\}$$
$$=\exp\left\{-\left[(-\ln 0.01)^{\frac{1}{0.01}}+(-\ln 0.02)^{\frac{1}{0.01}}\right]^{0.01}\right\}=0.01$$

According to Eq. (8.62), the reliability is

$$R_s=1-C(u,1)-C(1,v)+C(u,v)=1-u-v+C(u,v)=1-0.01-0.02+0.01=0.98$$

The reliability for the dependent series system is 0.98.

For comparison, if the two units are total independent from each other, the reliability is

$$R_s=0.99*0.98=0.9702$$

It can be seen that the dependent system has higher reliability than that of independent system.

The bound of the series system considering dependency can be derived by applying the Frechet-Hoeffding bounds in (8.58). The reliability for series system is then bounded by [17]

$$\prod_{i=1}^{n} R_i(t) \leq R_s(t) \leq \min(R_1(t), R_2(t), R_3(t)....) \tag{8.63}$$

For two-unit series system in Example 8.7, for the various θ in the Copula function, the system reliability calculated from Eq. (8.60) is decreasing with the θ. The lower bounder of reliability is the $\prod_{i=1}^{n} R_i(t)$ and the upper bound of system reliability is equal to the weakest unit in system. For the case presented in Example 8.7, the lower bound is 0.9702 when the two units are totally independent from each other. The upper bound of the system reliability is the reliability of the weakest unit which is 0.98. Figure 8.6 plots the system reliability for θ ranges from 0 to 1, where we can see the system reliability decreasing from upper bound 0.98 to lower bound 0.9702, which is exact as described by Eq. (8.63).

In Chap. 6, we have discussed for a large-scale system, the reliability approaches to 0 when the number of constituent units is large, and they are connected in series. The assumption is not realistic. The number of units in the aircraft is large but the reliability is still high, at least not as low as near 0. One issue is the dependency among the components is not considered. The Copula method can solve the problem by considering their dependency. An example here is illustrated to show how the

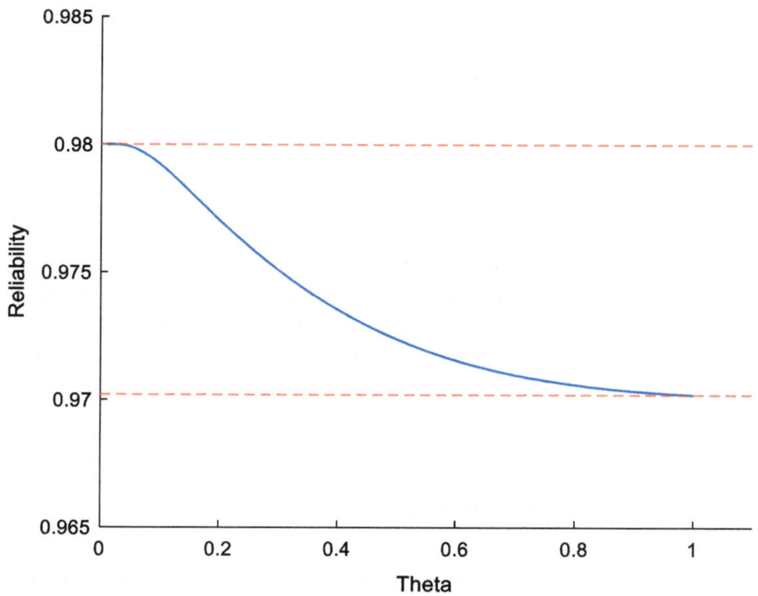

Fig. 8.6 System reliability for dependent series system with various θ

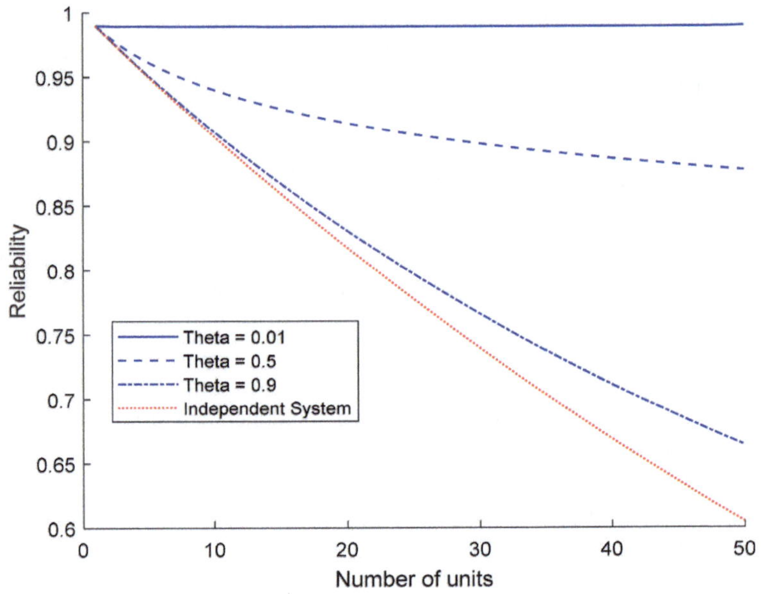

Fig. 8.7 System reliability with number of constituent units

reliability decreases with number of units when dependency considered. Suppose the units are identical and connected in series. The reliability function of Eq. (8.62) is simplified as follows when Copula function $C(u, v; \theta) = \exp\left\{-\left[(-\ln u)^{\frac{1}{\theta}} + (-\ln v)^{\frac{1}{\theta}}\right]^{\theta}\right\}$ are used.

$$R_s = 1 - \sum_{i=1}^{n} (-1)^{i+1} \binom{n}{i} \exp\left\{-\left[\mathrm{i} * (-\ln u)^{\frac{1}{\theta}}\right]^{\theta}\right\} \tag{8.64}$$

Suppose the $u = 0.01$, the system reliability is decreasing with number of units, but it varies with the dependency parameter θ. For $\theta = 0.01$, which implies a high dependency within the units, the system reliability does not drop significantly as shown in Fig. 8.7. For $\theta = 0.9$, which implies low dependency within the units, the system reliability exhibits close to the reliability of considering totally independence. The example demonstrates how dependency affects the whole system reliability.

8.4.3 Parallel System Dependency

Parallel system is defined as until all substituent units connected in parallel failed, the whole system will fail. In terms of probability of failure or the CDF function of the failure time distribution, the probability of whole system failure is simpler than that of series system with

$$F_s(t) = F_1(t) \cap F_2(t) \tag{8.65}$$

Write Eq. (8.65) in terms of Copula function as

$$F_s(t) = C(u, v) \tag{8.66}$$

For number of n substituent units connected in parallel, it is not hard to extend Eq. (8.66) as

$$F_s(t) = C(u_1, u_2, u_3, \ldots.) \tag{8.67}$$

where $u_i = F_i(t)$. Reliability is then

$$R_s(t) = 1 - C(u_1, u_2, u_3, \ldots.) \tag{8.68}$$

Example 8.8

A system with two-unit connected in parallel and the two units are dependent from each other. The reliability of the two units is 0.99, 0.98 respectively. Suppose the dependency of the two units follows the Copula function as $C(u, v; \theta) = \exp\left\{-\left[(-\ln u)^{\frac{1}{\theta}} + (-\ln v)^{\frac{1}{\theta}}\right]^{\theta}\right\}$. The $\theta = 0.01$. What is the reliability of the system?

Solution: The CDF for the two units is $u = 1 - 0.99 = 0.01$; $v = 1 - 0.98 = 0.02$. The

$$\begin{aligned} C(u, v; \theta) &= \exp\left\{-\left[(-\ln u)^{\frac{1}{\theta}} + (-\ln v)^{\frac{1}{\theta}}\right]^{\theta}\right\} \\ &= \exp\left\{-\left[(-\ln 0.01)^{\frac{1}{0.01}} + (-\ln 0.02)^{\frac{1}{0.01}}\right]^{0.01}\right\} = 0.01 \end{aligned}$$

According to Formula (8.68), the system reliability in parallel is

$$R_s(t) = 1 - C(u_1, u_2, u_3, \ldots.) = 1 - 0.01 = 0.99$$

If the two units are totally independent from each other, the system reliability is then

$$R_s(t) = 1 - uv = 1 - 0.01 * 0.02 = 0.9998 > 0.99$$

The system reliability of independent system is higher than that of dependent system.

Without providing proof, the bound of parallel system considering dependency is [17].

$$\max(R_1(t), R_2(t), R_3(t)....) \le R_s(t) \le 1 - \prod_{i=1}^{n} (1 - R_i(t)) \tag{8.69}$$

i.e. the system reliability for dependent system is lower than its counterpart of independent system for a parallel system, as shown in the Example 8.8.

8.4.4 Parameter Estimation for Copula Function

For any Copula function, the unknown parameters should be estimated to allow the practical use of the model. It is a challenge to estimate the parameters for a model with plenty number of parameters: parameters in the Copula function and the parameters in each marginal CDF. A popular method, which is implemented by Matlab applies a two-stage procedure: firstly, estimate the parameters in the Copula function, next step is to estimate the parameter for each marginal CDF. The joint distribution is

$$H(t_1, t_2, ...) = C(u_1 = F_1(t_1), u_1 = F_2(t_2), ...) \tag{8.70}$$

From Eq. (8.70), it is not hard to get the PDF of the joint distribution of number of n CDFs is [18]

$$\frac{\partial^n H}{\partial t_1 \partial t_2 \partial t_3 ....} = \frac{\partial^n C}{\partial u_1 \partial u_2 \partial u_3 ....} \frac{dF_1(t_1)}{dt_1} \frac{dF_2(t_2)}{dt_2} ... = \frac{\partial^n C}{\partial u_1 \partial u_2 \partial u_3 ....} \prod_{i=1}^{n} f_1(t_i) \tag{8.71}$$

Given observation data

$$t_1^j, t_2^j, t_3^j, ...t_n^j \quad j = 1, 2, 3, ...m$$

From Eq. (8.71), it is easy to obtain the log likelihood function

$$\ell(\boldsymbol{\theta}) = \sum_{j=1}^{m} \ln c(F_1(t_1^j), F_2(t_1^j), F_3(t_1^j)...) + \sum_{j=1}^{m} \sum\nolimits_{i=1}^{n} \ln f_i(t_i^j) \tag{8.72}$$

where $c(F_1(t_1^j), F_2(t_1^j), ...) = \frac{\partial^n C}{\partial u_1 \partial u_2 \partial u_3 ....}$.

Rewrite the (8.72) as

$$\ell(\boldsymbol{\theta}) = \sum_{j=1}^{m} \ln c\left(u_1{}^j, u_2{}^j, u_3{}^j \ldots\right) + \sum_{j=1}^{m} \sum\nolimits_{i=1}^{n} \ln f_i\left(t_i^j\right) \tag{8.73}$$

In Eq. (8.73), the $\sum_{j=1}^{m} \ln c(u_1{}^j, u_2{}^j, u_3{}^j \ldots)$ does not contain parameter of the marginal distribution. The second part does not contain the parameters of Copula. The maximum likelihood estimators from the maximum of (8.73) can thus be divided into two independent functions. Let α represents the parameters in the Copula function, the θ_i represents parameters in the marginal distribution function. Formula (8.73) is rewritten as

$$\ell(\boldsymbol{\theta}) = \sum_{j=1}^{m} \ln c\left(u_1{}^j, u_2{}^j, u_3{}^j \ldots; \alpha\right) + \sum\nolimits_{j=1}^{m} \ln f_1\left(t_1{}^j; \theta_1\right) + \sum\nolimits_{j=1}^{m} \ln f_2\left(t_2{}^j; \theta_2\right) + \ldots. \tag{8.74}$$

which means the parameter estimator on marginal distribution is divided to estimate its own parameters without mixing with others. The parameter estimation is hence significantly simplified.

The u_i^j is CDF, approximated by the empirical CDF of each variable t_i. As we discussed in Chap. 5, the empirical CDF can be approximated by median rank $\frac{i-0.3}{n+0.4}$, or by other ranks.

Take a Copula function defined as $C(u, v; \theta) = \exp\left\{ - \left[(-\ln u)^{\frac{1}{\theta}} + (-\ln v)^{\frac{1}{\theta}} \right]^{\theta} \right\}$ as example [4]. The PDF of the Copula is then [19]:

$$\frac{\partial'' C}{\partial u \partial v} = \mathrm{C}(u, v) \frac{1}{uv} (\ln u \ln v)^{\frac{1}{\theta}-1} \left[(-\ln u)^{\frac{1}{\theta}} + (-\ln v)^{\frac{1}{\theta}} \right]^{\theta-2} \left\{ \left[(-\ln u)^{\frac{1}{\theta}} + (-\ln v)^{\frac{1}{\theta}} \right]^{\theta} + \theta - 1 \right\} \tag{8.75}$$

This example demonstrates a complex expression for the Copula PDF function. As usual, for n-variable Copula function, computer software is required to accomplish the computation to find solutions.

Example 8.9

The Copula function is $(u, v; \theta) = \exp\left\{-\left[(-\ln u)^{\frac{1}{\theta}} + (-\ln v)^{\frac{1}{\theta}}\right]^{\theta}\right\}$. The data for the two variables are shown in Table 8.5:

Estimate the parameter value θ.

Solution: The parameter θ can be estimated by using Formula (8.75) as PDF. To solve the problem, we use the ksdensity() function in Matlab to obtain the empirical CDF. Other CDF approximation methods can also be used, for example, the median-rank method. The empirical CDF corresponds to the above table is shown in Table 8.6.

By using the maximum likelihood method using Formula (8.75) as PDF, this example uses the copulafit() in the Matlab, we obtain the

$$\widehat{\theta} = 0.8233$$

Table 8.5 Two variable data

x	0.30	0.12	3.06	0.10	0.36	2.43	2.28	0.80	0.05	0.14
y	2.70	0.16	0.10	2.45	0.35	3.61	1.93	1.18	1.47	0.18

Table 8.6 Empirical CDF for the two variable data

u	0.405	0.252	0.950	0.235	0.451	0.823	0.778	0.642	0.195	0.269
v	0.772	0.234	0.222	0.733	0.274	0.902	0.629	0.461	0.528	0.238

Appendix 8A

Table 8A.1 Failure data collected from field for Truck 1 and Truck 2 (Data Source: [15])

Mining Truck 1			Mining Truck 2		
ID	Time to failure (in hours)	Workload (in ton × km)	ID	Time to failure (in hours)	Workload (in ton × km)
1	18.55	2013	1	225.15	21997.8
2	31.15	3501.3	2	255.35	26485.8
3	67.05	6986.1	3	109.4	10840.5
4	10.4	1511.4	4	52.1	6754
5	26.55	2963.4	5	44.4	6664.9
6	292.5	29494.3	6	280.11	34886.5
7	10.3	1217.7	7	39.2	4527.6
8	20.4	2992	8	180.55	24619.1
9	101.15	10906.5	9	254.1	26753.1
10	85	6705.6	10	22.45	2821.5
11	46.05	5009.4	11	45.4	6695.7
12	266.29	31519.4	12	252.1	28967.4
13	21.3	5544	13	33.5	4664
14	35	3375.9	14	9.2	3483.7
15	11	1230.9	15	56.25	6496.6
16	243.15	26690.4	16	22.15	3465
17	120.55	13084.5	17	31.35	4048
18	34.1	5359.2	18	44.2	12698.4
19	42.05	4775.1	19	169.41	34491.6
20	136.35	16525.3	20	20.45	2079
21	10.2	2420	21	149.05	14813.7
22	19.3	10,351	22	326.3	47537.6
23	22.15	10997.8	23	61.35	11809.6
24	61.15	6755.1	24	33.55	4818
25	28.45	3177.9	25	17.55	1570.8
26	39.45	4492.4	26	37.35	3036
27	29.15	3831.3	27	203.2	29823.2
28	2.1	653.4	28	97.55	11267.3
29	10.4	2051.5	29	110.25	16124.9
30	5	1100	30	45.25	6211.7
31	22.05	2607	31	86.45	13818.2
32	39.45	5643	32	85.1	18,590
33	67.21	14032.7			
34	34.15	8911.1			
35	94.3	13212.1			
36	195.4	23,562			
37	32.4	5984			
38	69.5	16,434			

(continued)

Table 8A.1 (continued)

Mining Truck 1			Mining Truck 2
39	19.25	3410	
40	56.45	10609.5	
41	17.25	3421	
42	89.1	15446.2	
43	143.4	48605.7	
44	4.1	920.7	
45	3.2	742.5	
46	81.2	8434.8	
47	87	10611.7	
48	279.5	51966.2	
49	10.1	1007.6	

References

1. Johnson RA, Wichern DW (2003) Applied multivariate statistical analysis. Pearson Prentice Hall, Upper Saddle River, NJ
2. Genz A, Bretz F (2002) Comparison of methods for the computation of multivariate t probabilities. J Comput Graph Stat 11(4):950–971. https://doi.org/10.1198/106186002394
3. Bensimhoun ME (2009) N-dimensional cumulative function, and other useful facts about gaussians and normal densities
4. Marshall AW, Olkin I (1967) A multivariate exponential distribution. J Am Stat Assoc 62(317): 30–44
5. Hanagal DD (1996) A multivariate Weibull distribution. Econ Qual Contr 11:193–200
6. Roy D (1994) Classification of life distributions in multivariate models. IEEE Trans Reliab 43(3):445–445
7. Lee L (1979) Multivariate distributions having Weibull properties. J Multivar Anal 9(2): 267–277
8. Wen CKLM-J (2009) A multivariate Weibull distribution. Pak J Stat Oper Res 5(2):55–66
9. Lu JC, Bhattacharyya GK (1990) Some new constructions of bivariate Weibull models. Ann Inst Stat Math 42(3):543–559
10. Roy D, Mukherjee SP (1998) Multivariate extensions of univariate life distributions. J Multivar Anal 67(1):72–79
11. Hougaard P (1986) A class of multivariate failure time distributions. Biometrika 73(3):671–678
12. Kundu D, Gupta AK (2013) Bayes estimation for the Marshall-Olkin bivariate Weibull distribution. Comput Stat Data Anal 57(1):271–281
13. Lynch SM (2007) Introduction to applied Bayesian statistics and estimation for social scientists, Statistics for social science and public policy. Springer, New York
14. Yuan F (2018) Parameter estimation for bivariate Weibull distribution using generalized moment method for reliability evaluation. Qual Reliab Eng Int 34(4):631–640
15. Yuan F, Barabadi A, Lu J (2017) Reliability modelling on two-dimensional life data using bivariate weibull distribution: with case study of truck in mines. Eksploatacja i Niezawodnosc—Maint Reliab 19(4)
16. Nelsen RB (1999) An introduction to copulas, Lecture notes in statistics, no. 139. Springer, New York

17. Fang G, Pan R, Hong Y (2020) Copula-based reliability analysis of degrading systems with dependent failures. Reliab Eng Syst Saf 193
18. Bouy E (2000) Copulas for finance a reading guide and some applications. Groupe de Recherche Op'erationnelle, Paris
19. Gu YK, Fan CJ, Liang LQ, Zhang J (2019) Reliability calculation method based on the copula function for Merchanical systems with dependent failure. Ann Oper Res 311:99–116

Chapter 9
Proportional Hazard Model

Accounting for operating conditions that have a significant impact on failure time in the life distribution model can improve the accuracy of reliability analysis. The Proportional Hazard Model (PHM) is one of such popular models that can account for these operating conditions. The PHM was initially introduced by Cox as a regression method. It combines the hazard rate—also known as the failure rate in reliability engineering—with other variables, known as covariates, into a single model using a form of linear regression [1]. The strength of the PHM lies in its partial likelihood estimation approach, which allows estimation of covariate coefficients without requiring knowledge of the underlying failure rate. This advantage makes it useful in both medical and engineering applications. This chapter presents the fundamentals of the PHM and addresses the key issues relevant to reliability data analysis.

9.1 Introduction to PHM

We start with the gas turbine from the energy industry to introduce the necessity of considering operating condition. The turbine is a critical equipment in energy industries which has a variety of applications including power generation and mechanical drive. When turbine failures occur frequently, the resulting economic loss can be considerable. Figure 9.1 shows a figure of the factors that could affect the turbine in operation.

During operation, several types of failure could occur. Table 9.1 shows some failure modes and failure mechanisms of a turbine used in energy industry [2].

Table 9A.1 in the Appendix 9A shows some time to failure and the operating condition data. From these data, it demonstrates the time to failure (TTF) changing with temperature. If we ignore the temperature and just use the TTF data, we can just perform Weibull analysis or other distribution analysis. But the information about the temperature affecting the TTF is not used. But from our previous knowledge, the

F. Yuan, *Essentials of Reliability Engineering*, Springer Series in Reliability Engineering, https://doi.org/10.1007/978-3-032-15544-3_9

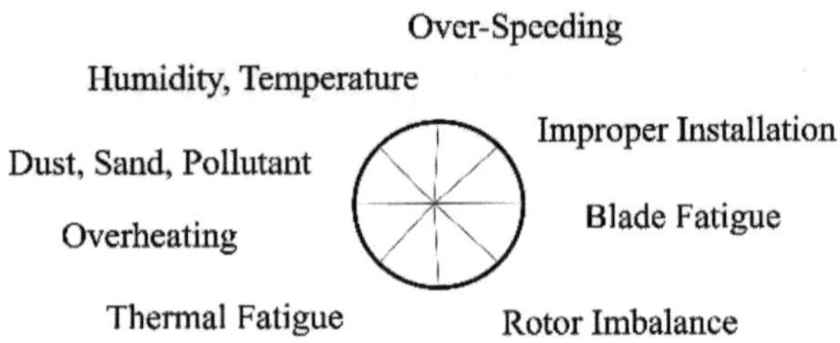

Fig. 9.1 Factors affecting turbine operation

Table 9.1 Failure modes and mechanisms for gas turbine [2]

Sub-System	Failure Mode	Possible Failure Mechanism
Fuel system	Unexpected stop or shutdown	Sticking caused by contamination, impurities
	Failed to start on demand	Pressure or flow sensor error, control logic error
High pressure turbine	Failure to start on demand	Sticking caused by contamination
Accessory drive system	Speed, torque out abnormal	Shaft misalignment
Exhaust system	Failure to start on demand	Debris buildup
Lubrication system	Chocked	Too much contamination
Control & Monitoring	Erratic output	Material fatigue
	Abnormal instrument reading	Sensor drift and aging
	Failure to start on demand	Too high ambient temperature

temperature might be a significant factor to affect the TTF. By plotting the temperature and TTF in figure, as shown in Fig. 9.2, no clear pattern is observed in how temperature changes with TTF; therefore, the effect of temperature on TTF remains unclear. Analytical methods need to be engaged to examine the effect quantitatively.

Take electronics as another example. The failure rate of a unit might have been obtained by tests in the lab where the operation condition is homogeneous. But in the field, the operating condition varies and thus the field failure rate will fluctuate. If the failure rate in the lab, which might be also found in the manual provided by supplier, is provided as λ. For the field, the failure then should be adjusted and one method is to multiply it by a constant factor π, which results to $\pi\lambda$ [3]. The factor π is evaluated by experience. For favoured operating condition, specify it a lower value for example $\pi = 0.9$. For adverse condition, specify it a higher value for example $\pi = 1.1$. Giving such adjustment is popular in industries. The idea behind it is the characteristic of the system failure rate is fixed, and it only relies on the inherent conditions for example the design of the system. The operating conditions, as external factors, will only affect the magnitude but not the characteristic of the failure rate. In general, we can write the inherent failure rate as $\lambda_0(t)$, same as the λ in the electronics case. The π is external factor and we replace it by a more general function as $g(z)$, where the z is the parameters of operating conditions. The z is a

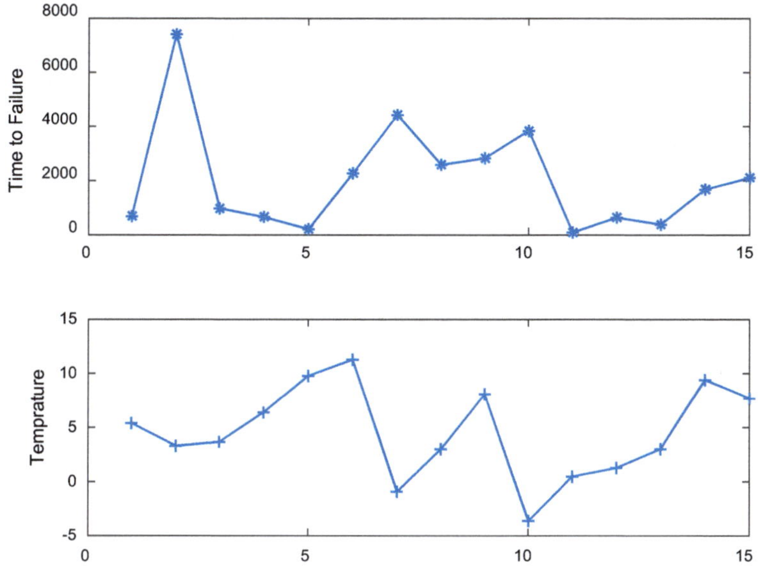

Fig. 9.2 Time to failure and averaging operating temperature

vector containing several operation factors. The modified failure rare considering operating condition is then

$$\lambda(t) = \lambda_0(t)g(z) \tag{9.1}$$

It is interesting to have: for failure rate $\lambda_1(t)$ and $\lambda_2(t)$ corresponding to two operating condition z_1 and z_2, the ratio of them

$$\frac{\lambda_1(t)}{\lambda_2(t)} = \frac{\lambda_0(t)g(z_1)}{\lambda_0(t)g(z_2)} = \frac{g(z_1)}{g(z_2)} \tag{9.2}$$

The ratio is independent of the inherent failure rate $\lambda_0(t)$. Therefore, we might be able to analyse the impact of operation factor without knowing the failure rate $\lambda_0(t)$. The model (9.1) is named **proportional hazard model**.

9.2 Proportional Hazard Model

The Proportional Hazards Model assumes that the inherent failure rate of a system remains unchanged, meaning the shape of the failure rate function is preserved. Operating condition factors affect the failure rate proportionally, only scaling it up or down but without altering its shape. We can demonstrate it graphically. Let the failure rate of system as follows, where the inherent failure rate is Weibull.

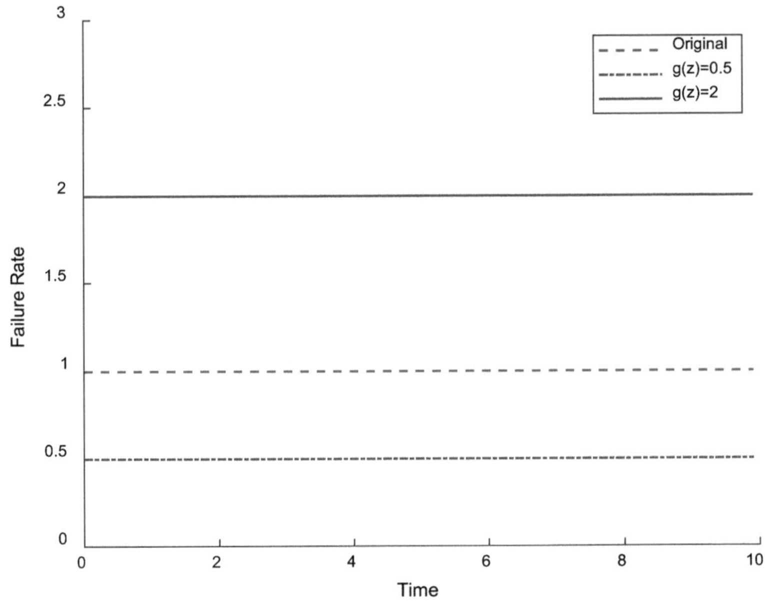

Fig. 9.3 Weibull failure rate function under various conditions

$$\lambda(t) = \frac{\beta}{\alpha}\left(\frac{t}{\alpha}\right)^{\beta-1} g(z); \tag{9.3}$$

Assume the $\alpha = 1$; $\beta = 1$ and operating condition function is set as $g(z) = 0.5$ and $g(z) = 2$ respectively, the plot of the failure rate function is shown in Fig. 9.3. We can readily find the failure rate are scaled by the $g(z)$. The original failure rate for this case is constant 1. When $g(z) = 0.1$, the failure rate decreases to constant 0.1. When $g(z) = 2$, the failure rate increases to constant 2. The proportional change is evident. Here we should emphasize that the proportion of failure changing with $g(z)$ is a fundamental assumption of PHM. This assumption needs to be validated when applying it for data analysis, as it might not hold for some situations.

Specifically, let the observation data present in a pair of time t and operating condition vector z, where z is a real valued row vector contain multiple dimensions. Each dimension corresponds to an operating condition such as temperature and humidity. The z is named **covariate** by convention in this book.

$$(t_1, z_1), (t_2, z_2), \ldots, (t_i, z_i), \ldots, (t_n, z_n) \tag{9.4}$$

The z_i in (9.4) are defined in multiple dimensions as row vector

$$z_i = (z_{1\mathrm{i}}, z_{2\mathrm{i}}, \ldots, z_{ji}, \ldots z_{pi}) \tag{9.5}$$

The inherent failure rate $\lambda_0(t)$ is named as baseline failure rate by convention. Operating at condition z, the resulting failure rate considering operating condition is denoted as $\lambda(t|z)$, while it is written as $\lambda(t)$ sometimes for simplicity unless specially specified. Furthermore, classically, the operating condition function $g(z)$ is defined as

$$g(z_i) = exp^{(\beta_1 z_{1i} + \beta_2 z_{2i} + \beta_3 z_{3i} + \ldots)} \tag{9.6}$$

The full PHM is then

$$\lambda(t) = \lambda_0(t)\, exp^{(\beta_1 z_{1i} + \beta_2 z_{2i} + \beta_3 z_{3i} + \ldots)} \tag{9.7}$$

The $\lambda(t)$ is the resulting failure rate, $\lambda_0(t)$ is the baseline failure rate without considering any covariates, and z_{ij} is the covariate value. Corresponding to failure rate written in Eq. (9.7), the reliability function is

$$R(t) = \mathrm{e}^{-\int_0^t \lambda(\tau) d\tau} = \mathrm{e}^{-\int_0^t \lambda_0(t) exp^{(\beta_1 z_{1i} + \beta_2 z_{2i} + \beta_3 z_{3i} + \ldots)} d\tau} = \left[e^{-\int_0^t \lambda_0(t) d\tau} \right]^{exp^{(\beta_1 z_{1i} + \beta_2 z_{2i} + \beta_3 z_{3i} + \ldots)}} = [R_0(t)]^{g(\mathbf{z})} \tag{9.8}$$

where $R_0(t) = e^{-\int_0^t \lambda_0(t) \mathrm{d}\tau}$ is the reliability function without considering covariate.

The proportional hazard model of Eq. (9.7) is the classical PHM, which is also named Cox model [4, 5]. The PHM model, if not specially specified, refers to the Cox model. The Cox model was not for engineering, but for survival analysis in medicine. Thereafter, the Cox model was extensively applied in reliability data analysis to study the influence of operating conditions.

9.3 Partial Likelihood

Same as Weibull analysis and Copula analysis that was discussed before, a key issue to apply the PHM on reliability data analysis is to find a proper parameter estimation method. Adding operation condition into a model can hold multiple factors, but meanwhile it complicates the parameter estimation. But fortunately, the PHM has an efficient parameter estimation method named partial likelihood method that takes advantage of the ratio present in Eq. (9.2). The partial likelihood solves the parameter estimation issue without needing to know the baseline failure rate of the PHM. The partial likelihood is one driving force of using the PHM in applications.

Given observation data present pairwise as Formula (9.4), the regular maximum likelihood data can still be applied to find the optimal parameters. However, for the proportional hazard model, directly using the maximum likelihood is not common as the number of parameters could be large and numerical computation should be involved. Moreover, the obtained solution from maximum likelihood might not be optimal. To estimate the coefficient β_i for each covariate, the most common method is the aforementioned partial likelihood method. As we described, the partial likelihood method can estimate the coefficient β_i without knowing the $\lambda_0(t)$, suggesting a great advantage over maximum likelihood. We can derive the partial likelihood here by specifying the covariate part in the PHM as a general function as $g(\boldsymbol{\beta}, z)$.

Order the failure time t_i ascendingly, i.e.

$$0 \le t_1 \le t_2 \le \ldots \le t_i \le \ldots \le t_n \le t_{*n+1} + \infty \tag{9.9}$$

The t_{*n+1} is the next failure after t_n which has not been observed. The introduction of the unobserved t_{*n+1} is for the ease of derivation of the partial likelihood.

The failure rate is then $\lambda_0(t)g(\boldsymbol{\beta}, z)$, and PDF of the PHM is given by

$$f(t) = \lambda_0(t)g(\boldsymbol{\beta}, z).\mathrm{e}^{-\int_0^t \lambda_0(\tau)g(\boldsymbol{\beta}, z)d\tau} \tag{9.10}$$

Assume the observations are independent from each other, the joint PDF function of the observations is

$$L = \prod_{i=1}^{n+1} f(t_i) \tag{9.11}$$

According to the order in (9.9), the probability of this order by using the joint function (9.11) is

$$\iiint\limits_{0 \le t_1 \ldots \le t_i \le \ldots \le t_n} L = \int_0^{\infty} \int_{t_1}^{\infty} \int_{t_2}^{\infty} \cdots \int_{t_{n-1}}^{\infty} \int_{t_n}^{\infty} \left\{ \prod_{i=1}^{n+1} f(t_i) \right\} dt_{n+1} dt_n \ldots dt_2 dt_1 \tag{9.12}$$

For simplicity of notation, we let

$$\Lambda(t; \boldsymbol{\beta}, z) = \int_0^t \lambda_0(\tau) g(\boldsymbol{\beta}, z) d\tau \tag{9.13}$$

The $f(t)$ of (9.10) is then rewritten as

$$f(t) = \lambda_0(t) g(\boldsymbol{\beta}, z).\mathrm{e}^{-\Lambda(\boldsymbol{\beta}, z)} \tag{9.14}$$

Equation (9.12) is equal to

$$\iiint\limits_{0\le t_1\ldots\le t_i\le\ldots\le t_n} L = \int\limits_0^{\infty}\int\limits_{t_1}^{\infty}\int\limits_{t_2}^{\infty}\cdots\int\limits_{t_{n-1}}^{\infty}\prod_{i=1}^{n-1} f(t_i).\int\limits_{t_n}^{\infty}\left\{\lambda_0(t_{n+1})g(\boldsymbol{\beta},z_n).\mathrm{e}^{-\Lambda(\boldsymbol{\beta},z_n)}\right\}dt_{n+1}\ldots dt_2 dt_1$$

It can be readily obtained the

$$\int\limits_{t_n}^{\infty}\left\{\lambda_0(t)g(\boldsymbol{\beta},z_n).\mathrm{e}^{-\Lambda(t;\boldsymbol{\beta},z_n)}\right\}dt_{t_{n+1}} = \mathrm{e}^{-\Lambda(t_n;\boldsymbol{\beta},z_n)} \tag{9.15}$$

and

$$\int\limits_{t_{n-1}}^{\infty}\left\{\lambda_0(t_n)g(\boldsymbol{\beta},z_{n-1}).\mathrm{e}^{-\Lambda(t_n;\boldsymbol{\beta},z_{n-1})}\mathrm{e}^{-\Lambda(t_n;\boldsymbol{\beta},z_n)}\right\}dt_n$$

$$= \frac{g(\boldsymbol{\beta},z_{n-1})}{g(\boldsymbol{\beta},z_{n-1})+g(\boldsymbol{\beta},z_n)}\mathrm{e}^{-\{\Lambda(t_{n-1};\boldsymbol{\beta},z_{n-1})+\Lambda(t_{n-1};\boldsymbol{\beta},z_n)\}} \tag{9.16}$$

Repeat the (9.15) in Eq. (9.12) recursively. Equation (9.12) is finally obtained as [6]

$$\frac{g(\boldsymbol{\beta},z_{n-1})}{g(\boldsymbol{\beta},z_{n-1})+g(\boldsymbol{\beta},z_n)}\cdot\frac{g(\boldsymbol{\beta},z_{n-2})}{g(\boldsymbol{\beta},z_{n-2})+g(\boldsymbol{\beta},z_{n-1})+g(\boldsymbol{\beta},z_n)}\cdots\cdot$$
$$\frac{g(\boldsymbol{\beta},z_1)}{(\boldsymbol{\beta},z_1)+\ldots+(\boldsymbol{\beta},z_{n-2})+g(\boldsymbol{\beta},z_{n-1})+g(\boldsymbol{\beta},z_n)}$$
$$\int\limits_0^{+\infty}\lambda_0(t_1)\sum_{i=1}^{n}\mathrm{g}(\boldsymbol{\beta},z_i)\mathrm{e}^{-\sum\limits_{i=1}^{n}\Lambda(t_1;\boldsymbol{\beta},z_i)}dt_1 \tag{9.17}$$

The $\int\limits_0^{+\infty}\lambda_0(t_1)\sum\limits_{i=1}^{n}\mathrm{g}(\boldsymbol{\beta},z_i)\mathrm{e}^{-\sum\limits_{i=1}^{n}\Lambda(t_1;\boldsymbol{\beta},z_i)}dt_1 = 1$, obviously. Equation (9.17) can be rewritten as

$$L(\boldsymbol{\beta}) = \prod_{i=1}^{n}\frac{g(\boldsymbol{\beta},z_i)}{\sum\limits_{j\ge i}g(\boldsymbol{\beta},z_j)} \tag{9.18}$$

Equation (9.18) is named partial likelihood, which, as suggested in the equation, does not contain any term related to the baseline function.

Equation (9.18) is based on the ordered failure time t_i. By convention of the PHM, we define the set D as the set of t_i that the failure has been observed. The set $Risk(t_i)$ is the set for a t_i in Set D where failure has not been observed at time t_i, i.e. the failure time that occurred at and after time t_i. Fig. 9.4 demonstrates the risk set of t_i.

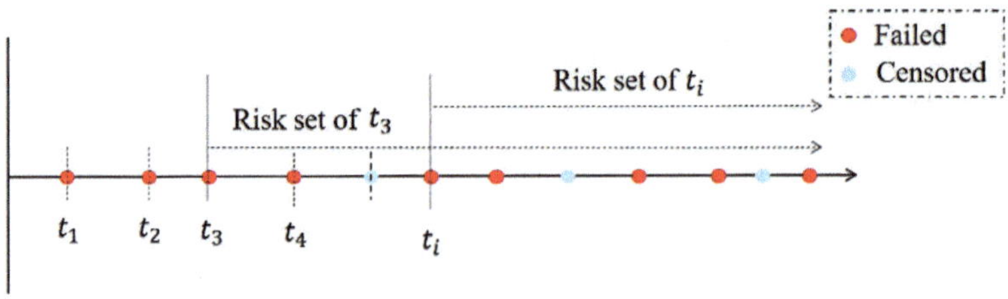

Fig. 9.4 Risk set of failure time

The partial likelihood function is then rewritten as

$$L(\boldsymbol{\beta}) = \prod_{i \in D} \left\{ \frac{g(\boldsymbol{\beta}, z_i)}{\sum\limits_{j \in Risk(t_i)} g(\boldsymbol{\beta}, z_j)} \right\} \tag{9.19}$$

For the Cox model, $g(\boldsymbol{\beta}, z) = \boldsymbol{\beta}' z$, Eq. (9.19) is rewritten as

$$L(\boldsymbol{\beta}) = \prod_{i \in D} \left\{ \frac{\exp(\boldsymbol{\beta}' z_i)}{\sum\limits_{j \in Risk(t_i)} \exp(\boldsymbol{\beta}' z_j)} \right\} \tag{9.20}$$

9.4 Partial Likelihood Estimation

Let the optimal estimate to $\boldsymbol{\beta}$ be denoted as $\widehat{\boldsymbol{\beta}}$. Same as the regular maximum likelihood method, the $\widehat{\boldsymbol{\beta}}$ is obtained by maximizing $L(\boldsymbol{\beta})$,which is an unconstraint optimization problem. To obtain the solution, as usual, we can just take the first derivative in respect to, and then equal the first derivative to zero. Take the logarithm of Eq. (9.20) as follows

$$\ell(\boldsymbol{\beta}) = \sum\nolimits_{i \in D} \left[\boldsymbol{\beta}' z_i - \ln \sum_{j \in Risk(t_i)} \exp(\boldsymbol{\beta}' z_j) \right] \tag{9.21}$$

Define a score vector as follows

$$\mathbf{U}(\boldsymbol{\beta}) = \left(\frac{\partial \ell}{\partial \beta_1}, \frac{\partial \ell}{\partial \beta_2}, \ldots, \frac{\partial \ell}{\partial \beta_i} \ldots, \frac{\partial \ell}{\partial \beta_p} \right)' \tag{9.22}$$

The p is the number of factors. Then the first derivative of Eq. (9.21) is given by

$$\mathbf{U}(\boldsymbol{\beta}) = \sum \left[z_i - \frac{\sum_{j \in Risk(t_i)} z_i \exp(\boldsymbol{\beta}' z_j)}{\sum_{j \in Risk(t_i)} \exp(\boldsymbol{\beta}' z_j)} \right] \tag{9.23}$$

Rewrite Eq. (9.23) as follows for easy understanding, and equal them to 0

$$\mathbf{U}(\boldsymbol{\beta}) = \begin{pmatrix} \sum \left(z_{1i} - \frac{\sum_{j \in Risk(t_i)} z_{1j} \exp(\boldsymbol{\beta}' z_j)}{\sum_{j \in Risk(t_i)} \exp(\boldsymbol{\beta}' z_j)} \right) \\ \cdots \\ \sum \left(z_{mi} - \frac{\sum_{j \in Risk(t_i)} z_{mj} \exp(\boldsymbol{\beta}' z_j)}{\sum_{j \in Risk(t_i)} \exp(\boldsymbol{\beta}' z_j)} \right) \\ \cdots \\ \sum \left(z_{pi} - \frac{\sum_{j \in Risk(t_i)} z_{pj} \exp(\boldsymbol{\beta}' z_j)}{\sum_{j \in Risk(t_i)} \exp(\boldsymbol{\beta}' z_j)} \right) \end{pmatrix} = \begin{pmatrix} 0 \\ \cdots \\ 0 \\ \cdots \\ 0 \end{pmatrix} \tag{9.24}$$

The analytical solution to Eq. (9.24) is unavailable. Numerical method should be employed to find the optimal. It is a typical optimization problem using numerical method. This book does not discuss the numerical algorithm. Software packages, such as SPSS and Matlab coxphfit() (version 2021a) provide the solution to Eq. (9.24) by using numerical method. The information matrix $(\boldsymbol{\beta}) = -\frac{\partial^2 \ell}{\partial \beta_i \beta_j}$, a $p \times p$ matrix with its element as

$$\boldsymbol{I}_{kl} = \sum \frac{\sum_{j \in Risk(t_i)} z_{kj} z_{lj} \exp(\boldsymbol{\beta}' z_j) \sum_{j \in Risk(t_i)} \exp(\boldsymbol{\beta}' z_j) - \sum_{j \in Risk(t_i)} z_{kj} \exp(\boldsymbol{\beta}' z_j) \sum_{j \in Risk(t_i)} z_{lj} \exp(\boldsymbol{\beta}' z_j)}{\left(\sum_{j \in Risk(t_i)} \exp(\boldsymbol{\beta}' z_j) \right)^2}; \quad k, l \in 1, 2, \ldots p \tag{9.25}$$

According to central limit theory, the $\widehat{\boldsymbol{\beta}}$ can be approximated by multivariate Normal distribution as [7],

$$\widehat{\boldsymbol{\beta}} \approx N_p\left(\boldsymbol{\beta}, \boldsymbol{I}^{-1}\left(\widehat{\boldsymbol{\beta}}\right)\right) \tag{9.26}$$

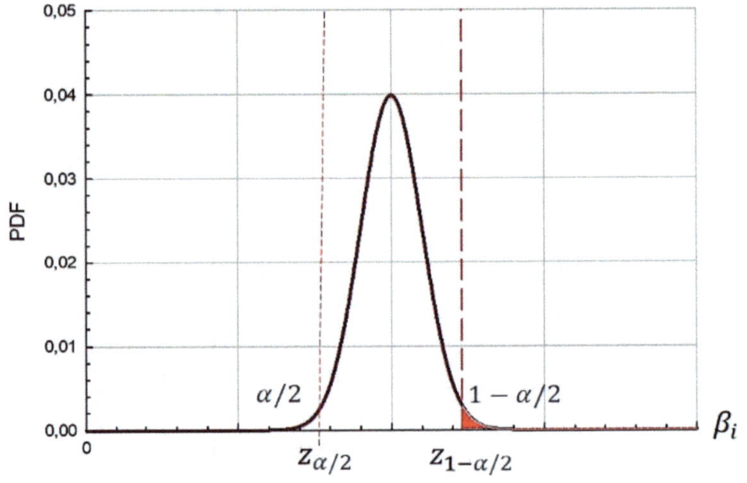

Fig. 9.5 Confidence interval of coefficient β_i

The variance of the parameter β_i corresponds to the diagonal element $\boldsymbol{I}_{ii}{}^{-1}$, written as V_{ii}. The confident interval can be obtained from

$$\frac{\widehat{\beta}_i - \beta_i}{\sqrt{V_{ii}}} \sim N(0, 1) \tag{9.27}$$

For a confidence interval $1 - \alpha$, it is not hard to obtain the confidence interval of β_i as

$$\left[\widehat{\beta}_i - z_{1-\frac{\alpha}{2}}\sqrt{V_{ii}}, \widehat{\beta}_i + z_{1-\alpha/2}\sqrt{V_{ii}}\right] \tag{9.28}$$

A plot of the confidence interval is shown in Fig. 9.5, where we can see the lower and higher bound of β_i at $z_{\frac{\alpha}{2}}$ and $z_{1-\frac{\alpha}{2}}$. The $z_{\frac{\alpha}{2}}$ is the value at percentage of $\frac{\alpha}{2}$ of standard Normal distribution.

Example 9.1
Using the accelerated life test data shown in Table 9A.3 to estimate the parameter of the temperature coefficient using PHM model. The data is only one covariate.

Solution: A rough solution can be obtained directly from Eq. (9.21) by plotting. If one unique solution exists, the curve should contain one and only one maximum. We plot the $\ell(\boldsymbol{\beta})$ ranging $\boldsymbol{\beta}$ from −0.2 to 0.4 as shown in Fig. 9.6. The function plot shows one and only one maximum existing between 0 and 0.1. To find the optimal point accurately, we use the first

(continued)

Example 9.1 (continued)

derivative of the $\ell(\boldsymbol{\beta})$, shown in Eq. (9.23) to obtain the solution. A plot of the $\mathbf{U}(\boldsymbol{\beta})$ is shown in Fig. 9.7, where you can see the curve $\mathbf{U}(\boldsymbol{\beta})$ crosses with horizontal line 0. Using numerical method, which uses fsolve() in this case, the optimal solution is 0.0321.

For comparison, we use the Matlab (version 2020b) built-in coxphfit() function to estimate the parameters for the same question, the obtained solution is 0.0310. The two obtains very close solution. The difference between them may be due to the use of different numerical methods.

Fig. 9.6 Empirical plot of $\ell(\boldsymbol{\beta})$

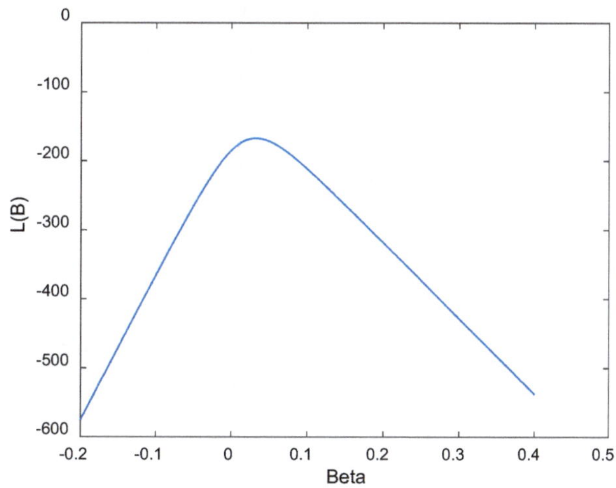

Fig. 9.7 Empirical plot of $U(\boldsymbol{\beta})$

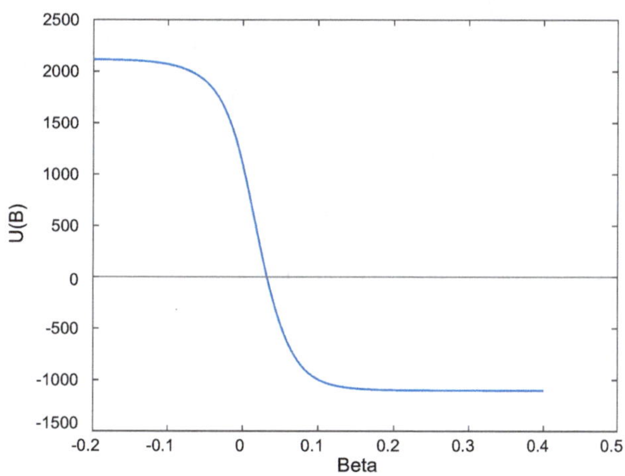

In Example 9.1, we use plots of Figs. 9.6 and 9.7 to verify the existence of the unique solution. But when the number of covariates is more than 2, the plot method is impossible. Several publications have discussed the proof of unique solution existing [8–11]. One can refer to them for details. We should be aware that the uniqueness, i.e. the only one maximum as shown in Fig. 9.6 is essential to numerical approach. Without knowing the maximal point is unique, the solution obtained from numerical approach might not be globally optimal.

9.4.1 Right Censored Situation

For situation where some lifetime data is right censored, which is very common in analysis, it is not hard to find the partial likelihood Eqs. (9.19) and (9.20) still hold. Set the censored indicator as δ_i and define $\delta_i = 1$ indicated uncensored (failure) time, $\delta_i = 0$ indicated censored time. For n observation data including censored data, the logarithm function of partial likelihood function is

$$\ell(\boldsymbol{\beta}) = \sum_{i=1}^{n} \delta_i \left[\boldsymbol{\beta}' z_i - \ln \sum_{j \in Risk(t_i)} \exp(\boldsymbol{\beta}' z_j) \right] \tag{9.29}$$

First derivative function is

$$\mathbf{U}(\boldsymbol{\beta}) = \begin{pmatrix} \sum_{i=1}^{n} \delta_i \left[z_{1i} - \dfrac{\sum_{j \in Risk(t_i)} z_{1j} \exp(\boldsymbol{\beta}' z_j)}{\sum_{j \in Risk(t_i)} \exp(\boldsymbol{\beta}' z_j)} \right] \\ \cdots \\ \sum_{i=1}^{n} \delta_i \left[z_{mi} - \dfrac{\sum_{j \in Risk(t_i)} z_{mj} \exp(\boldsymbol{\beta}' z_j)}{\sum_{j \in Risk(t_i)} \exp(\boldsymbol{\beta}' z_j)} \right] \\ \cdots \\ \sum_{i=1}^{n} \delta_i \left[z_{pi} - \dfrac{\sum_{j \in Risk(t_i)} z_{pj} \exp(\boldsymbol{\beta}' z_j)}{\sum_{j \in Risk(t_i)} \exp(\boldsymbol{\beta}' z_j)} \right] \end{pmatrix} = \begin{pmatrix} 0 \\ \cdots \\ 0 \\ \cdots \\ 0 \end{pmatrix} \tag{9.30}$$

Element for the information matrix is

$$I_{kl} = \sum_{i=1}^{n} \delta_i \frac{\sum_{j \in Risk(t_i)} z_{kj} z_{lj} \exp(\beta' z_j) \sum_{j \in Risk(t_i)} \exp(\beta' z_j) - \sum_{j \in Risk(t_i)} z_{kj} \exp(\beta' z_j) \sum_{j \in Risk(t_i)} z_{lj} \exp(\beta' z_j)}{\left(\sum_{j \in Risk(t_i)} \exp(\beta' z_j) \right)^2} \tag{9.31}$$

Example 9.2
For the data from paper by Kjevik and Yuan [2], present in Table 9A.1, which is with 5 covariates: Average temperature, Min temperature, Max temperature, Average precipitation, Max precipitation. Using the proportional hazard model to estimate the coefficients of covariates, their variance, and confidence intervals of 95%.

Solution: Matlab is used to perform calculation. Firstly, the TTFs are ordered ascendingly as following Table 9.2.

Apply the method described in Sect. 9.4. Using the fslove() function in the Matlab to find the solution of equation system Eq. (9.30), the estimates of efficient $\boldsymbol{\beta}$ are obtained as

$$10.88057, \ -3.94648, \ -6.44016, \ -0.15708, \ -0.11432$$

According to Eq. (9.31), the Fisher matrix $\boldsymbol{I}(\boldsymbol{\beta})$ is as shown in Table 9.3.

Take the inverse of matrix $\boldsymbol{I}(\boldsymbol{\beta})$. We obtain $\boldsymbol{I}^{-1}\left(\widehat{\boldsymbol{\beta}}\right)$ as in Table 9.4.

The variance of $\widehat{\beta}_i$, corresponding to the diagonal values in the matrix $\boldsymbol{I}^{-1}\left(\widehat{\boldsymbol{\beta}}\right)$, are

$$41.13335, 12.94869, 8.488238, 0.298245, 0.004141$$

The $z_{1-\frac{\alpha}{2}} = z_{0.975} = 1.96$ according to the standard Normal distribution. According to Eq. (9.28), the 95% CI of $\widehat{\boldsymbol{\beta}}$ is as shown in Table 9.5.

We have also verified the inverse information matrix, which is the covariance of the $\boldsymbol{\beta}$ is shown in the Table 9.4, and the estimates obtained from partial likelihood function reaches the same that has obtained from the Matlab built in function coxphfit().

9.5 Parametric Proportional Hazard Model

The classical proportional hazard model does not require the implicit expression of the underlying baseline function. However, for reliability analysis, to evaluate reliability, the baseline function needs to be known. We can let the underlying

Table 9.2 Ordered TTF data

	TTF	Ave temp	Min temp	Max temp	Average precipitation	Max precipitation
i	t	β_1	β_2	β_3	β_4	β_5
1	96	0.5	−2.3	2.7	0	0
2	216	9.8	7.5	12.1	0.8	2.4
3	384	3	0.9	4.9	4.72	14.2
4	648	1.3	−1.1	3.5	0.49	3.5
5	672	6.4	4.4	8.4	2.73	14.5
6	696	5.4	3.3	7.4	4.35	19.3
7	984	3.7	1.4	5.8	3.3	13.3
8	1680	9.4	7	12.1	3.3	29.6
9	2112	7.7	5.4	10.5	2	17.6
10	2280	11.3	8.4	14.6	1.13	14.7
11	2592	3	0.6	5.2	3.08	22.4
12	2832	8.1	5.3	11.3	2.97	28.3
13	3840	−3.6	−6.5	−0.9	2.01	13.2
14	4440	−0.9	−3.5	1.7	2.3	11.1
15	7416	3.3	0.6	6.1	2.29	45.9

Table 9.3 Information matrix

130.4198	127.8151	140.5051	−11.3702	39.15707
127.8151	125.7446	137.1543	−9.80195	42.55781
140.5051	137.1543	152.1079	−13.9257	36.59522
−11.3702	−9.80195	−13.9257	10.23389	36.61802
39.15707	42.55781	36.59522	36.61802	435.8277

Table 9.4 Inverse of information matrix

41.13335	−22.3216	−17.8493	0.134845	−0.02855
−22.3216	12.94869	8.903272	−0.37101	0.024662
−17.8493	8.903272	8.488238	0.242405	0.001187
0.134845	−0.37101	0.242405	0.298245	−0.0213
−0.02855	0.024662	0.001187	−0.0213	0.004141

Table 9.5 CI of estimates

CI	β_1	β_2	β_3	β_4	β_5
Lower bound	10.16445	−4.64965	−7.21354	−0.35768	−1.42341
Upper bound	11.59669	−3.24331	−5.66679	0.043524	1.194782

base line function as Weibull distribution, Exponential or other. This section discusses the method to perform reliability using PHM with failure rate of Weibull distribution and Exponential as baseline function. The model is named parametric PHM as the entire model is parametrized.

9.5.1 Weibull Proportional Hazar Model

The failure rate function for Weibull distribution as baseline function is

$$\lambda(t_i) = \frac{\beta}{\alpha}\left(\frac{t_i}{\alpha}\right)^{\beta-1} e^{(\beta_1 z_{1i} + \beta_2 z_{2i} + \ldots)} \tag{9.32}$$

The likelihood function considering right censored situation, where the complete data situation is special of case of, is given by

$$L(\boldsymbol{\beta}) = \prod_{i=1}^{n} \lambda(t_i)^{\delta_i} S(t_i) \tag{9.33}$$

where $S(t_i) = e^{-\int_0^{t_i} \lambda(\tau)d\tau}$. And let $\Lambda(t_i) = \int_0^{t_i} \lambda(\tau)d\tau$. Take the logarithm of (9.33). We obtain

$$\ln L(\boldsymbol{\beta}) = \sum_{i=1}^{n} \{\delta_i \ln \lambda(t_i) - \Lambda(t_i)\} \tag{9.34}$$

We rewrite the failure rate function (9.32) as follows

$$\lambda(t_i) = \alpha^{-\beta} \beta t_i^{\beta-1} g(z) = \alpha^{-\beta} \beta t_i^{\beta-1} e^{(\beta_1 z_{1i} + \beta_2 z_{2i} + \ldots)} \tag{9.35}$$

For further simplification, let the constant $\alpha^{-\beta}$ be a covariate $\mathbf{z_0}$ which is a constant unit vector so the parameter α can be as a coefficient in the PHM, which is

$$\mathbf{z_0} = (1, 1, 1, \ldots) \tag{9.36}$$

The (9.35) is thus rewritten as

$$\lambda(t_i) = \beta t_i^{\beta-1} e^{(\beta_0 z_{0i} + \beta_1 z_{1i} + \beta_2 z_{2i} + \ldots)} \tag{9.37}$$

The corresponding cumulative failure rate function is then

$$\Lambda(t_i) = t_i^{\beta} e^{(\beta_0 z_{0i} + \beta_1 z_{1i} + \beta_2 z_{2i} + \ldots)} \tag{9.38}$$

Rewrite the (9.34) in a full expression as

$$\ln L(\boldsymbol{\beta}) = \sum_{i=1}^{n} \{\delta_i \ln \lambda(t_i) - \Lambda(t_i)\} = \sum_{i=1}^{n}$$
$$\times \left\{\delta_i(\ln \beta + (\beta - 1)\ln t_i + (\beta_0 z_{0i} + \beta_1 z_{1i} + \beta_2 z_{2i} + \ldots)) - t_i^{\beta} e^{(\beta_0 z_{0i}+\beta_1 z_{1i}+\beta_2 z_{2i}+\ldots)}\right\} \tag{9.39}$$

The 1st derivative of ln$L$ is

$$\frac{\partial \ln L}{\partial \beta} = \sum_{i=1}^{n} \left\{\delta_i\left(\frac{1}{\beta} + \ln t_i\right) - \ln t_i . t_i^{\beta} e^{(\beta_0 z_{0i}+\beta_1 z_{1i}+\beta_2 z_{2i}+\ldots)}\right\} = \sum \delta_i \frac{1}{\beta}$$
$$+ \sum \delta_i \ln t_i - \sum \ln t_i . t_i^{\beta} e^{(\beta_0 z_{0i}+\beta_1 z_{1i}+\beta_2 z_{2i}+\ldots)} \tag{9.40}$$

$$\frac{\partial \ln L}{\partial \beta_j} = \sum_{i=1}^{n} \left\{\delta_i z_{ji} - z_{ji} t_i^{\beta} e^{(\beta_0 z_{0i}+\beta_1 z_{1i}+\beta_2 z_{2i}+\ldots)}\right\} = \sum \delta_i z_{ji}$$
$$- \sum z_{ji} t_i^{\beta} e^{(\beta_0 z_{0i}+\beta_1 z_{1i}+\beta_2 z_{2i}+\ldots)} \tag{9.41}$$

The 2nd derivative is

$$\frac{\partial^2 \ln L}{\partial \beta \partial \beta} = -\sum \delta_i \frac{1}{\beta^2} - \sum \ln^2 t_i . t_i^{\beta} e^{(\beta_0 z_{0i}+\beta_1 z_{1i}+\beta_2 z_{2i}+\ldots)}$$
$$\frac{\partial^2 \ln L}{\partial \beta \partial \beta_j} = -\sum \ln t_i . t_i^{\beta} z_{ji} . e^{(\beta_0 z_{0i}+\beta_1 z_{1i}+\beta_2 z_{2i}+\ldots)}$$
$$\frac{\partial^2 \ln L}{\partial \beta_j \partial \beta_k} = -\sum z_{ji} z_{ki} t_i^{\beta} e^{(\beta_0 z_{0i}+\beta_1 z_{1i}+\beta_2 z_{2i}+\ldots)} \tag{9.42}$$

The estimates for β and β_j can be obtained by letting (9.40) and (9.41) equal to zero. The corresponding estimates for α in Eq. (9.32) is $e^{\beta_0} = \alpha^{-\beta}$. The

$$\alpha = e^{-\beta_0/\beta} \tag{9.43}$$

The coefficient β_i for covariates is directly obtained from (9.40) and (9.41) without changing.

Example 9.3
Using the accelerated life test data presented in Table 9A.3 to estimate the parameter of the temperature coefficient, and the parameters of failure rate when the underlying failure rate function is from Weibull distribution.

Solution: By using numerical methods to solve the equations Eq. (9.40) and (9.41), we obtain the full parameters are:

$$\beta = 0.5574; \beta_0 = -8.9398, \beta_1 = 0.0303.$$

The β is shape parameter of the Weibull distribution. The $\beta_1 = 0.0303$ is coefficient of the single temperature covariate. According to Eq. (9.43), the scale parameter is:

$$\alpha = e^{-\beta_0/\beta} = e^{8.9398/0.5574} = 9.24 \times 10^6$$

The whole Weibull failure rate function with covariate is then

$$\lambda(t_i) = \frac{\beta}{\alpha}\left(\frac{t_i}{\alpha}\right)^{\beta-1} e^{(\beta_1 z_{1i} + \beta_2 z_{2i} + \dots)} = \frac{0.5574}{9.24 \times 10^6}\left(\frac{t_i}{9.24 \times 10^6}\right)^{0.5574-1} e^{0.0303 * z_1}$$

The plot of the failure rate function when $z_1 = 200$, as shown in Fig. 9.8.

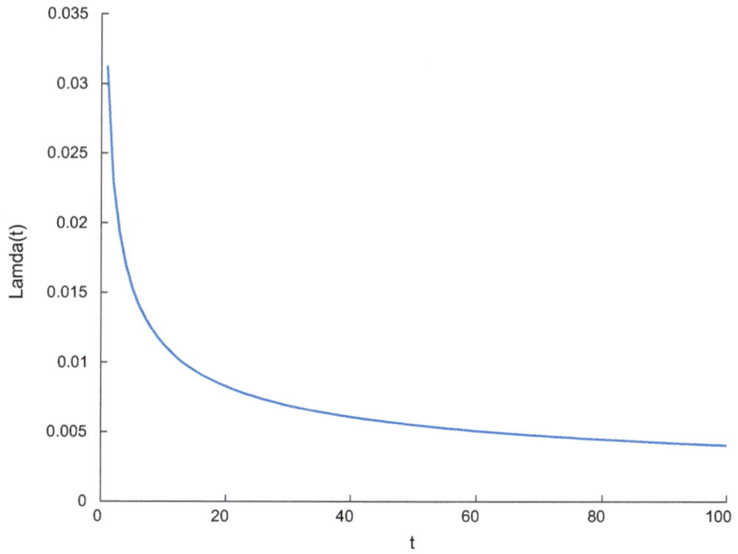

Fig. 9.8 Failure rate function from full parameter method

In the model (9.32), the linear function $\beta_1 z_{1i} + \beta_2 z_{2i} + \ldots$ does not have a constant part, letting us say β_c. If it contains constant part, the linear regression is then $\beta_c + \beta_1 z_{1i} + \beta_2 z_{2i} + \ldots$. In the model (9.37), the$\lambda(t) = \beta t^{\beta-1} e^{(\beta_0 z_{0i} + \beta_1 z_{1i} + \beta_2 z_{2i} + \ldots)}$, if the constant β_c contains in the model, the β_0 and β_c are then mixed. The estimate of scale parameter from β_0 will then not be accurate as it absorbs the β_c, which is a drawback of the model (9.32).

A simple method to check if constant is there or not is to use direct regression analysis. We write a linear regression function as

$$\ln TTF = \beta_c + \beta_1 z_{1i} + \beta_2 z_{2i} + \ldots$$

which is also an accelerated life test model widely used in the reliability. For the data in Table 9A.3, the linear regression is estimated as follows

$$\ln TTF = 14.6783 - 0.00532 * \text{Temprature}$$

The constant $\beta_c = 14.6783$, a significant value. If the model (9.37) used, the β_c will be merged with the scale parameter of Weibull distribution without differentiation. This is a drawback of the full model.

9.5.2 Two-Stage Parameter Estimation for Weibull PHM

The two-stage method firstly use the partial likelihood method to estimate the coefficients of the covariate. Suppose the estimators of covariate coefficient are correct and the estimators are considered as the true parameters. The parameter estimation in the baseline function is then a simple task. Without losing the generality, we let the failure rate function of Weibull with covariates be written as

$$\lambda(t_i) = \alpha^{-\beta} \beta t_i^{\beta-1} g(z) = \alpha^{-\beta} \beta t_i^{\beta-1} B(z_i) \tag{9.44}$$

Now, the $g(z)$ is a known quantity without any unknown parameter. Let the quantity $e^{(\beta_1 z_{1i} + \beta_2 z_{2i} + \ldots)} = B(z_i)$. The first derivative of $\ln L$ using Eq. (9.44) as PDF is then

$$\frac{\partial \ln L}{\partial \beta} = \sum_{i=1}^{n} \left\{ \delta_i \left(\frac{1}{\beta} + \ln t_i \right) - \ln t_i . t_i^{\beta} e^{\beta_0} B(z_i) \right\} = \sum \delta_i \frac{1}{\beta} + \sum \delta_i \ln t_i - \sum \ln t_i . t_i^{\beta} e^{\beta_0} B(z_i) \tag{9.45}$$

$$\frac{\partial \ln L}{\partial \beta_0} = \sum_{i=1}^{n} \left\{ \delta_i - t_i^{\beta} e^{\beta_0} B(z_i) \right\} = \sum \delta_i - e^{\beta_0} \sum t_i^{\beta} B(z_i) \tag{9.46}$$

Let the first derivative equal to 0, from Eq. (9.46), the

$$e^{\beta_0} = \frac{\sum \delta_i}{\sum t_i^{\beta} B(z_i)}; \tag{9.47}$$

Equation (9.45) is rewritten as

$$\sum \delta_i \frac{1}{\beta} + \sum \delta_i \ln t_i - \frac{\sum \delta_i . \sum \ln t_i . t_i^{\beta} B(z_i)}{\sum t_i^{\beta} B(z_i)} = 0 \tag{9.48}$$

It has been known the equation to solve scale parameters of classical Weibull distribution is as follows Eq. (9.49), referring to Chap. 5. Comparing Eq. (9.48) with Eq. (9.49), it finds they are of same form, except the extra $B(z_i)$.

$$\frac{n}{\widehat{\beta}} - n \frac{\sum t_i^{\widehat{\beta}} \ln t_i}{\sum t_i^{\widehat{\beta}}} + \sum \ln t_i = 0 \tag{9.49}$$

Equations (9.48) and (9.49) might share some common mathematical properties.

Merging Eq. (9.43) with (9.47), the α parameter in the Weibull distribution is

$$\alpha = e^{-\left(\ln \frac{\sum \delta_i}{\sum t_i^{\widehat{\beta}} B(z_i)}\right) / \widehat{\beta}} \tag{9.50}$$

Example 9.4

For the data from paper by Kjevik and Yuan [2], present in Table 9A.1, which is with 5 covariates: Average temperature, Min temperature, Max temperature, Average precipitation, Max precipitation. Suppose the underlying baseline failure rate function is Weibull failure rate. Estimate all the unknown parameters for the Weibull PHM.

Solution: The estimated coefficients are $\beta_1 = 10.88$, $\beta_2 = -3.95$, $\beta_3 = -6.44$, $\beta_4 = -0.16$, $\beta_5 = -0.11$. β_1: Ave temp, β_2:Min temp, β_3:Max temp, β_4:Average precipitation, β_5:Max precipitation. The estimates are obtained by using the partial likelihood method Eq. (9.24).

From Eq. (9.48), we obtain the shape parameter β as 2.6426. Substitute the $\widehat{\beta} = 2.6426$ into Eq. (9.50), the scale parameter is $\alpha = 147.38$. The full parameters for the Weibull distribution with 5 covariates are:

$$\lambda(t) = \frac{\beta}{\alpha}\left(\frac{t}{\alpha}\right)^{\beta-1} e^{(\beta_1 z_{1i} + \beta_2 z_{2i} + \ldots)} = \frac{2.6426}{147.38}\left(\frac{t}{147.38}\right)^{2.6426-1} e^{(10.88 z_{1i} - 3.95 z_{2i} - 6.44 z_{3i} - 0.16 z_{4i} - 0.11 z_{5i})}$$

The computation was performed by the Matlab (version 2020b), using the fsolve() function to solve Eq. (9.48).

Example 9.5

Using the accelerated life test data shown in Table 9A.3 to estimate the parameter of the temperature coefficient in PHM model. The underlying failure rate function is from Weibull distribution.

Solution: The $\beta_1 = 0.0321$, from the partial likelihood estimation using Eq. (9.20).

By solving Eq. (9.48), we obtained the shape parameter β as 0.5718. Substitute the $\widehat{\beta} = 0.5718$ into Eq. (9.50), we then obtained the scale parameter as $\alpha = 1.37 \times 10^7$.

The Weibull failure rate function with covariate is then

$$\lambda(t) = \frac{\beta}{\alpha}\left(\frac{t}{\alpha}\right)^{\beta-1} e^{(\beta_1 z_{1i} + \beta_2 z_{2i} + \ldots)} = \frac{0.5718}{1.37 \times 10^7}\left(\frac{t}{1.37 \times 10^7}\right)^{0.5718-1} e^{0.0321 z_{1i}}$$

The computation was performed by the Matlab (version 2020b) by using the fsolve() function to solve the Eq. (9.48). The plot of the failure rate function when $z_{1i} = 200$ is shown in Fig. 9.9, which shows the failure is decreasing with time.

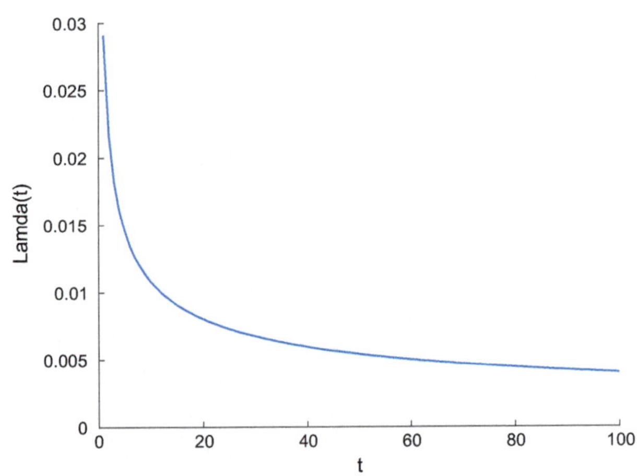

Fig. 9.9 Failure rate function from two-stage parameter estimating method

9.5.3 Exponential Proportional Hazar Model

If the failure rate function in the baseline function is a constant λ_0, then the failure time follows Exponential distribution, and the PHM failure rate is as follows

$$\lambda = \lambda_0 e^{(\beta_1 z_{1i} + \beta_2 z_{2i} + \ldots)} \tag{9.51}$$

Similar to the (9.33), the likelihood function considering right censored situation, where the complete data situation is special of case, is given by

$$L(\boldsymbol{\beta}) = \prod_{i=1}^{n} \lambda^{\delta_i} S(t_i) \tag{9.52}$$

where $S(t_i) = e^{-\lambda t_i}$. And let $\Lambda(t_i) = \lambda t_i$. Take logarithm of Eq. (9.52), we obtain

$$\ln L(\boldsymbol{\beta}) = \sum_{i=1}^{n} \{\delta_i \ln \lambda - \Lambda(t_i)\} \tag{9.53}$$

Substitute Eq. (9.51) into (9.53). The logarithm likelihood function is then written as

$$\begin{aligned} \ln L(\boldsymbol{\beta}) &= \sum_{i=1}^{n} \{\delta_i \ln \lambda - \Lambda(t_i)\} \\ &= \sum_{i=1}^{n} \left\{\delta_i (\ln \lambda_0 + \beta_1 z_{1i} + \beta_2 z_{2i} + \ldots) - t_i . \lambda_0 e^{(\beta_1 z_{1i} + \beta_2 z_{2i} + \ldots)}\right\} \end{aligned} \tag{9.54}$$

The 1st derivative of Eq. (9.54) is as follows

$$\begin{cases} \dfrac{\partial \ln L}{\partial \lambda_0} = \displaystyle\sum_{i=1}^{n} \left\{\delta_i \frac{1}{\lambda_0} - t_i e^{(\beta_1 z_{1i} + \beta_2 z_{2i} + \ldots)}\right\} \\ \dfrac{\partial \ln L}{\partial \beta_j} = \displaystyle\sum_{i=1}^{n} \left\{\delta_i z_{ji} - z_{ji} t_i . \lambda_0 e^{(\beta_1 z_{1i} + \beta_2 z_{2i} + \ldots)}\right\}, j = 1, 2, \ldots p \end{cases} \tag{9.55}$$

Let Eq. (9.55) equal to 0, then

$$\widehat{\lambda_0} = \frac{\sum_{i=1}^{n} \delta_i}{\sum_{i=1}^{n} t_i e^{(\beta_1 z_{1i} + \beta_2 z_{2i} + \ldots)}} \tag{9.56}$$

The 2nd derivative is

$$\frac{\partial^2 \ln L}{\partial \lambda_0 \partial \lambda_0} = -\sum \delta_i \frac{1}{\lambda_0{}^2}$$

$$\frac{\partial^2 \ln L}{\partial \lambda_0 \partial \beta_j} = -\sum t_i z_{ji}.e^{(\beta_1 z_{1i} + \beta_2 z_{2i} + \ldots)}$$

$$\frac{\partial^2 \ln L}{\partial \beta_j \partial \beta_k} = -\sum z_{ji} z_{ki} t_i e^{(\beta_1 z_{1i} + \beta_2 z_{2i} + \ldots)} \tag{9.57}$$

Alternative, Eq. (9.51) can be rewritten as follows by letting the constant failure rate written as a covariate factor.

$$\lambda = e^{(\beta_0 z_{0i} + \beta_1 z_{1i} + \beta_2 z_{2i} + \ldots)} \tag{9.58}$$

$$\lambda_0 = e^{\beta_0};\, z_{0i} = 1 \; for \; i = 1, 2, .., 0 \tag{9.59}$$

The logarithm likelihood function is then written as

$$\ln \boldsymbol{\beta} = \sum_{i=1}^{n} \left\{ \delta_i (\beta_0 z_{0i} + \beta_1 z_{1i} + \beta_2 z_{2i} + \ldots) - t_i e^{(\beta_0 z_{0i} + \beta_1 z_{1i} + \beta_2 z_{2i} + \ldots)} \right\} \tag{9.60}$$

The 1st derivative is

$$\frac{\partial \ln L}{\partial \beta_j} = \sum_{i=1}^{n} \left\{ \delta_i z_{ji} - z_{ji} t_i e^{(\beta_0 z_{0i} + \beta_1 z_{1i} + \beta_2 z_{2i} + \ldots)} \right\}, j = 0, 2, \ldots p \tag{9.61}$$

By letting Eq. (9.61) equal to 0, the estimator to β_j can be obtained.

Example 9.6
Using the accelerated life test data presented in Table 9A.3 to estimate the parameter of the temperature coefficient and the parameters of failure rate. The underlying failure rate function is supposed from Exponential distribution.

Solution: According to Eq. (9.61), we obtain

$$\beta_0 = -16.169, \beta_1 = 0.0530$$

The failure rate of the baseline function is

$$\lambda_0 = e^{\beta_0} = 9.50 * 10^{-8}$$

According to Eq. (9.51), the full failure rate function is

$$\lambda = 9.50 * 10^{-8} e^{0.0530 * z}$$

At temperature z = 200 ° C, the failure rate is $= 9.50 * 10^{-8} e^{0.0530 * 200} = 0.0038$. The corresponding negative $L(\boldsymbol{\beta})$ is 368.8940.

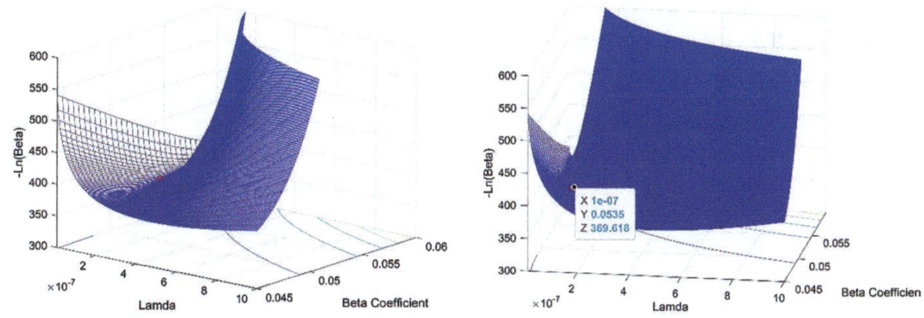

Fig. 9.10 Empirical plot for the Example 9.6

The $-L(\boldsymbol{\beta})$ is plotted with various β_0 and λ_0, as shown in Fig. 9.10 to check if the optimal solution obtained for the Example 9.6. The function $-L(\boldsymbol{\beta})$ is of complex relation with β_0 and λ_0.The lowest point is the red dot as shown in Fig. 9.10. The lowest point as shown on the right figure of 9.10 is $\lambda_0 = 10 * 10^{-8}$, $\beta_1 = 0.0535$, which is very near the parameters obtained from the maximum likelihood $\lambda_0 = 9.50 * 10^{-8}$, $\beta_1 = 0.0530$. it verifies the parameter estimation method for this case works well.

9.5.4 Two-Stage Parameter Estimator for Exponential

If the coefficients of covariates are known, the remaining unknown parameter for the Exponential proportional hazard model is the λ_0. We can also employ a two-stage method. The first stage uses the partial likelihood method to obtain the coefficients of covariates. The second stage is to evaluate the failure rate by

$$\widehat{\lambda_0} = \frac{\sum_{i=1}^{n} \delta_i}{\sum_{i=1}^{n} t_i e^{(\beta_1 z_{1i} + \beta_2 z_{2i} + \ldots)}} \tag{9.62}$$

Equation (9.62) is the same as Eq. (9.56). The derivation is also similar to the derivation to Eq. (9.56).

Example 9.7
Using the accelerated life test data presented in Table 9A.3 to estimate the parameter of the temperature coefficient, and the parameters of failure rate when the underlying failure rate function is from Exponential distribution. Use the two-stage approach.

(continued)

Example 9.7 (continued)

Solution: The β_1=0.0321, from the partial likelihood estimation using Eq. (9.20). According to Eq. (9.62), the failure rate is

$$\widehat{\lambda_0} = \frac{\sum_{i=1}^{n} \delta_i}{\sum_{i=1}^{n} t_i e^{(\beta_1 z_{1i} + \beta_2 z_{2i} + \ldots)}} = 6.9269 * 10^{-6}$$

According to Eq. (9.51), the full failure rate function is

$$\lambda = 6.9269 * 10^{-6} e^{0.0321 * z}$$

At temperature z = 200 ° C, the failure rate is =6.9269 * $10^{-6} e^{0.0321 * 200}$ = 0.0034. The corresponding negative $L(\boldsymbol{\beta})$ is 383.3492. The two-stage method is not the optimal, as the $-L(\boldsymbol{\beta})$ is higher than that from the full maximum likelihood method shown in Example 9.6. This example shows the two-stage method has lower accuracy than the one stage method that uses full likelihood. But the two-stage method has much lower computational cost.

9.6 Test Zero Covariate Coefficient

A practical issue for using proportional hazard models is the selection of proper covariates. Some covariates might be noise covariate that does not carry any information about the failure rate. In statistics, it is an issue of testing the significance of the covariate. Equation (9.27) describes a method to compute the estimator's confidence interval. Here we can use the same Eq. (9.27) to test the significance. A non-significant covariate is with coefficient near 0. This implies we can test if the covariate coefficient $\beta_i = 0$. The hypothesis is then proposed as

$$H_0: \ \beta_i = 0$$

Let the real $\beta_i = 0$. According to Eq. (9.27),

$$\frac{\widehat{\beta_i}}{\sqrt{V_{ii}}} \sim N(0, 1) \tag{9.63}$$

For significance level α, if the H_0 does not hold, the quantity $\frac{\widehat{\beta_i}}{\sqrt{V_{ii}}}$ will locate outside of the region $1 - \alpha$, as shown in Fig. 9.11.

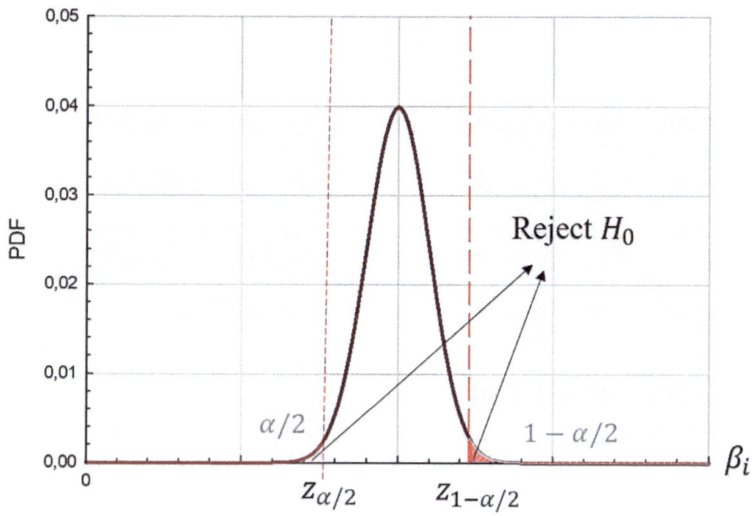

Fig. 9.11 Reject region of zero coefficient

Example 9.8
For the data from paper by Kjevik and Yuan [2], present in Table 9A.1, which is with 5 covariates: Average temperature, Min temperature, Max temperature, Average precipitation, Max precipitation. Test if the 5 covariates are not significant, choosing the significance level $\alpha = 0.05$.

Solution: From Example 9.2, we have obtained the estimators of efficient $\boldsymbol{\beta}$ are obtained as

$$10.88057, -3.94648, -6.44016, -0.15708, -0.11432$$

And the variance of $\widehat{\beta}_i$ is

$$41.13335, 12.94869, 8.488238, 0.298245, 0.004141$$

From the standard Normal distribution, we know the $z_{\frac{\alpha}{2}} = -1.96$; $z_{1-\frac{\alpha}{2}} = 1.96$. The covariate will be rejected if $\frac{\widehat{\beta}_i}{\sqrt{V_{ii}}} \geq 1.96$ or $\frac{\widehat{\beta}_i}{\sqrt{V_{ii}}} \leq -1.96$.

Ave temperature: $\frac{\widehat{\beta}_1}{\sqrt{V_{11}}} = \frac{10.88057}{\sqrt{41.1335}} = 1.6965.$ Accepted H_0

Min temperature: $\frac{\widehat{\beta}_2}{\sqrt{V_{22}}} = \frac{-3.94648}{\sqrt{12.94869}} = -1.0967.$ Accepted H_0

Max temperature: $\frac{\widehat{\beta}_3}{\sqrt{V_{33}}} = \frac{-6.44016}{\sqrt{8.488238}} = -2.2105 \leq -1.96.$ Rejected H_0

(continued)

Example 9.8 (continued)

Average precipitation: $\frac{\widehat{\beta_4}}{\sqrt{V_{44}}} = \frac{-0.15708}{\sqrt{0.298245}} = -0.2876.$ Accepted H_0

Max precipitation: $\frac{\widehat{\beta_5}}{\sqrt{V_{55}}} = \frac{-0.11432}{\sqrt{0.004141}} = -1.7765.$ Accepted H_0

Only the max temperature covariate shows significantly not zero, suggesting only the max temperature is a covariate should be considered in the PHM model.

9.7 Goodness of Fit Test

The PHM relies on several implicit assumptions, which may render it unreasonable in some cases:

- The failure rate relates to the covariates should be in a form of exponential.
- The covariates affect the failure rate proportionally.
- The covariate function is in a linear form.

Several methods have been developed to check the adequacy of PHM, either graphical method or statistical test using Cox-Snell residual [12], Schoenfeld residual [13] or martingale residuals [14]. This chapter selects the popular Grambsch and Therneau method [15].

9.7.1 *Grambsch and Therneau Test*

The method assesses the proportionality of the proportional hazard model by analysing the Schoenfeld residual. Suppose the observed failure times are ordered ascendingly as

$$t_1, t_2, \ldots t_k, \ldots, t_d \tag{9.64}$$

The trick of the method is introducing auxiliary time-dependent coefficient $g_j(t)$. If the coefficient is constant, the auxiliary coefficient, shown as θ_j in Eq. (9.65) will be zero.

$$\beta_j(t) = \beta_j + \theta_j g_j(t) \tag{9.65}$$

Usually, the $g_j(t)$ is specified as ln$t$. The derivation of this statistic test is complicated and lengthy, involving plenty knowledge in stochastic processing. This Chapter only presents the results from Grambsch and Therneau [15], without elaborating the derivation process. The null hypothesis is all the auxiliary θ_j coefficient zero

$$H_0 : \boldsymbol{\theta} = \mathbf{0} \tag{9.66}$$

The Schoenfeld residuals are defined as Eq. (9.67), using the covariates $\boldsymbol{z_k}$ and the ordered failure times (9.64).

$$\mathbf{r}(\boldsymbol{\beta}, \boldsymbol{z_k}) = \boldsymbol{z_k} - \overline{\boldsymbol{z_k}} \tag{9.67}$$

where $\overline{\boldsymbol{z_k}} = \frac{\sum_{j \in Risk(t_k)} z_j \exp(\beta' z_j)}{\sum_{j \in Risk(t_k)} \exp(\beta' z_j)}$, the $\boldsymbol{z_k}$ is the covariate vector corresponding to the failure time t_k. When the true covariate coefficients are unknown, we can use the maximum likelihood estimator of $\widehat{\boldsymbol{\beta}}$ to calculate $\mathbf{r}\left(\widehat{\boldsymbol{\beta}}, \boldsymbol{z_k}\right)$, which is simply written as $\widehat{\mathbf{r}_k}$. The statistics for the goodness of fit test are [15, 16].

$$T(G) = \left(\sum_{k=1}^{d} G_k \widehat{\mathbf{r}_k}\right)' H^{-1} \left(\sum_{k=1}^{d} G_k \widehat{\mathbf{r}_k}\right) \tag{9.68}$$

where

$$G_k = \begin{pmatrix} g_1(t_k) & 0 & 0 \\ 0 & g_j(t_k) & 0 \\ 0 & 0 & g_p(t_k) \end{pmatrix}_{p \times p} \tag{9.69}$$

$$H = \sum_{k=1}^{d} G_k V_k G_k' - \left(\sum_{k=1}^{d} G_k V_k\right) \left(\sum_{k=1}^{d} V_k\right)^{-1} \left(\sum_{k=1}^{d} G_k V_k\right)' \tag{9.70}$$

$$V_k = \frac{\sum_{i=1}^{n} Y(\mathrm{t}_k) \exp\{\beta' z_i\} \{z_i - \overline{z_i}(\beta, \mathrm{t}_k)\} \{z_i - \overline{z_i}(\beta, \mathrm{t}_k)\}'}{\sum_{i=1}^{n} Y(\mathrm{t}_k) \exp\{\beta' z_i\}} \tag{9.71}$$

$$Y(\mathrm{t}_k) = \mathrm{I}(\mathrm{t}_k \le \mathrm{T}_i). \tag{9.72}$$

The test statistics $T(G)$ of Eq. (9.68) is supposed to follow the $\chi^2(p)$ where the degree of freedom p is equal to the number of covariates. Given a significance level α, if the $T(G)$ is greater than the value obtained from the inverse function of the CDF $\chi^2(p)$ at α, reject the null hypothesis H_0 of Eq. (9.66). Fig. 9.12 plots the PDF function of $\chi^2(10)$, the rejection area is shown as the shadowed.

The computation of the statistics is heavy, impossible by manual calculation. R or Matlab software can be used to perform the computation. This chapter uses MATLAB to perform the computation for the test.

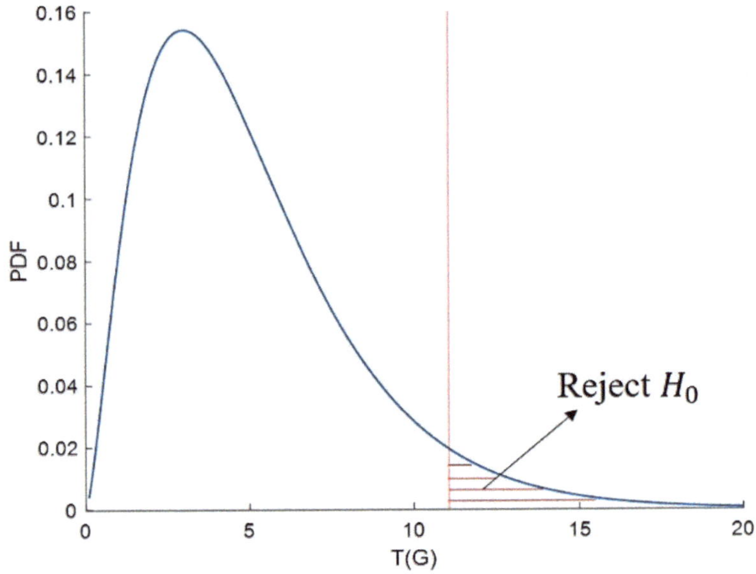

Fig. 9.12 Reject region of proportionality

Example 9.9
For the data from paper by Kjevik and Yuan [2], present in Table 9A.1, which is with 5 covariates: Ave temp, Min temp, Max temp, Average precipitation, Max precipitation. Test if the coefficients of the covariates are constant, choosing the significance level $\alpha = 0.05$.

Solution: From Example 9.2, we have obtained the partial likelihood estimators of efficient $\boldsymbol{\beta}$ as

$$10.88057, -3.94648, -6.44016, -0.15708, -0.11432$$

The test statistics of Eq. (9.68) is 14.3561 where the $g_j(t) = \ln t$. The distribution of the $T(G)$ is $\chi^2(p)$ with degree of freedom 5, equal to the number of covariates. At significance level $\alpha = 0.05$, the inverse CDF function of $\chi^2(p)$ at $1 - \alpha = 0.95$ is 11.0705. Obviously, the result from Eq. (9.68) which is $14.3561 > 11.0705$. The proportionality assumption does not hold. But this test result is for the situation where all the 5 covariates are considered together. If any one of the covariates violates the proportionality, the proportionality assumption will fail. Alternatively, we can choose covariate one by one to test. The results are shown in the following Table 9.6.

The table shows Average precipitation Max precipitation rejected the $H_0 : \boldsymbol{\theta} = \mathbf{0}$, which means the two factors does not meet the proportionality assumption. The PHM does not work if the two factors are considered. Be aware this test only validates the proportionality assumption, not the adequacy of the whole PHM model.

Table 9.6 Test results for individual factor

Covariate	Partial Likelihood Estimator	T(G)	$inv(\chi^2(1))$	Test Results
Ave temp	0.0646	1.3527	3.8415	Not reject
Min temp	0.0736	1.4195	3.8415	Not reject
Max temp	0.0465	1.5640	3.8415	Not reject
Average precipitation	−0.1216	4.1924	3.8415	Reject
Max precipitation	−0.0777	8.6379	3.8415	Reject

9.7.2 Exponential Goodness of Fit Test

This section discusses the test of the fitting of PHM to the given data. The first method uses the speciality of the PHM structure. By Eq. (9.13), the reliability function of the PHM can be written as follows, regardless of covariate function

$$R(t) = e^{-\Lambda(t)} \tag{9.73}$$

The function $\Lambda(t)$ is the cumulative hazard function. Conditioned on the PDF function Eq. (9.13) is correct, the (9.73) holds. Considering the $\Lambda(t)$ as a variable, Eq. (9.73) is a function of the form

$$R(t) = e^{-x}, \tag{9.74}$$

suggesting the $\Lambda(t)$ follows Exponential distribution with failure rate 1. We can use this property to perform goodness of fit test on the proportional hazard model. For a parametric PHM, the $\Lambda(t)$ is explicitly written as

$$\Lambda(t_i) = H_0(t_i)e^{(\beta_1 z_{1i} + \beta_2 z_{2i} + \ldots)} \tag{9.75}$$

It is not hard to find out, for Weibull baseline function, Eq. (9.75) is

$$\Lambda(t_i) = \left(\frac{t_i}{\alpha}\right)^{\beta} e^{(\beta_1 z_{1i} + \beta_2 z_{2i} + \ldots)} \tag{9.76}$$

For Exponential baseline function, Eq. (9.75) is

$$\Lambda(t_i) = \lambda t_i e^{(\beta_1 z_{1i} + \beta_2 z_{2i} + \ldots)} \tag{9.77}$$

By taking both sides of logarithm, Eq. (9.74) is rearranged as a linear function as follows

$$\ln R(t) = -x \tag{9.78}$$

By substituting Eq. (9.77) into (9.78), Eq. (9.78) is rewritten as

$$\ln R(t) = -\Lambda(t) \tag{9.79}$$

The remaining task to find out the empirical R(t). As discussed in Chap. 5, the CDF can be approximated by media rank $\frac{i-0.3}{n+0.4}$. The reliability is then approximated as $1 - \frac{i-0.3}{n+0.4}$. Equation (9.79) by using the approximated cumulative function is then

$$\ln\left(1 - \frac{i-0.3}{n+0.4}\right) = -\Lambda(t_i) \tag{9.80}$$

Specially, if the covariate function contains only one covariate, and at each covariate value, there are several failures, then we can group the data by each discrete value and we calculate the values in Eq. (9.80) only for each group. In the group case, the $\ln R(t_i) = \ln\left(1 - \frac{i-0.3}{n+0.4}\right)$ still hold but the n used for each group is not the total number of whole data sets, but the total number of data sets in this group. We then put all the group data together and consider them as the empirical data. The desirable line is $y = -x$. If empirical data sets fit the $y = -x$ well, the model is acceptable; otherwise, reject the model. We can demonstrate it by an example as follows

Example 9.10
Using the accelerated life test data presented in Table 9A.3 to estimate the parameter of the temperature coefficient, and the parameters of failure rate when the underlying failure rate function is from Weibull distribution. It is known the estimated parameters are

$$\alpha = 1.3741 \times 10^7, \beta = 0.5718; \beta_1 = 0.0321$$

Does the model fit the data?

Solution: Group the data by the three levels $z = 170,200,250$. According to Eq. (9.76), the cumulative hazard function, acting as x in the linear function, is

$$\Lambda(t_i) = \left(\frac{t_i}{\alpha}\right)^{\beta} e^{(\beta_1 z_{1i} + \beta_2 z_{2i} + \ldots)}$$

By Eq. (9.80), the $y = \ln R(t_i) = \ln\left(1 - \frac{i-0.3}{n+0.4}\right)$. For each group, computing the (x, y), the detailed results are shown in the following Table 9.7.

Plot the data sets in the above table and the describable line $y = -x$ together, the result is shown in the following Fig. 9.13.

It can be seen the data sets fit the desirable line well. The model

$$\lambda(t) = \frac{\beta}{\alpha}\left(\frac{t}{\alpha}\right)^{\beta-1} e^{(\beta_1 z_1 + \beta_2 z_2 + \ldots)} = \frac{0.5718}{1.37 \times 10^7}\left(\frac{t}{1.37 \times 10^7}\right)^{0.5718-1} e^{0.0321 z_1}$$

is acceptable.

Table 9.7 Empirical linear points for Example 9.10

z=170				z=200				z=250			
i	TTF	$x=\Lambda(t_i)=\left(\frac{t_i}{\alpha}\right)^{\beta}e^{\beta_1 z_1}$	$y=\ln R(t_i)$	i	TTF	x	y	i	TTF	x	y
1	0.20	0.01	−0.03	1	0.10	0.01	−0.03	1	0.10	0.07	−0.04
2	5.00	0.05	−0.09	2	0.20	0.02	−0.09	2	0.20	0.10	−0.09
3	27.00	0.13	−0.14	3	0.20	0.02	−0.14	3	0.20	0.10	−0.15
4	51.00	0.18	−0.20	4	2.00	0.08	−0.20	4	0.20	0.10	−0.21
5	103.50	0.28	−0.26	5	5.00	0.13	−0.26	5	0.50	0.17	−0.28
6	192.00	0.39	−0.33	6	5.00	0.13	−0.33	6	2.00	0.38	−0.35
7	192.00	0.39	−0.40	7	25.00	0.32	−0.40	7	5.00	0.64	−0.42
8	429.00	0.62	−0.47	8	52.50	0.49	−0.47	8	5.00	0.64	−0.51
9	429.00	0.62	−0.56	9	52.50	0.49	−0.56	9	5.00	0.64	−0.60
10	954.00	0.99	−0.65	10	52.50	0.49	−0.65	10	10.00	0.95	−0.69
11	2495.00	1.71	−0.74	11	52.50	0.49	−0.74	11	10.00	0.95	−0.80
12	2495.00	1.71	−0.85	12	123.50	0.80	−0.85	12	23.00	1.53	−0.92
13	2495.00	1.71	−0.97	13	123.50	0.80	−0.97	13	23.00	1.53	−1.06
14	2495.00	1.71	−1.11	14	123.50	0.80	−1.11	14	23.00	1.53	−1.22
15	2495.00	1.71	−1.28	15	219.00	1.12	−1.28	15	23.00	1.53	−1.42
16	2495.00	1.71	−1.47	16	219.00	1.12	−1.47	16	23.00	1.53	−1.66
17	2495.00	1.71	−1.71	17	524.00	1.84	−1.71	17	50.00	2.39	−1.97
18	2495.00	1.71	−2.02	18	524.00	1.84	−2.02	18	50.00	2.39	−2.43
19	2948.00	1.88	−2.48	19	1145.00	2.87	−2.48	19	120.00	3.94	−3.32
20	2948.00	1.88	−3.37	20	1145.00	2.87	−3.37				

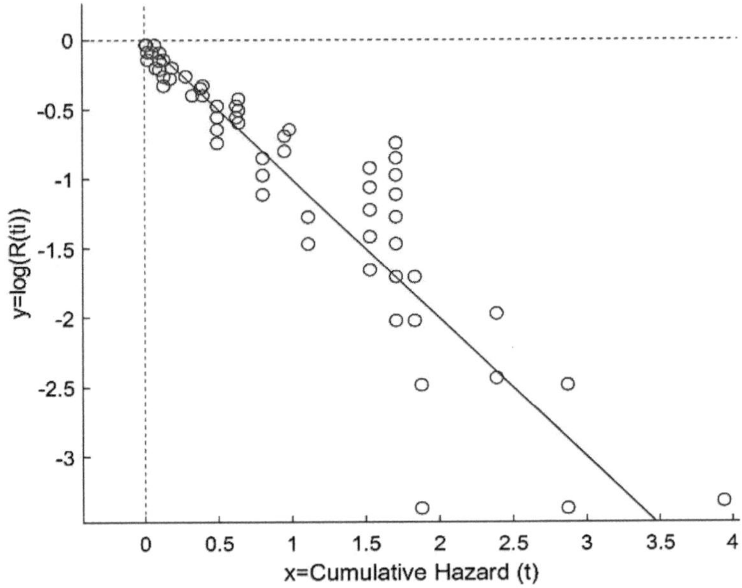

Fig. 9.13 Exponential goodness of fit test for Example 9.10

9.7.3 Goodness of Fit Test Using Cumulative Hazard Function

Section 9.7.1 discusses the Grambsch and Therneau approach to test the proportionality of PHM numerically, which is named formal test in statistics. This section discusses another unformal approach using graphical approach. This approach is simple but effective. According to Eq. (9.79) in the preceding section, the CDF function can be estimated by $\frac{i-0.3}{n+0.4}$. On the other hand, the

$$R(t_i) = 1 - F(t_i) = e^{-H_0(t_i)e^{(\beta_1 z_{1i} + \beta_2 z_{2i} + \ldots)}} \tag{9.81}$$

where $H_0(t_i)e^{(\beta_1 z_{1i}+\beta_2 z_{2i}+\ldots)}$ is the cumulative hazard function written as $\Lambda(t_i)$. It is readily to obtain

$$\Lambda(t_i) \approx -\ln\left(1 - \frac{i-0.3}{n+0.4}\right) \tag{9.82}$$

The cumulative hazard function $H_0(t_i)e^{(\beta_1 z_{1i}+\beta_2 z_{2i}+\ldots)}$ is a function proportional to the constant $e^{(\beta_1 z_{1i}+\beta_2 z_{2i}+\ldots)}$. If the covariate is constant, the $\Lambda(t_i)$ will be strictly proportional from each other at various covariate values. Therefore, it can be used to test the proportionality graphically, if we can plot Eq. (9.82) against time.

Test Proportionality for Weibull PHM

For the parametric PHM where the baseline function is known, for example a Weibull failure rate, the cumulative hazard function is written as

$$\Lambda(t_i) = \left(\frac{t_i}{\alpha}\right)^{\beta} e^{(\beta_1 z_{1i} + \beta_2 z_{2i} + \ldots)} \tag{9.83}$$

The plot of Eq. (9.83) for three covariate levels is demonstrated in Fig. 9.14. The three cumulative hazard function are all originating from origin point, and without any crossing other than at origin point. The empirical cumulative data should show the similar pattern if the model is able to fit the data.

However, checking the proportionality visually from Fig. 9.14 is not convenient. Further transformation is necessary to make the curve linear. Take logarithm at both sides of (9.83),

$$\ln \Lambda(t_i) = \beta \ln t - \beta \ln \alpha + \beta_1 z_{1i} + \beta_2 z_{2i} + \ldots \tag{9.84}$$

which is a linear function in form of $y = ax + b$, where

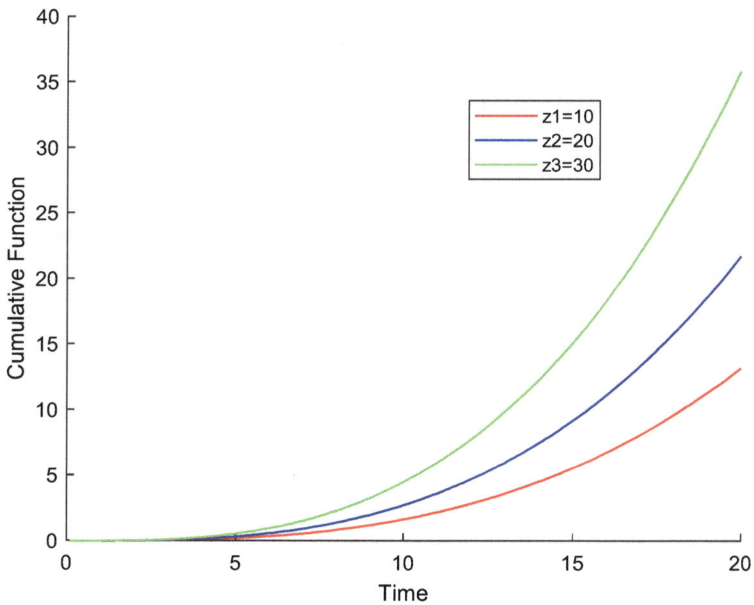

Fig. 9.14 Cumulative Hazard function plot against time

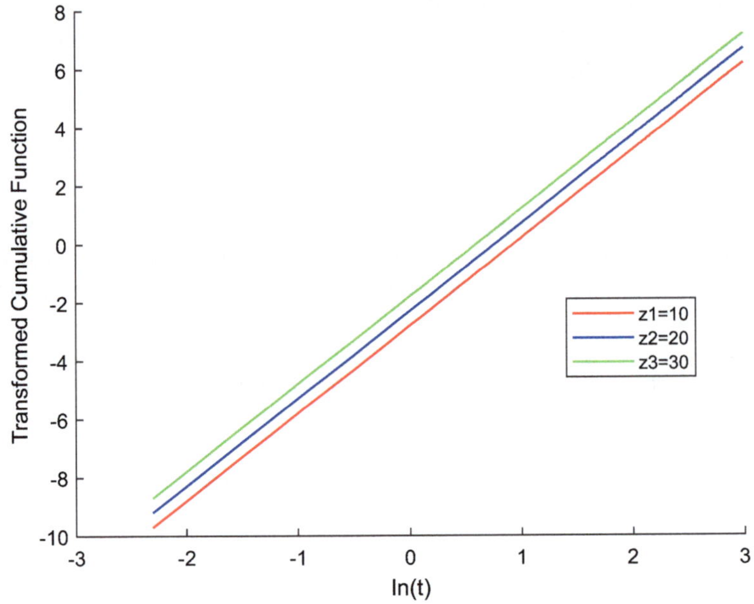

Fig. 9.15 Linearized cumulative Hazard function against time

$$\begin{cases} a = \beta \\ b = -\beta \ln \alpha + \beta_1 z_{1i} + \beta_2 z_{2i} + \ldots \\ x = \ln t \\ y = \ln \Lambda(t) \end{cases} \tag{9.85}$$

The b is the constant threshold. For different covariate values, the line will only differ in its threshold, i.e. visually, the line is only shifted up and down with the changing of covariate values as shown in Fig. 9.15.

Checking the proportionality from the linear function is very simple. If the empirical data is parallel as in Fig. 9.15, the proportionality assumption holds.

The procedure to perform the test is as **Algorithm 9.1** as follows:

Step **1**: Group by the covariate level. For each level, using Eq. (9.82) to calculate the empirical cumulative function value.

Step **2**: Plot the empirical data in two dimensions, where $y = \ln \Lambda(t_i)$; $x = \ln t_i$.

Step **3**: Plot the linear line $\ln\Lambda(t_i) = \beta \ln t_i - \beta \ln \alpha + \beta_1 z_{1i} + \beta_2 z_{2i} + \ldots$ by Eq. (9.85) for each covariate level.

Step **4**: Visually check the plot for various covariate levels. If the empirical $\ln\Lambda(t_i)$ for various levels shows proportional pattern, which means at least no crossing among them, the proportionality assumption holds and the PHM can be used to the data; if no, the covariate does not affect the cumulative hazard function proportionally. It is because eighter the PHM does not fit the data, or the data should be transformed. It can also be used to validate the estimated parameters, if all the straight lines for each covariate level fit their corresponding empirical data, accept the estimates; otherwise, reject the estimates.

Example 9.11
Using the accelerated life test data presented in Table 9A.3 to estimate the parameter of the temperature coefficient, and the parameters of failure rate when the underlying failure rate function is from Weibull distribution. It is known the estimated parameters are

$$\alpha = 1.3741 \times 10^7, \beta = 0.5718; \beta_1 = 0.0321$$

Does the model fit the data?

Solution: Group the data by the three levels $z = 170,200,250$. For each group, compute the (x, y) by the Algorithm 9.1, the detailed results are shown in the following Table 9.8.

The plot of the emperical data exhibits the emperial cumulative hazard function shown in parallel, despite not perfectly, as shown in Fig. 9.16 as follows.The proportionality assumption can be roguthly accpeted.

The PHM using the estimators are acceptabel since their lines can fit their emperical cumulative hazard function generally, as shown in Fig. 9.17 as follows.

Test Proportionality for Exponential PHM
Despite the Exponential distribution is special case of Weibull distribution, the PHM with failure rate of Exponential distribution, a constant failure rate, and the PHM with Weibull failure rate could have some differences in the proportionality test. The cumulative hazard function of constant failure rate is written as

$$\Lambda(t_i) = \lambda t_i e^{(\beta_1 z_{1i} + \beta_2 z_{2i} + \ldots)} \tag{9.86}$$

Take logarithm on both sides further. Equation (9.86) is written as

$$\ln \Lambda(t) = \ln t + \ln \lambda + \beta_1 z_{1i} + \beta_2 z_{2i} + \ldots \tag{9.87}$$

The function is again of linear function of form $y = x + b$, where

$$\begin{cases} y = \ln \Lambda(t) \\ x = \ln t \\ b = \ln \lambda + \beta_1 z_{1i} + \beta_2 z_{2i} + \ldots \end{cases} \tag{9.88}$$

Table 9.8 Empirical linear points for Example 9.11

z=170				z=200				z=250			
i	TTF	$y = \ln \Lambda(t_i)$	$x = \ln t_i$	i	TTF	y	x	i	TTF	y	x
1	0.20	−3.35	−1.61	1	0.10	−3.35	−2.30	1	0.10	−3.30	−2.30
2	5.00	−2.44	1.61	2	0.20	−2.44	−1.61	2	0.20	−2.39	−1.61
3	27.00	−1.95	3.30	3	0.20	−1.95	−1.61	3	0.20	−1.90	−1.61
4	51.00	−1.61	3.93	4	2.00	−1.61	0.69	4	0.20	−1.55	−1.61
5	103.50	−1.34	4.64	5	5.00	−1.34	1.61	5	0.50	−1.28	−0.69
6	192.00	−1.12	5.26	6	5.00	−1.12	1.61	6	2.00	−1.06	0.69
7	192.00	−0.92	5.26	7	25.00	−0.92	3.22	7	5.00	−0.86	1.61
8	429.00	−0.75	6.06	8	52.50	−0.75	3.96	8	5.00	−0.68	1.61
9	429.00	−0.59	6.06	9	52.50	−0.59	3.96	9	5.00	−0.52	1.61
10	954.00	−0.44	6.86	10	52.50	−0.44	3.96	10	10.00	−0.37	2.30
11	2495.00	−0.30	7.82	11	52.50	−0.30	3.96	11	10.00	−0.22	2.30
12	2495.00	−0.16	7.82	12	123.50	−0.16	4.82	12	23.00	−0.08	3.14
13	2495.00	−0.03	7.82	13	123.50	−0.03	4.82	13	23.00	0.06	3.14
14	2495.00	0.11	7.82	14	123.50	0.11	4.82	14	23.00	0.20	3.14
15	2495.00	0.24	7.82	15	219.00	0.24	5.39	15	23.00	0.35	3.14
16	2495.00	0.38	7.82	16	219.00	0.38	5.39	16	23.00	0.50	3.14
17	2495.00	0.53	7.82	17	524.00	0.53	6.26	17	50.00	0.68	3.91
18	2495.00	0.70	7.82	18	524.00	0.70	6.26	18	50.00	0.89	3.91
19	2948.00	0.91	7.99	19	1145.00	0.91	7.04	19	120.00	1.20	4.79
20	2948.00	1.22	7.99	20	1145.00	1.22	7.04				

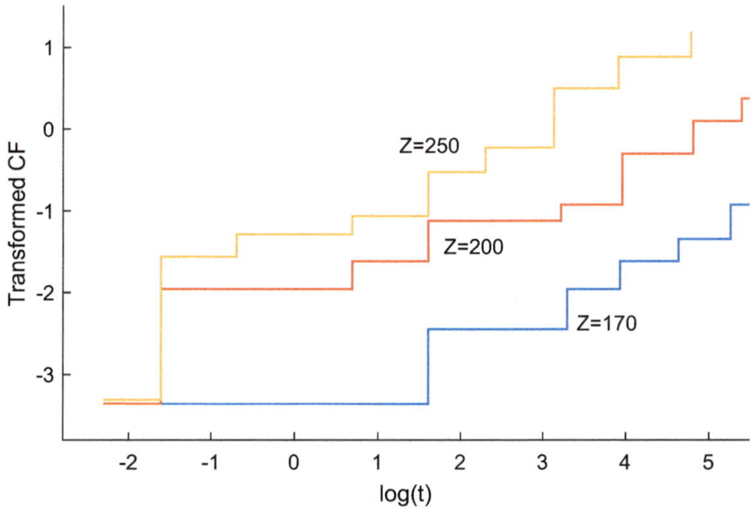

Fig. 9.16 Proportionality check for Example 9.11

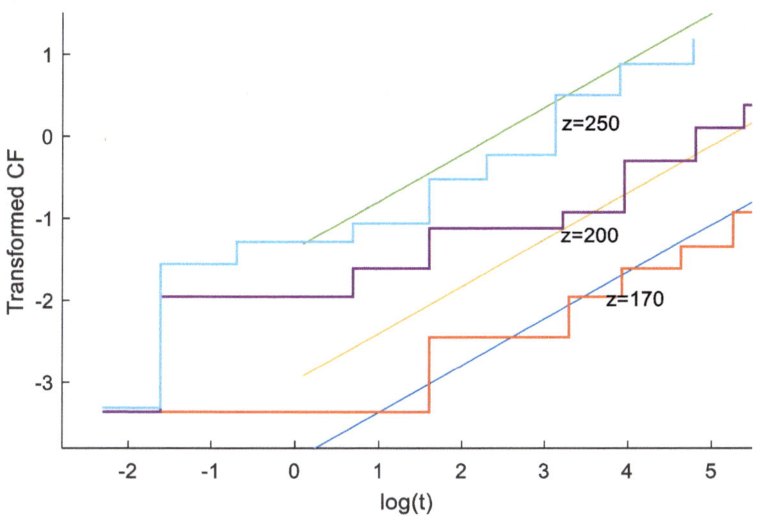

Fig. 9.17 Fitting check for Example 9.11

The lines for different covariate level differs at its threshold, and visually, they are present as linear lines in parallel. Same as the Weibull PHM, the cumulative hazard function is approximated by

$$\Lambda(t_i) \approx -\ln\left(1 - \frac{i - 0.3}{n + 0.4}\right) \tag{9.89}$$

If the Exponential PHM holds, the plot of the empirical data sets will be linear, and their line are parallel for the various covariate level. The **Algorithm 9.1** still applies to the Exponential PHM, except the Step 3, which is changed as follows:

Step **3**: Plot the linear line $\ln\Lambda(t) = \ln t + \ln\lambda + \beta_1 z_{1i} + \beta_2 z_{2i} + \ldots$ by Eq. (9.88) for each covariate level.

Example 9.12

Using the accelerated life test data presented in Table 9A.3 to estimate the parameter of the temperature coefficient, and the parameters of failure rate when the underlying failure rate function is from Exponential distribution. It is known the estimated parameters are

$$\lambda = 6.9269 * 10^{-6}; \beta_1 = 0.0321$$

Does the model fit the data?

Solution: It has been discussed in the **Example 9.11** the data can be modelled by PHM as the proportionality assumption holds. The remaining question is if the Exponential PHM can fit the data or not. We plot the linear line by Eq. (9.88) as shown in Fig. 9.18. It is obvious all the three lines cannot fit the empirical cumulative hazard function. The Exponential PHM using the estimates $\lambda = 6.9269 * 10^{-6}$; $\beta_1 = 0.0321$ is thus not acceptable.

Concerning the empirical cumulative hazard function, there are other alternatives other than the method used in this Section, for example the popular Nelson-Aalen method [7]. There are certainly some differences among those methods in their

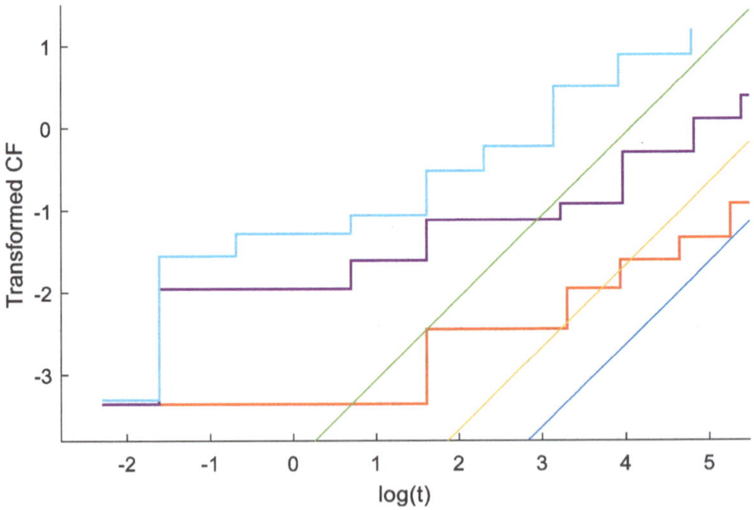

Fig. 9.18 Plots of Example 9.12

performance; while for practical purpose, their difference might not be significant as the data might be not accurate as well. This chapter puts many efforts on the proportionality test as this is a point ignored usually in many state-of-art applications. One should be aware the proportionality is very necessary to properly apply the PHM. On more theoretical discussions on PHM, one can refer the book by Lawless for further reading [7].

9.8 Generalized PHM

The physical interpretation of PHM is desired in reliability data analysis. Using the covariates to accommodate some physical factors and using the failure rate function to cope the randomness, the PHM in this perspective is advantageous over other simple distribution for data analysis. Nevertheless, the classical PHM model is restricted as an exponential form and the covariates are linked linearly. Given an example where the covariates are the simple but most common factor temperature and humidity, the classical PHM model is then

$$\lambda(t) = \lambda_0(t)e^{(\beta_1 \mathrm{Temp} + \beta_2 \mathrm{Humidity})} \tag{9.90}$$

Will the temperature and humidity be proper to be linearly connected and is it reasonable to use the exponential form? Surely it is not certain as it depends on the failure mechanism. The goodness of fit test approach can check the adequacy of the model. While for failure data analysis, we might need some other options to relax the linearity and exponentiality assumption.

9.8.1 Simple Transformation of Covariate

The first relaxation is to write the covariate to a general function, let say $w(z)$, the PHM model is then written as

$$\lambda(t) = \lambda_0(t)e^{\beta_1 w_1(\mathrm{Temp}) + \beta_2 w_2(\mathrm{Humidity})} \tag{9.91}$$

The transform function $w(z)$ is preferably monotonic and continuous. The selection of the transform function can be empirical, or one can chose a function by referring the undergoing failure physics. The temperature, for instance, is preferable to be of the form as follows in physics

$$w(\text{Temp}) = \frac{1}{\text{Temp}} \tag{9.92}$$

instead of the $w_1(\text{Temp}) = Temp$ shown in the classical PHM. Chapters 11 and 12 in this book will interpret it in detail from failure physics perspective. The transformation will not add any complexity to the mathematics, as the parameter estimation, goodness of fit test method remains the same. The only extra work is to change the numerical temperature value to $\frac{1}{Temp}$ for analysis. Other transformation such as $w(z) = z^2$, $w(z) = z^3$, $w(z) = a^z$, $w(z) = \ln z$ can be also options. The following example demonstrates the difference between the transformed and the linear PHM.

Numerical Example 9.13

The data was adopted from an accelerated life test with only one temperature covariate, as shown in the Table 9A.4 [17]. The analysis was performed by using both the original data and the transformed data by $w(\text{Temp}) = \frac{1}{Temp}$. The baseline function chooses the Weibull failure rate. The two-stage parameter estimation was used. For classical PHM, where the covariant is the original temperature T, from the partial likelihood method, the coefficient was estimated as $\beta_T = 0.0541$. The baseline function for Weibull failure rate is with parameter $\alpha = 5.7866 \times 10^6$; $\beta = 1.46333$. The full PHM model is written as

$$\lambda(t) = \frac{1.4633}{5.7866 \times 10^6}\left(\frac{t}{5.7866 \times 10^6}\right)^{0.4633} \exp(0.0541 * T) \tag{9.93}$$

The second PHM model uses the $1/T$ as the covariant. Applying the two-stage parameter estimation, the coefficient for $1/T$ is $\beta_{1/T} = -522$. The parameter for the Weibull failure rate is, after estimating the coefficient of covariate, is $\alpha = 2.6199$; $\beta = 1.3783$. The full model is then

$$\lambda(t) = \frac{1.3783}{2.6199}\left(\frac{t}{2.6199}\right)^{0.3783} \exp\left(-522 * \frac{1}{T}\right) \tag{9.94}$$

By plotting at temperature 50 °C, as shown in Fig. 9.19, Eqs. (9.93) and (9.94) exhibits some difference.

The difference in the two PHM models is visually significant. It implies sometimes transformation is necessary if the knowledge of failure physics can be used.

Shift of Data

The second data transformation is the shift of observed values. For example, the temperature unit might be changed from Celsius to Kelvin. The temperature is transformed as

$$\text{Kelvin} = \text{Celsius} + 273 \tag{9.95}$$

The Kelvin is more common used than the Celsius used in the failure physics.

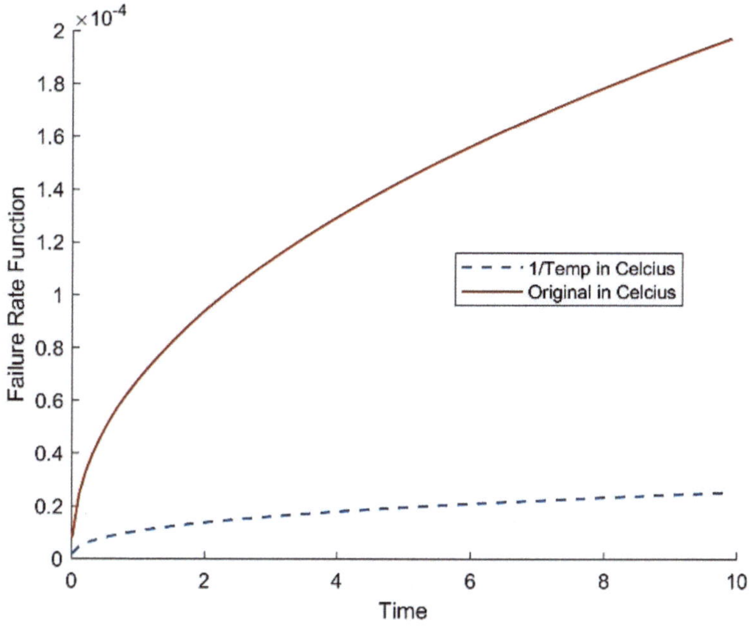

Fig. 9.19 Comparison of original and transformed PHM

Numerical Example 9.14
For the case used in the Example 9.13, change the temperature unit from Celsius to Kelvin. The covariate is then

$1/K = 1/(\text{Celsius}+273)$. Applying the two-stage parameter estimation, the coefficient for 1/K is $\beta_{1/K} = 8.9 \times 10^{-5}$. The parameter for the Weibull failure rate is $\alpha = 7.68 \times 10^3$; $\beta = 1.4469$. The plot of failure rate function also shows notable differences as shown in Fig. 9.20.

9.8.2 *Nonexponential Covariate Function*

The transformation discussed in the previous section retains in a form of exponentiality for the covariate function. But there are some situations it does not follow the exponentiality. We can try to transform the non-exponential functions into the exponential form. For example, the failure time of some electronic units depends on the voltage of a form z^{β_j}. The failure rate function for the PHM is not of exponential form but as follows

$$\lambda(t) = \lambda_0(t) z^{\beta_j} \tag{9.96}$$

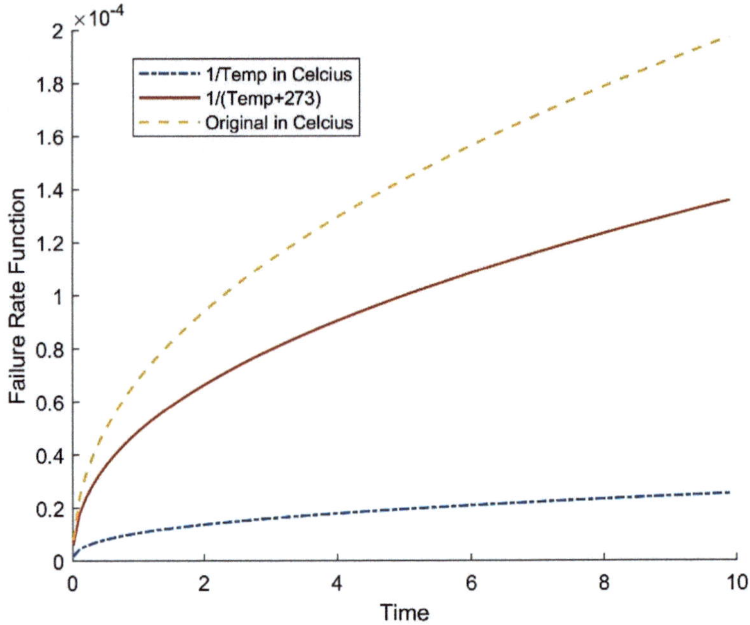

Fig. 9.20 Comparison of original and two others transformed PHMs

In order to use the approach developed in the classical PHM, one trick is to rewrite Eq. (9.96) in an exponential form

$$\lambda(t) = \lambda_0(t)e^{\beta_j \ln z} \tag{9.97}$$

Then the PHM can be used as normal. Special to the $\lambda_0(t)z^{\beta_j}$, it is not hard to find the failure rate is still proportional to the covariate value. The Appendix 9B of this Chapter proves even without transformation as to Eq. (9.97), the partial likelihood estimation approach still applies.

9.8.3 General Covariate Function PHM

The explicit expression of the covariate can be any form from physical perspective. In order to apply the PHM for more general failure mechanisms, with a least modification on the classical PHM, we can have a failure rate function as follows

$$\lambda(t/z) = \lambda_0(t)g(\beta, z) \tag{9.98}$$

Here we limit the case as a univariate case despite it loses generality, while the applicability doesn't compromise, as it can extended to the multivariate case easily. For two covariate level z_1, z_2, the ratio of the $\lambda(t/z)$ of Eq. (9.98) is straightforward as

$$\frac{\lambda(t/z_1)}{\lambda(t/z_2)} = \frac{g(\beta, z_1)}{g(\beta, z_2)} \tag{9.99}$$

Equation (9.99) suggests the failure rate ratio is proportional regardless of the time, implying it is still a proportional hazard model. But some necessary restrictions on the general function $g(\beta, z_1)$ should be:

(a) It is monotonic.
(b) Continuous and the first and 2nd derivative exist.

As derived in the beginning of this chapter, the partial likelihood function is as follows

$$L(\boldsymbol{\beta}) = \prod_{i \in D} \left\{ \frac{g(\boldsymbol{\beta}, z_i)}{\sum\limits_{j \in Risk(t_i)} g(\boldsymbol{\beta}, z_j)} \right\} \tag{9.100}$$

It is great if the partial likelihood estimation can still apply, since it is a unique advantage of the PHM against other multivariate models. For the classical PHM, the partial likelihood estimation was totally accepted just until the justification of the partial likelihood was proved. The Appendix 9B shows an example where the partial likelihood estimation still applies for $g(\beta, z) = z^{\beta}$. But it is not clear if other functions also apply. This is an open question that needs to be further investigated in the future.

Example 9.15
This example uses the data in Table 9A.5. The time to failure of the resistor is related to the applied stress, namely the volage. The dependency of the TTF and stress is unknown. This example uses both the classical PHM and the PHM with non-exponential covariate function z^{β_j}. For the classical PHM, the covariate is the voltage stress. A plot of the empirical cumulative hazard function, discussed in Sect. 9.7.3, are shown in Fig. 9.21.

The empirical cumulative function exhibits some proportionality, as they are roughly in parallel. The PHM is thus applicable for such case. Firstly, we use the classical PHM to fit the data. The two-stage PHM estimation method was used, supposing the baseline function is Weibull distribution. The parameters are:

$$\alpha = 2.77109 \times 10^3, \beta = 2.3237; \beta_1 = 0.7217$$

The classical PHM is as follows

(continued)

Example 9.15 (continued)

$$\lambda(t) = \frac{2.3237}{2.77109 \times 10^3}\left(\frac{t}{2.77109 \times 10^3}\right)^{1.3237} \exp(0.7217 * z) \qquad (9.101)$$

The next PHM using covariate function z^{β_j}. The $z^{\beta_j} = e^{\beta_j \ln z}$ just for parameter estimation. We can transform the covariate, the voltage stress by $\ln z$ in this example, the others remain unchanged. The parameters are

$$\alpha = 4.08 \times 10^6, \beta = 2.3716; \beta_1 = 10.4953$$

The PHM model is as follows

$$\lambda(t) = \frac{2.3716}{4.08 \times 10^6}\left(\frac{t}{4.08 \times 10^6}\right)^{1.3716} z^{10.4953} \qquad (9.102)$$

A plot for Eqs. (9.101) and (9.102) for voltage stress 15v does exhibit some difference, not even significantly, as shown in Fig. 9.22.

For this case, the power PHM is from the physical background, stating the time to failure is related to the applied in a form of power. While when an application concerns less on the physical background, the classical PHM is preferable as its statistical properties have been intensively investigated and there are commercial software or package, such as in Matlab, is available for the classical PHM.

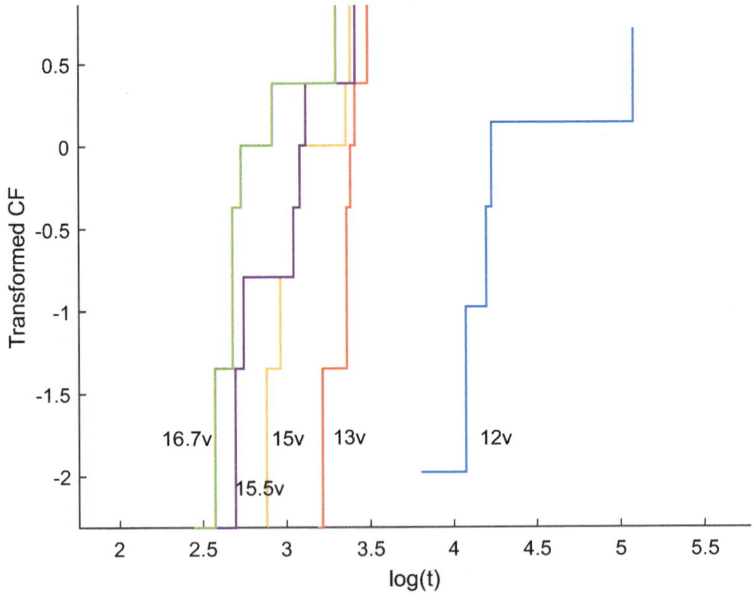

Fig. 9.21 Proportionality of cumulative Hazard function

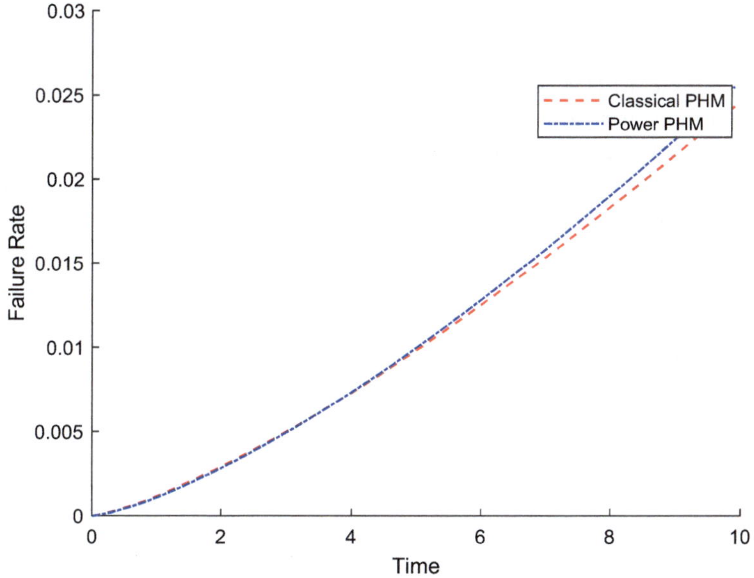

Fig. 9.22 Failure rate function of classical and power PHMs

9.8.4 Another General Covariate Function PHM

If we let the time to failure (TTF) be proportional to a function of the factors such as temperature$f(T)$ in a form as

$$TTF \propto f(T) \tag{9.103}$$

The corresponding failure rate is then inverse proportional to $f(T)$ as

$$\lambda \propto \frac{1}{f(T)} \tag{9.104}$$

For simplicity of notation, let $G(T) = \frac{1}{f(T)}$. A proportional hazard model can be written as

$$\lambda(t/T) = \lambda_0(t)G(T) \tag{9.105}$$

The $f(T) > 0$ since the TTF is positive. Equation (9.105) can be rearranged in an exponential form

$$\lambda(t/T) = \lambda_0(t)e^{\ln G(T)} \tag{9.106}$$

An artificial coefficient β_j is added as

$$\lambda(t/z) = \lambda_0(t)e^{\beta_j \ln G(T)} \tag{9.107}$$

which is a typical PHM with transformed covariate value. The $G(T)$ is a physical model and the significance of the factor T is reflected by the β_j. When $\beta_j = 0$, the factor has no impact on the failure rate. Equation (9.107) retains all the properties of classical PHM. Nevertheless, to be an eligible PHM, at least a constraint should be imposed: the $\ln G(T)$ should be a monotone function in the range of the questions of interest.

Appendix 9A

Table 9A.1 Time to failure with related temperature values (Data source: [2])

TTF hours	Average temp	Min temp.	Max temp	Average precipitation	Max precipitation
696	5.4	3.3	7.4	4.35	19.3
7416	3.3	0.6	6.1	2.29	45.9
984	3.7	1.4	5.8	3.3	13.3
672	6.4	4.4	8.4	2.73	14.5
216	9.8	7.5	12.1	0.8	2.4
2280	11.3	8.4	14.6	1.13	14.7
4440	−0.9	−3.5	1.7	2.3	11.1
2592	3	0.6	5.2	3.08	22.4
2832	8.1	5.3	11.3	2.97	28.3
3840	−3.6	−6.5	−0.9	2.01	13.2
96	0.5	−2.3	2.7	0	0
648	1.3	−1.1	3.5	0.49	3.5
384	3	0.9	4.9	4.72	14.2
1680	9.4	7	12.1	3.3	29.6
2112	7.7	5.4	10.5	2	17.6

Table 9A.2 Experimental data (Data source [19])

150 °C	Censor	170 °C	Censor	190 °C	Censor
2582	1	2077	1	667	1
2801	1	2787	1	1323	1
6015	0	3050	1	1376	1
3666	1	3055	1	1758	1
3834	1	3153	1	1781	1
6015	0	3163	1	1956	1
3966	1	3222	1	1963	1
6015	0	5856	1	1997	1
6015	0	3307	1	2094	1
6015	0	3550	1	2334	1
4050	1	3695	1	2381	1
4116	1	5856	0	2435	1
6015	0	3789	1	2491	1
4180	1	3975	1	2540	1
4212	1	4046	1	2919	1
4263	1	4221	1	3101	1
6015	1	4277	1	3205	1
6015	0	4303	1	3249	1
4626	1	5856	0	3257	1
4929	1	4640	1	3281	1
6015	0	5856	0	3339	1
4951	1	5856	0	3434	1
5211	1	4712	1	3587	1
6015	0	5856	0	3653	1
6015	0	4635	1	4323	1

Table 9A.3 Accelerated life test data (Data source [18])

170 °C	200 °C	250 °C
0.2	0.1	0.1
5	0.2	0.2
27	0.2	0.2
51	2	0.2
103.5	5	0.5
192	5	2
192	25	5
429	52.5	5
429	52.5	5
954	52.5	10
2495	52.5	10
2495	123.5	23
2495	123.5	23
2495	123.5	23
2495	219	23
2495	219	23
2495	524	50
2495	524	50
2948	1145	120
2948*	1145	

2948*-Censored Data

Table 9A.4 TTF at various Temperature levels (Data source [17, 20])

Time T	70 °C	90 °C	115 °C	140 °C
TTF (in hours)	39	6	1	1
	83	45	1	3
	208	65	44	7
	336	82	46	7
	373	87	65	8
	462	157	71	18
	475	177	95	21
	552	208	99	21
		259	108	23
		298	109	24
		364	114	26
			118	36
			118	37
			131	37
			143	44
			180	66
				78
				78

Table 9A.5 TTF at various temperature levels

Run Stress	12 V	13 V	15 V	15.5 V	16.7 V
1	66.5	32.7	17.9	21.0	14.6
2	58.73	28.8	21.8	22.6	14.6
3	68.8	28.8	19.4	21.8	13.1
4	160.60	30.3	17.9	14.8	18.5
5	44.8	24.1	28.7	15.6	11.5
6	/	24.9	21.0	13.2	15.4
7	/	29.5	29.5	30.3	27.0

Appendix 9B: Proof of the Asymptotic Consistency for $g(\beta, z) = z^{\beta}$

The partial likelihood estimate function is written in a modified manner as

$$L(\beta) = \prod_{i \in D} \left\{ \frac{\exp(\beta z_i)}{\sum_{i \in R(t_i)} \exp(\beta z_i)/n} \right\} \tag{A.1}$$

The corresponding log partial likelihood function is

$$l(\beta) = \frac{1}{n} \log\{\exp(\beta z_i)\} - \frac{1}{n} \log\left\{ \sum_{j \in R(t_i)} \exp(\beta z_j)/n \right\} \tag{A.2}$$

Writing the exponential term as a general function $g(\boldsymbol{\beta}, z_i)$, the (A.2) is then rewritten as

$$l(\beta) = \frac{1}{n} \log\{g(\beta, z_i)\} - \frac{1}{n} \log\left\{ \sum_{j \in R(t_i)} g(\beta, z_j)/n \right\} \tag{A.3}$$

Again, rewrite the (A.3) into a more general form as

$$l(\boldsymbol{\beta}) = \frac{1}{n} \log\{g(\boldsymbol{\beta}, z_i)\} - \frac{1}{n} I(t > t_j) . \log\left\{ \sum_{j=1}^{j=n} g(\boldsymbol{\beta}, z_j)/n \right\} \tag{A.4}$$

When $n \to \infty$,

$$\frac{1}{n}\log\{g(\boldsymbol{\beta}, z_i)\} \to \int_0^{+\infty} \log\{g(\boldsymbol{\beta}, z)\}f(z)dz \tag{A.5}$$

The derivation is similar to the [10]. Let the probability of a product that can survive to time t at the operating environment z, as $R(t/z)$. Then

$$R(t) = \int_0^{+\infty} R(t/z)f(z)dz \tag{A.6}$$

$$\frac{1}{n}I(t > t_j).\log\left\{\sum_{j=1}^{j=n} g(\boldsymbol{\beta}, z_j)/n\right\} \to \int_0^{+\infty} -dR(t)\int_t^{+\infty} \times \left[\log\int_0^{+\infty} exp^{-\int_0^{\tau}\lambda(s)g(\boldsymbol{\beta},z)ds} g(\boldsymbol{\beta}, z)f(z)dz\right]d\tau dt \tag{A.7}$$

The likelihood function (A.4) for the infinite n is

$$H(\boldsymbol{\beta}) = \int_0^{+\infty} \log\{g(\boldsymbol{\beta},z)\}f(z)dz + \int_0^{+\infty}\int_t^{+\infty}\left[\log\int_0^{+\infty} \mathrm{R}(\tau/z)g(\boldsymbol{\beta},z)f(z)dz\right]d\tau.dR(t) \tag{A.8}$$

The 1st derivative is

$$\frac{\partial \mathrm{H}(\boldsymbol{\beta})}{\partial \boldsymbol{\beta}} = \int_0^{+\infty} \frac{g'(\boldsymbol{\beta}, z)}{g(\boldsymbol{\beta}, z)} f(z)dz + \int_0^{+\infty}\int_t^{+\infty} \frac{\int_0^{+\infty} \mathrm{R}(\tau/z)g'(\boldsymbol{\beta}, z)f(z)dz}{\int_0^{+\infty} \mathrm{R}(\tau/z)g(\boldsymbol{\beta}, z)f(z)dz} d\tau.dR(t) \tag{A.9}$$

Let $E(g(\boldsymbol{\beta}, z)/t) = \int_0^{+\infty} \mathrm{R}(\tau/z)g(\boldsymbol{\beta}, z)f(z)dz$ and $E(g'(\boldsymbol{\beta}, z)/t) = \int_0^{+\infty} \mathrm{R}(\tau/z)g'(\boldsymbol{\beta}, z)f(z)dz$. The (A.9) is rewritten as

$$\frac{\partial \mathrm{H}(\boldsymbol{\beta})}{\partial \boldsymbol{\beta}} = \int_0^{+\infty} \frac{g'(\boldsymbol{\beta}, z)}{g(\boldsymbol{\beta}, z)} f(z)dz + \int_0^{+\infty}\int_t^{+\infty} \frac{E(g'(\boldsymbol{\beta}, z)/\tau)}{E(g(\boldsymbol{\beta}, z)/\tau)} d\tau.dR(t) \tag{A.10}$$

The 2nd derivative is

$$\frac{\partial^2 \mathrm{H}(\boldsymbol{\beta})}{\partial \boldsymbol{\beta}^2} = \int_0^{+\infty} \frac{g''(\boldsymbol{\beta},z).g(\boldsymbol{\beta},z) - g'^2(\boldsymbol{\beta},z)}{g^2(\boldsymbol{\beta},z)} f(z)dz + \int_0^{+\infty} \int_t^{+\infty} \frac{E(g''(\boldsymbol{\beta},z)/\tau)E(g(\boldsymbol{\beta},z)/\tau) - E^2(g'(\boldsymbol{\beta},z)/\tau)}{E^2(g(\boldsymbol{\beta},z)/\tau)} d\tau.dR(t) \tag{A.11}$$

When the covariant function is in a form of

$$g(\boldsymbol{\beta},z) = z^{\beta} \tag{A.12}$$

The (A.8) to (A.11) is specified as

$$H(\beta) = \int_0^{+\infty} \beta \ln z f(z)dz + \int_0^{+\infty} \log E\left(z^{\beta};t\right)dR(t) \tag{A.13}$$

$$\frac{\partial \mathrm{H}(\beta)}{\partial \beta} = \int_0^{+\infty} \ln z f(z)dz + \int_0^{+\infty} \log E\left(\ln z.z^{\beta};t\right)dR(t) \tag{A.14}$$

$$\frac{\partial^2 \mathrm{H}(\beta)}{\partial \beta^2} = \int_0^{+\infty} \left[\frac{E\left(z^{\beta}\ln^2 z;t\right)}{E(z^{\beta};t)} - \left(\frac{E\left(z^{\beta}\ln z;t\right)}{E(z^{\beta};t)}\right)^2\right] dR(t) \tag{A.15}$$

The (A.15) can be rewritten as

$$\frac{\partial^2 \mathrm{H}(\beta)}{\partial \beta^2} = \int_0^{+\infty} \left[\frac{E\left(\left(\ln z - \frac{E\left(z^{\beta}\ln z;t\right)}{E(z^{\beta};t)}\right)^2 z^{\beta};t\right)}{E(z^{\beta};t)}\right] dR(t) < 0 \tag{A.16}$$

The above formulas suggest the function $H(\beta)$ has a local maximum. Let the true parameter be β. There exists $\delta > 0$ when the $n \geq n_1$.

$$\begin{cases} l(\beta+\delta) - l(\beta)^{a.s.}{}_{a.s.} H(\beta+\delta) - H(\beta) < 0, \\ l(\beta-\delta) - l(\beta)^{a.s.}{}_{a.s.} H(\beta-\delta) - H(\beta) < 0; \end{cases} \tag{A.18}$$

There exists a point β^* with local maximum between $[\beta - \delta, \beta + \delta]$, since $l(\beta)$ is a continuous function and it is differentiable, i.e. the $\frac{\partial l(\beta)}{\partial \beta} = 0$ at β^*. Therefore, the $\frac{\partial l(\beta)}{\partial \beta} = 0$ has solution β^* with probability 1 and $|\beta - \beta^*| < \delta$. Since the δ is arbitrary small, the β^* converges to the true β asymptotically.

Problems and Questions

1. Using the data present in Table 9A.2, check if the proportional hazard model can fit the data. If yes,
   a. Estimate the coefficient of the temperature covariate using partial likelihood estimation method.
   b. Choose the baseline function as Weibull failure rate and estimate the parameters of the failure rate. Validate if the model can fit the data.
2. The failure rate function of logistic distribution is given by

$$\lambda(t) = \frac{e^t}{\sigma(1 + e^t)}$$

   Use this failure rate as baseline function in the PHM, derive a parameter estimation method for all the parameters.
3. What are the assumptions made in the proportional hazard model and how to validate these assumptions?

References

1. Cox DR (1972) Regression models and life-tables. J R Stat Soc Ser B Stat Methodol 34(2): 187–220
2. Johanne K, Yuan F (2020) The determination of the influences of operation environment on the reliability of oil and gas turbine in Arctic area. Int J COMADEM 22(4):1–6
3. USA. Department of Defense (1998) MIL-HDBK-338B Electronic reliability design handbook
4. Cox DR (1975) Partial likelihood. Biometrika 62(2):269–276. https://doi.org/10.1093/biomet/62.2.269
5. Cox DR (1981) Survival times - aspects of partial likelihood - discussion. Int Stat Rev 49(3): 258–258
6. Kalbfleisch JD, Prentice RL (1973) Marginal likelihoods based on Cox's regression and life model. Biometrika 60(2):267–278
7. J. F. Lawless, Statistical models and methods for lifetime data, 2 Hoboken, NJ: Wiley-Interscience, 2003
8. Næs T (1982) The asymptotic distribution of the estimator for the regression parameter in Cox's regression model. Scand J Stat 9(2):107–115
9. Oakes D (1977) The asymptotic information in censored survival data. Biometrika 64(3): 441–448. https://doi.org/10.2307/2345319
10. Tsiatis AA (1981) A large sample study of Cox's regression model. Ann Stat 9(1):93–108
11. Lin DY, Wei LJ (1989) The robust inference for the Cox proportional hazards model. J Am Stat Assoc 84(408):1074–1078. https://doi.org/10.2307/2290085

12. Cox DR, Snell EJ (1968) A general definition of residuals. J Royal Stat Soc Series B Methodol 30(2):248–275
13. Schoenfeld D (1982) Partial residuals for the proportional hazards regression model. Biometrika 69(1):239–241
14. Lin DY, Wei LG (1991) Goodness-of -fit tests for the general cox regression mode. Stat Sin 1: 1–17
15. Grambsch PM, Therneau TM (1994) Proportional hazards tests and diagnostics based on weighted residuals. Biometrika 81(3):515–526. https://doi.org/10.2307/2337123
16. Xue Y, Wang H, Yan J, Schifano ED (2020) An online updating approach for testing the proportional hazards assumption with streams of survival data. Biometrics 76(1):171–182
17. Yuan F, Lu J, Batalden B, Yao J (2017) The investigation of physical explanation for proportional hazard model (PHM) for typical failure mechanisms, presented at the 2017 annual reliability and maintainability symposium (RAMS)
18. Elsayed EA, Chan CK (1990) Estimation of thin-oxide reliability using proportional hazards models. IEEE Trans Reliab 39(3):329–335
19. Rodríguez-Borbón I, Medina MAR, Rodríguez-Picón L, Alvarado-Iniesta A (2017) Reliability estimation for accelerated life tests based on a Cox proportional Hazard model with error effect: reliability estimation via proportional Hazard model with error. Qual Reliab Eng 33(7)
20. Shisong M, Yincai T, Lingling W (2008) Reliability statistics. Higher Education Press, Beijing

Chapter 10
Reliability Simulation

Obtaining analytical solutions to complex problems is often preferable; however, it requires substantial mathematical effort, and in some cases, analytical solutions may not be available. Simulation methods, by contrast, are relatively simple yet highly effective, especially for solving complex problems, as they usually require fewer assumptions to obtain solutions. With the widespread availability of computers and significant advances in their processing power, simulation has become a popular approach in engineering analysis, for example, in modeling dynamic systems, performing uncertainty analyses, and analyzing failure data. This chapter introduces the fundamentals of simulation, highlights its limitations, and discusses various methods for applying simulation to reliability data analysis.

10.1 Pseudo Random Number Generating

Simulation of problems involving probabilistic behavior usually requires the generation of random numbers using computer software. In reliability engineering, for example, Exponential, Weibull, or Normal random numbers are often generated for simulation analysis. These simulated numbers are called pseudo-random numbers. Regardless of the target distribution, all such pseudo-random numbers can be generated from the simplest uniform distribution over the range [0,1]. Therefore, this chapter begins with pseudo-random numbers. Since uniform random numbers are the foundation for generating random numbers from other distributions, their credibility and accuracy have a significant impact on the accuracy of simulation results.

One common method to generate random number for uniform distribution uses a formula as follows [1]

F. Yuan, *Essentials of Reliability Engineering*, Springer Series in Reliability Engineering, https://doi.org/10.1007/978-3-032-15544-3_10

$$x_{n+1} = (ax_n + b) \bmod m \tag{10.1}$$

The a and b are pre-defined constants, the x_n is the number at the nth step. The *mod* is the Modulo operation. For an example of $m = 2$, the modulo operation is performed as

$$0 \ mod\ 2 = 0;$$

$$1 \ mod\ 2 = 1;$$

$$2 \ mod\ 2 = 0.$$

In computer hardware, all integer numbers are expressed in binary. A 8 bit binary number can have integers 0, 1, 2, …, $2^8 - 1$. If we choose the mod $m = 2^8 - 1 + 1 = 2^8$, for any integer, the operation *mod m* confines the integers ranged between 0 and $m - 1$, so the integer can be stored in the 8 bit. The mod seems not related to the uniform distribution. But if we let the x_{n+1}/m, then the x_{n+1}/m is confined between 0 and 1. Each x_{n+1}/m can be seen a number drawn from the uniform distribution $[0, 1]$. Therefore, the question of random drawn from $[0, 1]$ is equivalent to draw an integer from integers from 0 to $m - 1$. The Formula (10.1) is named linear congruential generator, a basic generator that can be extended to many other generators.

We can illustrate an example to demonstrate how to generate random number by the modulo operation. The first integer number, saying the x_0 in the modulo is named *seed*. Given an example letting the $m = 9$; $a = 4$; $b = 1$. Choose the seed as 0.

$$x_1 = (4x_0 + 1) \bmod 9 \rightarrow (1) \bmod 9 = 1$$

$$x_2 = (4x_1 + 1) \bmod 9 \rightarrow (4 + 1) \bmod 9 = 5$$

$$x_3 = (4x_2 + 1) \bmod 9 \rightarrow (20 + 1) \bmod 9 = 3$$

$$x_4 = (4x_3 + 1) \bmod 9 \rightarrow (12 + 1) \bmod 9 = 4$$

$$x_5 = (4x_4 + 1) \bmod 9 \rightarrow (16 + 1) \bmod 9 = 8$$

$$x_6 = (4x_5 + 1) \bmod 9 \rightarrow (32 + 1) \bmod 9 = 6$$

$$x_7 = (4x_6 + 1) \bmod 9 \rightarrow (24 + 1) \bmod 9 = 7$$

$$x_8 = (4x_7 + 1) \bmod 9 \rightarrow (28 + 1) \bmod 9 = 2$$

$$x_9 = (4x_8 + 1) \bmod 9 \rightarrow (8 + 1) \bmod 9 = 0$$

$$\cdots$$

Apparently, if the next x_{10} will be same as the x_1 and the integer sequences will repeat as a period as 0 1 5 3 4 8 6 7 2, as shown in Fig. 10.1. This period contains all the possible integer, which is desirable. But it might not apply to other seed. For example, a generator such as $m = 9$; $a = 2$; $b = 0$, for seed 3, the period sequence is

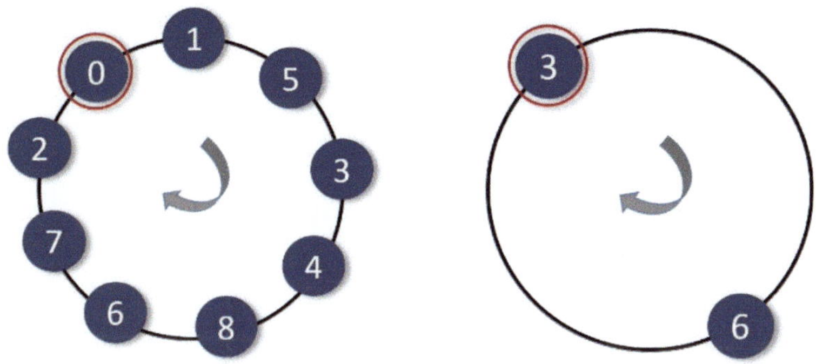

Fig. 10.1 Pseudo random number sequences

3 6 3. A longer sequence period is preferable, since longer sequence implies a higher resolution of pseudo numbers.

One simple generator named Kiss random generator is a congruential generator of the form of Eq. (10.1) with

$$x_{n+1} = (69069x_n + 23606797) \bmod 2^{32}$$

The period of Kiss generator is of order 2^{95} [1], the popular Mersenne-Twisters generator in simulation is of order $2^{19937} - 1$ [2]. An obvious problem is: When the pseudo numbers are used for approximation on the continuous uniform $U[0, 1]$, where the approximation is essentially a discrete distribution, there will be very small interval between two consecutive discrete numbers. It implies the generated number is not continuous but with gap. The real continuous distribution is not possible to be generated by a principle upon Eq. (10.1). Although other principles may offer improvements, they also have similar inherent limitations. The generation of truly continuous random numbers is still an open challenge in simulation.

A Simulation Experiment

To demonstrate the limitation of current generator further, we use the most popular Mersenne-Twisters generator to generate the random number targeting the uniform distribution $U[0, 1]$. The $U[0, 1]$ is with a constant density function 1, the CDF function is:

$$F(x) = x; x \in [0, 1] \tag{10.2}$$

The results using the popular Mersenne-Twisters generator as shown in Fig. 10.2. The empirical CDF corresponding to Eq. (10.2) is constructed from pseudo-random numbers generated for various sample sizes.

From the above figures, we can find when the number of random number size n increases, it approaches to the ideal cumulative distribution function visually. As shown in the last figure in Fig. 10.2, the simulated one seems exactly a line described

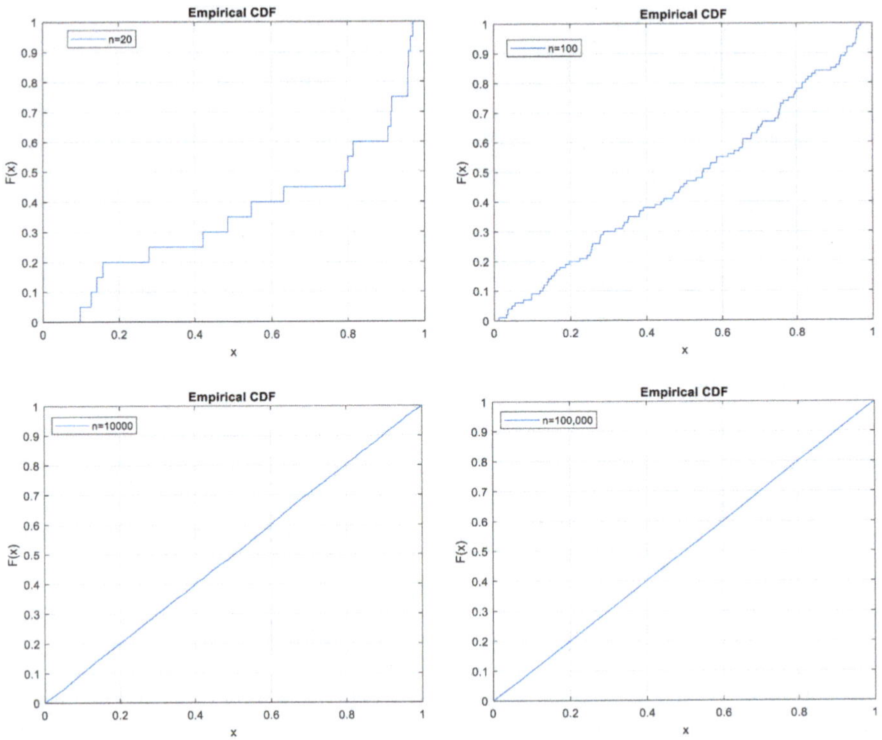

Fig. 10.2 Convergence of simulation with sample size

Fig. 10.3 Resolution demonstration

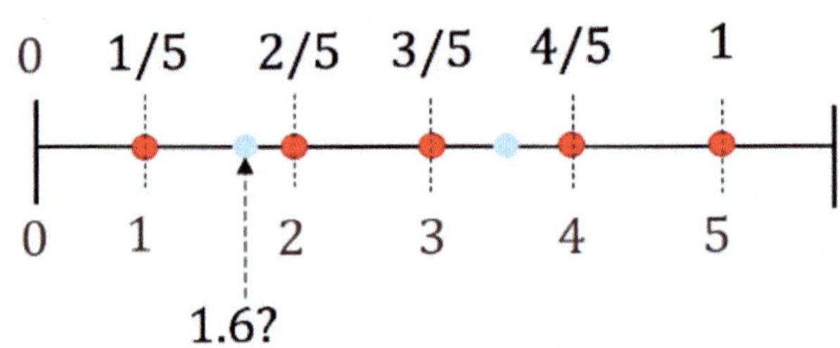

by Eq. (10.2), which means the accuracy of the random number is high. However, as we mentioned earlier, since the pseudorandom numbers uses discrete values to approximate the continuous function, simulation error is then inevitable. The following Fig. 10.3 demonstrates an example. Given the 5 equally spaced number to approximate the continuous interval [0, 1]. Only the 0, 1/5, 2/5,…, 1, which mean only the 6 numbers, has no error. All the other numbers are approximated by these 6 numbers. For example, the number 1.6 is between 1 and 2, it can be approximated by either 1 or 2. The maximum error is then 1/5 for the continuous interval [0, 1].

For m bit data, for example 32 bit, the total discrete number is 2^m, the maximum error is then

$$\frac{1}{2^m - 1} \approx \frac{1}{2^m} \tag{10.3}$$

The Mersenne-Twisters generator is said to have resolution 53 bits. The maximum error for a random number drawn from the Mersenne-Twisters is then $\frac{1}{2^{53}} = 1.11 \times 10^{-16}$, an extremely small value. For reliability data analysis, such small error can be ignorable.

10.2 Probability and Reliability Simulation

It is normal in reliability to state, e.g. the reliability of a laptop is 0.7. How to simulate reliability 0.7? One common method is based on the pseudo random number drawn from $U[0, 1]$. In uniform distribution $U[0, 1]$, the probability of 0.7 is a range from 0.3 to 1, or a range from 0 to 0.7, or from 0.1 to 0.8, any range with length 0.7 is with probability 0.7, as illustrated in Fig. 10.4. For easy of understanding and easy of coding, usually, we take the range [0, 0.7] or [0.3, 1]. When we take the range [0.3, 1], if the pseudo random number is within the range, the laptop is simulated as reliable; otherwise, as unreliable.

We can evaluate the accuracy of the simulation by statistical analysis. Suppose the system is only with two states: failed or functional. By simulation, we obtained the number of failure states, saying, with total number of iterations n. The probability of failure can be estimated as i/n, which can be proved by statistic theory. Suppose the real probability of failure is p. The probability of number of i failures observed is then as follows, according to Binomial distribution

$$b(i) = \binom{n}{i} p^i (1 - p)^{n-i} \tag{10.4}$$

The PDF of the Binomial is discrete, as shown in Fig. 10.5. The mean and variance of the Binomial distribution is as following Table 10.1.

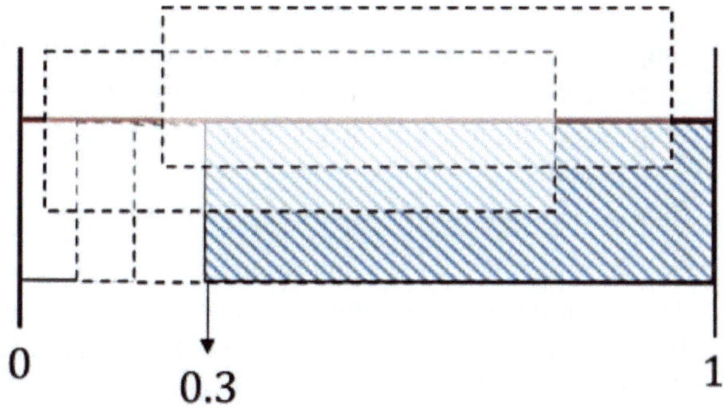

Fig. 10.4 Simulation a probability value

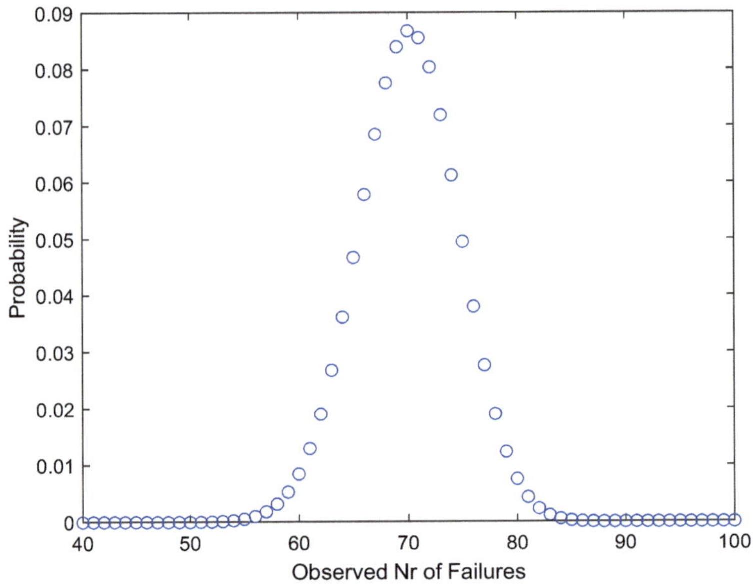

Fig. 10.5 PDF of binomial distribution

Table 10.1 Statistic of Binomial distribution

Mean	np
Variance	$np(1-\text{p})$
MLE	$\widehat{p}=\frac{i}{n}$

Then the expectation of $E\left(\frac{i}{n}\right)=p$, $Var\left(\frac{i}{n}\right)=\frac{\text{p}(1-\text{p})}{n}$, implying when $n\rightarrow\infty$, the $\frac{i}{n}$ can reach the true parameter p.

Example 10.1

Simulating Probability 0.7. Check the convergence when the n increases.

Solution: We choose the interval [0.3, 1] for the probability 0.7. Generate random number for uniform distribution ranged in [0,1] using the Mersenne-Twisters generator. The algorithm is as follows:

Step 1: Generate pseudo random number U_i. If the U_i is within range [0.3, 1], add the number by 1 and $i=i+1$; otherwise, do nothing.
Step 2: Continue the Step1 until the predefined number of simulations n reaches.
Step 3: The simulated probability is then i/n.

For the example, we set the number of iterations ranged from 10 to 10,000. Then plotting mean of the iterations against the number of iterations. From Fig. 10.6, it shows the estimated probability fluctuates in the beginning, then it decreases and stabilizes to the probability 0.7 around n =4000.

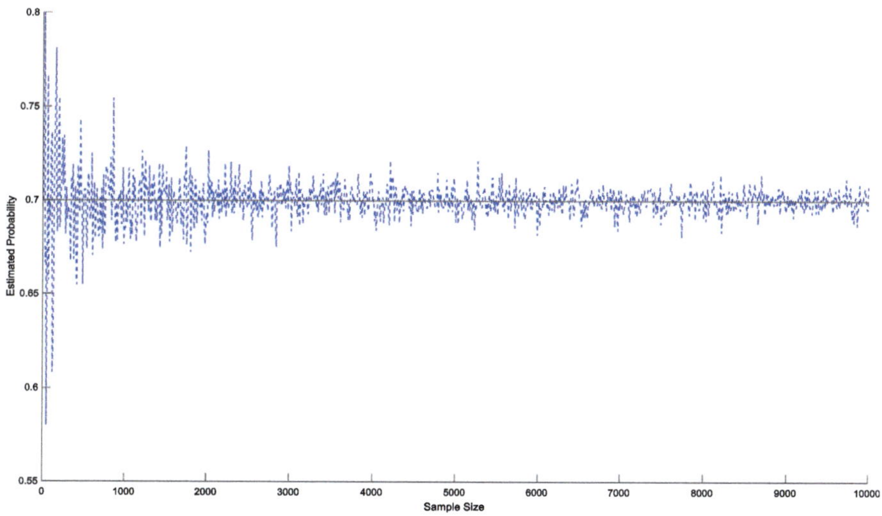

Fig. 10.6 Estimated probability converges

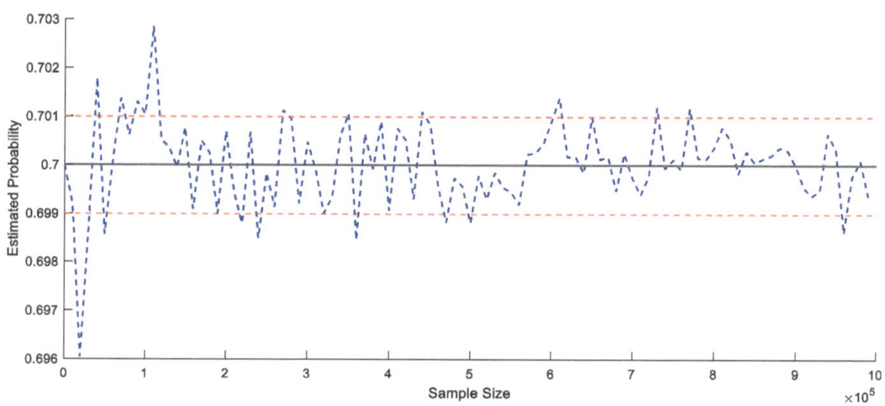

Fig. 10.7 Estimated probability not converges to zero

10.2.1 A Problem Concerning

Let us to have another plot shown in Fig. 10.7, where we can see the estimated probability ends with randomness, which is not desirable. The error for a large $n = 10^6$ is still with error around 0.001, shown from Fig. 10.7. This is a fundamental problem for this simulation. But fortunately, the error is explainable and controllable to some extent. We can use central limit theorem to investigate the error of the i/n by using the central limit theorem.

According to the central limit theorem, the

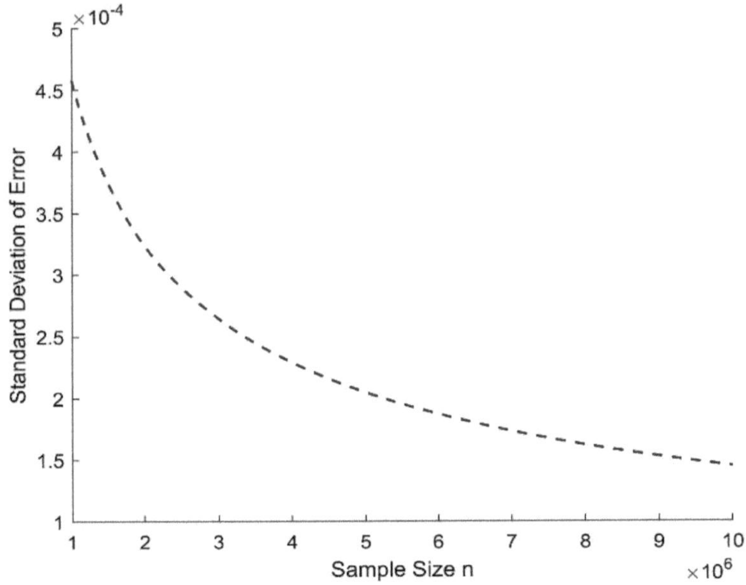

Fig. 10.8 Standard deviation decreases with *n*

Table 10.2 Standard deviation value with large *n*

$n = 1 \times 10^6$	4.5826e-4	$n = 2 \times 10^6$	3.2404e-4	$n = 3 \times 10^6$	2.6458e-4
$n = 4 \times 10^6$	2.2913e-4	$n = 5 \times 10^6$	2.0494e-4	$n = 6 \times 10^6$	1.8708e-4
$n = 7 \times 10^6$	1.7321e-4	$n = 8 \times 10^6$	1.6202e-4	$n = 9 \times 10^6$	1.5275e-4

$$\frac{i}{n} \sim N\left(p, \frac{p(1-\mathrm{p})}{n}\right) \tag{10.5}$$

The variance $\frac{p(1-\mathrm{p})}{n}$ is absolutely decreasing with *n*, as shown in Fig. 10.8. Table 10.2 tabulates the numerical value of the $\sqrt{\frac{p(1-\mathrm{p})}{n}}$, where we can see the numerical decreasing. Figure 10.8 demonstrates the decreasing trend.

Further, we have some plots to show how the sample distributed. For each sample size *n* below, we conducted 5000 iterations, the empirical plot of histograms is shown in Fig. 10.9. we can see the empirical PDF tends to normality as *n* increases.

The numerical values can show the theoretical STD computed from $\sqrt{\frac{p(1-\mathrm{p})}{n}}$ is close the empirical STD as shown in Table 10.3. Equation (10.5) works and the nonconvergence shown in Fig. 10.7 is inevitable but measurable.

The empirical mean can result to a mean very close to true probability at the cost of running the 5000 iterations. We can conduct another simulation to validate: the mean and variance by number of iterations with 5000*n is same as that by averaging the mean and variance from divided 5000 groups with each group having iteration

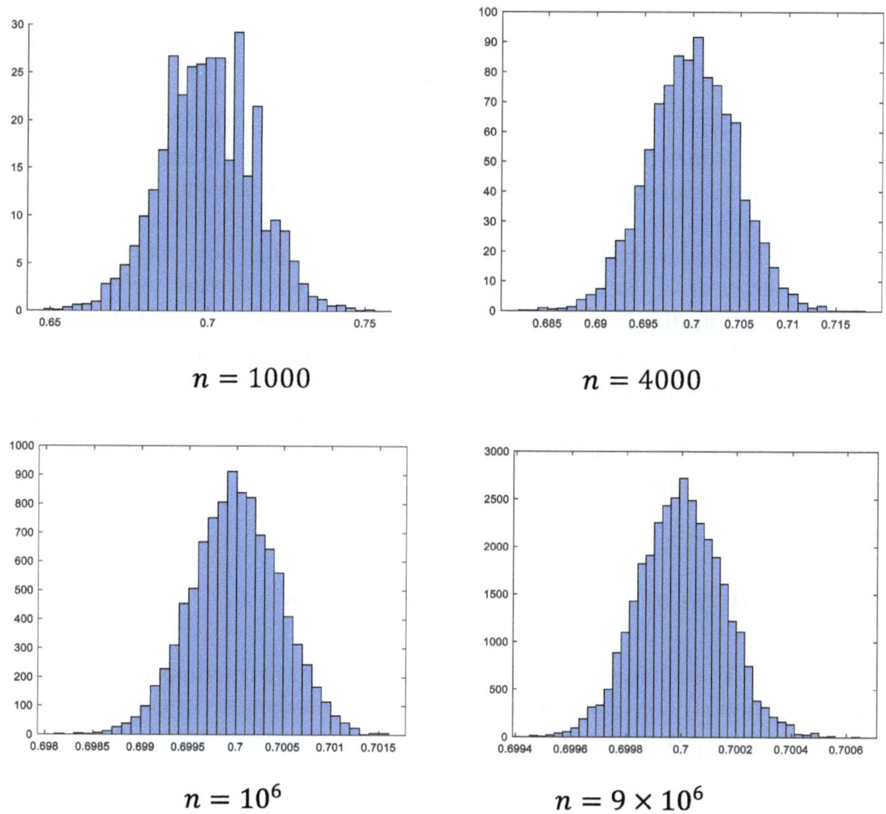

Fig. 10.9 Empirical PDF for various sample size

Table 10.3 Standard deviation value with large n

Sample Size	Theoretical STD	Empirical STD	Empirical Mean
$n = 10^3$	0.0145	0.0146	0.700054400000000
$n = 4 \times 10^3$	7.2e-3	7.2e-3	0.699999000000000
$n = 1 \times 10^4$	4.6e-3	4.5e-3	0.700026100000000
$n = 1 \times 10^6$	4.5826e-4	4.6203e-4	0.6999993246
$n = 9 \times 10^6$	1.5275e-4	1.5282e-4	0.699998997088889

number n. Suppose the number of iterations for each division is i_m; $m = 1, 2, \ldots,$ 5000. For the first, the empirical mean

$$A = \frac{i_1 + i_2 + i_3 \ldots + i_{5000}}{5000 * n} \tag{10.6}$$

The $E(A) = p$; $Var(A) = \frac{p(1-p)}{n * 5000}$. For the case with 5000 groups, i.e. dividing the 5000*n samples equally to 5000 groups.

$$B = \frac{\frac{i_1}{n} + \frac{i_2}{n} + \frac{i_3}{n} \ldots + \frac{i_{5000}}{n}}{5000} \tag{10.7}$$

The $E(B) = \frac{p+p+\ldots p}{5000} = p$; $Var(AB) = \frac{\frac{p(1-\mathrm{p})}{n} + \frac{p(1-\mathrm{p})}{n} \ldots}{5000^2} = \frac{p(1-\mathrm{p})}{5000*n}$. The statistics A and B are equal in its mean and the variance, implying dividing the sample size equally will not improve the estimation performance. This result is useful for simulation strategy designing.

Back to the nonconvergence issue, which is no matter how the sample size increases, the error of the estimated probability does not decrease to 0. From the theoretical analysis, this is not expected. As $Var\left(\frac{i}{n}\right) = \frac{p(1-\mathrm{p})}{n} \rightarrow_{n=\infty} 0$, the mean of the iterations should surely converge to 0. But the reality in simulation is not. One reason is attributed to the pseudo randoms generated which are discussed in the beginning of this Chapter. The theoretical random number should be continuous from 0 to 1, while the real ones are discrete with gap. This issue should bear in mind when we conduct simulation.

10.2.2 Criteria of Sample Size Determination

The variance will decrease with the sample size. If we have determined the maximal standard deviation is c, the number of iterations can then be calculated without difficulty. The standard deviation is $\sqrt{\frac{p(1-\mathrm{p})}{n}}$. The

$$\sqrt{\frac{p(1-\mathrm{p})}{n}} \leq c \tag{10.8}$$

Then

$$n \geq \frac{p(1-\mathrm{p})}{c^2} \tag{10.9}$$

Equation (10.9) requires the numerical value p ranged between 0 and 1. It is not hard to find the $p(1-\mathrm{p})$ has its maximum value at $p = 0.5$. Therefore, a conservative calculation is the

$$n \geq \frac{1}{4c^2} \tag{10.10}$$

Example 10.2
Determine the minimal sample size when simulating probability of 0.7, to have a standard deviation within 0.001.

Solution: According to Eq. (10.10),

$$\frac{1}{4c^2} = \frac{1}{4 * 0.001^2} = 250{,}000.$$

The sample size should be at least 250,000. For comparison, if the real probability is 0.7, a more accurate calculation is according to Eq. (10.9).

$$\frac{p(1-\mathrm{p})}{c^2} = \frac{0.3 * 0.7}{0.001^2} = 210{,}000$$

By simulation with 5000 iterations and each iteration having simple size $n = 210{,}000$, the empirical standard deviation is found around 0.0010075, which is very close to the desired 0.001, implying Eq. (10.9) is applicable. Figure 10.10 shows the empirical PDF of the estimated probability, where we can see the estimated probability centring around the real probability 0.7, as expected.

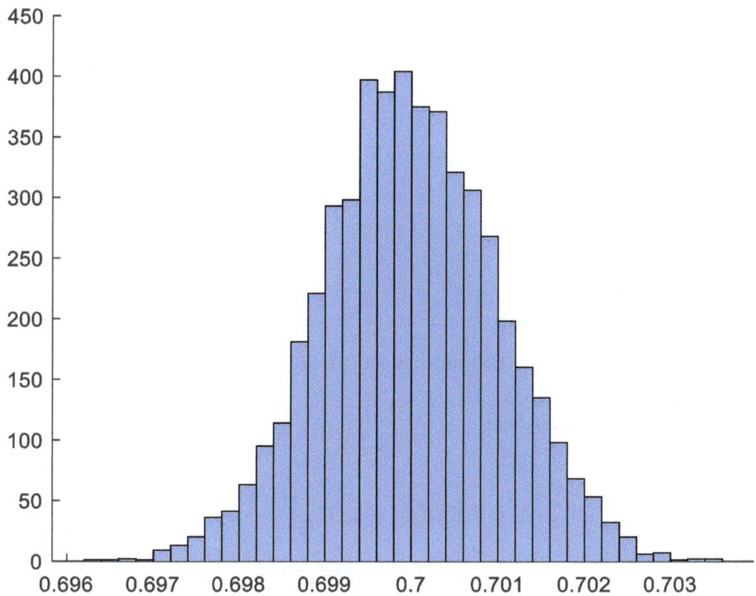

Fig. 10.10 Empirical PDF for sample size n = 210,000 for 5000 iterations

10.2.3 Confidence Interval of Estimated Probability

In Fig. 10.9, it has shown the estimated probability, written as $\widehat{p}_n$ approaches to normality when the sample size n increases. The $\widehat{p}_n$ exhibits normality even when the $n = 1000$. In simulation, the sample size problem is not a concerning issue as we can usually increase it at our will. The sample size n in simulation is very rare to have a size less than 1000. According to the central limit theorem as shown in Eq. (10.5), the

$$\widehat{p}_n = \frac{i}{n} \sim N\left(p, \frac{p(1-\text{p})}{n}\right) \tag{10.11}$$

Therefore,

$$\frac{\widehat{p}_n - p}{\sqrt{\frac{p(1-\text{p})}{n}}} \sim N(0, 1) \tag{10.12}$$

When the real p is unknown, one can use the $\widehat{p}_n$ as the p, Eq. (10.12) is rewritten as

$$\frac{\widehat{p}_n - p}{\sqrt{\frac{\widehat{p}_n(1-\widehat{p}_n)}{n}}} \sim N(0, 1) \tag{10.13}$$

The confidence interval for $1 - \alpha$ is then

$$\widehat{p}_n \pm z_{\alpha/2}\sqrt{\frac{\widehat{p}_n(1-\widehat{p}_n)}{n}} \tag{10.14}$$

Equation (10.14) uses a property $z_{\alpha/2} = 1 - z_{\alpha/2}$ due to symmetry of PDF of the Normal distribution.

Example 10.3
Find out a confidence interval of 95% for a simulation with sample size $n = 1000$ if $\widehat{p}_n = 0.7000544$.

Solution: The 95% confidence interval corresponds to a $\alpha = 0.025$. $z_{\alpha/2} = -1.96$.

According to Eq. (10.14),

$$\widehat{p}_n \pm z_{\alpha/2}\sqrt{\frac{\widehat{p}_n(1-\widehat{p}_n)}{n}} = 0.7000544 \pm 1.96\sqrt{\frac{0.7000544(1-0.7000544)}{1000}}$$

$$\widehat{p}_n \in [0.6717, 0.7285]$$

(continued)

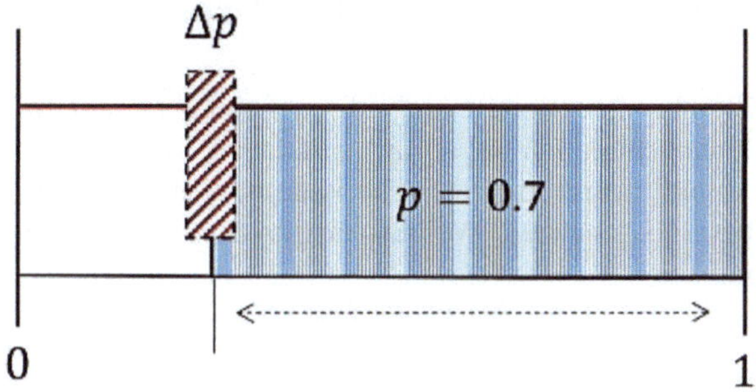

Fig. 10.11 Error due to generator

Example 10.3 (continued)

In the case, the real p is known as 0.7, a more accurate CI is

$$\widehat{p}_n \pm z_{\alpha/2}\sqrt{\frac{p(1-p)}{n}} = 0.7000544 \pm 1.96\sqrt{\frac{0.7(1-0.7)}{1000}}$$

The CI is,

$$\widehat{p}_n \in [0.6717, 0.7285]$$

which is same as CI using $\widehat{p}_n = p$, which is not universal but accidentally, as the $\widehat{p}_n$ is numerically close the real p in this special case. The CI between them is not different at precision of 4 decimal digits.

Analysis of Effect of Generator Imperfection

The simulation of the probability assumes the pseudo random generator is perfect, meaning the random number is a continuous uniform distribution. However, as we discussed before, the simulated uniform distribution is discrete. For example, for the probability simulation, the error denoted as Δp shown in Fig. 10.11, which is the resolution of the generator, and it means each pseudo random number has this error. The estimated probability is $\frac{i}{n}$ according to Eq. (10.5). The observed i contains error with a mean number of $n\Delta p$, implying the $\frac{i}{n}$ is $\frac{n\Delta p}{n} = \Delta p$. The Δp equates to $\frac{1}{2^m}$ by Eq. (10.3). For 32 bits, the $\frac{1}{2^m} = 2.3283 \times 10^{-10}$. For the popular Mersenne-Twisters generator, the $m = 53$, $\frac{1}{2^m} = 1.11 \times 10^{-16}$.

The error from the imperfection of generator is thus of very small magnitude. We can compare this error with the error induced from the sample size. As shown from Table 10.3, the standard deviation from 9,000,000 of simulations, which is an extremely large number of iterations, is 1.5282×10^{-4}. This value is still significantly larger than the error from the imperfection of generator 1.11×10^{-16}. Therefore, the effect of the imperfection of generator can be ignored practically in some cases.

10.3 Continuous Random Generating

The above sections mainly discuss the generator for uniform distribution, as it is fundamental. For reliability simulation, the common distribution is Exponential, Weibull distribution etc. The issue now is how to generate the random number for these Exponential, Weibull distributions etc. using the uniform pseudo random number.

10.3.1 Inverse Transform Method

The CDF of a distribution is defined between 0 and 1. An important property of distribution is the CDF $F(t)$ follows the uniform distribution $U[0, 1]$. An idea to generate random number is then first generating random number from $U[0, 1]$.

$$U_1, \ldots, U_i, \ldots, U_n$$

Thereafter let the $F(t) = U_i$ and obtain the random number t_i for the specified distribution. The method is straightforward and easy to be implemented by computer languages.

Exponential Random Number
The CDF of Exponential distribution is

$$F(t) = 1 - e^{-\lambda t} \tag{10.15}$$

Let the $F(t) = U_i$. We obtain $1 - e^{-\lambda t} = U_i$. Rearrange it and obtain an Exponential random number

$$t_i = -\ln(1 - U_i)/\lambda \tag{10.16}$$

Figure 10.12 shows the distribution of simulated Exponential random numbers (100,000 samples) with $\lambda = 5$ by Eq. (10.16). It shows the simulated random numbers fits the real PDF very well.

Weibull Random Number
The CDF of Weibull distribution is given by

$$F(t) = 1 - e^{-\left(\frac{t}{\alpha}\right)^{\beta}} \tag{10.17}$$

The random number t_i is written as

$$t_i = \alpha(-\ln(1 - U_i))^{1/\beta} \tag{10.18}$$

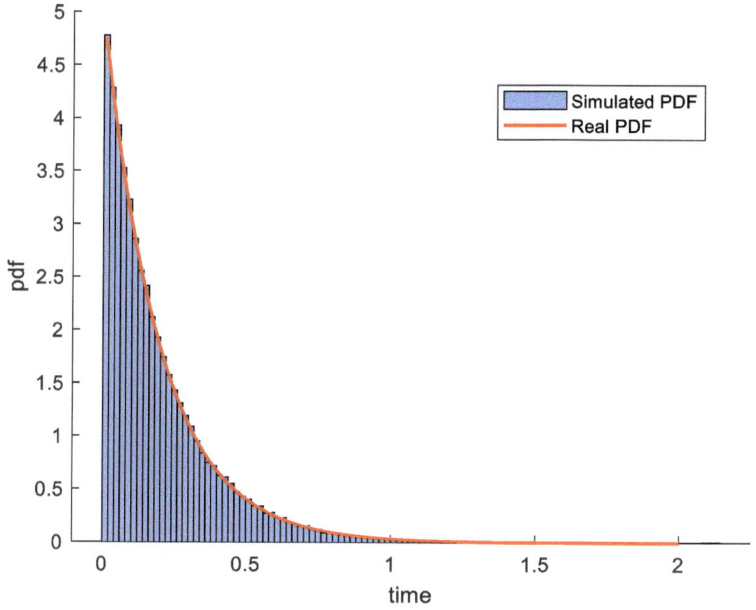

Fig. 10.12 Distribution of exponential random numbers

Apply Eq. (10.18) to generate 100,000 random numbers for Weibull distribution with $\alpha = 5; \beta = 2$. The empirical PDF is seen fitting with the real PDF very well, as shown in Fig. 10.13.

The inverse transform method requires an explicit expression of CDF function $F(t)$ and its inverse function $F^{-1}(U)$. For reliability simulation, the explicit expression of $F(t)$ and $F^{-1}(U)$ normally exist. The inverse transform method is thus very popular in the reliability simulation. The Matlab (2020b) also uses the inverse transform method to generate random number for Exponential and Weibull.

10.3.2 Acceptance-Rejection Method

The Acceptance-Rejection (AR) method is more complex than the inverse method but can be applied more broadly. The acceptance-rejection method defines a principle for the region where the target PDF function $f(x)$ locates is acceptance region; the other is rejection region [1]. If we sample from a 2-dimensional region as shown in the rectangle region in Fig. 10.14. The sample below the function $f(x)$ is the acceptance region, above the $f(x)$ is rejection region. The issue now is to find the acceptance region by sampling. Firstly, we sample from the horizontal region $g(x)$ and suppose the sampled point is x_i, then the vertical line crosses the x_i is also divided by two, one in acceptance and one in rejection region. Suppose the vertical line (the bold dashed one) in Fig. 10.14 is uniformly distributed. When the function $cg(x)$ is

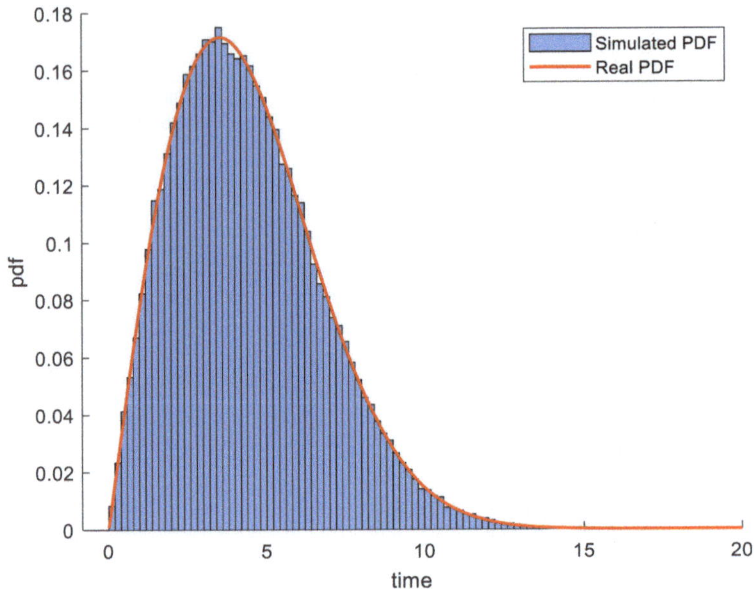

Fig. 10.13 Distribution of Weibull random numbers

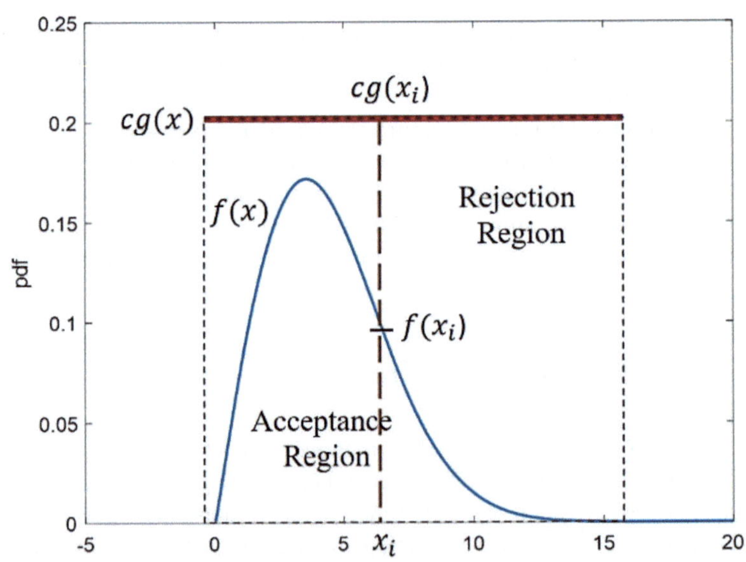

Fig. 10.14 Principle of acceptance-rejection method

above the function $f(x)$, the probability of the sample in the acceptance region, i.e. under $f(x_i)$, is $\frac{f(x_i)}{cg(x_i)}$. Let the uniformly distributed point denoted by u_i. In mathematical term, when the $u_i < \frac{f(x_i)}{cg(x_i)}$, the point is in acceptance region, otherwise in the rejection region. Repeat the sampling from the $g(x)$ (In Fig. 10.14 is the horizonal line) and check if it is located in acceptance region for many times. If it is in acceptance region, the sample x_i is then considered from PDF $f(x)$. We can thus obtain many samples in the acceptance region, which is the random number from PDF $f(x)$. The sample from the horizonal line can be accomplished by sampling from PDF $cg(x_i)$.

In the $cg(x_i)$, the c is constant, which scaling the PDF function $g(x)$ up and down, mainly scaling up, to assure the function $cg(x_i)$ above function $f(x)$ so the acceptance and rejection is clearly divided. The $g(x)$ in Fig. 10.14 is uniform function, while it can be any other PDF function. But the sampling from function $g(x)$ should be easy to facilitate the sampling from the horizonal line. The $g(x)$ is called envelope PDF function. The $f(x)$ is called the targe PDF function. Since the probability should no more than 1, the $\frac{f(x_i)}{cg(x_i)} \leq 1$, then it is not hard to deduce the c should satisfy

$$c \geq \max\left(\frac{f(x)}{g(x)}\right) < \infty \tag{10.19}$$

Moreover, the $\frac{f(x_i)}{cg(x_i)}$ is preferable to be bigger, to ensure the probability of a sample is higher in the acceptance region. The c should be as small as possible to reduce the sample region. In some cases, the c can take the value as $\max\left(\frac{f(x)}{g(x)}\right)$, which is the smallest c. The shape of envelope function prefers to being as close as the target function to reduce rejection region. As shown in Fig. 10.15, we can see the function $g_1(x)$ has a much smaller rejection region than $g_2(x)$.

The algorithm of performing the acceptance-rejection method is as follows [1].

Algorithm 10.1: Acceptance-Rejection Method
Step 1: Choose a reference distribution $g(x)$. Find out the optimal c.
Step 2: Sample a random number u from $U[0, 1]$.
Step 3: Sample a random number x_i from the reference distribution $g(x)$.
Step 4: Computing $\frac{f(x)}{c.g(x)}$. If $\frac{f(x)}{c.g(x)} \geq u$, accept the x_i; otherwise, go to **Step 2**.

Example 10.4
Use the acceptance-rejection method to generate random numbers between [0,1] for Exponential distribution $\lambda = 5$. Choose the $c = 10$.

Solution: The reference distribution $g(x)$ is chosen as the uniform distribution [0,1], and the $c = 10$. Using Algorithm 10.1 to simulate 100,000 random numbers, the empirical distribution of the random numbers are shown in Fig. 10.16.

The inverse transform requires the explicit expression of CDF, while the Acceptance-Rejection (AR) method requires the PDF even without an explicit expression. But the AR method is not as straightforward as the inverse transform approach, and AR needs another reference distribution and the finding of the value c.

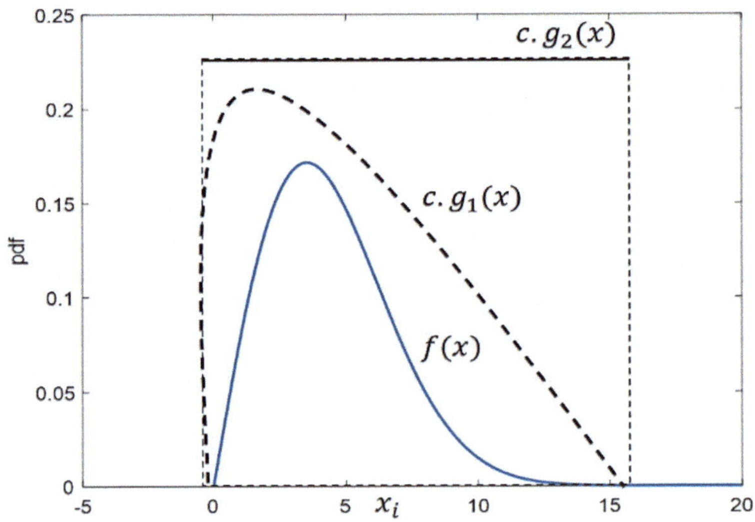

Fig. 10.15 Selection of envelope function

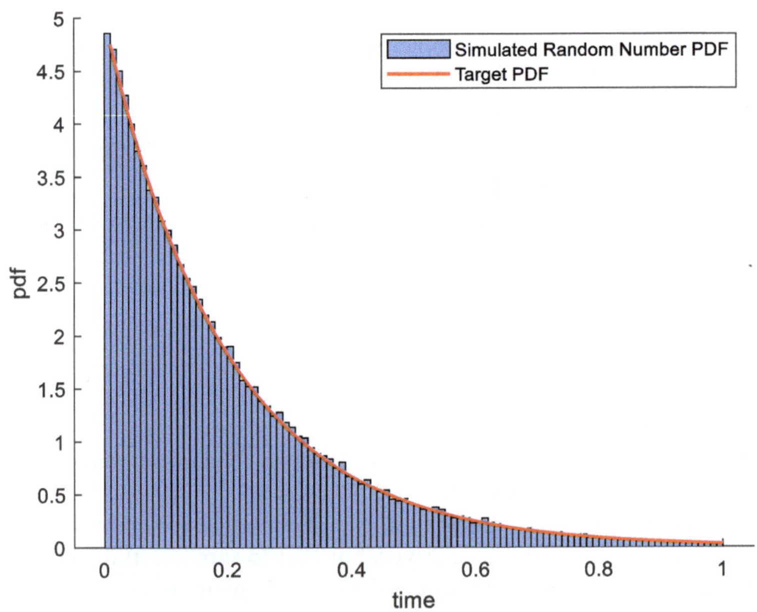

Fig. 10.16 Distribution of exponential random numbers from AR method

The AR is not commonly used for sampling in reliability simulation, but it is a good alternative to the inverse method, and it is useful when the CDF expression of the data distribution is not analytic. From the computational cost perspective, the AR method should sample from a whole 2-D region, but the inverse transform method only sample from a curve $f(x)$, the inverse method is thus with much light computational cost.

Example 10.5
A mixed Weibull distribution is written as

$$f(t) = \frac{1}{2}\frac{\beta_1}{\alpha_1}\left(\frac{t}{\alpha_1}\right)^{\beta_1 - 1} e^{-(t/\alpha_1)^{\beta_1}} + \frac{1}{2}\frac{\beta_2}{\alpha_2}\left(\frac{t}{\alpha_2}\right)^{\beta_2 - 1} e^{-(t/\alpha_2)^{\beta_2}}$$

$$\alpha_1 = 0.2; \beta_1 = 1.4$$

$$\alpha_2 = 0.3; \beta_2 = 6$$

Use the acceptance-rejection method to generate random numbers between [0,1], selecting the uniform distribution as envelope distribution.

Solution: The uniform distribution has PDF distribution $g(x) = 1$, the maximal $\frac{f(x)}{c.g(x)} = \frac{f(x)}{c}$. The minimal c is then the maximal $f(x)$. We plot the mixed density function as follows Fig. 10.17.

The maximal $f(x)$ value is around 4.5. We can choose the $c = 5$. After sampling 20,000 random numbers, the plot of the empirical distribution is as follows Fig. 10.18, where it can be seen the empirical distribution can fit the target PDF well. The acceptance-rejection method works. In this example, the inverse transform method does not work, as the analytical solution to $t_i = F^{-1}(U_i)$ is not available.

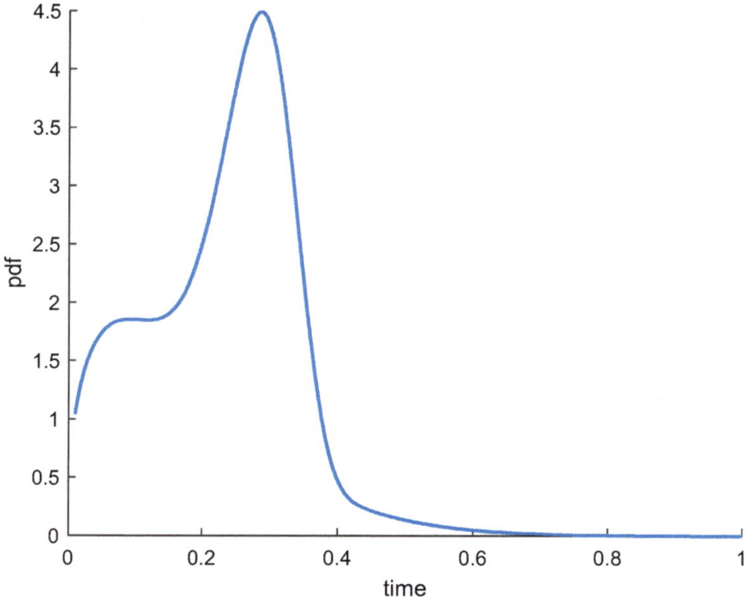

Fig. 10.17 Mixed density function

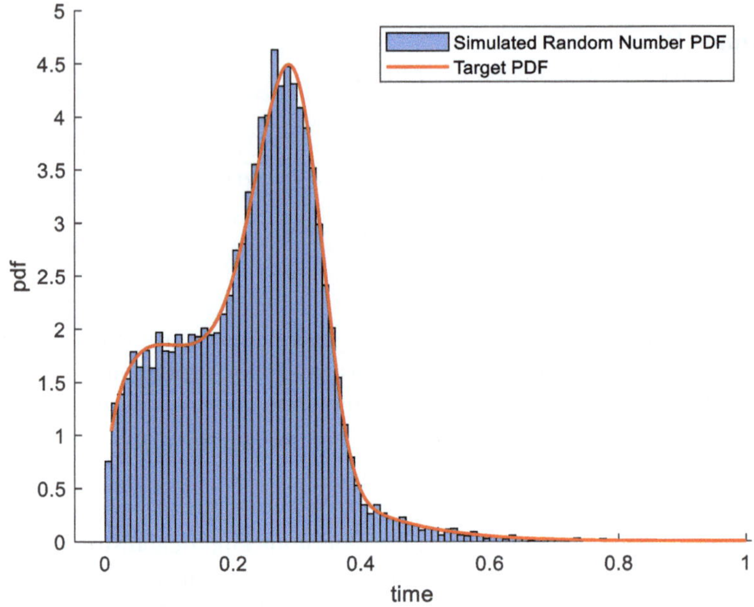

Fig. 10.18 Sample histogram for the mixed density function

10.4 Markov Chain Monte Carlo Method

Markov chain is a stochastic process modeling the dynamic of an event development when uncertainty is presented. The simulated probability as shown in Fig. 10.19 is an example of stochastic process where the estimated probability is of random process. Markov chain is a widely used method to model the simulation process.

10.4.1 Basics of Markov Chain

The idea of Markov chain is relatively simple. Given the example of particle movement, we know the particle is moving constantly and suppose the particle can have only three positions, denoted by A, B, C as shown in Fig. 10.19. The probability of the particle A moves from $A \to B$ is denoted as p_{AB}. From $A \to C$ is denoted as p_{AC}. If not moving, it is denoted by p_{AA}.

The probability matrix of moving is as follows

$$p = \begin{pmatrix} p_{AA} & p_{AB} & p_{AC} \\ p_{BA} & p_{BB} & p_{BC} \\ p_{CA} & p_{CB} & p_{CC} \end{pmatrix} = \begin{pmatrix} 0.24 & 0.5 & 0.26 \\ 0 & 0.5 & 0.5 \\ 0.3 & 0.3 & 0.4 \end{pmatrix} \tag{10.20}$$

Fig. 10.19 Transition in Markov chain

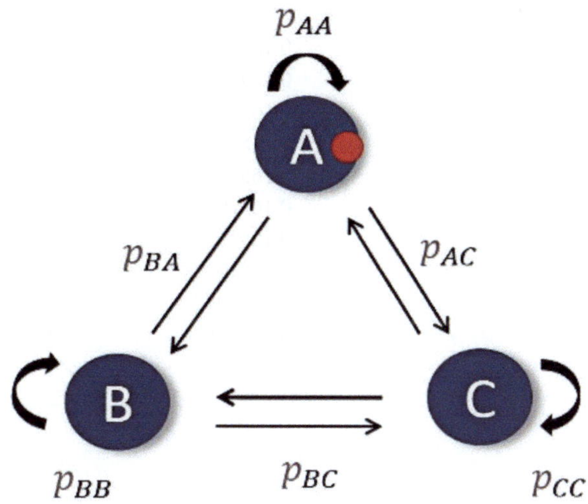

It is very hard to speculate the location of a specific particle at given time as it can be in any position. We can have a probability analysis on the location. Suppose the particle locates the position A, B, C with probability of $q_0 = (0.33, 0.33, 0.34)$ respectively initially. The next step, the probability of the next position is then

$$q_1 = q_0 p = (0.33, 0.33, 0.34)\begin{pmatrix} p_{AA} & p_{AB} & p_{AC} \\ p_{BA} & p_{BB} & p_{BC} \\ p_{CA} & p_{CB} & p_{CC} \end{pmatrix} = (0.1812, 0.4320, 0.3868)$$

$$q_2 = q_1 p = (0.1812, 0.4320, 0.3868)\begin{pmatrix} p_{AA} & p_{AB} & p_{AC} \\ p_{BA} & p_{BB} & p_{BC} \\ p_{CA} & p_{CB} & p_{CC} \end{pmatrix} = (0.1595, 0.4226, 0.4178)$$

$$\cdots$$

$$q_9 = q_8 p = (0.1652, 0.4162, 0.4185)\begin{pmatrix} p_{AA} & p_{AB} & p_{AC} \\ p_{BA} & p_{BB} & p_{BC} \\ p_{CA} & p_{CB} & p_{CC} \end{pmatrix} \approx (0.1652, 0.4162, 0.4185)$$

$$q_{10} = q_9 p \approx (0.1652, 0.4162, 0.4185)$$

$$\cdots$$

The probability for each step are shown in Table 10.4.

To illustrate the process, we plot the probability of particle A as Fig. 10.20, where we can see the probability converges when the iteration reaches $N = 4$.

If we change the initial state of probability $q_0 = (0.1, 0.2, 0.7)$, the

Table 10.4 Iteration of probability

A	B	C
0.18120000	0.43200000	0.38680000
0.15952800	0.42264000	0.41783200
0.16363632	0.41643360	0.41993008
0.16525174	0.41601398	0.41873428
0.16528070	0.41625314	0.41846615
0.16520721	0.41630677	0.41848602
0.16519554	0.41630280	0.41850167
0.16519743	0.41629967	0.41850290
0.16519825	0.41629942	0.41850233
0.16519828	0.41629953	0.41850219

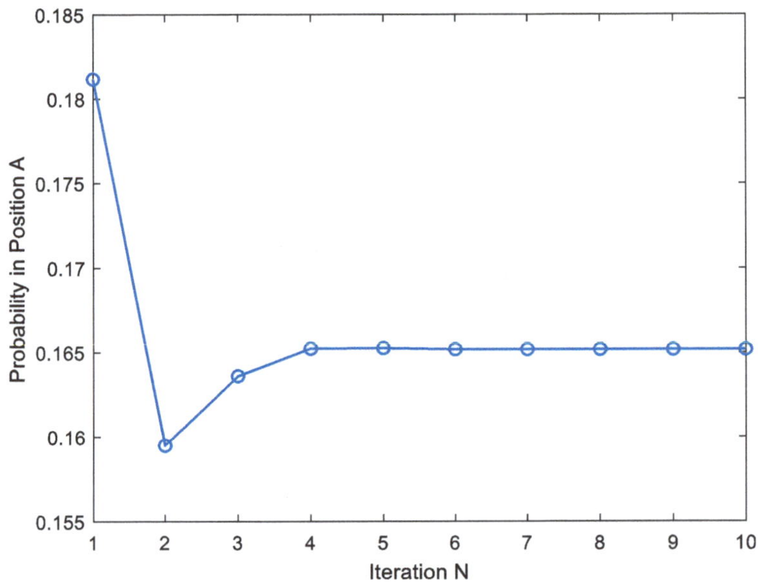

Fig. 10.20 Probability of transition in Markov chain

$$q_{10} \approx (0.1652, 0.4163, 0.4185)$$

If let initial probability $q_0 = (0.2, 0.3, 0.5)$, the q_{10} is still

$$q_{10} \approx (0.1652, 0.4163, 0.4185)$$

which means, experimentally, the $q_n = q_0 p^n$ tends to have a same final probability, which is not case specific. For a given transition matrix p, the same state limit will reach regardless of the initial probability of state, as shown in the above calculation and more examples can be shown in the following Fig. 10.21. The probability will

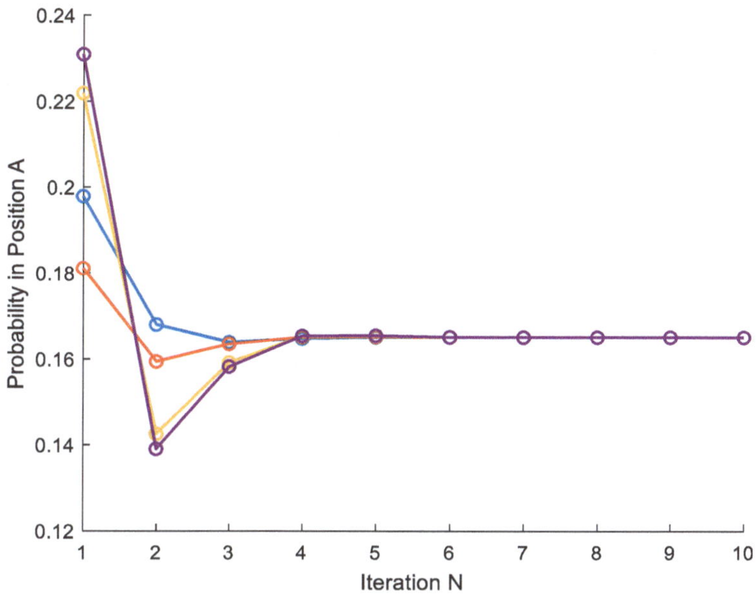

Fig. 10.21 Probability of transition in Markov chain

converge to a same value when they are starting from different values. This is a desirable property of Markov Chain. The particle movement is an example of Markov chain.

In Markov chain, the A, B, C is called *states*, the matrix p in Eq. (10.20) is called *transition* matrix. When the iteration continues, i.e. $N \to \infty$, the $q_n \ ^{n\to\infty} \to (\pi_A, \pi_B, \pi_C)$. The

$$\pi_A + \pi_B + \pi_C = 1 \tag{10.21}$$

10.4.2 Convergence of Markov Chain

The convergence of the Markov chain in the particle movement example is not special. In general, if the Markov chain is irreducible and aperiodic, the Markov chain can converge as same as shown in Fig. 10.21. A Markov chain is irreducible is defined as there is possibility to transit from one to another position for any state, ensuring no state will vanish during the iteration process. The aperiodicity further ensures the Markov chain will not go cyclically when iteration increases. Intuitively, we can imagine if a periodic Markov chain will iterate cyclically, then it is sure that cyclic chain is not able to converge to stationary value.

In more general, we can denote the state by number 1, 2, 3, … . An irreducible and aperiodic Markov chain will have a unique stationary distribution with the transition matrix of p. The stationary distribution can be written as,

$$\pi_j = \sum_{i=1}^{N} \pi_i p_{ij} \tag{10.22}$$

In the particle example shown in Fig. 10.19, $\pi_1 = 0.1652$; $\pi_2 = 0.4163$; $\pi_3 = 0.4185$. The numerical value of $\sum_{i=1}^{N} \pi_i p_{ij}$ in Eq. (10.22) is

$$\pi_1 p_{11} + \pi_2 p_{21} + \pi_3 p_{31} = 0.1652 = \pi_1$$

$$\pi_1 p_{12} + \pi_2 p_{22} + \pi_3 p_{32} = 0.4163 = \pi_2$$

$$\pi_1 p_{13} + \pi_2 p_{23} + \pi_3 p_{33} = 0.4185 = \pi_3$$

Equation (10.22) thus holds. Eq. (10.22) can be written as $\boldsymbol{\pi} = \boldsymbol{\pi}\ p$ where $\boldsymbol{\pi} = (\pi_1, \pi_2, \ldots \pi_j, \ldots)$. The

$$\boldsymbol{\pi} = \boldsymbol{\pi}\ p = \boldsymbol{\pi}\ p^2 = \ldots = \boldsymbol{\pi}\ p^n$$

implying the $\boldsymbol{\pi}$ is the unique stationary distribution.

The limit of the probability π_j can be also obtained from the limit of p^n, for

$$p^n \xrightarrow{n \to \infty} (\pi_1, \pi_2, \ldots, \pi_j, \ldots) \tag{10.23}$$

For the particle example, the $p^{1000} = \begin{pmatrix} 0.1652 & 0.4163 & 0.4185 \\ 0.1652 & 0.4163 & 0.4185 \\ 0.1652 & 0.4163 & 0.4185 \end{pmatrix}$. The $\pi_1 = 0.1652$; $\pi_2 = 0.4163$; $\pi_3 = 0.4185$, same as from $q_1 = q_0 p^n$ in Sect. 10.4.1. Note the initial probability does not affect the stationary distribution as expected, since the Markov is independent on the initial probability.

Detailed Balance Equations (10.22) and (10.23) can be used to find the stationary distribution. When for the construction of a converged Markov chain, a more tractable but stronger condition is to let the [1]

$$\pi_i p_{ij} = \pi_j p_{ji} \text{ for any } i, j.$$

which is named detailed balance in Markov chain. We can prove the detailed balance can ensure a converged Marko chain:

$$\sum_i \pi_i p_{ij} = \sum_i \pi_j p_{ji} = \pi_j \sum_i p_{ji} = \pi_j$$

Therefore Eq. (10.22) holds and it is equivalent to the $\boldsymbol{\pi} = \boldsymbol{\pi}\ p$. The chain can converge to $\boldsymbol{\pi}$.The detailed balance of Markov plays a key condition in Markov chain for simulation. We will discuss it in the next section.

10.4.3 Markov Chain for Simulation: Metropolis-Hastings

Simulation based on the Markov chain takes advantage of the dynamic process of the Markov Chain will approach to the stationary distribution. For sampling from a PDF $f(x)$, the stationary distribution can be the $f(x)$, named target function [1]. When the iteration goes sufficiently large and the distribution stabilizes, the samples from the Markov chain can be considered as the sample from the $f(x)$ with error very small and ignorable. One may doubt why not sampling from the $f(x)$ directly since it is straightforward and simple, and why use the Markov chain to complicate the case? In some cases, sampling directly from the $f(x)$, e.g. using acceptance-rejection method is preferable. However, for some examples, the dimension of the $f(x)$ is high, sampling directly from the $f(x)$ is not easy. Markov chain can simplify the sampling, one example for which will be discussed later in this chapter.

In Markov chain, the stationary distribution π_j and the transition matrix p in Eq. (10.20) can uniquely determine the chain. For simulation purpose, the stationary distribution can select the target function $f(x)$. One foremost approach to constructing the Markov chain to estimate PDF is the Metropolis-Hastings algorithm.

Metropolis-Hastings (MH) Algorithm

The target function $f(x)$ is selected as the stationary distribution and the remaining issue is for the transition matrix. The transition probability is a conditional probability. We can propose a conditional PDF $q(y/x)$ and the conditional PDF should be easy to sample from, and the explicit expression is preferable to be available, which is same as the envelope distribution in the acceptance-rejection method. But the MH algorithm by convention names the envelop distribution as proposal distribution. The proposal PDF $q(y/x)$ is not related to the targe function and it is arbitrary so finding a good $q(y/x)$ is not hard. Back to the Markov chain, in order to have a converged chain, the detailed balance condition [1]

$$f(x)q(y/x) = f(y)q(x/y)$$

should be met. Since the $q(y/x)$ is selected arbitrarily, mostly, the above equation is not able to meet. A trick is to add another function $a(x, y) = a(y, x)$ to let the equation hold.

$$f(x)q(y/x)a(y,x) = f(y)q(x/y)a(x,y) \tag{10.24}$$

Simply let the

$$a(y,x) = f(y)q\left(\frac{x}{y}\right). \tag{10.25}$$

Substitute it to Eq. (10.24) and it is not hard to find Eq. (10.24) hold. The new proposal function for the Markov chain is

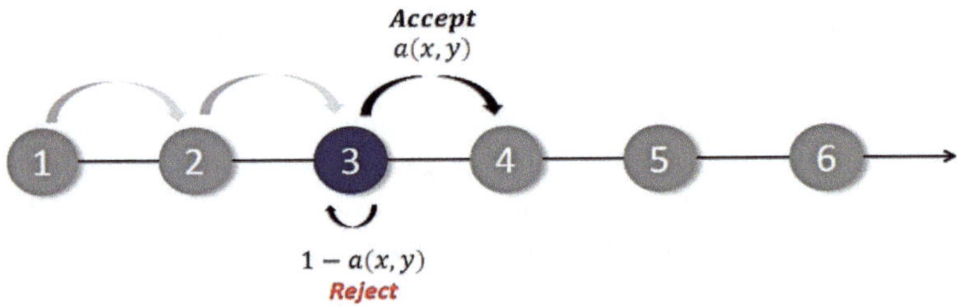

Fig. 10.22 Presence of probability for transition

$$q'(y/x) = q(y/x)a(x,y)$$

$$\frac{f(x)\boxed{q(y/x)a(y,x)}}{q'(y/x)} = \frac{f(y)\boxed{q(x/y)a(x,y)}}{q'(x/y)}$$

The new proposal function ensures Eq. (10.24) holds, and it functions to be as a probability $a(y,x)$ to accept the transition, as demonstrated in Fig. 10.22. If $a(y,x)$ function is not used, the transition will surely occur. In this sense, the MH algorithm is also an acceptance-rejection algorithm.

The Markov chain defined for MH can surely converge. But if the acceptance probability $a(y,x)$ is too small and the rejection rate is too high, the sampling will be of very low efficiency. Equation (10.24) can be further multiplied by a constant C,

$$f(x)\mathrm{q}(y/x)a(y,x)C = f(y)q(x/y)a(x,y)C \tag{10.26}$$

And let the bigger one of $a(y,x)C$ or $a(x,y)C$ to be 1, i.e.

$$\max(a(y,x)C, a(x,y)C\,) = 1;$$

Thus,

$$C = \frac{1}{\max(a(y,x), a(x,y)\,)}$$

Then the

$$a(y,x)C = \frac{a(y,x)}{\max(a(y,x), a(x,y)\,)} = \frac{1}{\max\left(1, \frac{a(x,y)}{a(y,x)}\right)}$$

which is equivalent to

$$a(y,x)C = \min\left(\frac{a(y,x)}{a(x,y)}, 1\right) \tag{10.27}$$

According to Eq. (10.25), the $a(y,x) = f(y)q(x/y)$. Equation (10.27) is equal to

$$a(y,x)C = \min\left(\frac{f(y)q(x/y)}{f(x)q(y/x)}, 1\right)$$

Meaning the new acceptance probability is

$$\min\left(\frac{f(y)q(x/y)}{f(x)q(y/x)}, 1\right). \tag{10.28}$$

According to the derivation procedure, we know Eq. (10.28) has been scaled up to a maximum to 1, by which the acceptance can be increased significantly, while the Markov chain still converges. Hitherto all the issues have been resolved. The MH algorithm, similar to the acceptance-rejection in Algorithm 10.1, is then proposed as follows.

Algorithm 10.2: Metropolis-Hastings Algorithm [1]
Step 1: Choose the proposal distribution $q(y/x)$.
Step 2: Generate a sample $\widehat{y}$ from $q(y/x_i)$.
Step 3: Sample a random number u from $U[0, 1]$.
Step 3: If $u < \min\left(\frac{f(\widehat{y})q(x_i/\widehat{y})}{f(x_i)q(\widehat{y}/x_i)}, 1\right)$, accept the $x_{i+1} = \widehat{y}$; otherwise, let $x_{i+1} = x_i$.
Step 4: Go back to **Step 2** if the number samples is not sufficient.

Example 10.6
The example demonstrates the implementation of Algorithm 10.2. The target PDF function $f(x)$ is the standard Normal distribution

$$f(t) = \frac{1}{\sqrt{2\pi}} e^{-\frac{1}{2}t^2}$$

Draw the random numbers from the Normal distribution using Metropolis-Hastings algorithm.

Solution: We choose the simple Uniform distribution as the proposal distribution. Let the random variable x, y are with a joint distribution density function defined as follows

$$y - x \sim U[-b, b] = \begin{cases} \frac{1}{2b}, (y-x) \in [-b, b] \\ 0, \quad \textit{Others} \end{cases}$$

The CDF function is

(continued)

Example 10.6 (continued)

$$y-x \sim \begin{cases} \frac{y-x+b}{2b}, (y-x) \in [-b, b] \\ 0, \qquad\qquad Others \end{cases}$$

It is defined in a symmetric range from $-b$ to b where $b > 0$ since the Normal distribution is also symmetric and ranges from $[-\infty, +\infty]$. The conditional transition distribution $q(x/\widehat{y})$ and $q(y/x_i)$ are

$$q(x/\widehat{y}) = \frac{-x+b+\widehat{y}}{2b}$$

$$q(y/x_i) = \frac{y-x_i+b}{2b}$$

Obviously, $q(x_i/\widehat{y}) = q(\widehat{y}/x_i)$. The proposal distribution is symmetric. The

$$\min\left(\frac{f(y)q(x/y)}{f(x)q(y/x)}, 1\right) = \min\left(\frac{f(y)}{f(x)}, 1\right)$$

Draw a random number from $q(y/x_i)$ by letting $\frac{-x_i+b+\widehat{y}}{2b} = u$, where u is Uniform random number from 0 to 1. Then

$$\widehat{y} = 2bu + x_i - b$$

The algorithm is implemented by coding of Matlab. Setting the number of samples 20,000, run the simulation, the samples distribution and the target function are plotted as follows Fig. 10.23. The b chooses as 0.5.

It is not a perfect fitting. The mean of the samples is -0.0186. Standard deviation is 0.9690, showing the fitting is roughly acceptable with some error. The acceptance rate for the simulation is around 0.90, a high rate, attributing to an improvement by Eq. (10.28).

Example 10.7

This example is to demonstrate the MH algorithm. Improvement is needed either by using better proposal distribution, or by selecting only part of the sampling. Another problem of Algorithm 10.2 is when the magnitude of the $f(x)$ is extremely small. Algorithm 10.2 will fall with extremely long computational time for even just hundreds of samples. One practical improvement is necessary to cope the extreme low magnitude of the $f(x)$. We can take the $\ln f(x)$ instead of $f(x)$, then the problem can be solved. The improved algorithm is as follows.

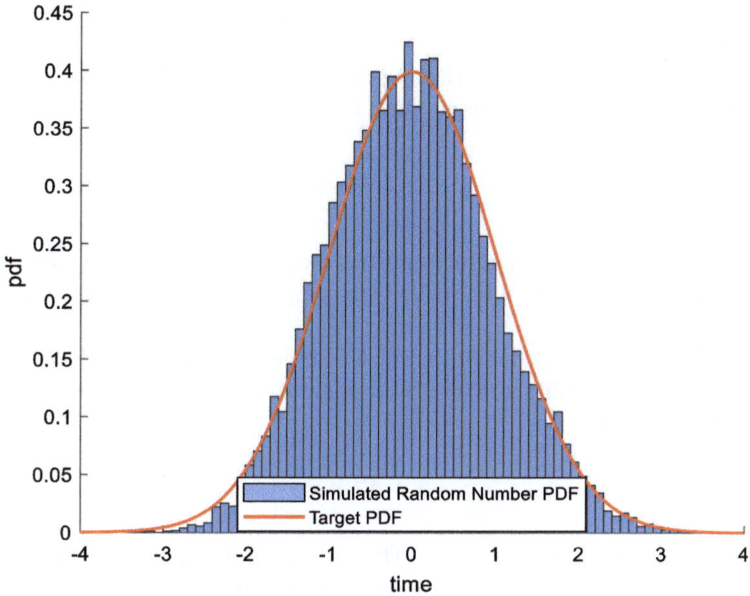

Fig. 10.23 Samples histogram from Metropolis-Hastings algorithm

Algorithm 10.3: Modified Metropolis-Hastings Algorithm [1]
Step 1: Choose the proposal distribution $q(y/x)$.
Step 2: Generate a sample $\widehat{y}$ from $q(y/x_i)$.
Step 3: Sample a random number u from $U[0, 1]$.
Step 3: If $\ln u < \min\left(\ln \frac{f(\widehat{y})q(x_i/\widehat{y})}{f(x_i)q(\widehat{y}/x_i)}, 0\right)$, accept the $x_{i+1} = \widehat{y}$; otherwise, let $x_{i+1} = x_i$.

For symmetric proposal function, $\min\left(\ln \frac{f(\widehat{y})q(x_i/\widehat{y})}{f(x_i)q(\widehat{y}/x_i)}, 0\right) = \min(\ln f(\widehat{y}) - \ln f(x_i), 0)$.
Step 4: Go back to **Step 2** if the number samples is not sufficient.

The modified Metropolis-Hastings Algorithm just modified the **Step 3**.

Example 10.8

This example uses Metropolis-Hastings algorithm for reliability analysis. The real Weibull distribution is known with PDF function $f(x)$ as

$$f(t) = \frac{\beta}{\alpha}\left(\frac{t}{\alpha}\right)^{\beta-1} e^{-\left(\frac{t}{\alpha}\right)^{\beta}}$$

The $\alpha = 1$; $\beta = 2$. Draw random numbers from the Weibull distribution using Metropolis-Hastings algorithm. Given the sample is in unit of hours, estimate the reliability for $t = 1$.

Solution: We choose the Normal distribution as the proposal distribution. Let the random variable x, y are with a joint distribution density function defined as follows

$$y = x + N\left(0, \sigma^2\right)$$

The $q(y, x_i)$ is then $N(x_i, \sigma^2)$. The $q(\widehat{y}, x)$ is then $N(\widehat{y}, \sigma^2)$. The $q(y/x_i) = \frac{q(y, x_i)}{\int q(y, x_i)dy}$. Since $q(y, x_i)$ is PDF of $N(x_i, \sigma^2)$, the $\int q(y, x_i)dy$ is then unity. We obtain $q(y/x_i) = q(y, x_i)$. Similarly, we can obtain the $q(x/\widehat{y}) = q(x, \widehat{y})$. Then it is straightforward the $q(x_i/\widehat{y}) = q(\widehat{y}/x_i)$. The proposal distribution is symmetric. The $\min\left(\ln \frac{f(\widehat{y})q(x_i/\widehat{y})}{f(x_i)q(\widehat{y}/x_i)}, 0\right) = \min(\ln f(\widehat{y}) - \ln f(x_i), 0)$.

The random number drawn from $q(y/x_i)$ is then equal to draw from $N(x_i, \sigma^2)$. The algorithm was implemented by Matlab. There is total 20,000 samples obtained from the simulation. The $\sigma = 0.4$. The sample distribution is shown in following Fig. 10.24.

It is not a perfectly fitting, while it is acceptable. If we use the 20,000 samples as observation data, the maximum likelihood estimate from the sample is $\widehat{\alpha} = 1.0192$; $\widehat{\beta} = 2.0820$. Compared with the real parameter $\alpha = 1$; $\beta = 2$, the simulated parameters are quite close. For reliability at $t = 1$, we can estimate it from the simulated samples. If the variable in the Weibull is time to failure, the simulated samples are then simulated time to failure samples. For reliability of time t, the reliability corresponds to the area right of the vertical line $t = 1$, as shown on the left figure in Fig. 10.25. We can estimate the area from the samples. A simper method, as discussed in Chap. 5, for observed time to failure data, the reliability can be estimated by

$$\widehat{R} = 1 - \frac{i}{n+1} \tag{10.29}$$

(continued)

Example 10.8 (continued)

The i is the order of the samples which have been ascendingly ordered. Reliability can be approximated from the order, as we discussed in Chap. 3. In this example, we can use the order of the sample i which is nearest to the t, thereafter using the order i in Eq. (10.29) to estimate reliability. The n is the total number of samples. In this Example, the resulting estimated reliability is

$$\widehat{R} \approx 0.3751$$

from the simulated samples, while the real reliability is

$$R(t) = e^{-(t/\alpha)^{\beta}} = e^{-(1/1)^2} = 0.3679$$

The acceptance rate in this simulation is 74.12%, a very high rate. Unfortunately, the high acceptance rate does not mean high quality samples. In this example, the quality of this simulation example is not high. We should point out: the parameter in the proposal distribution affects the simulation performance. The following Fig. 10.25 presents figures for two σ. The $\sigma = 5$ exhibits an even worse fitting, implying the simulation performance is worse. Selecting the σ for the Normal distribution of the proposal distribution is challenging issue, which is a drawback of the MH algorithm.

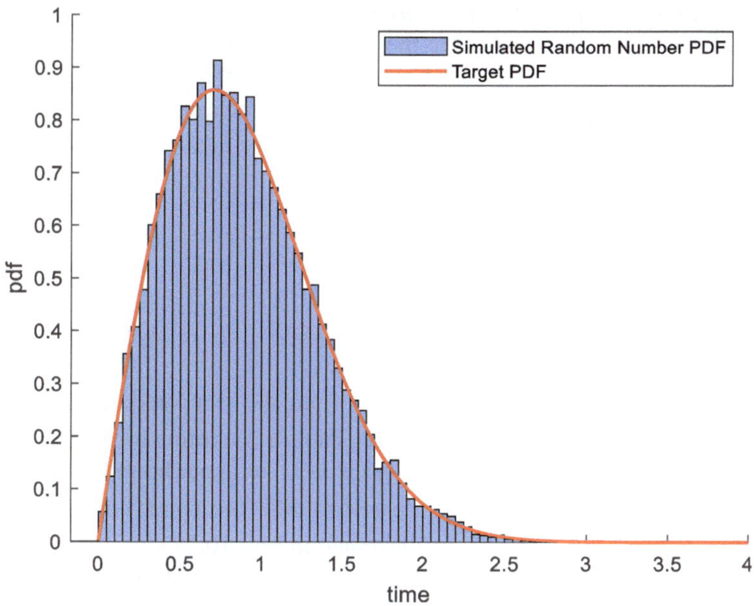

Fig. 10.24 Samples histogram for Weibull distribution

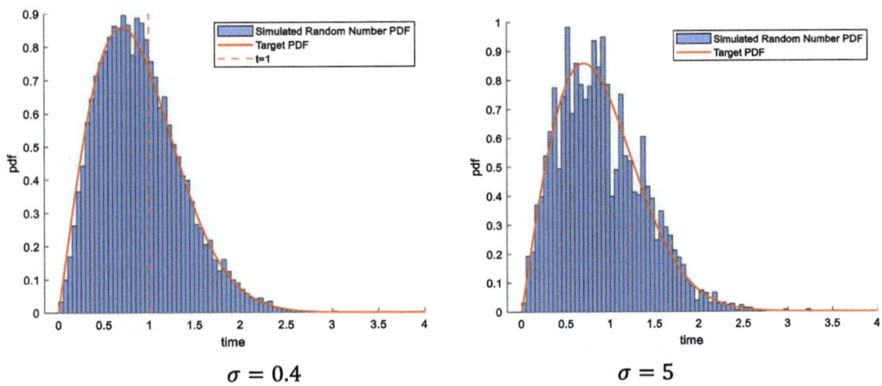

Fig. 10.25 Presence of probability for transition

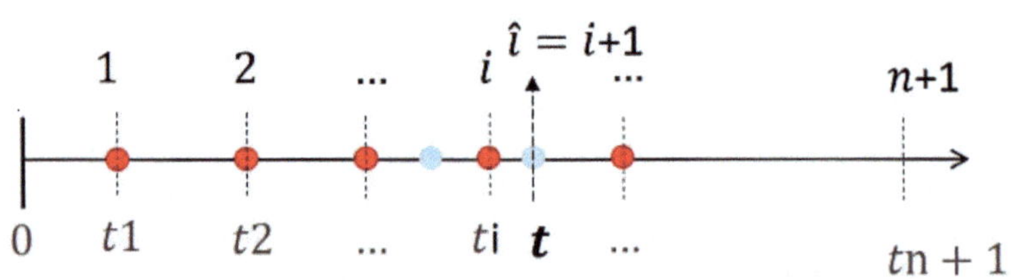

Fig. 10.26 Ordered simulated samples

Reliability Estimation from Simulated Samples

In Example 10.8, we have stated the method to estimate the reliability from simulated samples. The method is based on the order of the descending samples, as shown in Fig. 10.26 as follows. Given an arbitrary time t, another method adds the time t into the ordered samples. Suppose the order of t in the simulated sample is $\widehat{i}$, the reliability is then

$$\widehat{R} = 1 - \frac{\widehat{i}}{n+2}$$

where n is the size of original samples. This method is very simple and easy to code in computer languages.

The next issue is to deduce an interval from the simulated samples. We can estimate the reliability, let us say $\widehat{R}_i$, for each sample according to Eq. (10.29). For a confidence interval, given the lower bound is $\alpha/2$, the upper bound is $1 - \alpha/2$. The lower bound chooses time whose $\widehat{R}_i$ nearest to $\alpha/2$. The upper bound chooses the time whose $\widehat{R}_j$ nearest to $1 - \alpha/2$, as illustrated in Fig. 10.27.

For the case in Example 10.8, the 95% confidence interval from simulated samples are [0.1812, 1.9367]. The real 95% confidence interval by

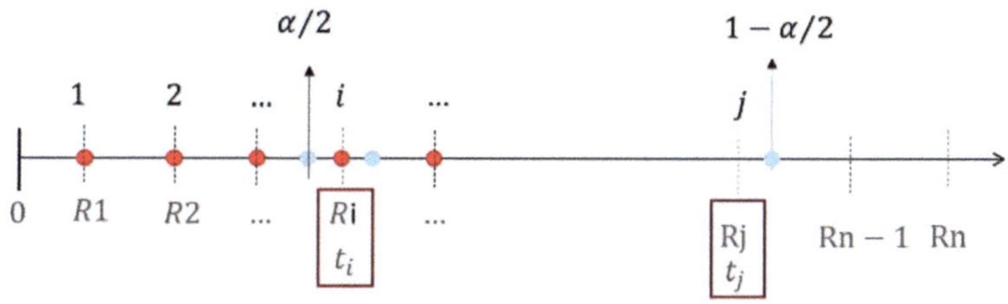

Fig. 10.27 Confidence interval from simulated samples

$$t_L = \alpha[-\ln(1-\alpha/2)]^{\frac{1}{\beta}}; \quad t_U = \alpha[-\ln(\alpha/2)]^{\frac{1}{\beta}}$$

The real CI confidence interval [0.1591, 1.9206], where we can see the interval from simulated sample is close but with a certain error.

10.4.4 Gibbs Sampling

The Gibbs sampling is another sampling approach which is simple and popular. The Gibbs is especially effective for multivariate target distribution $f(\boldsymbol{X})$. The $\boldsymbol{X}$ is with d dimensions $\boldsymbol{X} = (x_1, x_2, \ldots, x_d)$. The ith sample is denoted by $\boldsymbol{X}^i$. The algorithm of the Gibbs sampling is simpler than the acceptance-rejection and the Metropolis-Hastings algorithm [1]. Letting the derivation of the Gibbs and the usage of the sampling to be discussed later, we would like to present the algorithm firstly as it is easy to be understood.

Algorithm 10.4: Gibbs Algorithm [1]
Step 1: Provide an arbitrary initial point $\boldsymbol{X}^0$.
Step 2: Start the iterative procedure. The new value $\boldsymbol{X}^{i+1}$ is generated for each dimension of $\boldsymbol{X}$.

For the 1st dimension, the random number x_1^{i+1} is generated from the conditional distribution

$$x_1^{i+1} \sim f\left(x_1/x_2^i, x_3^i, \ldots, x_j^i, \ldots x_d^i\right)$$

The 2nd dimension, the random number x_2^{i+1} is generated from the conditional distribution

$$x_2^{i+1} \sim f\left(x_2/x_1^i, x_3^i, \ldots, x_j^i, \ldots x_d^i\right)$$

The 3rd dimension, the random number x_3^{i+1} is generated from the conditional distribution

$$x_3^{i+1} \sim f\left(x_3/x_1^i, x_2^i, x_4^i \ldots, x_j^i, \ldots x_d^i\right)$$

$$\cdots$$

The jth dimension, the random x_j^{i+1} is generated from the conditional distribution

$$x_j^{i+1} \sim f\left(x_j/x_1^i, x_2^i, \ldots, x_{j-1}^i, x_{j+1}^i, , \ldots x_d^i\right)$$

$$\cdots$$

The dth dimension, the random x_d^{i+1} is generated from the conditional distribution

$$x_d^{i+1} \sim f\left(x_d/x_1^i, x_2^i, \ldots, x_j^i, , \ldots x_{d-1}^i\right)$$

The new $\boldsymbol{X}^{i+1}$ is thus generated by above procedure.

Step 3: Iterate the procedure by letting the $\boldsymbol{X}^{i+1}$ as the old $\boldsymbol{X}^i$. A sequence of random number is than obtained

$$\boldsymbol{X}^1, \boldsymbol{X}^2, \boldsymbol{X}^3, \ldots, \boldsymbol{X}^n$$

Example 10.9
This example illustrates the sampling from a two-dimensional Normal distribution with parameter

$$\boldsymbol{u} = (0, 0); \quad \boldsymbol{\Sigma} = \begin{pmatrix} 1 & 0.6 \\ 0.6 & 1 \end{pmatrix}$$

(Please refer the Chap. 8 for the detail of multivariate Normal distribution). Use the Gibbs approach to draw 5000 samples from the above distribution.

Solution: It is known the $x_1^{i+1} \sim f(x_1/x_2^i)$ and $x_2^{i+1} \sim f(x_2/x_1^i)$ still follow Normal distribution of [3]

$$f(x_1/x_2^i) = N\left(u_1 + \frac{\sigma_{12}}{\sigma_{22}}(x_2 - u_2), \sqrt{\sigma_{11} - \sigma_{12}{}^2/\sigma_{22}}\right)$$

(continued)

Example 10.9 (continued)

$$f\left(x_2/x_1^i\right)=N\left(u_2+\frac{\sigma_{12}}{\sigma_{11}}(x_1-u_1),\sqrt{\sigma_{22}-{\sigma_{12}}^2/\sigma_{11}}\right)$$

The random number drawing for each dimension is thus from a univariate Normal distribution. The algorithm is run by using Matlab. The simulated sample for 5000 are plotted as follows Fig. 10.28.

The mean of the sample is

$$\widehat{\boldsymbol{u}}=[0.022,0.0125].$$

The standard deviation is

$$\hat{\boldsymbol{\Sigma}}=\begin{pmatrix}0.9848 & 0.6264\\ 0.6264 & 1.0145\end{pmatrix}$$

The estimated parameters from the simulated samples are roughly closed to the real parameters.

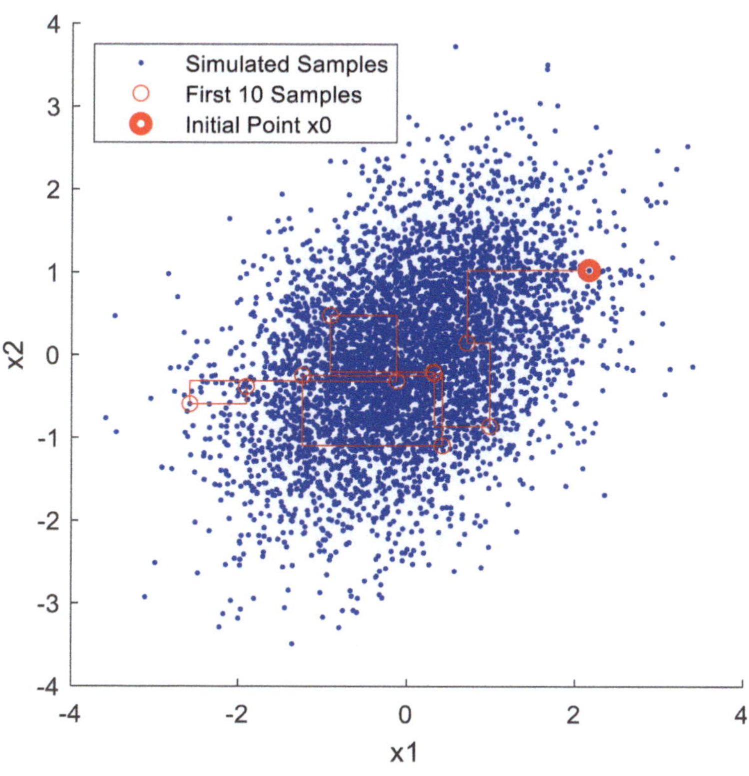

Fig. 10.28 Sample distribution for Example 10.9

The Gibbs with Metropolis-Hastings Algorithm

The Gibbs appears a big difference with the Metropolis-Hastings Algorithm, while essentially, the Gibbs is a special case of MH algorithm [1]. The target distribution is $f(\boldsymbol{X})$ where $\boldsymbol{X}$ is a random vector for general case. For each drawing in Algorithm 10.4, if the ith dimension is selected, the next iterated value X^{t+1} is written as

$$X^{t+1} = \left(x_1{}^t, x_2{}^t, \ldots, x_i{}^{t+1}, \ldots., x_m{}^t\right)$$

For dimensions i, the transition distribution for the Gibbs is the

$$\frac{1}{m} f\left(x_i{}^{t+1} / \boldsymbol{X}_{-i}{}^t\right) \tag{10.30}$$

where the $\boldsymbol{X}_{-i}$ denotes the $\boldsymbol{X}$ value without the ith dimension. The $\frac{1}{m}$ is due to each dimension is equally selected. The probability of the ith dimension is selected is then $\frac{1}{m}$. Let us rewrite in a more explicit form as

$$f\left(\boldsymbol{X}^{t+1}\right) = f\left(x_i{}^{t+1} / \boldsymbol{X}_{-i}{}^t\right)$$

For MH algorithm, the transition probability is dependent on $\frac{f(\widehat{y}) q(x_i/\widehat{y})}{f(x_i) q(\widehat{y}/x_i)}$. For Gibbs, the $\widehat{y}$ is $\boldsymbol{X}^{t+1}$, x_i is $\boldsymbol{X}^t$. According to Eq. (10.30), the transition probability $q\left(\boldsymbol{X}^{t+1}/\boldsymbol{X}^t\right) = \frac{1}{m} f\left(x_i{}^{t+1}/\boldsymbol{X}_{-i}{}^t\right)$. The $q\left(\boldsymbol{X}^t/\boldsymbol{X}^{t+1}\right) = \frac{1}{m} f\left(x_i{}^t/\boldsymbol{X}_{-i}{}^{t+1}\right) = \frac{1}{m} f\left(x_i{}^t/\boldsymbol{X}_{-i}{}^t\right)$ due to the $\boldsymbol{X}^{t+1}$ differs with $\boldsymbol{X}^t$ only at x_i and $\boldsymbol{X}_{-i}{}^t = \boldsymbol{X}_{-i}{}^{t+1}$.

$$\begin{aligned}\frac{f(\widehat{y}) q(x_i/\widehat{y})}{f(x_i) q(\widehat{y}/x_i)} &= \frac{f\left(\boldsymbol{X}^{t+1}\right) q\left(\boldsymbol{X}^t/\boldsymbol{X}^{t+1}\right)}{f\left(\boldsymbol{X}^t\right) q\left(\boldsymbol{X}^{t+1}/\boldsymbol{X}^t\right)} = \frac{f\left(x_i{}^{t+1}, \boldsymbol{X}_{-i}{}^{t+1}\right) q\left(\boldsymbol{X}^t/\boldsymbol{X}^{t+1}\right)}{f\left(x_i{}^t, \boldsymbol{X}_{-i}{}^t\right) q\left(\boldsymbol{X}^{t+1}/\boldsymbol{X}^t\right)} \\ &= \frac{f\left(x_i{}^{t+1}, \boldsymbol{X}_{-i}{}^{t+1}\right) . \frac{1}{m} f\left(x_i{}^t/\boldsymbol{X}_{-i}{}^t\right)}{f\left(x_i{}^t, \boldsymbol{X}_{-i}{}^t\right) . \frac{1}{m} f\left(x_i{}^{t+1}/\boldsymbol{X}_{-i}{}^t\right)}\end{aligned} \tag{10.31}$$

According to the conditional probability definition, the

$$f\left(x_i{}^t/\boldsymbol{X}_{-i}{}^t\right) = \frac{f\left(x_i{}^t, \boldsymbol{X}_{-i}{}^{t+1}\right)}{\int \boldsymbol{X}_{-i}{}^t dx_i};$$

$$f\left(x_i{}^{t+1}/\boldsymbol{X}_{-i}{}^t\right) = \frac{f\left(x_i{}^{t+1}, \boldsymbol{X}_{-i}{}^t\right)}{\int \boldsymbol{X}_{-i}{}^t dx_i}$$

Equation (10.30),

$$\frac{f\left(x_i^{t+1},\boldsymbol{X}_{-i}^{t+1}\right).\frac{1}{m}f\left(x_i^t/\boldsymbol{X}_{-i}^t\right)}{f\left(x_i^t,\boldsymbol{X}_{-i}^t\right).\frac{1}{m}f\left(x_i^{t+1}/\boldsymbol{X}_{-i}^t\right)}=\frac{f\left(x_i^{t+1},\boldsymbol{X}_{-i}^{t+1}\right).\frac{f\left(x_i^t,\boldsymbol{X}_{-i}^{t+1}\right)}{\int \boldsymbol{X}_{-i}^t dx_i}}{f\left(x_i^t,\boldsymbol{X}_{-i}^t\right).\frac{f\left(x_i^{t+1},\boldsymbol{X}_{-i}^t\right)}{\int \boldsymbol{X}_{-i}^t dx_i}}=\frac{f\left(x_i^{t+1},\boldsymbol{X}_{-i}^t\right).\frac{f\left(x_i^t,\boldsymbol{X}_{-i}^t\right)}{\int \boldsymbol{X}_{-i}^t dx_i}}{f\left(x_i^t,\boldsymbol{X}_{-i}^t\right).\frac{f\left(x_i^{t+1},\boldsymbol{X}_{-i}^t\right)}{\int \boldsymbol{X}_{-i}^t dx_i}}=1$$

Therefore, the acceptance probability for MH algorithm according to Eq. (10.28) is

$$\min\left(\frac{f(y)q(x/y)}{f(x)q(y/x)},1\right)=\min(1,1)=1$$

implying the acceptance probability is 1. It thus verifies the Gibbs is a special case of Metropolis-Hastings Algorithm.

10.5 Bayesian Inference

The Gibbs has many other variants. Algorithm 10.4 is the simplest [1]. The Gibbs is not a common approach for sampling of the multivariate distribution such as the case in Example 10.9, but very popular for the Bayesian inference. Essentially, the famous Markov Chain Monte Carlo (MCMC) related to Bayesian inference uses the Gibbs as one of its main methods for computing. The MCMC refers to simulation using Markov Chain for sampling. The Gibbs, Metropolis-Hastings discussed here are all MCMC approach. For Bayesian approach, if we have understood the Markov Chain, the Metropolis-Hastings and the Gibbs algorithm, understanding the Bayesian approach is then easy. MCMC can be applied to solve many statistical issues. This book only concerns the Bayesian inference for parameter estimation.

If a PDF function with multivariate random vector $\boldsymbol{X}$, representing for example the observed data, the parameter in the distribution is a vector $\boldsymbol{\Theta}$. The PDF can be written as

$$f(\boldsymbol{X}/\boldsymbol{\Theta}) \tag{10.32}$$

meaning given parameters $\boldsymbol{\Theta}$, the X is observed. One special point for Bayesian approach is the parameters in the model are considered as random variables. It is not hard to understand the parameter estimation is based on the data to obtain the parameters $\boldsymbol{\Theta}$. In mathematical term, it can be written in a term of conditional probability as

$$\pi(\boldsymbol{\Theta}/\mathbf{X}) \tag{10.33}$$

By the definition of conditional probability, Eq. (10.33) is equal to

$$\pi(\boldsymbol{\Theta}/\mathbf{X}) = \frac{\pi(X, \boldsymbol{\Theta})}{m(\mathbf{X})}; \quad m(\mathbf{X}) = \iiint f\left(\frac{\boldsymbol{X}}{\boldsymbol{\Theta}}\right) d\boldsymbol{\Theta} \tag{10.34}$$

The $\pi(\boldsymbol{\Theta}, \mathbf{X})$ is the joint distribution of parameter $\boldsymbol{\Theta}$ and the observed data $\mathbf{X}$. The joint distribution is a joint of the data distribution $f(X/\boldsymbol{\Theta})$ and the parameter $\boldsymbol{\Theta}$ distribution $\pi(\boldsymbol{\Theta})$ as follows

$$\pi(\boldsymbol{\Theta}, \mathbf{X}) = f(X/\boldsymbol{\Theta})\pi(\boldsymbol{\Theta}) \tag{10.35}$$

Substitute Eq. (10.35) into Eq. (10.34), the full Bayesian expression is

$$\pi(\boldsymbol{\Theta}/\mathbf{X}) = \frac{\pi(X, \boldsymbol{\Theta})}{m(\mathbf{X})} = \frac{f(X/\boldsymbol{\Theta})\pi(\boldsymbol{\Theta})}{m(\mathbf{X})} \tag{10.36}$$

For purpose of parameter estimation, the aim is to find the distribution of $\pi(\boldsymbol{\Theta}/\mathbf{X})$. The $\pi(\boldsymbol{\Theta}/\mathbf{X})$ is the stationary distribution of our interest, which is named as Posterior distribution in Bayesian approach. The $\pi(\boldsymbol{\Theta})$ is the density function of distribution for parameter $\boldsymbol{\Theta}$, named as prior distribution.

10.5.1 Bayesian Parameter Estimation

Let the observed failure times be $t_1, t_2, t_3, \ldots, t_n$. Assume the failures are statistically independent from each other. The parameter is θ. The likelihood function be

$$L(\boldsymbol{t}/\theta) = f(t_1/\theta) f(t_2/\theta) f(t_3/\theta) \ldots f(t_n/\theta) \tag{10.37}$$

The likelihood function is a joint density function for the observed data, i.e. considered as the $f(X/\boldsymbol{\Theta})$ in Eq. (10.36). The Bayesian posterior distribution is rewritten as

$$\pi(\boldsymbol{\Theta}/\mathbf{X}) = \frac{f(X/\boldsymbol{\Theta})\pi(\boldsymbol{\Theta})}{m(\mathbf{X})} = \frac{L(\boldsymbol{t}/\theta)\pi(\theta)}{\int L(t/\theta)\pi(\theta)d\theta} \tag{10.38}$$

When the data observed at $\boldsymbol{t}$ are fixed, the $\int L(t/\theta)\pi(\theta)d\theta$ is a quantity unrelated to the parameter θ. This quantity is a numerical value, denoted as M. The posterior distribution is then

$$\pi(\boldsymbol{\Theta}/\mathbf{X}) = \frac{L(\boldsymbol{t}/\theta)\pi(\theta)}{M} \tag{10.39}$$

Example 10.10
Derive the posterior distribution for Exponential distribution when the prior distribution given by a Uniform distribution $U(a, b)$.

Solution: For observed time to failure t_i, the PDF is $f(t_i) = \lambda e^{-\lambda t_i}$. The likelihood function is

$$L(\lambda) = \lambda^n e^{-\lambda \sum_{i=1}^{i=n} t_i}$$

The $\pi(\theta) = \frac{1}{b-a}$. According to Eq. (10.39), the posterior distribution is

$$\pi(\mathbf{\Theta}/\mathbf{X}) = \frac{\lambda^n e^{-\lambda \sum_{i=1}^{i=n} t_i}}{M(b-a)}; \quad M = \int_0^{\infty} \lambda^n e^{-\lambda \sum_{i=1}^{i=n} t_i} d\lambda \tag{10.40}$$

We can find the $M = \int_0^{\infty} \lambda^n e^{-\lambda \sum_{i=1}^{i=n} t_i} d\lambda$ is a numerical value, but the integration is complex, for which we have to use numerical integration method to obtain the distribution of $\pi(\mathbf{\Theta}/\mathbf{X})$.

A difficulty has encountered in Example 10.10, the posterior distribution is not able to be obtained if the M is unknown. But this difficulty can be overcome by the Metropolis-Hastings Algorithm or Gibbs algorithm. The Metropolis-Hastings Algorithm, of which Gibbs is a special case of, is dependent on the target distribution by the Formula

$$\min\left(\frac{f(y)q(x/y)}{f(x)q(y/x)}, 1\right)$$

In Bayesian, the target distribution is the posterior distribution $\pi(\mathbf{\Theta}/\mathbf{X})$. The proposal $q(x/y)$ distribution is independent from the $\pi(\mathbf{\Theta}/\mathbf{X})$. Be aware the variable in this case is the $\mathbf{\Theta}$. Our focus is only on the $\frac{f(y)}{f(x)} = \frac{\pi(\mathbf{\Theta}_1/\mathbf{X})}{\pi(\mathbf{\Theta}_2/\mathbf{X})}$. It is straightforward to find out the M has been cancelled out in the ratio $\frac{f(y)}{f(x)}$. The Metropolis-Hastings Algorithm and Gibbs algorithm don't require the M for sampling, which is give a great advantage of the computation. Therefore, for MCMC Bayesian, we only concern the

$$\pi(\mathbf{\Theta}/\mathbf{X}) \propto f(X/\mathbf{\Theta})\pi(\mathbf{\Theta}) \tag{10.41}$$

Example 10.11
Sample from the posterior distribution for Exponential distribution when the prior distribution given by a Uniform distribution $U(a, b)$. Apply the Metropolis-Hastings Algorithm. The time to failure data is

1.59 4.96 2.95 0.18 3.76 0.38 1.24 0.01 5.10 1.63

The posterior distribution is $\pi(\mathbf{\Theta}/\mathbf{X}) \propto \lambda^n e^{-\lambda \sum_{i=1}^{i=n} t_i}$.

Solution: The proposal distribution for the Metropolis-Hastings Algorithm uses Normal distribution *N*(0,0.1). The samples size is set as 20,000. The algorithm is implemented by Matlab. The histogram for the simulated samples is shown as follows Fig. 10.29.

For comparison, the distribution of $\lambda^n e^{-\lambda \sum_{i=1}^{i=n} t_i}$ is plotted. It can be seen the empirical simulated samples, and the real distribution is with same shape, differing with the scale. The difference of the scale is due to the M in Eq. (10.40).

Since the distribution of the $\widehat{\lambda}$ is obtained from the simulated samples, the parameter estimation and its confidence interval derivation is a very easy task. The CI can use the same approach shown in Fig. 10.27. The results are tabulated in following Table 10.5 where the CI values is at significance interval 0.95.

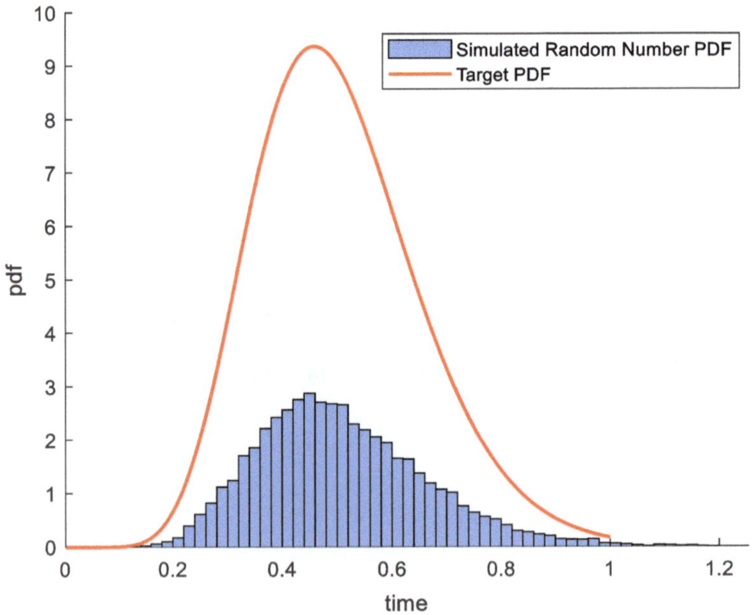

Fig. 10.29 Sample distribution for Example 10.11

Table 10.5 Statistics of simulated estimates

Mean	Std	Max	Min	CI Lower (0.95)	CI Upper (0.95)
0.5086	0.1537	1.1985	0.1443	0.2560	0.8597

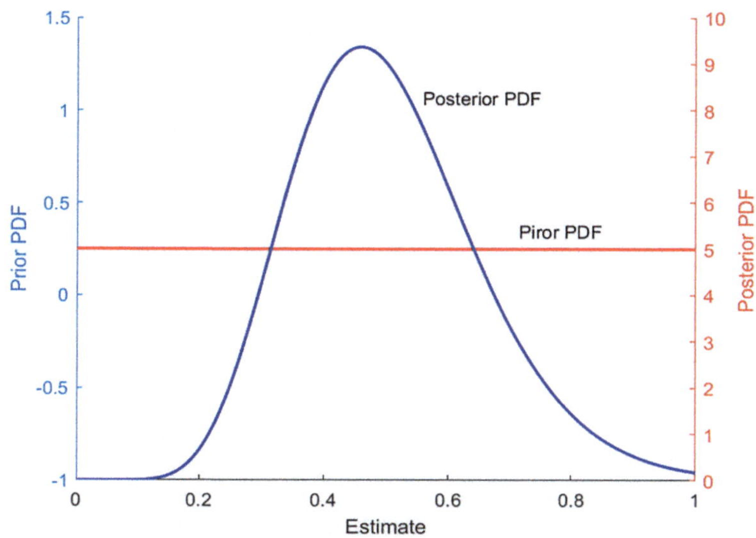

Fig. 10.30 Prior and posterior distribution for uniform prior

10.5.2 Prior Distribution and Posterior Distribution

In Example 10.11, the parameter to be estimated is assumed to initially follow a Uniform distribution. This initial distribution is referred to as the prior distribution. Uniform distribution implies all the candidate parameter values are equally likely to be the true parameter. The prior distribution is thereafter improved by absorbing the information of time to failure t_1, t_2, t_3, …, t_n. Figure 10.30 shows the prior distribution where no observed data used and the posterior distribution after absorbing the data information. We can see the information of posterior contains much more information of parameter λ, as the posterior sharper curvature implies more information.

If we choose the Normal distribution $N(3, 4)$ as the prior distribution, the posterior distribution using Normal as prior can be seen within a much-concentrated, less spreading region as shown in Fig. 10.31 than the Uniform distribution prior. The less-spreading PDF suggests higher accuracy of the estimate.

The uniform prior distribution has all values are equal, meaning it contains no information on the parameter. Such prior is named non-informative prior. The Normal distribution is informative prior since candidate estimates are not equal with probability. But a Normal distribution with a larger deviation is with flatter PDF, corresponding a less parameter information has been known in advance.

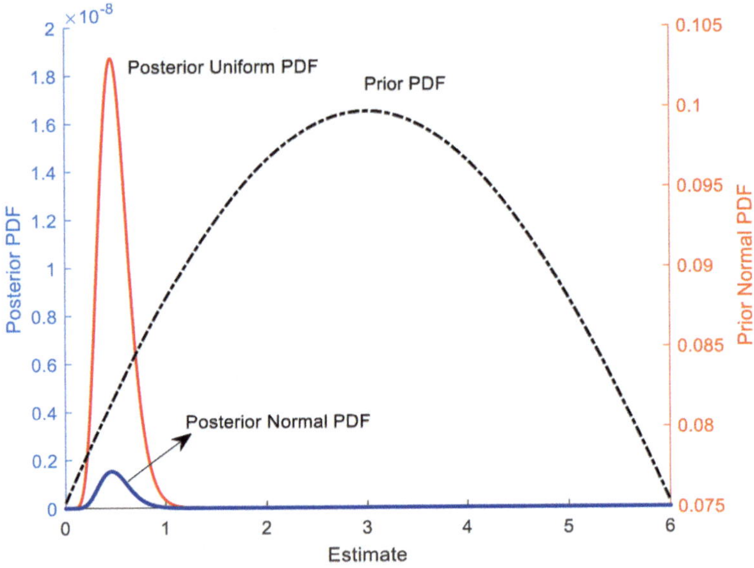

Fig. 10.31 Prior and posterior distribution for normal prior

10.5.3 Example Using Bayesian Inference for Parameter Estimation

The probability density function (PDF) is as follows for two-dimensional distribution.

$$f(s,t)=\gamma^2\alpha_1^{-1}\beta_1\left(\frac{t}{\alpha_1}\right)^{\beta_1-1}\alpha_2^{-1}\beta_2\left(\frac{s}{\alpha_2}\right)^{\beta_2-1}\mathrm{B}^{\gamma-2}\left(\frac{1}{\gamma}-1+\mathrm{B}^{\gamma}\right)\exp(-\mathrm{B}^{\gamma}) \tag{10.42}$$

where $\mathrm{B}=(t/\alpha_1)^{\beta_1}+(s/\alpha_2)^{\beta_2}$. For given observed data (s_i, t_i); $i = 1, 2, 3\ldots.n$, the likelihood function $L(Data)=\prod_{i=1}^{n} f(s_i,t_i)$. The prior distribution for each α_1, α_2, β_1, β_2 uses Gamma distribution. The γ is assumed to be Beta distribution since it confined in (0,1). The full Bayesian expression is then

$$\begin{aligned}\pi(\alpha_1,\alpha_2,\beta_1,\beta_2,\gamma|Data) \propto L(Data) &\times \frac{1}{\Gamma(a_1)b_1^{a_1}}\alpha_1{}^{a_1-1}e^{-\frac{\alpha_1}{b_1}} \times \frac{1}{\Gamma(a_2)b_2^{a_2}}\alpha_1{}^{a_2-1}e^{-\frac{\alpha_2}{b_2}}\\ &\times \frac{1}{\Gamma(a_3)b_3^{a_3}}\beta_1{}^{a_3-1}e^{-\beta_1/b_3} \times \frac{1}{\Gamma(a_4)b_4^{a_4}}\beta_2{}^{a_4-1}e^{-\beta_2/b_4}\\ &\times \frac{\Gamma(a_5)}{\Gamma(a_5)\Gamma(b_5)}\gamma^{a_5-1}(1-\gamma)^{b_5-1}\end{aligned} \tag{10.43}$$

If we use the Uniform distribution for each parameter as prior, the

Table 10.6 Statistics of estimate samples from Bayesian MCMC for Truck 2

Truck	Parameter	Mean of Estimates	Standard Deviation	Lower	Upper
Truck 2	α_1	14,950	2066	11,040	19,130
	β_1	2.922	0.446	2.085	2.909
	α_2	112.9	19.28	78.66	153.6
	β_2	2.526	0.4194	1.749	3.375
	γ	0.4233	0.07326	0.2949	0.5803

$$\pi(\alpha_1, \alpha_2, \beta_1, \beta_2, \gamma|Data) \propto L(Data) \tag{10.44}$$

The data used here are shown in Table 8A.1 in Chap. 8. We only present the results by using the Uniform prior. For the non-informative prior, the parameter used are $\alpha_1 \sim U(1,\ 3000)$, $\beta_1 \sim U(0.1,10)$ $\alpha_2 \sim U(1,\ 3000)$, $\beta_2 \sim U(0.1,10)$. The $\gamma \sim U(0,\ 1)$. A number of 10,000 of iterations has been ran in the MCMC computation using Gibbs sampling. The results are shown in Table 10.6. The lower and upper for intervals derived from the posterior distribution (10.44).

In this example, the parameter estimates, and their confidence intervals have been easily obtained using the MCMC approach, even for a complex PDF as shown in Eq. (10.42). For the same PDF, estimating the parameters using the maximum likelihood method is challenging due to the complex form of its likelihood function. Moreover, applying numerical methods to find the solution is uncertain, as the presence of local optima in the function cannot be analytically verified. In such cases, using MCMC within a Bayesian framework provides a stable and effective approach.

10.6 MCMC Convergence Diagnosis

The samples obtained from the MCMC is not independent from each other and from the presented example, we can find the samples drawn by MCMC is not as good as the inverse transform method. For the inverse transform, the empirical PDF from the samples can fit the target distribution perfectly, while the MCMC tends to have more error. Other than the sample independence issue, diagnosing if the MCMC has reach stationarity is another issue. A simple but effective way to cope with the stationary issues is graphic based method, despite there are plenty other numerical method available [4]. Diagnosis convergence is still an open question in state of art.

10.6.1 Stationarity Diagnostics

In the beginning of the Markov chain, we have seen the chain will take some time before stabilizing. As shown in the discrete Markov chain case in Sect. 10.4.1, the

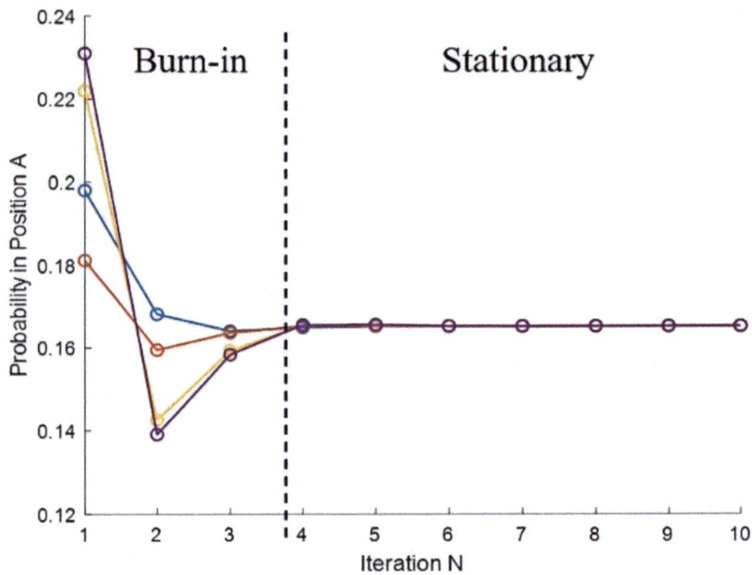

Fig. 10.32 Burn-in stage of Markov chain

probability stabilizes from step 4. Before the step 4, it is chaotic, and this stage is called the burn-in phase, as shown in Fig. 10.32. The sample obtained from the burn-in phase should be excluded as the real samples. The stage after step 4 becomes flat with small and unnoticeable fluctuation, which we can consider as stationary. The samples are preferable to be taken from the stationary phase.

Diagnosing the stationarity by plot is subjective, but it is simple and very popular in using the MCMC method.

10.6.2 Autocorrelation and Thinning

For the samples in the stationary phase, the samples are still correlated between the consecutive samples. To demonstrate the autocorrelation, we use the simulation results from **Example 10.6** for demonstration. The sample (named traces in MCMC) don't show significant burn-in stage, as shown in Fig. 10.33.

The autocorrelation can evaluate the dependency between the consecutive samples. For $k-$Lag of autocorrelation, the autocorrelation is defined as [5]

$$r_k = \frac{\frac{1}{n}\sum_{i=1}^{n-k}(x_i - \overline{x})(x_{i+k} - \overline{x})}{\frac{1}{n}\sum_{i=1}^{n}(x_i - \overline{x})^2} \tag{10.45}$$

For the samples in Fig. 10.33, the autocorrelation is shown as

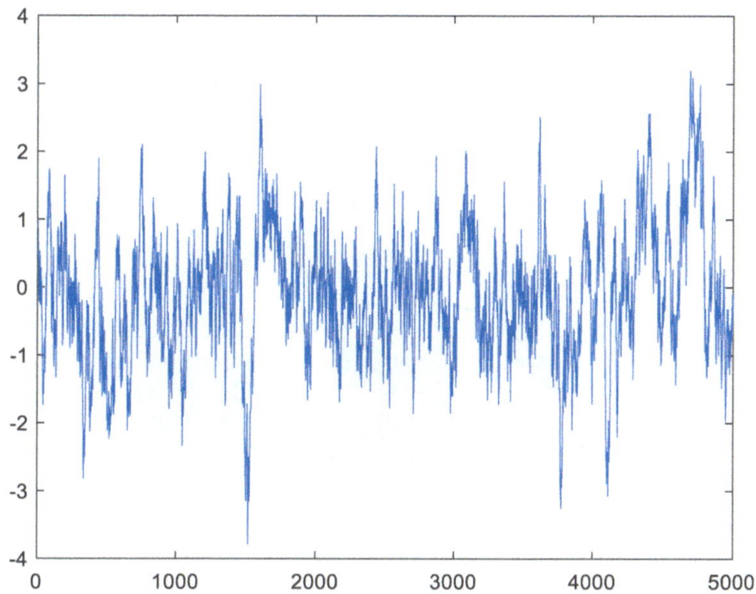

Fig. 10.33 Traces plot of Example 10.6

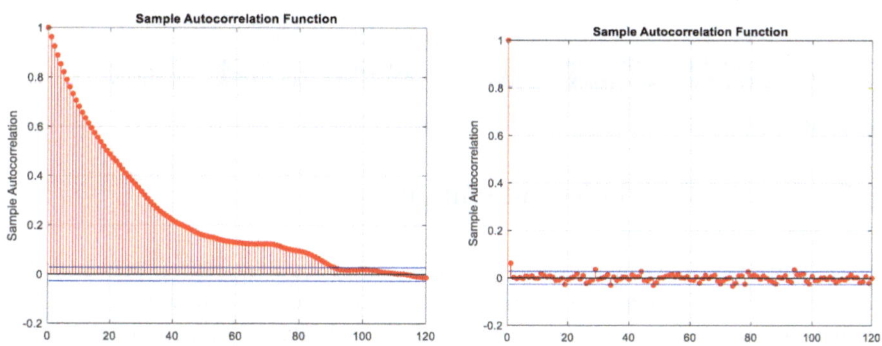

Before thinning After thinning

Fig. 10.34 Autocorrelation plot

To obtain the independent samples as final samples, we can take the sample for every m sample, since they are not significantly correlated. This method is called "thinning". From the left of Fig. 10.34, we can find the lag k = 80–90, the autocorrelation becomes very low. We therefore can take the thinning step k = 80, i.e. taking one from every 80 samples, the autocorrelation plot is then as the right of Fig. 10.34, where we can see no significant autocorrelation. The mean and STD can be seen an improvement using the thinning, as shown in Table 10.7. The results from "thinning" are closer to real value than that of "no thinning".

Table 10.7 Statistics of simulated samples with and without thinning

	Mean	STD
Real value	0	1
No thinning	−0.0921	0.9733
Thinning	0.0369	1.0070

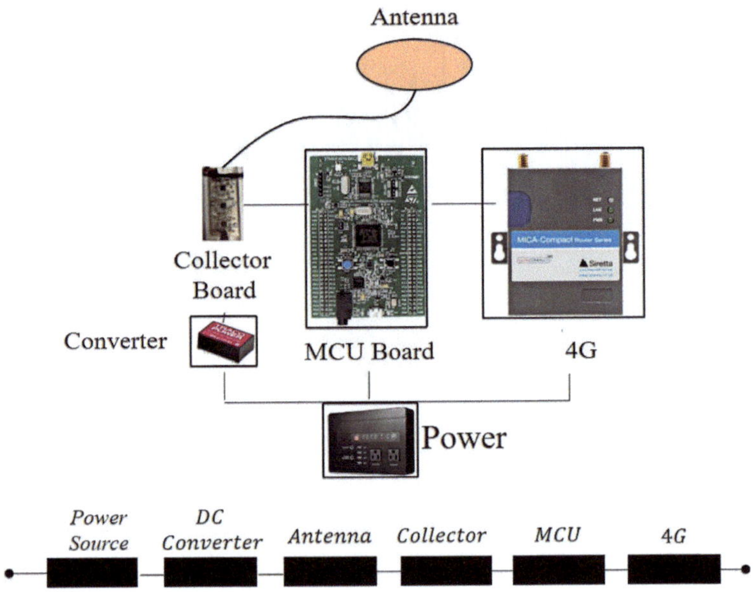

Fig. 10.35 RBD for a IOT device

10.7 System Reliability Simulation

For system connected with basic parallel, series, cold standby, network. The methods discussed in Chap. 6 is sufficient to be used for reliability evaluation. For practical purpose, for a situation where the reliability of some components is dependent on others, the simulation is a preferable technique can be used to evaluate the reliability.

Example 10.12 Reliability Simulation on IOT Device
The aim of this example is to illustrate the performance of simulation. An internet of thing system contains 6 major components with reliability block diagram is as follows Fig. 10.35.

The $R_1(Power\ Source) = 0.9$; $R_2(DC\ Converter) = 0.9$; $R_3(Antenna) = 0.9$; $R_4(Collector) = 0.9$; $R_5(MCU) = 0.99$; $R_6(4G) = 0.8$.

The real reliability can be computed analytically as

(continued)

Table 10.8 Statistics of simulated system reliability

Mean	Std	Max	Min
0.5204	0.0112	0.5630	0.4775

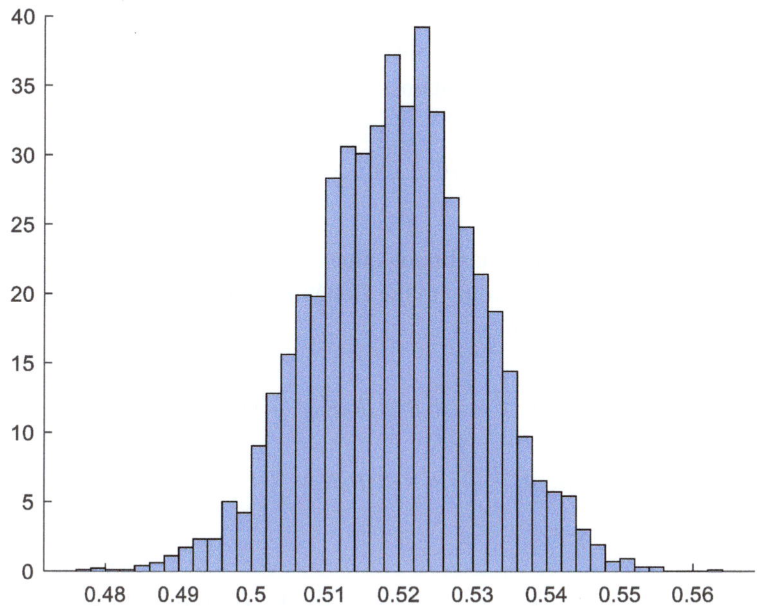

Fig. 10.36 Histogram of the $\widehat{R}_s$ for 5000 times

Example 10.12 (continued)

$$R_s = R_1R_2R_3R_4R_5R_6 = 0.9 * 0.9 * 0.9 * 0.9 * 0.9 * 0.99 * 0.8 = 0.5196$$

Using the random generating method of Mersenne-Twister to generate a random number between 0 and 1, if random number, for example the $R_1 = 0.9$, is greater than 0.1, then the component is considered as 1. If all the components are *on*, then the system is *on*. Repeat it for number of n=2000 times, the system reliability is the total number of *on* system state, let us say n_{on}, divided by the n=2000.

$$\widehat{R}_s = \frac{n_{on}}{n}$$

The algorithm is implemented by Matlab. Repeat the procedure for 5000 times. The $\widehat{R}_s$ differs at each of the 5000 iterations and the results are shown in Table 10.8, and its histogram is shown in Fig. 10.36.

Theoretical Standard Deviation

The system reliability is explicitly $R_s = R_1R_2R_3R_4R_5R_6$. The variance of each R_i is nearly identical since all of them using the same generating algorithm and with the same sample size. According to the discussion in Sect.10.2.2, for simulation on one probability, the standard deviation of simulation for sample size n is $\sqrt{\frac{p(1-p)}{n}}$. If the p is unknown, an upper bound is obtained from

$$\sqrt{\frac{p(1-p)}{n}} \leq \frac{0.5}{\sqrt{n}} \tag{10.46}$$

by using the $p(1-p)$ with the maximum value at $p = 0.5$. Then all the $Var(R_i) = \frac{1}{4n}$ according to Eq. (10.46). According to the Delta method (see in Appendix 10A), the variance of the system reliability is

$$\begin{aligned} Var(R_s) = & \Big((R_2R_3R_4R_5R_6)^2 + (R_1R_3R_4R_5R_6)^2 + (R_1R_2R_4R_5R_6)^2 \\ & + (R_1R_2R_3R_5R_6)^2 + (R_1R_2R_3R_4R_6)^2 + (R_1R_2R_3R_4R_5)^2 \Big) * Var(R_i) \\ = & \, 2.53 \times 10^{-4} \end{aligned}$$

The standard deviation of R_s is then $\sqrt{2.53 \times 10^{-4}} = 0.0159$ for sample size 2000. The real measured STD for the system reliability, shown in Table 10.8 is 0.0112, which is close but preferably less than the theoretically calculated STD 0.0159. The standard deviation for other sample size n can be computed by

$$\begin{aligned} & STD(R_s) \\ & = \frac{0.5}{\sqrt{n}} \sqrt{\left((R_2R_3R_4R_5R_6)^2 + (R_1R_3R_4R_5R_6)^2 + (R_1R_2R_4R_5R_6)^2 + (R_1R_2R_3R_5R_6)^2 + (R_1R_2R_3R_4R_6)^2 + (R_1R_2R_3R_4R_5)^2 \right)} \end{aligned} \tag{10.47}$$

Figure 10.37 plots the STD against the sample size where we can see the STD decreases with the sample size.

Simulation has been carried for other sample size. The results are shown in Table 10.9, where we can see the theoretical variance calculated from Eq. (10.47) in general agrees with the simulation results. Therefore, practically, we can use the Delta method along with Eq. (10.46) to control the simulation accuracy for the system reliability assessment.

The example presents here is just to demonstrate the simulation method. For reliability analysis such as the problem in this example, the simulation approach is however not recommended since the analytical method using $R_s = R_1R_2R_3R_4R_5R_6$ is more accurate and more computationally efficient than the simulation method.

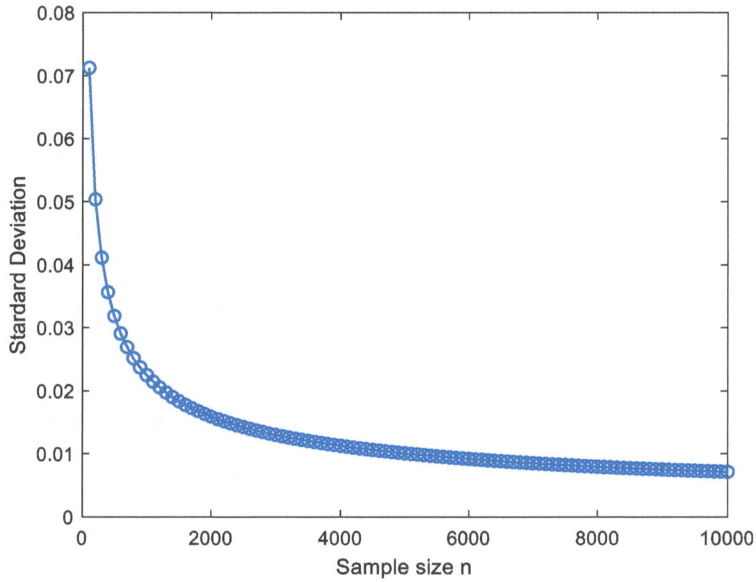

Fig. 10.37 Standard deviation with increasing sample size *n*

Table 10.9 Theoretical STD and simulated STD (5000 iterations)

n	Theoretical STD	Simulated STD
2000	0.0159	0.0112
5000	0.0101	0.0071
8000	0.0080	0.0056
10,000	0.0071	0.0050

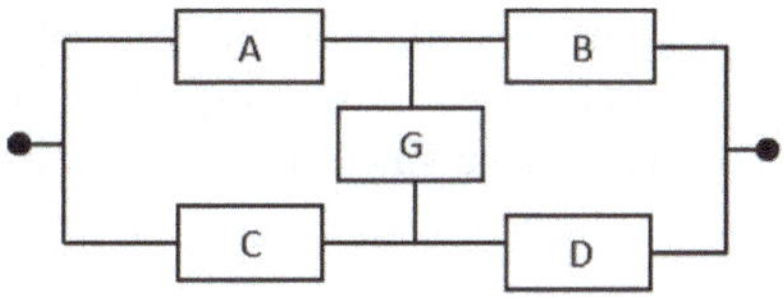

Fig. 10.38 Bridge RBD network

Example 10.13 Reliability Simulation on Bridge System

In Sect. 6.8 of Chap. 6, we illustrated an example of bridge network as following Fig. 10.38.

The detailed simulation method was presented in Sect. 6.9. This chapter we continue that example and present the method to assess the variance. The reliability of the RBD is the sum of all the probability corresponding to system state 1. For each state, the probability is of form

$$P(i) = R_A(1 - R_B)\ldots.$$

(continued)

Table 10.10 Theoretical STD and simulated STD for system reliability (5000 iterations)

n	Theoretical STD	Simulated STD
2000	0.05	0.0032
5000	0.0316	0.0021
8000	0.0250	0.0016
10,000	0.0224	0.0014

Example 10.13 (continued)

$$\frac{\partial P(i)}{\partial R_A} = (1 - R_B)R_c\ldots; \quad \frac{\partial P(i)}{\partial R_B} = -R_A R_c\ldots;$$

For any probability $P_A * P_B \leq 0.5 * 0.5$. Then $\frac{\partial P(i)}{\partial R_A} = (1 - R_B)R_c\ldots \leq 0.5 * 0.5 * 0.5\ldots$. i.e. we can obtain an upper bound of the $\left|\frac{\partial P(i)}{\partial R_A}\right| \leq 0.5^{m-1}$, where m is the number of components. According to the delta method, the variance is

$$Var(R_s) = m\left(\frac{2^m}{2^{m-1}}\right)^2 Var(R_i) = 4mVar(R_i)$$

$$4mVar(R_i) \leq \frac{4m}{4n} = \frac{m}{n}$$

we obtain

$$Var(R_s) \leq \frac{m}{n} \tag{10.48}$$

The theoretical STD and the simulated STD is shown in Table 10.10, where we can see the theoretical STD is higher than the simulated STD. The theoretical STD uses Eq. (10.47). The example shows the theoretical STD is conservative since it is an upper bound of the real STD.

Example 10.14 Reliability Simulation on Stress-Strength System

This example is to estimate the reliability using simulation approach for the problem. Assume the strength follows Normal distribution $N(4{,}000, 400^2)$.

$$\mu_{Strength} = 4000; \quad \sigma_{Strength} = 400$$

Stress is also Normal Distribution as $N(3{,}000, 300^2)$.

(continued)

Table 10.11 Stress-strength simulation results

No.	No of Iterations	Reliability
1	50	0.98
2	50	0.96
3	5000	0.9768
4	5000	0.976
5	500,000	0.9771
6	500,000	0.9771
7	2,000,000	0.9772
8	2,000,000	0.9774

Example 10.14 (continued)

$$\mu_{Stress} = 3000; \quad \sigma_{Stress} = 300$$

What is the reliability?

Solution: The limit function for this problem is

$$G(Strength, Stress) = Strength - Stress$$

The procedure for the simulation is

Step 1. Generate Normal random numbers from the Stress x(stress) and Strength distribution y(strength).

Step 2. if Strength > Stress, system functions. Repeat the procedure for n iterations.

$$\text{Failure Probability} = \frac{\text{Total number of results with functional trial}}{\text{Number of Simulations}}$$

The results for various iterations are shown as follows Table 10.11.

The real exact reliability has been known as $R = 0.977$. When the number of iterations reaches 500,000, the simulated reliability will be very close to the real reliability. One may wonder since the reliability for this example can be obtained analytically, why we need the simulation? We can have an example. An equation for a simple circuit is as simple as follows

$$R = U/I$$

when U and I follow the simple Normal distribution, the analytical distribution of R is not easy to be obtained. While the simulation approach is very simple. We can just draw random number as U_i, I_i and obtain a sample of R by U_i/I_i. Repeat the procedure, we can obtain the empirical distribution of the R.

Remark Simulation approach needs relatively simple mathematical background but with heavy computational cost. The uncertainty in the results is also concerning, despite the uncertainty is small and controllable for most cases. For practical purpose, when the analytical approach is available, we recommend using the analytical approach since analytical method can be more accurate and more computationally efficient. Moreover, in this Chapter, some advanced statistics methods and algorithms are introduced. One can avoid some of them if it is unnecessary. For the simulation algorithms described in this Chapter, more detailed discussion can be found in reference [1].

Appendix 10A: A Revisit of Delta Method

The delta method is a simple method to estimate the distribution of a complex continuous function. For a univariate variable x_n, given its mean is μ and variance as $Var(x)$, suppose the x_n following the central limit theorem

$$(x_n - \mu) \sim N(0, Var(x_n)). \tag{10A.1}$$

For a continuous function $g(.)$ and has at least the 2nd continuous derivative, the following equation follows

$$g(x_n) - g(\mu) \sim N\left(0, |g'(\mu)|^2 Var(x_n)\right) \tag{10A.2}$$

From the (10A.1) and (10A.2), we can derive the variance of $g(x)$

$$Var(g(x_n)) = |g'(\mu)|^2 Var(x_n) \tag{10A.3}$$

Proof of (10A.2):

According to Taylor expansion, the continuous function $g(x)$ is expanded at point μ as follows

$$g(x_n) \approx g(\mu) + g'(\theta)(x_n - \mu) \tag{10A.4}$$

Rearrange the (10A.4) as

$$g(x) - g(\mu) \approx g'(\mu)(x - \mu)$$

Since the $(x_n - \mu)$~$N(0, Var(x_n))$ and the $g^{'}(\mu)$ is a constant, it is straightforward to obtain

$$g(x) - g(\mu) \sim N\left(0, |g'(\mu)|^2 Var(x_n)\right)$$

For the multivariate delta method, we can use the multivariate central limit theorem. It is not hard to obtain by using the multivariate Taylor expansion as follows

$$g(y_1, y_2, y_3, \ldots y_m) - g(\mu_1, \mu_2, \mu_3, \ldots \mu_m) \sim N\left(0, \nabla \mathrm{g}.\boldsymbol{\Sigma}.(\nabla \mathrm{g})'\right) \qquad (10A.5)$$

$\nabla \mathrm{g} = \left(\frac{\partial g}{\partial y_1}, \frac{\partial g}{\partial y_2}, \frac{\partial g}{\partial y_3}, \ldots \frac{\partial g}{\partial y_m}\right)_{m \times 1}$; $\boldsymbol{\Sigma}$ is covariance with element i, j as Cov(y_i, y_j). If the multi random variable y_1, y_2, y_3, $\ldots y_m$ are independent from each other, then the following has been obtained.

$$Var(g(y_1, y_2, y_3, \ldots y_m)) \sim \sum_{i=1}^{m} \left|\frac{\partial g(\mu_i)}{\partial y_i}\right|^2 Var(y_i) \qquad (10A.6)$$

Problems and Questions

1. Calculate the least number of iterations to simulate the probability 0.99, given a standard deviation as maximal 0.001.
2. Use the inverse transform method to draw 1000 random numbers for an Exponential distribution with unity failure rate. Compute the mean and standard deviation for the 1000 random numbers.
3. A mixed Weibull distribution is written as

$$f(t) = \frac{1}{3}\frac{\beta_1}{\alpha_1}\left(\frac{t}{\alpha_1}\right)^{\beta_1 - 1} e^{-(t/\alpha_1)^{\beta_1}} + \frac{2}{3}\frac{\beta_2}{\alpha_2}\left(\frac{t}{\alpha_2}\right)^{\beta_2 - 1} e^{-(t/\alpha_2)^{\beta_2}}$$

$$\alpha_1 = 2; \beta_1 = 1$$

$$\alpha_2 = 3; \beta_2 = 2$$

Use the Acceptance-Rejection method to generate random numbers between [0,1] for this mixed distribution, selecting the Uniform distribution as envelope distribution.

4. For discrete Markov chain, the transition matrix is as follows

$$p = \begin{pmatrix} p_{AA} & p_{AB} & p_{AC} \\ p_{BA} & p_{BB} & p_{BC} \\ p_{CA} & p_{CB} & p_{CC} \end{pmatrix} = \begin{pmatrix} 0.25 & 0.5 & 0.25 \\ 0 & 0.25 & 0.75 \\ 0.2 & 0.3 & 0.5 \end{pmatrix}$$

For two $q_0 = (0.33, 0.33, 0.34)$ and $q_0 = (0.2, 0.4, 0.4)$, compute the limit probability for $n = 10{,}000$. Can you achieve equal results?

5. For a reliability system with RBD as Fig. 10.39 and reliability for each component is equal as 0.9.

Use simulation method to compute the system reliability and its std. error of the simulation is limited to be less than 0.05.

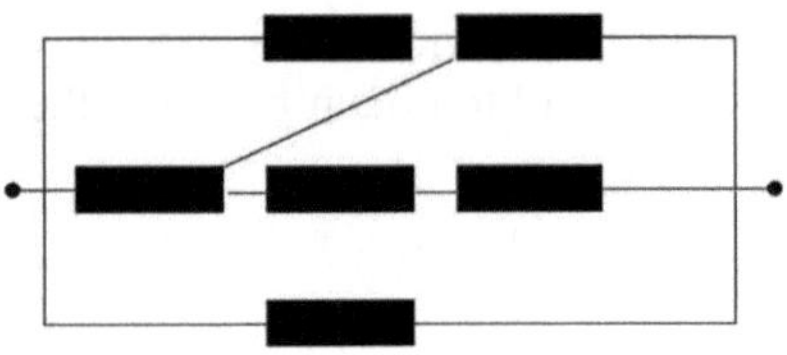

Fig. 10.39 A RBD network

References

1. Robert CP, Casella G (2004) Monte Carlo statistical methods. Springer, New York
2. Matsumoto M, Nishimura T (1998) A 623-dimensionally equidistributed uniform pseudo-random number generator. ACM Trans Model Comput Simul 8(1):3–30
3. Johnson RA, Wichern DW (2003) Applied multivariate statistical analysis. Pearson Prentice Hall, Upper Saddle River, NJ
4. Roy V (2020) Convergence diagnostics for Markov chain Monte Carlo. Ann Rev Stat Appl 7: 387–412
5. Box GEP, Jenkins GM, Reinsel GC, Ljung GM (2015) Time series analysis: forecasting and control. Willey

Chapter 11
Reliability Physics

Failure is undesirable in engineering. Explaining the root causes of failure and predicting them through mathematical models based on physical principles provides a convincing and solid basis for decision making. Using physics to explain failure can help overcome the shortcomings of life data analysis, which is sometimes criticized for lacking a solid physical foundation. In general, engineering systems consist of mechanical, electrical, electronic, and software components. This chapter addresses fundamental physical principles related only to mechanical and electronic systems, while omitting software, as it is not governed by physical laws. The concepts discussed in this chapter may be less familiar to some engineers and may not directly explain specific failures, but they can enhance the understanding of the fundamental nature of failures.

11.1 Mechanical Failure Physics

The mechanical properties of material are closely related to its atomic structure at the microscopic level. The constituent atoms in material interact with each other, and when subjected to external forces, these atoms respond to the external force in some manners. At the macroscopic level, this response is reflected in the material's mechanical properties.

11.1.1 Atomic Interaction

We can start with a highly simplified scenario. In a solid, the constituent atoms follow specific structures in space. These atoms interact with one another, behaving similarly to springs as following Fig. 11.1.

F. Yuan, *Essentials of Reliability Engineering*, Springer Series in Reliability Engineering, https://doi.org/10.1007/978-3-032-15544-3_11

Fig. 11.1 Atomic interaction

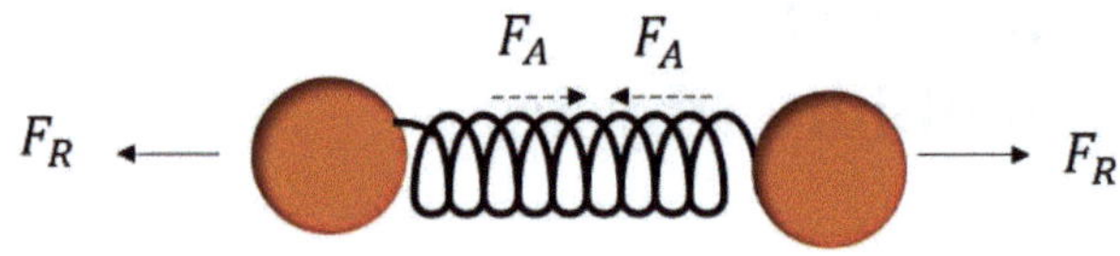

The atom constituting the solid subjects to both attractive F_A and repulsive force F_R. The net force is

$$F_{Net} = F_A + F_R \tag{11.1}$$

The force is directional. If we define the F_A as positive, the F_R is then opposite as negative. The net force F_{Net} can be zero at some positions. The attractive force plays the critical role to bond the atom together. The attractive force originates from the electrostatic of the positive charges of the nuclei and the negative charges of the electrons. The two opposite charges attract each other. The repulsive force is complex and can be explained by quantum physics in some ways. When two atoms approach closer, the electrons, which locate outside of the nuclei will overlap. But the capacity of the space to hold the number of electrons is limited, repulsive force is to resist this overlapping. The outermost electrons of the atom are the main participant in the constitution of the solid and those electrons are called **valance electrons**. Atoms are held together by a balance of attractive and repulsive forces, which maintain a certain equilibrium distance between them.

Suppose the distance between the two atoms is r. Two functions that can represent the two forces are as [1, 2]

$$F_A = \frac{mA}{r^{m+1}}; F_R = -\frac{nB}{r^{n+1}} \tag{11.2}$$

The net force F_{Net} is shown as in Fig. 11.2, where we can see there is a point with $F_{Net} = 0$, which is the equilibrium, corresponding to the interatomic distance r_0, which is considered as the distance between the two atoms. For some atoms, the r_0 is around 0.3 nm [3]. Atomic forces play a key role in responding to external mechanical forces and determining mechanical properties such as tensile strength and yield strength, which will be discussed later in this chapter.

In analysis of atomic interaction, we can use potential energy instead of the force for analysis. The potential energy for mechanical force F is of unit ($N.\ m$) and for the atomic system, at position r, potential energy is defined as [3]

$$E(r) = \int_{+\infty}^{r} F dr \tag{11.3}$$

And the potential energy at infinite distance is defined as 0 for atomic interaction, which is $\underset{r \to +\infty}{E(r)} = 0$. From Eq. (11.3), we can obtain

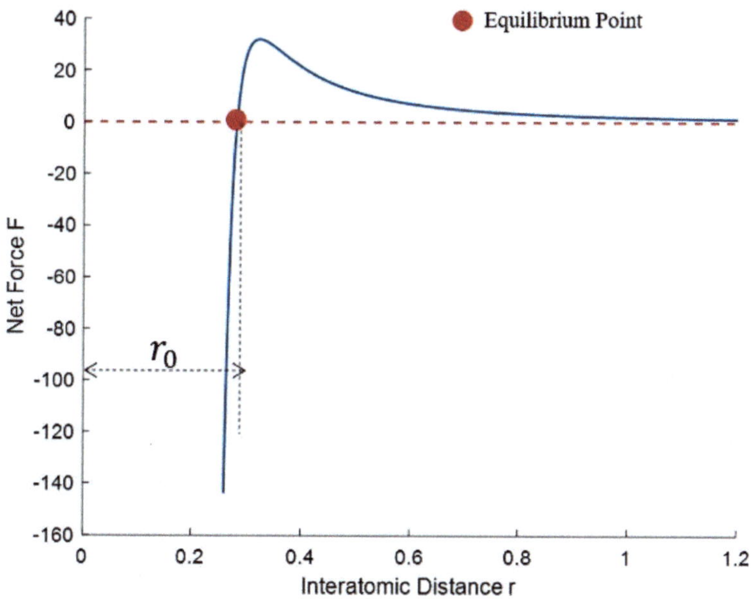

Fig. 11.2 Net force of atomic structure

Fig. 11.3 Potential energy

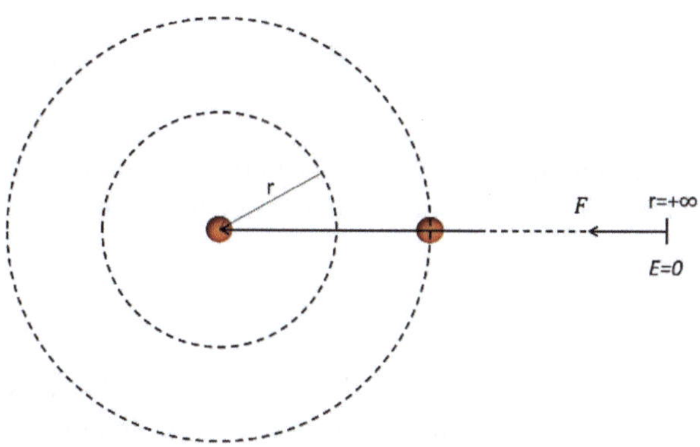

$$F = \frac{dE(r)}{dr} \tag{11.4}$$

Figure 11.3 illustrates the potential energy for atomic interaction. The force F is directed to the center, which is the direction of attractive force. The potential energy can be calculated by the above Eq. (11.4).

For example, if the attractive force is $F = \frac{10}{r^2}$, according to Eq. (11.4), the potential energy at point r_0 is then

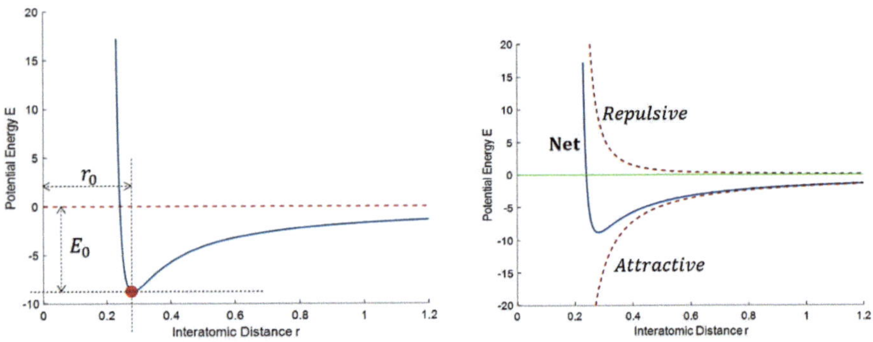

Fig. 11.4 Atomic potential energy

Table 11.1 Bond energy [3]

Materials	Bond Energy (kJ/mol)	Melting Temperature (°C)
NaCl	640	801
MgO	1000	2800
Si	450	1410
Al	330	660

$$E(r) = \int_{+\infty}^{r_0} F dr = \int_{+\infty}^{r_0} \frac{10}{r^2} dr = -\frac{10}{r}\Big|_{+\infty}^{r_0} = -\frac{10}{r_0} < 0 \tag{11.5}$$

The potential energy can be negative, depending on the definition of 0 potential energy. A widely used formula of potential energy for atomic bonds is as follows

$$E_{Net} = E_A + E_R = -\frac{A}{r^m} + \frac{B}{r^n}; n > m \tag{11.6}$$

where $E_A = -\frac{A}{r^m}; E_R = \frac{B}{r^n}$. When $r \rightarrow +\infty$, the potential energy of Eq. (11.5) is 0, agreeing with the 0 potential energy definition. The lowest energy E_{Net} in Eq. (11.6) is achieved at the point of equilibrium, meaning the point of net force 0 is the point with lowest potential energy E_0. This is why this energy E_0 is considered as the bond energy for material.

A plot of potential energy as shown in Fig. 11.4, where it can be seen the deepest potential energy obtained at r_0, which is the bond energy of the two atoms. It is obvious that deeper E_0 implying a stronger bond.

Table 11.1 lists bond energy of some substances [3].

Example 11.1
Suppose the total potential energy between two ions be the sum of attractive and repulsive energies E_A and E_R as

$$E_{Net} = E_A + E_R = -\frac{A}{r} + \frac{B}{r^n}$$

a. Find out the equilibrium distance r_0 and the bond energy E_0 given the A, B and n are known. The bond energy is the lowest energy of E_{Net}.
b. For a K^+-Cl^- ion pair, it has been known the attractive and repulsive energies E_A and E_R (in eV) (r is in nm) are given by [3]

$$E_A = -\frac{1.436}{r}; \quad E_R = \frac{5.86 \times 10^{-6}}{r^9}$$

Find the equilibrium distance r_0 and bonding energy E_0.

Solution:

a. Lowest energy point E_0 calculation:

$$E_{net} = -\frac{A}{r} + \frac{B}{r^n};$$

Find the r_0 by taking the first derivative of E_{net} as follows

$$\frac{\partial E_{net}}{\partial r} = \frac{A}{r^2} + (-n)Br^{-n-1};$$

We obtain $r^{1-n} = \frac{A}{nB}$. Thus

$$r_0 = \left(\frac{A}{nB}\right)^{\left(\frac{1}{1-n}\right)};$$

The corresponding lowest energy is

$$E_0 = -\frac{A}{r_0} + \frac{B}{r_0{}^n};$$

b. In this case, the $A = 1.436$; $B = 5.86 \times 10^{-6}$; $n = 9$. The

(continued)

Example 11.1 (continued)

$$r_0 = \left(\frac{A}{nB}\right)^{\left(\frac{1}{1-n}\right)} = \left(\frac{1.436}{9 \times 5.86 \times 10^{-6}}\right)^{\left(\frac{1}{1-9}\right)} = 0.2790 \text{ nm}$$

The bond energy

$$E_0 = -\frac{A}{r_0} + \frac{B}{r_0{}^n} = -\frac{1.436}{0.2790} + \frac{5.86 \times 10^{-6}}{0.2790^9} = -4.57 \text{ eV}$$

11.1.2 Linear Region of Atomic Interaction

For an elastic spring, meaning the spring can be fully recovered after the external force is released. The force F and its incremental Δx is linear as

$$F = k\Delta x \tag{11.7}$$

For atomic interaction, the incremental $\Delta x = r - r_0$, where r_0 is the equilibrium spacing. The net force F and the incremental Δx is approximately linear only in a

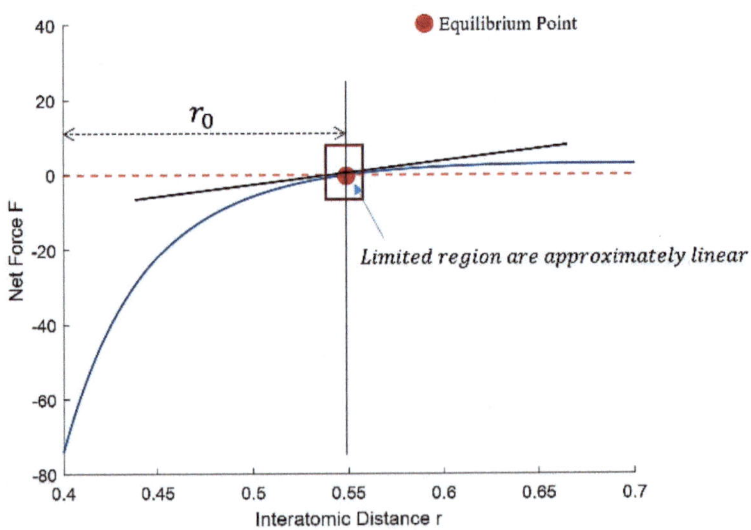

Fig. 11.5 Limited region are linear

region near the r_0, as demonstrated in Fig. 11.5. Outside of this valid region, linearity is obviously violated. The bonding, when upon the application of external force and the net force that is out of the valid region, would be damaged. At macroscopic level, the material would deform or fail as it cannot recover fully.

11.1.3 Defect in Solid Structure

The previous sections have discussed the two atomic interactions, which have been simplified. The real solid is much more complex, and some kinds of solid material are present in crystalline. A crystalline material contains atoms that are periodic and regular. The atoms are bounded by the force discussed in the previous sections. Figure 11.6 exhibits three examples of crystalline type which is the structure of some solids [4]. Aluminum and coper metal are face-centered structure. Tantalum is body centered cubic.

Highly relevant to reliability, the defect in the solid is relevant to many failures. In general, the defect in solid can classified as point defect and line defect. Point defect is an imperfection of the crystalline containing vacancy or extra atom occupying interstitial site, as shown in Fig. 11.7.

The number of vacancies can be approximated by a formula as [1, 3]

$$N_v = N.\exp\left(-\frac{Q}{kT}\right) \tag{11.8}$$

The N_v is the number of vacancies. The Q is activation energy. The N is total number of atomic sites. The k is Boltzmann's constant which is given by $8.62 \times 10^{-5} eV/atom.\ K$. The T is temperature in unit of Kelvin. The form of Eq. (11.8) shows in a

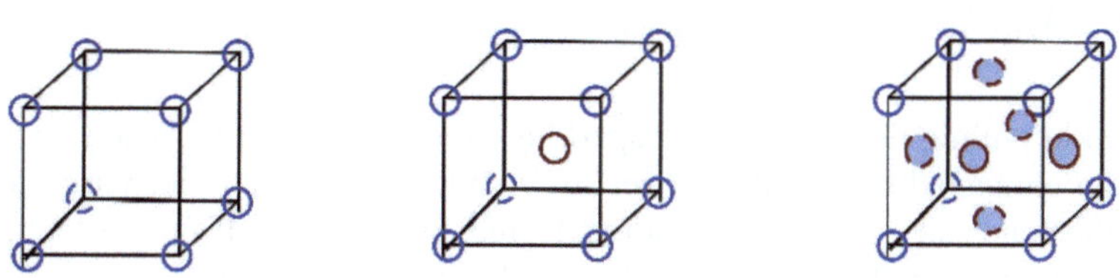

Fig. 11.6 Simple cubic crystalline, body centered cubic crystalline, and face centered cubic crystalline

Fig. 11.7 Point defect in a structure

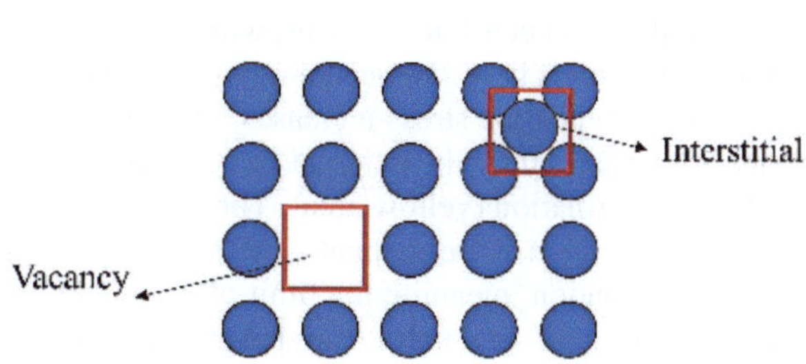

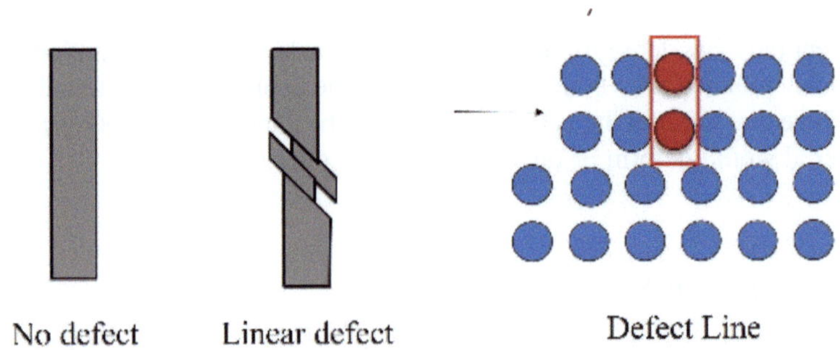

Fig. 11.8 Linear defect in a structure

form of $\exp\left(-\frac{Q}{kT}\right)$ which is very common in failure physics models related to temperature. Some failure models are derived from the model Eq. (11.8).

Linear defects illustrate the defect presents in a line, for example a line of atoms in the structure are positioning in the wrong place. Figure 11.8 demonstrates a linear defect of a rod, left one is perfect, right one contains linear defect. The rod has dislocated in a specific direction and all the atoms in the direction are mismatched, suggesting defects. The right most figure in Fig. 11.8 illustrates the defect line marked within square area at atomic level in two dimensions. It can be seen the atoms in the defect line mismatches (presenting as extra) with their nearby atoms. The linear defect, named dislocation in material science, plays a critical role in the strength of the materials.

When external force is applied on the material, as discussed in Sect. 11.1.2, when its magnitude reaches some limit, the bonding will be damaged and the point defect, linear defect might occur. Some mechanical failures are induced due to these defects.

11.1.4 Mechanical Properties

Image a kid pulling the rubber band apart, as shown in Fig. 11.9. When his force is weak, the elongated band will recover fully. When the force is increased to some extent, the band can still recover to some degree, but not fully. If further let the force be larger enough, the band will stretch to its limit and then break.

Analogous to the rubber band, the solid shows three stages, in terms of stress and strain. In the first stage, the applied stress (force) will elongate or compresses the solid, but the solid can fully recover, which is termed as elastic deformation. At that moment, the bonds have not been destroyed, and it is in the elastic deformation stage (green area). When the stress increases, the solid then cannot recover fully, as some bonds are destroyed by the external force permanently. This phenomenon is termed as plastic deformation (yellow area). The elastic deformation is temporary, while the plastic deformation is permanent. The stress between the elastic deformation and plastic deformation, meaning the limit of elastic deformation is **yield strength**. The maximal stress of the solid can tolerate against breaking is **tensile strength**.

Fig. 11.9 Rubber band can respond to external force elastically

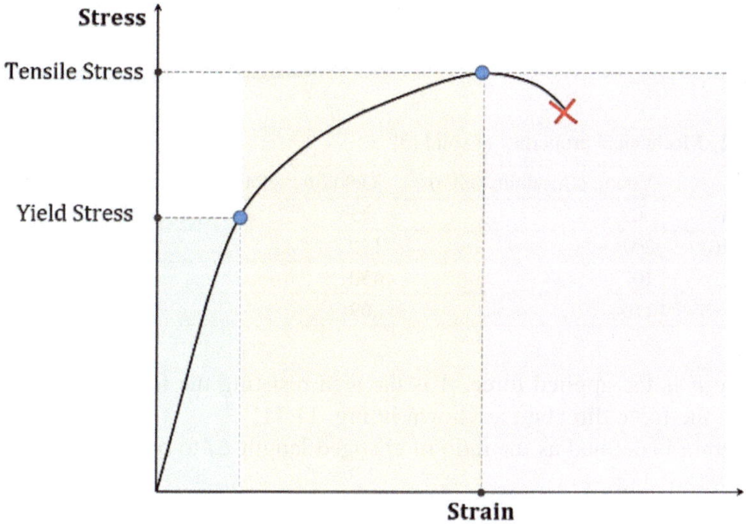

Fig. 11.10 Stress-strain curve of a typical solid

Figure 11.10 shows a stress-strain curve for typical solid material, where the yield strength is the limit of the linear part of the curve, the highest point of the curve is the tensile strength [3].

We have not yet defined stress and strain, which are two fundamental terms in mechanics. The stress, normally denoted by σ is defined as

$$\sigma = \frac{F}{A} \tag{11.9}$$

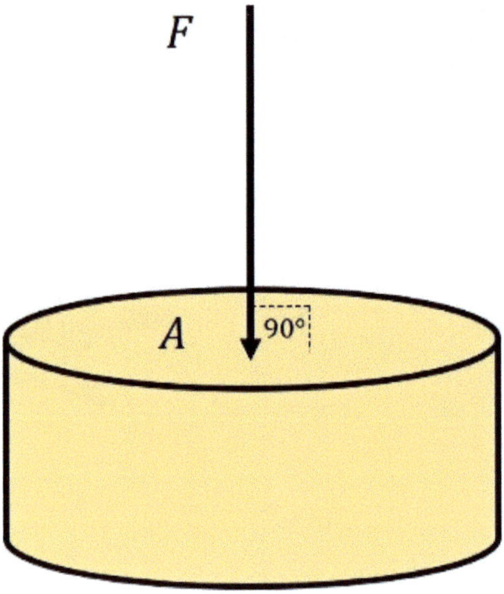

Fig. 11.11 Stress definition

Table 11.2 Mechanical properties of solid [3]

Materials	Young's Modulus (GPa)	Yield Strength(MPa)	Tensile Strength (MPa)
Aluminum	69	35	90
Steel (1020)	207	180	380
Titanium	107	450	520
Copper	110	69	200

where the F is the applied force, A is the area resisting the force which is perpendicular to the force direction as shown in Fig. 11.11.

The strain is defined as the ratio of changed length Δl to original length l.

$$\epsilon = \frac{\Delta l}{l} \tag{11.10}$$

From physical perspective, stress and strain are defined from macroscopic level. The elastic deformation is equivalent to the linear relationship between stress and strain, implying the

$$\sigma = E\epsilon \tag{11.11}$$

The E, here is the slope of the elastic line in Fig. 11.10 is defined as the **Young's modulus**, or **modulus of elasticity**. The Young's modulus is normally a value of great magnitude, some of them shown in Table 11.2.

Example 11.2
We know a steel rod is originally 100 mm long. The tensile stress on it is 200 MPa. The steel rod has been extended to 100.5 mm after applying stress. What is the Young's Modulus of the steel.

Solution: The $\Delta l = 100.5 - 100 = 0.5mm$. The $\epsilon = \frac{0.5mm}{100mm} = 0.005$. Since $\sigma = E\epsilon$, the Young's modulus is then

$$E = \frac{\sigma}{\epsilon} = \frac{200MPa}{0.005} = 4.0 \times 10^4 MPa = 40GPa$$

11.2 Fatigue Failure

Fatigue failure is a common failure mechanism in mechanical engineering. It has been claimed fatigue contributes to 90% of metallic failure and 50% of mechanical failure [3, 5].

11.2.1 Cyclic Stress

The fatigue occurs at an applied stress that is even lower than the tensile stress. The stress applied on the material is cyclic with alternating compressive and tensile stress, as shown in Fig. 11.12, where we can see the stress is not constant but cyclic.

The stress range $\Delta\sigma$ is defined as [3]

$$\Delta\sigma = \sigma_{max} - \sigma_{min} \tag{11.12}$$

11.2.2 Physics of Fatigue

Fatigue failure is induced by the crack growing in weak regions of the material, such as the cavity, notch, damaged surface. The fatigue is generally subjected to three phases until failure [3]:

- I: Crack initiation.
- II: Crack growing.
- III: Ultimate failure.

The crack initiation can be examined from microstructural perspective. Suppose the microstructure of the material is organized as a perfect cubic as shown in Fig. 11.13. The material is subjected to cyclic tensile and compressive stress. During each cycle, the tensile stress stretches the structure as b in Fig. 11.13, then compressive stress recovers the structure to the original fully as shown in Figure c of Fig. 11.13.

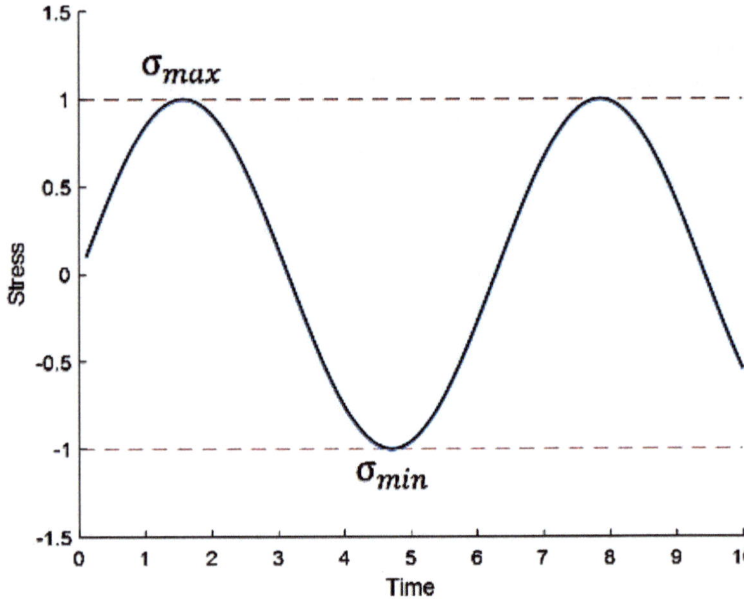

Fig. 11.12 Stress cycle

Fig. 11.13 Cubic structure material

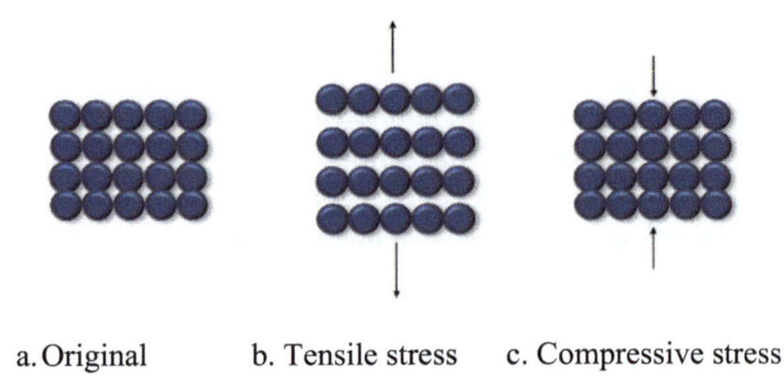

However, when imperfection is present in the structure, as shown in Fig. 11.14 where vacancies are present. After one cycle of tensile and compressive stress, the structure cannot fully recover to its original. The imperfection can be of point-defect with vacancies in the material, or dislocation, which is a linear defect that causes fatigue failure. The edge and screw dislocation are two basic types of linear defect present in the material that causes the microstructural change, and the change cannot fully recover.

Once the crack has initialized, the material is then more vulnerable to failure. At this stage, the slip band grows significantly after each cycle. The slip band is the area where defect solid slides. The slide usually occurs in a favored direction which is the easiest direction to slide. Nevertheless, at this stage, the crack growth rate is still stable and linear. When the crack grows to certain level, e.g. a clear cleavage appears, the final failure will come in a sudden. This final stage is relatively in a shorter period [3]. Figure 11.15 illustrates the fatigue failure phases.

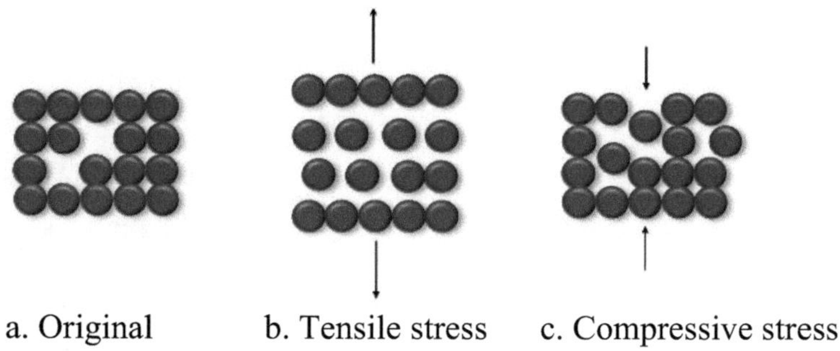

Fig. 11.14 Imperfect cubic structure material

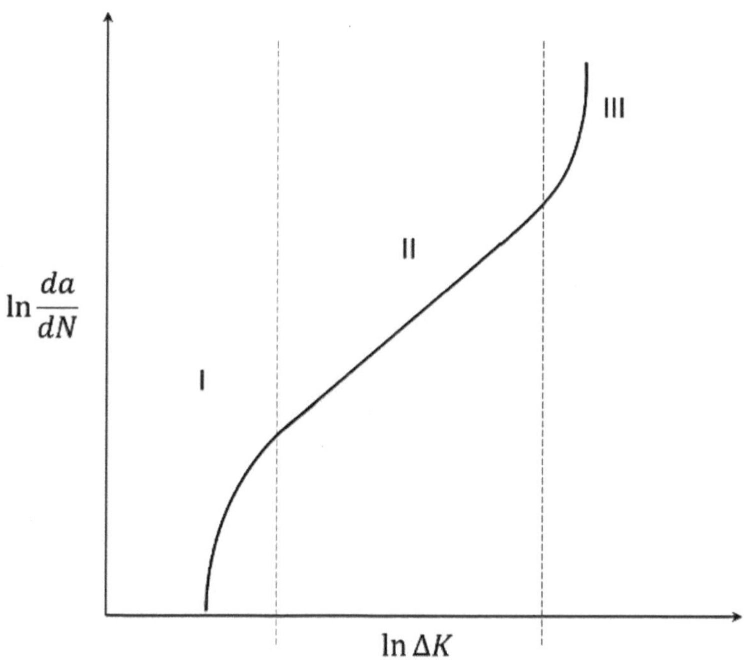

Fig. 11.15 The phase of fatigue failure development

Stress Amplification

Cyclic stress applied to the material induces voids, openings, notches, etc., where stress can be amplified. For a small openness as shown in Fig. 11.16, even when the force F is small, if the A is small, the σ can be very big. This is why the needle can penetrate an object even using a small magnitude of force. If the amplified stress reaches some limit to which the material cannot withstand, the crack then initializes [3].

Fig. 11.16 Stress amplified

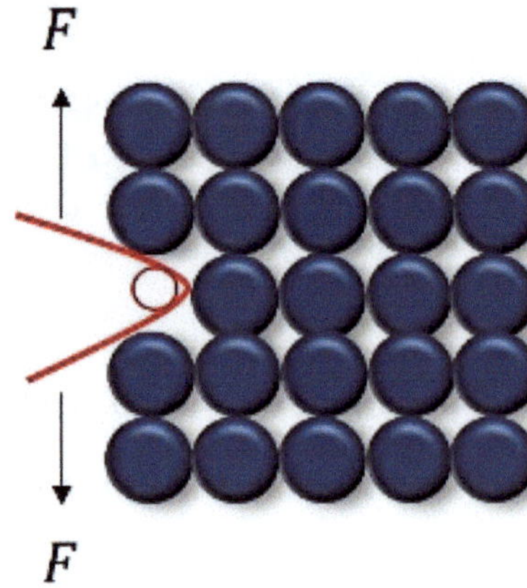

Fig. 11.17 Crack in solid material

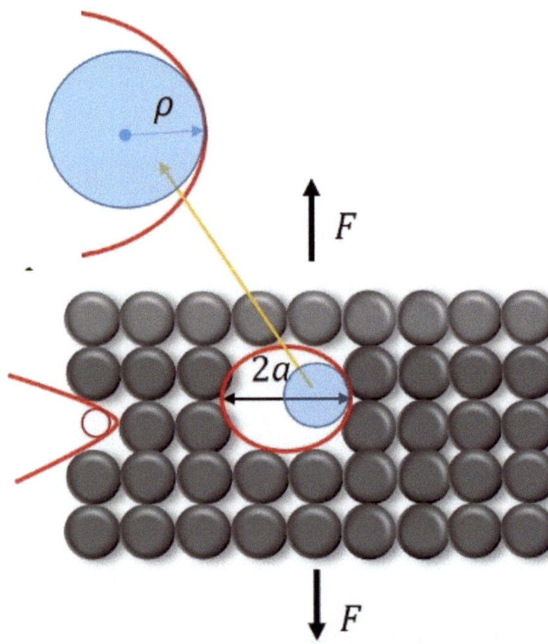

Normally, a crack is concentrated at the crack tip, which is the sharpest point, as shown in Fig. 11.17 where the crack has already initialized.

Given ρ as the radius of the crack tip, a as the crack size. σ_0 as the applied external stress. The stress at the crack tip is then given by [3]

$$\sigma_m = 2\sigma_0 \left(\frac{a}{\rho}\right)^{1/2} \tag{11.12}$$

The stress concentration factor K, the magnitude of the stress has been amplified, is given by [3]

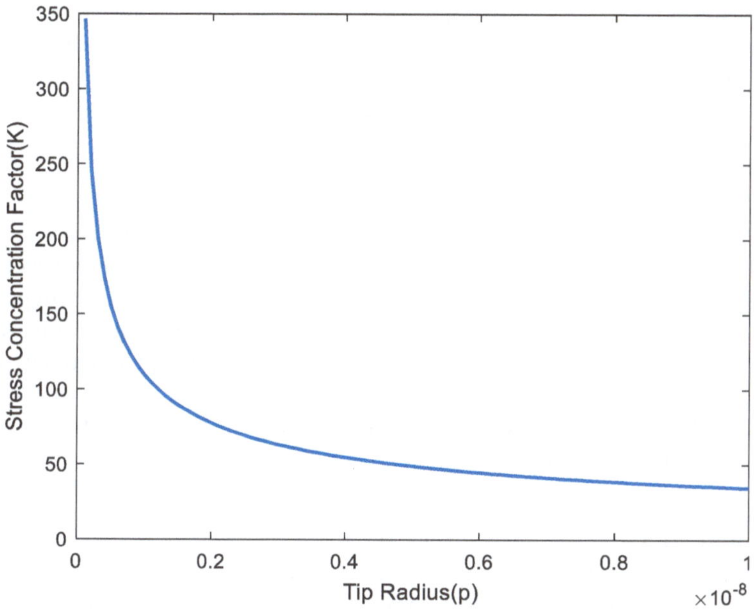

Fig. 11.18 Stress concentration decreases with crack tip radius

$$K_t = \frac{\sigma_m}{\sigma_0} = 2\left(\frac{a}{\rho}\right)^{1/2} \tag{11.13}$$

Figure 11.18 plots the stress concentration factor decreasing with the tip radius ρ at crack size fixed at $a = 3 \times 10^{-6}$.

Example 11.3
A defective brittle material contains a crack size $a = 5 \times 10^{-6}$m on the surface. The radius of the tip is 5×10^{-10}m. Suppose the external tensile stress applied on the material is 100 MPa. Evaluate stress concentration at the crack tip.

Solution: According to Formula (11.13), the stress concentration factor is given by

$$K_t = 2\left(\frac{a}{\rho}\right)^{1/2} = 2 \times \left(\frac{5 \times 10^{-6}\text{m}}{5 \times 10^{-10}\text{m}}\right)^{1/2} = 200$$

The concentration factor K as expected to be dimensionless as 200. The actual tip stress is 200 times more than the applied external stress. The stress at the tip is then

$$200^{*} 100 \text{ MPa} = 20{,}000 \text{ MPa}$$

Critical Stress

All material has a limit of maximal stress that will lead to fracture. For the crack tip, a critical stress at which the material will fracture is given by [3]

$$\sigma_c = \left(\frac{2E\gamma}{\pi a}\right)^{1/2} \tag{11.14}$$

where E is modulus of elasticity, γ is the surface energy. The surface energy is a concept related to the microstructure of the material and is positively correlated to the strength of atomic bond. Macroscopically, high surface energy suggests the surface is hard. For example, the surface energy for liquid is low, but for hard solids such as glasses, hard metal is high. The surface energy can be quantitatively defined as the energy required to increase the area size of the surface, where we can easily understand the practical implication. The crack, a procedure to create bigger surface, certainly depends on the surface energy.

Example 11.4

A hard material is with elasticity modulus 60 GPa and surface energy 0.6 J/m^2. Suppose the external tensile stress applied on the material is 50 MPa. Evaluate the maximal surface flaw that the material can be free of crack.

Solution: Rearrange Eq. (11.14) to let

$$a = \frac{2E\gamma}{\pi {\sigma_c}^2}$$

The $E = 60GPa$, $\gamma = 0.6$ J/m^2, σ_c is specified as 50 MPa.

$$a = \frac{2E\gamma}{\pi {\sigma_c}^2} = \frac{2 \times (60 \times 10^9 N/m^2) \times (0.6N/m)}{\pi \times (50 \times 10^6 N/m^2)^2} = 9.2 \times 10^{-6}\text{m}$$

11.2.3 Fracture Toughness

In material science, a measure to represent the toughness of the material to resist the fracture when crack is present is defined as [3, 6]

$$K_c = \sqrt{2E(\gamma + \gamma_p)} \tag{11.15}$$

For highly brittle material where no significant plastic deformation present, the toughness is simplified as [3]

Table 11.3 Toughness of some materials [3]

Materials	$K_c(\text{MPa}\sqrt{m})$
Alloy of Aluminum	24–44
Alloy of Steel	50–88
Alloy of Titanium	55
Other Common Metals: Cu, Ni, Ag	100–350

$$K_c = \sqrt{2E\gamma} \tag{11.16}$$

The K_c is defined from microscopic perspective. Since $\sigma_c = \left(\frac{2E\gamma}{\pi a}\right)^{1/2}$ where the σ_c is measurable from macroscopic perspective, it is not hard to define the fracture toughness as follows [6].

$$K_c = \sigma_c \sqrt{\pi a}$$

The toughness of some typical materials is shown in Table 11.3.

11.2.4 Fatigue Life Model

Once the crack is initialized, the crack will propagate exponentially. A simple but widely applied model regarding fatigue is the Paris-Erdogan fatigue model. This model assumes the crack growth rate $\frac{da}{dN}$ depends on a parameter named stress intensity factor defined as [3]

$$K_I = \sigma \sqrt{\pi a} \tag{11.17}$$

The Paris-Erdogan crack growth *da* per cycle *dN* is written as [3, 6]

$$\frac{da}{dN} = C(\Delta K_I)^m \tag{11.18}$$

where $\Delta K_I = (\sigma_{max} - \sigma_{min})\sqrt{\pi a} = \Delta\sigma\sqrt{\pi a}$. The a is the size of the openness, the N is the number of cycles, $\Delta\sigma$ is the applied cyclic load, the m is a constant with a typical value between 1 and 6, and C is a constant coefficient [3]. The Paris-Erdogan model under the stress range is written as

$$\frac{da}{dN} = C\left(\Delta\sigma\sqrt{\pi a}\right)^m \tag{11.19}$$

Formula (11.19) suggests the growth of openness for each cycle is not constant but increasing with the size of a. In another word, the openness grows faster and faster with number of cycles. Figure 11.19 shows the growing rate with the openness a,

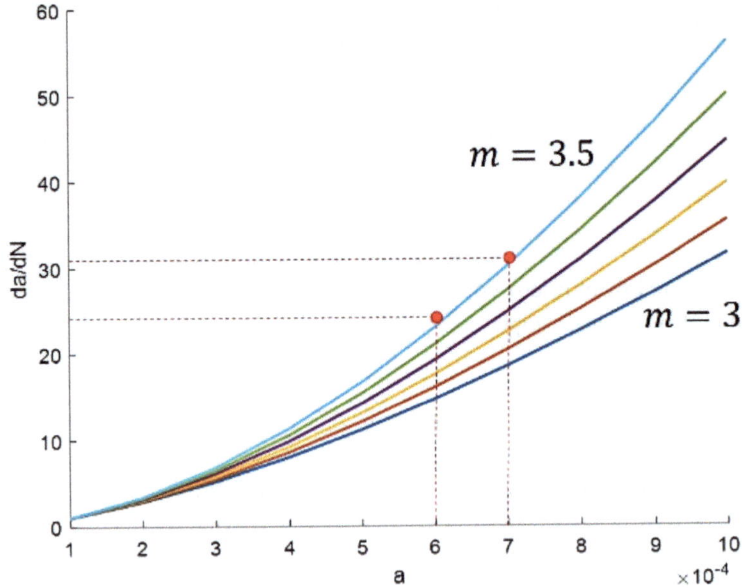

Fig. 11.19 Openness growing rate

where for example, at the $a = 6 \times 10^{-4}$, the $\frac{da}{dN} \approx 24$ per cycle, while for $a = 7 \times 10^{-4}$, the $\frac{da}{dN} \approx 31$ per cycle.

Rewrite the Formula (11.19),

$$dN = \frac{1}{C}(\Delta\sigma)^{-m}(\sqrt{\pi a})^{-m}da \tag{11.20}$$

Take integral of both sides of (11.20). We obtain

$$N_{life} = \left[\int_{a_0}^{a_{Failed}} \frac{1}{C}(\Delta\sigma)^{-m}(\sqrt{\pi a})^{-m}da\right](\Delta\sigma)^{-m} \tag{11.21}$$

For simplicity, Eq. (11.21) is rewritten as

$$N_{life} = A_0(\Delta\sigma)^{-m} \tag{11.22}$$

In case where the area subject to the stress is fixed, meaning in the $\sigma = \frac{F}{A}$, the A is constant, Eq. (11.22) is then rewritten to be a function of force F instead of stress σ, i.e.

$$N_{life} = A_0(\Delta F)^{-m} \tag{11.23}$$

where the parameter area A has been absorbed into the constant A_0. Equation (11.23) is referred to as Coffin-Manson model [5].

11.2.5 Data Analysis of Fatigue Life

In this section, we assume the measured data at macroscopical level is the total number of cycles experienced, and the stress is the stress difference $\Delta\sigma$. The purpose of the data analysis is to find the parameters of A_0 and m. For the convenience of the estimation, as many other data analysis methods do, we linearize Eq. (11.22) by taking logarithm on both sides of Eq. (11.22) as follows

$$\ln N = m \ln\Delta\sigma + \ln A_0 \tag{11.24}$$

which is of linear form of $y = ax + b$.

Given the observed data is N_i and $S_i = \Delta\sigma_i$, the corresponding x_i, y_i is as shown in Table 11.4.

Perform the least squares estimation method, we can obtain unbiased estimates of $\widehat{a}$, $\widehat{b}$. The corresponding

$$m = -\widehat{a}; \quad A_0 = e^{\widehat{b}} \tag{11.25}$$

Refer to Sect. 3.3.4 in Chap. 3 for the least squares estimation. The $\widehat{a}$ corresponds to $\widehat{\beta}_j$ with $j = 1$. The $\widehat{b}$ corresponds to $\widehat{\beta}_j$ with $j = 2$. The confidence interval is

$$\widehat{\beta}_j \pm t_{\frac{\alpha}{2},n-p} S_n \sqrt{D_{jj}} \tag{3.62}$$

Please refer to Sect. 3.3.4 in Chap. 3 for the S_n and D_{jj}. The n is the number of data sets. The $p = 2$ for this case. The confidence interval for m is then

$$-\widehat{\beta}_1 \pm t_{\frac{\alpha}{2},n-p} S_n \sqrt{D_{11}} \tag{11.26}$$

For A_0, the corresponding CI is

Table 11.4 Data transformation

$S_i = \Delta\sigma_i$	N_i	$y_i = \ln N_i$	$x_i = \ln \Delta\sigma_i$
10 MPa	50,000	10.8198	2.2306
9 MPa	56,000	10.9331	2.1972
…	…	…	…

$$\left[e^{\widehat{\beta_2}-t_{\frac{\alpha}{2},n-p}S_n\sqrt{D_{22}}}, e^{\widehat{\beta_2}+t_{\frac{\alpha}{2},n-p}S_n\sqrt{D_{22}}}\right] \quad (11.27)$$

For linear regression, the $\widehat{y}$ value for observed x_0 is $y=\widehat{a}x_0+\widehat{b}$. The confidence interval for the $\widehat{y}$ is

$$\widehat{y}(\boldsymbol{x_0}) \pm t_{\frac{\alpha}{2},n-p}S_n\sqrt{1+\boldsymbol{x_0}(X'X)^{-1}\boldsymbol{x_0}'} \quad (3.61)$$

The X in this case is the row vector $(\ln \boldsymbol{S}\ \boldsymbol{1})_{n\times 2}$ where $\boldsymbol{S}$ is the first column in Table 11A.1. Since $y=\ln N$, the confidence interval for the N of $\boldsymbol{x_0}=(\ln S_0\ 1)$ given the stress level S_0 is then

$$\left[e^{\widehat{y}(\boldsymbol{x_0})-t_{\frac{\alpha}{2},n-p}S_n\sqrt{1+\boldsymbol{x_0}(X'X)^{-1}\boldsymbol{x_0}'}}, e^{\widehat{y}(\boldsymbol{x_0})+t_{\frac{\alpha}{2},n-p}S_n\sqrt{1+\boldsymbol{x_0}(X'X)^{-1}\boldsymbol{x_0}'}}\right] \quad (11.28)$$

Note that the $(X'X)^{-1}=D$ referring to Sect. 3.3.4 in Chap. 3.

11.2.6 Fatigue Life Test Instrument

The fatigue life of hard metals is usually long under normal usage. To measure fatigue life, experiments must be conducted using a special machine that can shorten the test period. Figure 11.20 shows a fatigue strength test machine, WP 140, developed by Gunt in Germany. The load force generated by this type of machine can range from 0 to 300 N, with the corresponding alternating stress ranging from 0 to $600 N/mm^2$.

The test machine requires special specimen for the fatigue test. The force and the corresponding stress for the special specimen are numerically related as

$$\sigma=2F \quad (11.29)$$

Table 11.5 illustrates the test results as an example.

11.2.7 Example Case

In this example, we tested 21 metal samples, and the original results are tabulated in Table 11A.1. To estimate the parameter in A_0, m in equation $N=A_0(\Delta\sigma)^{-m}$, we transformed the data for S and N by taking the logarithm. The results are shown in Table 11.6.

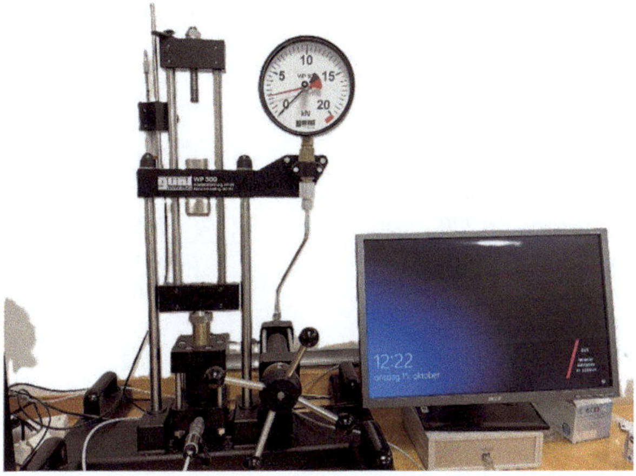

Fig. 11.20 Fatigue strength test machine

Table 11.5 Example data from fatigue strength test machine (Data source: [7])

No.	Load in Newton	Stress in N/mm^2	Endurance N in Cycles	Duration in min for speed 2800 per min
1	200	400	14,030	5
2	170	340	48,800	17
3	150	300	167,000	60
4	130	260	455,000	162
5	120	240	1,280,800	457

By performing the least squares estimation discussed in Sect. 11.2.5, and computing using Matlab code, we obtain the

$$\widehat{a} = -5.0254; \widehat{b} = 35.9409$$

According to Eq. (11.25), the

$$m = 5.0254;$$
$$A_0 = e^{35.9409} = 4.0638 \times 10^{15}$$

The full SN formula is then

$$N = 4.0638 \times 10^{15} (\Delta\sigma)^{-5.0254}$$

The R^2 value for this regression reaches 0.9623. A graphical plot of the observed SN data and the prediction are shown in Fig. 11.21, demonstrating a good fitting, consistent with the high R^2 value.

Table 11.6 Data transformation for example case

No.	S (Force Newton)	N	$x_i = \ln \Delta\sigma_i$	$y_i = \ln N_i$
1	230	5401	5.4381	8.5943
2	230	5015	5.4381	8.5202
3	230	3912	5.4381	8.2718
4	220	8217	5.3936	9.0140
5	220	7654	5.3936	8.9430
6	220	6912	5.3936	8.8410
7	210	11,202	5.3471	9.3238
8	210	9005	5.3471	9.1055
9	210	7702	5.3471	8.9492
10	200	12,400	5.2983	9.4255
11	200	11,401	5.2983	9.3415
12	200	9819	5.2983	9.1921
13	190	16,124	5.2470	9.6881
14	190	14,212	5.2470	9.5618
15	190	14,599	5.2470	9.5887
16	170	24,500	5.1358	10.1064
17	170	25,000	5.1358	10.1266
18	170	25,900	5.1358	10.1620
19	160	32,290	5.0752	10.3825
20	160	32,719	5.0752	10.3957
21	160	32,225	5.0752	10.3805

Confidence Interval

For this example, the $S_n = 0.1255$. The $D = \begin{pmatrix} 3.1302 & -16.5169 \\ -16.5169 & 87.1979 \end{pmatrix}$. The $n = 21$ and $p = 2$ in this example. The $t_{\frac{\alpha}{2},n-p} = t_{\frac{0.05}{2},1} = -2.0930$. The confidence interval for m is then

$$-\widehat{\beta}_1 \pm t_{\frac{\alpha}{2},n-p} S_{n\sqrt{D_{11}}} = 5.0254 \pm (-2.0930) * 0.1255 * \sqrt{3.1302}$$

The interval is thus [4.5605, 5.4903]. According to Eq. (11.27), the CI for A_0 is then

$$\left[e^{\widehat{\beta}_2 - t_{\frac{\alpha}{2},n-p} S_{n\sqrt{D_{22}}}}, e^{\widehat{\beta}_2 + t_{\frac{\alpha}{2},n-p} S_{n\sqrt{D_{22}}}}\right]$$
$$= \left[e^{35.9409 + (-2.0930) * 0.1255 * \sqrt{3.1302}}, e^{35.9409 - (-2.0930) * 0.1255 * \sqrt{3.1302}}\right]$$
$$= \left[3.4939 \times 10^{14}, 4.7267 \times 10^{16}\right]$$

Predict Fatigue Life

After the model parameters obtained, the model $N = A_0(\Delta\sigma)^{-m}$ can then be used to predict the fatigue life given stress $\Delta\sigma$. For stress level S_0=240, the

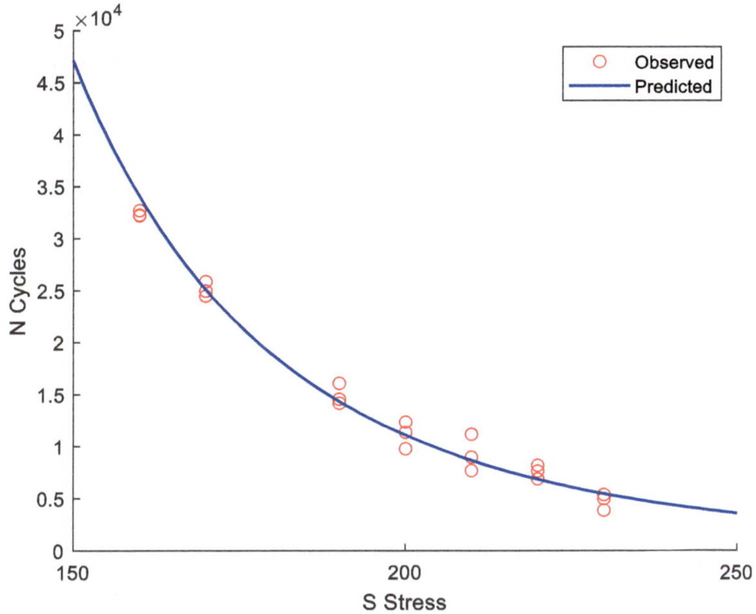

Fig. 11.21 Fitting of SN model

$$N = A_0(\Delta\sigma)^{-m} = 4.0638 \times 10^{15} \times (240)^{-5.0254} = 4.4404 \times 10^3 \approx 4440$$

Choose the CI as 0.95, according to Eq. (11.27), the CI for the $S_0 = 240$ is

$$\left[e^{\widehat{y}(\boldsymbol{x_0}) - t_{\frac{\alpha}{2},n-p} S_n \sqrt{1+\boldsymbol{x_0}(X'X)^{-1}\boldsymbol{x_0}'}}, e^{\widehat{y}(\boldsymbol{x_0}) + t_{\frac{\alpha}{2},n-p} S_n \sqrt{1+\boldsymbol{x_0}(X'X)^{-1}\boldsymbol{x_0}'}} \right] = [3338, 5905]$$

where the $\boldsymbol{x_0} = (\ln 240 \ 1)$, $(X'X)^{-1} = D = \begin{pmatrix} 3.1302 & -16.5169 \\ -16.5169 & 87.1979 \end{pmatrix}$.

A prediction for stress levels from 150 to 250 (in Newton) are plotted in the following Fig. 11.22. It can be seen the predicted N can fit the observed data well. And all the values observed fall within the 95% confidence interval. The prediction and CI were computed using software. In this example, Matlab code was developed to accomplish this task.

11.3 Some Electronic Failure Physics

Electronic devices perform their desired functions by manipulating the flow of electrons. The electrostatic forces, electric fields, and electromagnetic properties within the device are controlled or detected for specific purposes. Since electrons

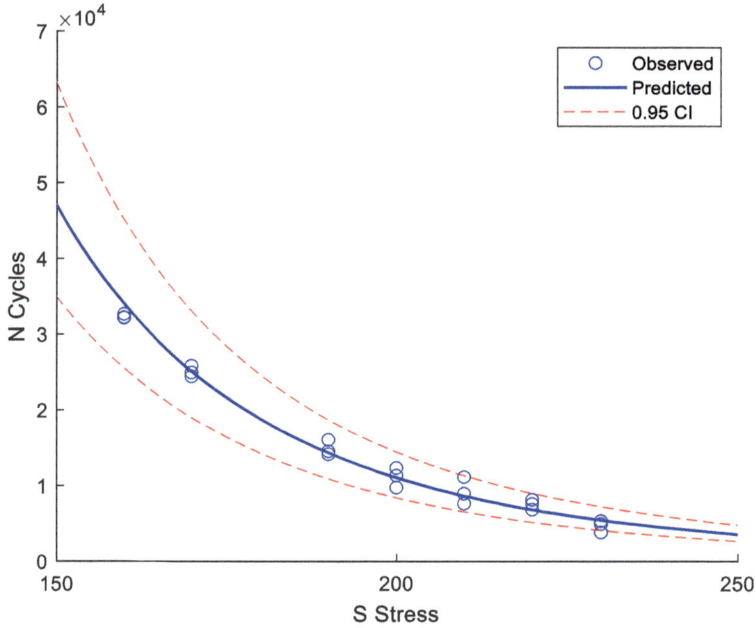

Fig. 11.22 A plot for the example case

are invisible to our naked eyes, controlling and detecting them in electronic devices is more challenging compared to mechanical devices. Some electronic failures are thus hidden and not visible, making failure analysis costly and complex. Figure 11.23 shows several old electronic devices used in my past, most of which have malfunctioned, but their failures are invisible.

This section discusses the kinetic energy and energy bands first since that is essential to electronic failure.

11.3.1 Kinetic Energy

Temperature plays a crucial role in electronic failure. It is connected to the equipartition theorem in physics, which links the macroscopic properties of a system to its microscopic behavior. According to classical physics, the kinetic energy of a moving object with speed v, mass m contains energy

$$E = \frac{1}{2}mv^2 \tag{11.30}$$

Fig. 11.23 Waste of electronics

For a particle in the system, each particle has the same kinetic energy as Eq. (11.30). When the system reaches to ideal equilibrium, the average energy, as described by the equipartition theorem [8] is

$$\frac{1}{2}mv^2 = \frac{1}{2}k_BT \tag{11.31}$$

The k_B is constant, the T is temperature in Kelvin. For ideal gas, where the particles move independently in all direction in a space. Mathematically, we can convert the moving to three orthogonal directions, the total kinetic energy is then

$$E = \frac{1}{2}m\left(v_x^2 + v_y^2 + v_z^2\right) \tag{11.32}$$

As each direction is $\frac{1}{2}k_BT$, the total energy is then $\frac{3}{2}k_BT$. If there is total number of particles N, the total energy of the system is then $\frac{3}{2}N.k_BT$. The k_B is Boltzmann constant with value [8]

$$1.380658 \times 10^{-23} JK^{-1} \text{ or } 8.617385 \times 10^{-5} eVK^{-1}$$

Obviously, the term k_BT represents the intensity of the particles' dynamic motion. A higher temperature corresponds to greater kinetic energy. Many failures relate to the undesired motion of the particle. Therefore, the term is closely related to the failure of electronic devices.

11.3.2 Energy Band

As mentioned before, the electron plays a critical role in electronics. The electron as a kind of particle at the microscopic level exbibits non-continuum state, to which classical physics cannot explain, while the quantum physics can explain it well. The energy state, as illustrated by the circles in Fig. 11.24, is discrete and only certain energy levels are allowed, i.e. the electron can only be located at a specific energy level.

In atomic structure, the outmost electrons participate in the conduction. The electrons that participate in the conduction are named free electrons since they are less confined by the atomic nucleus. As the solid is organized as crystalline, when multiple atoms are organized, the energy band will overlap. The overlapping energy level will affect the free electrons.

The overlapping will not merge the energy states in a way of numerical adding or subtracting, and it is still discrete. Each atom contributes to an energy level to the overall energy level, as illustrated in Fig. 11.25. Same as the single atom, the free electron will fill the lower energy state first. Filling the lower energy first is analogous to push balls into a well, as illustrated on the right of Fig. 11.25. The ball will fill the well from bottom to top.

The overlapping of energy states yields to three energy bands, as shown in Fig. 11.26. The middle is forbidden band where electrons are not allowable. The top is conduction band where the electron is relatively free. The lowest band is valence band where the electrons will occupy first. When the forbidden band is small or even doesn't exist, the electron in the valence band will be easy to be excited to the conduction band to participate conduction. But if the forbidden gap is big, the likelihood of electron excitation to the conduction band is much lower, and the material behaves as an electrical insulator [9].

Electrons remain in their energy states stably only at absolute zero temperature, i.e., 0 Kelvin. At temperatures above 0 Kelvin, there is always a possibility that some electrons will jump to higher energy states.

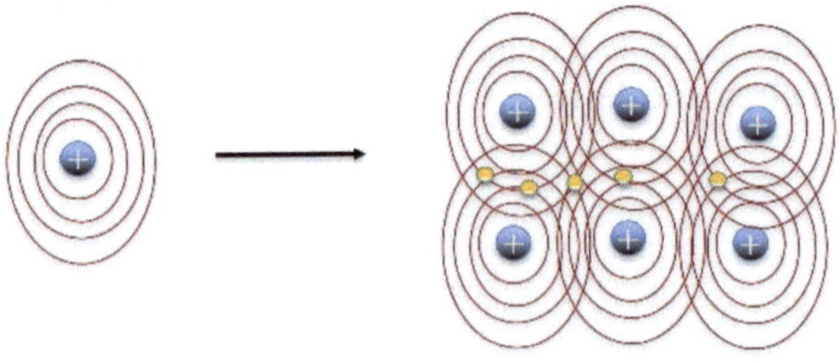

Fig. 11.24 Energy levels for single atoms and an atom lattice

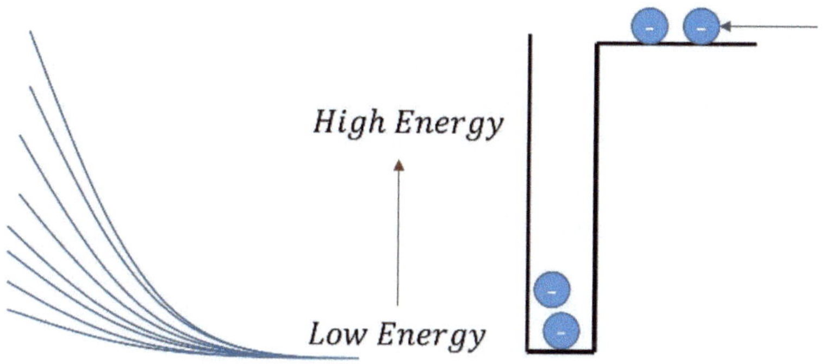

Fig. 11.25 Overlapping energy levels

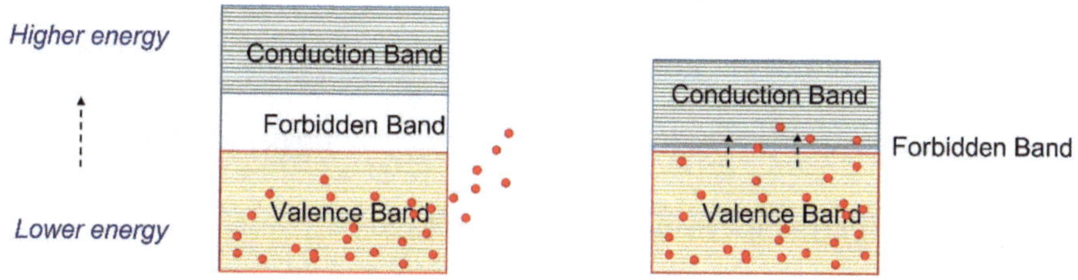

Fig. 11.26 Energy band

11.3.3 Fermi Energy

For free electrons, they will fill the orbitals of lower energy level first. For the higher energy states, if below a energy state, it is filled with electrons, upper than this state, it is empty. This boundary limit energy state is then named Fermi energy E_f. Like filling the electrons to a cubic box with length L, without elaborating the detail, the highest occupied energy level for total number of N electrons can be calculated from [10]

$$E(N) = \frac{\hbar^2}{2m_e}\left(\frac{3\pi^2 N}{V}\right)^{2/3} \tag{11.33}$$

The $V = L^3$, corresponding to the volume of the cubic box. The $\hbar$ is Plank's constant with $\hbar = 1.05459 \times 10^{-34} J\ s$. The m_e is mass of the electron with $m_e = 9.1093897 \times 10^{-31} kg$. The $\frac{N}{V}$ is the density of the free electron in the material.

Example 11.5
It has been known the number of free electrons for copper atom is around 1.48. The density of the copper is known as 8.9 g/cm^3. The weight per mol is known as 63.55 g/mol. (1 mol = 6.0221367×10^{23}) [3]

a. What is the density of free electrons?
b. What is the fermi energy of the copper?

Solution:

a. For 1 cm^3 copper, the number of atoms is

$$\frac{8.9\ g/cm^3}{63.55g/mol} \times 6.0221367 \times 10^{23} = 8.438 \times 10^{22}$$

The density of free electrons is then

$$1.48 * 8.438 \times 10^{22}/cm^3 = 1.2482 \times 10^{23}/cm^3 = 1.2482 \times 10^{29}/m^3$$

b. The Fermi energy is calculated from Eq. (11.33).

$$E(N) = \frac{\hbar^2}{2m_e}\left(\frac{3\pi^2 N}{V}\right)^{2/3} = \frac{(1.05459 \times 10^{-34} Js)^2}{2 * 9.1093897 \times 10^{-31} kg}(3\pi^2 * 1.2482 \times 10^{29}/m^3)^{2/3} = 1.4582 \times 10^{-22} J$$

If the 1 eV is $1.602176634 \times 10^{-19} J$. The Fermi energy is then

$$E_f = \frac{1.4582 \times 10^{-22}}{1.602176634 \times 10^{-19}} = 9.1016 \times 10^{-4}\ eV$$

The Fermi energy in this example uses the free electron density at room temperature, which is not accurate as the Fermi energy considers the density at absolute zero temperature. If the free electronic density at absolute zero temperature is $8.47 \times 10^{28}/m^3$ [10]. Then

$$E(N) = \frac{\hbar^2}{2m_e}\left(\frac{3\pi^2 N}{V}\right)^{2/3} = \frac{(1.05459 \times 10^{-34} Js)^2}{2 * 9.1093897 \times 10^{-31} kg}(3\pi^2 * 8.47 \times 10^{28}/m^3)^{2/3} = 1.1260 \times 10^{-18} J$$

$$E_f = \frac{1.1260 \times 10^{-18}}{1.602176634 \times 10^{-19}} = 7.0282eV$$

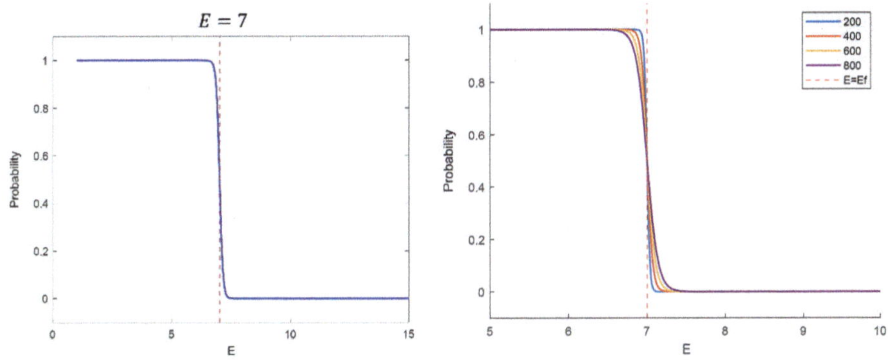

Fig. 11.27 A plot of Fermi-Dirac distribution

The Fermi energy concerns the free electrons at absolute zero temperature. Above zero temperature, the electron can be excited from one energy state to another state. For metal, the probability of a state is occupied can be modelled by Fermi-Dirac distribution [8] which is given by

$$f(E,T) = \frac{1}{e^{\left(E-E_f\right)/k_B T} + 1} \tag{11.34}$$

The function $f(E, T)$ is decreasing from 1 to 0 with a transition area. This transition is centered at the $E = E_f$. The left of Fig. 11.27 plots a figure for $E_f = 7$ and the $T = 800K$, where we can clearly see the transition area. The right of Fig. 11.27 shows the probability for various temperatures, where we can see the higher temperature, the flatter slope of the transition area.

The transition area suggests a phenomenon: the free electron will fill the very low energy state at probability of almost 1 and the probability to occupy the very high energy is very low near 0. While near the fermi energy level, it will distribute at a probability between 0 and 1. Higher probability is at lower energy state, lower probability is at higher energy state. Fermi Energy is critical to the stability of electronic material. In the Fermi-Dirac distribution, the form $1/k_BT$, named thermodynamic beta [11], can be seen in many failure models of electronic device.

Appendix 11A: Data

Table 11A.1 Experimental results

S (Force Newton)	*N*
230	5401
230	5015
230	3912
220	8217
220	7654
220	6912
210	11,202
210	9005
210	7702
200	12,400
200	11,401
200	9819
190	16,124
190	14,212
190	14,599
170	24,500
170	25,000
170	25,900
160	32,290
160	32,719
160	32,225

Problems and Questions

1. A tensile stress is applied on the long axis with force $F1 = 100N$ perpendicular to the area A as shown in Fig. 11.28. The force $F2 = 200N$ in the direction of $45\ ^\circ C$. Calculate the stress induced by the force $F1$ and $F2$.
2. A mechanical system suffers crack failure mode, the maximum stress the tip of the crack can withstand 10GPa as shown in Fig. 11.29 (1GPa = 1.0×10^9N/m^2). It has been estimated the stress at the tip of the crack $\sigma_m = 5\sigma_0$.

   a. If the applied stress σ_0 follows Weibull distribution with scale parameter $\alpha = 1\ GPa$ (1.0×10^9N/m^2) and shape parameter $\beta = 1.5$.

   $$f(\sigma_0) = \frac{\beta}{\alpha}\left(\frac{\sigma_0}{\alpha}\right)^{\beta-1} e^{-\left(\frac{\sigma_0}{\alpha}\right)^\beta}$$

   What is the distribution of the σ_m at the tip of the crack?

   b. What is the probability of failure and what is reliability of the system from the crack failure mode?

Fig. 11.28 Force of two directions

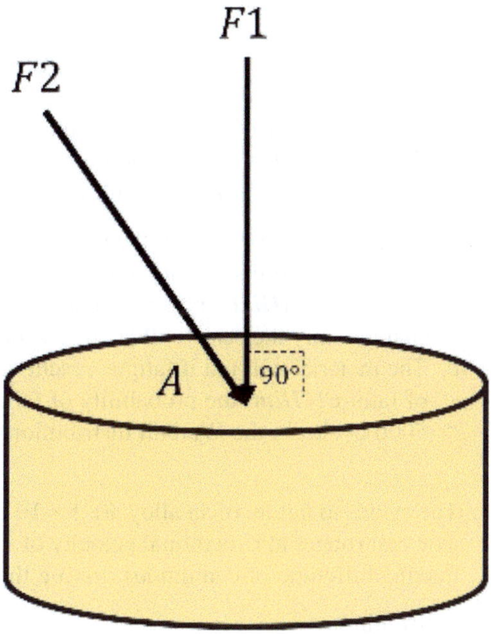

Fig. 11.29 Stress tip crack

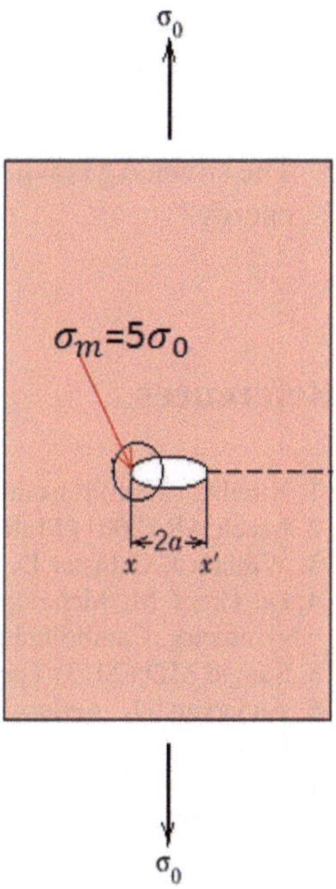

3. A mechanical system suffers fatigue failure. The fatigue model is

$$N = A.S^{-m}$$

(N is total number of cycles to failure. S the stress value. A is constant).

For a material, the m is known as 2. When the applied stress is 10 Pa, the number of cycles follows Weibull distribution $\alpha = 10{,}000$ and $\beta = 1.2$.

   a. What is the distribution of the number of cycles for applied stress at 20 Pa? if it is still Weibull distribution under the new stress, what is the parameter values of α and β? (***Hint***: the shape parameter in the Weibull distribution will not change, only the scale will change with the stress)
   b. The material will fail if fatigue reaches to 5000 cycles, what is the probability of failure? (***Hint***: the probability of failure is the right area that is bigger than 5000 cycles in the Weibull distribution)

4. The cycles to failure of an alloy are 1×10^7. A rod of this alloy is to be used for an axle that rotates at a rotational velocity of 10,000 cycles per minute. Calculate the maximal lifetime of continuous driving that is allowable?

   a. 10 minutes
   b. 10,000 minutes
   c. 1000 minutes
   d. None of the above

5. The silver Ag is with density of free electrons as $5.86 \times 10^{22}/m^3$. What is its Fermi energy?

References

1. Kittel C (2005) Introduction to solid state physics, 8th edn. Wiley
2. Leach AR (2001) Molecular modelling: principles and applications. Pearson Education
3. William J, Callister D, Rethwisch DG (2015) materials science and engineering. Wiley
4. De Graef M, Mchenry M (eds) (2012) An introduction to crystallography, diffraction and symmetry. Cambridge University Press, Cambridge
5. Sangid MD (2013) The physics of fatigue crack initiation. Int J Fatigue 57:58–72
6. Anderson TL, Anderson TL (2005) Fracture mechanics: fundamentals and applications. CRC Press
7. Mittasch P (2016) Experiment instructions WP 140 Fatigue testing machine. Gunt, Germany
8. Hofmann P (2015) Solid state physics an introduction. Wiley-VCH, Germany
9. Neamen DA, Biswas D (2011) Semiconductor physics and devices. McGraw-Hill Higher Education, New York
10. Ashcroft NW, Mermin ND (1976) Solid state physics. Harcort College Publishers
11. Meixner J (1975) Coldness and temperature. Arch Ration Mech Anal 57(3):281–290

Chapter 12
Accelerated Life Test

In industrial reliability analysis, life data obtained solely from field information may not be sufficient for a reliable evaluation of system reliability. Consequently, reliability engineers have increasingly focused on experimental approaches. The Accelerated Life Test (ALT) is one of the major experimental methods used in reliability analysis. This chapter discusses the principles of ALT and demonstrates several ALT models for mechanical and electronic systems.

12.1 Introduction to ALT

Estimating system life is crucial for system designers, or end users and customers to make decisions. Owing to technological improvement, the life expectancy of many products is very long. The mean life of a hard disk driver for example can reach 50 years. The life of some chip-type product is claimed even be 50,000 years [1]. For these durable products, evaluating life from field data is almost impossible as failure is very rare. Accelerated Life Test (ALT) is an approach for estimating product life by applying elevated conditions, such as higher stress, temperature, voltage, or humidity, to shorten the time to failure. ALT is an important reliability activity in industry today. Products such as LCD screens, adhesive joints, keyboards, and automobile tires have been seen using ALT to assess their expected lifetimes.

ALT was originally developed for electronic devices and can be applied throughout all phases of product design. During the design phase, ALT is used to verify the lifespan of components supplied by external vendors, as suppliers may provide inaccurate information or consider operating conditions that differ from the actual conditions in which the product will operate. Conducting ALT allows designers to obtain an accurate estimate of component life under specific operating conditions. Once prototypes are available, another ALT is performed to assess the lifespan of the new product. Development can continue when the evaluated life from ALT can meet

F. Yuan, *Essentials of Reliability Engineering*, Springer Series in Reliability Engineering, https://doi.org/10.1007/978-3-032-15544-3_12

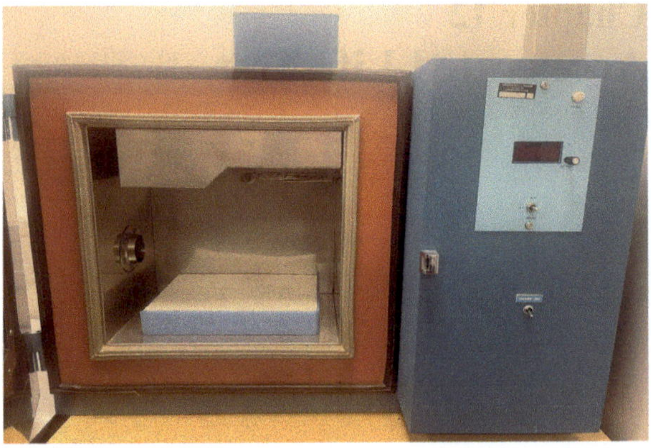

Fig. 12.1 A test chamber with temperature ranged from −50 °C to 200 °C

requirements. In new product development, ALT plays a crucial role in ensuring product reliability and performance.

ALT is failure-mechanism specific; there is no universal method because it depends heavily on the particular failure mechanism. Consequently, the ALT differs for different components, materials, and structures. Nevertheless, in general, ALT can be considered a model for assessing system life by adjusting certain operating conditions, typically temperature and humidity for electronic devices. Accordingly, the ALT model can be expressed in a general form that links the time to failure (TTF) with the relevant operating conditions.

$$TTF = f(Temprature, Humidity, Stress, \ldots) \tag{12.1}$$

The environmental chamber is a common apparatus used for ALT. The chamber can adjust operating conditions such as temperature and humidity. Some advanced chambers also integrate vibration and salt spray functionality. Figure 12.1 shows a simple environmental chamber that can adjust only temperature.

12.2 Classical ALT Model

Although ALT models differ on the failure mechanisms, there are some general principles that apply across various failure mechanisms. ALT models assume that the failure mechanism remains unchanged under extreme operating conditions, known as accelerated conditions. The following sections of this chapter will discuss the typical ALT models. Before introducing the ALT models, we define a key term in ALT: Acceleration Factor (AF), which is the ratio of the expected time-to-failure under normal operating conditions to that under accelerated stress conditions

$$AF = \frac{TTF_{normal}}{TTF_{accelerated}} \tag{12.2}$$

If a product life at accelerated condition is 10 hours, the life at normal condition is 1000 hours. The AF is then 1000/10 = 100, a dimensionless quantity. The use of AF in the ALT could simplify the computation. We will see the convenience of using AF in the Example 12.4.

12.2.1 Arrhenius Model

Arrhenius Model is one popular ALT model, probably the mostly popular model, when the accelerated condition is temperature. Suppose the failure of the system results from the accumulation of particles, heat, pressure, or other factors. The rate of the accumulation is

$$\frac{\partial P}{\partial t} = A_0 e^{-Q/k_b T} \tag{12.3}$$

Q: Activation energy of the failure
k_b: Boltzmann constant $8.61 \times 10^{-5} eV/k$.
A_0: Constant
T: Absolute temperature in Kelvin
P: Amount of particles, heat or others

At time t_1 with amount of P_1, t_2 is with amount of P_2. According to Arrhenius model (12.3),

$$\int_{P_1}^{P_2} dP = \int_{t_1}^{t_2} A_0 e^{-Q/k_b T} dt \tag{12.4}$$

Then,

$$P_2 - P_1 = A_0 e^{-\frac{Q}{k_b T}} (t_2 - t_1) \tag{12.5}$$

Suppose the amount reaches to P_2 results to failure, then the $t_2 - t_1$ is the time to failure. Rewrite Eq. (12.5),

$$TTF = t_2 - t_1 = \frac{P_2 - P_1}{A_0} e^{\frac{Q}{k_b T}} = A e^{\frac{Q}{k_b T}} \tag{12.6}$$

The $TTF = Ae^{\frac{Q}{k_b T}}$ is the Arrhenius model. The data TTF and temperature T can be measured from experiment. The unknown parameters in the model are then activation energy Q and A.

Table 12.1 ALT experiment

T	24 °C	55 °C	60 °C	65 °C
TTF	36 days	5 days	3 days	2 days

Linear regression method can be used to estimate the two parameters. Linearize Eq. (12.6) as

$$\ln TTF = \ln A + \left(\frac{Q}{k_b}\right) \cdot \frac{1}{T} = a + \frac{b}{T}, \tag{12.7}$$

which is a linear function.

Example 12.1
The life of an electronic device obtained from ALT is as following Table 12.1.
Evaluate the parameter Q and A for the Arrhenius model.
Solution: Linearize the Arrhenius model in a form of $y = ax + b$,

$$\ln TTF = \ln A + \left(\frac{Q}{k_b}\right) \cdot \frac{1}{T} = b + \frac{a}{T + 273}$$

$$y = \ln TTF; x = \frac{1}{T + 273};$$

The temperature in Table 12.1 should be converted to Kelvin degree by adding 273. The a, b values can be computed by (Refer Chap. 3),

$$\begin{cases} \widehat{a} = \dfrac{n \sum x_i y_i - \sum x_i \sum y_i}{n \sum x_i^2 - (\sum x_i)^2} \\ \widehat{b} = \overline{y} - \widehat{a}\overline{x} \end{cases}$$

$$\overline{x} = \frac{\sum x_i}{n}; \quad \overline{y} = \frac{\sum y_i}{n}$$

$$n = 4; \sum x_i y_i = 0.0223; \sum x_i = 0.0123774; \sum y_i = 6.9847; \sum x_i^2 = 3.8402996 \times 10^{-5}$$

$$\overline{x} = \frac{\sum x_i}{n} = \frac{0.0123774}{4} = 0.0031; \overline{y} = \frac{\sum y_i}{n} = \frac{6.9847}{4} = 1.7462$$

$$\widehat{a} = \frac{4 * 0.0223 - 0.0124 * 6.9847}{4 * 3.8402996 * 10^{-5} - 0.0124^2} = \frac{0.0026}{4.1195 * 10^{-7}} = 6.3114 \times 10^3$$

$$\widehat{b} = \overline{y} - \widehat{a}\overline{x} = 1.7462 - 6.3114 \times 10^3 * 0.0031 = -17.8191$$

(continued)

Example 12.1 (continued)

$$A = e^b = 1.8250 \times 10^{-8};$$

$$Q = k_b * a = 6.3114 \times 10^3 * 8.61 \times 10^{-5}\text{eV} = 0.5434 \text{ eV}$$

The result significantly depends on the precision of the value $\sum x_i y_i$, $\sum x_i^2$ etc. If we use Matlab code to calculate the $\widehat{a} = \frac{n\sum x_i y_i - \sum x_i \sum y_i}{n\sum x_i^2 - \left(\sum x_i\right)^2}$, where the value precision is much higher than the above calculation by hand.

We obtain slightly different answer as

$$\begin{cases} \hat{a} = 6.8736 \times 10^3 \\ \hat{b} = -19.5230; \end{cases}$$

From $\ln A = b; a = \frac{Q}{k_b}$, we obtain

$$A = e^b = 3.321 \times 10^{-9};$$

$$Q = k_b * a = 6.8736 \times 10^3 * 8.61 \times 10^{-5} \text{ eV} = 0.5918 \text{ eV}$$

We can how the precision affects the results from this example.

Example 12.2

An electronic device has Time to Failure 20 years at temperature 20 °C, what is the life at temperature 200 °C?

$$TTF = A_0 . \exp\left(\frac{E_\alpha}{k_b T}\right)$$

The E_α is 0.6 eV.

Solution:

$$A_0 = TTF_{20} / \exp\left(\frac{E_\alpha}{k_b T}\right) = 20 * 365 / \exp\left(\frac{E_\alpha}{k_b (20 + 273.15)}\right) \approx 3.46 \times 10^{-7}$$

$$TTF_{200} = A_0 . \exp\left(\frac{E_\alpha}{k_b T}\right) = 3.46 \times 10^{-7}$$

$$\times \exp\left(\frac{0.6}{8.61 \times 10^{-5} \times (200 + 273.15)}\right) = 0.8626 \text{ days}$$

The life decreases with temperature very sharply, as shown in Fig. 12.2.

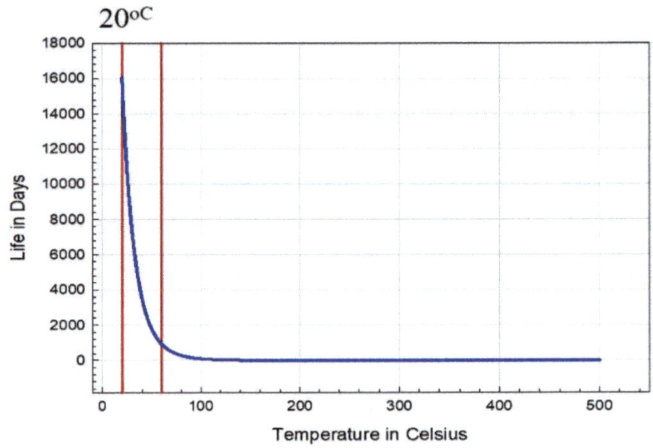

T in Celsius	Life
20	44 years
30	20 years
40	9.6 years
60	2.5 years
100	0.26years
120	37 days
150	11 days
180	3.6 days
.	
.	
.	
200	1.87 days

Fig. 12.2 Life decreases with temperature

12.2.2 Eyring Model

The Arrhenius model only considers temperature as a factor and thus it does not fit multiple factors. The Eyring model, as an alternative, is a more general approach that accounts for multiple factors. This model is based on Henry Eyring's theory of chemical reaction rates, which is derived from principles of chemistry and quantum mechanics. The model can be expressed as follows [2]:

$$TTF = A.T^{\alpha} e^{\left[\frac{Q}{k_b T} + \left(B + \frac{C}{T}\right) S_1 + \left(D + \frac{E}{T}\right) S_2 + \ldots\right]} \tag{12.8}$$

Q: Activation energy of the failure
k_b: Boltzmann constant $8.61 \times 10^{-5} eV/k$
A, α, B, C, D, E: Constants needed to be estimated
T: Absoluate temprature
S_1: Non-thermal stress values, such as voltage, applied force …

The Eyring model can accommodate more factors, but the multiple unknown parameters also complicate the parameter estimation.

12.2.3 Power-Law Model

According to failure physics, the time to failure could be a function of power form related to the applied condition in, which is written as

$$TTF = A_0(\xi)^{-n} \tag{12.9}$$

A_0: Constant
ξ: Stress (e.g. pressure)
n: Material dependent parameter

The model is with two unknown parameters. The stress on TTF follows the power-law for mechanical unit. The voltage stress on TTF of resistor also follows power-law.

Example 12.3
A product has a TTF Model of the form

$$TTF = A_0(\xi)^{-n} \exp\left(\frac{Q}{K_B T}\right)$$

If the temperature T is same under the normal condition and accelerated conditon, and $n = 2$. The applied stress is voltage $\xi_{Accelerated}$is 100 V and$\xi_{Normal}$is 12 V. What is the Acceleration Factor (AF)?

Solution: According to TTF model,

$$TTF_{Normal} = A_0(\xi_{Normal})^{-n} \exp\left(\frac{Q}{K_B T}\right)$$

$$TTF_{Accelerated} = A_0(\xi_{Accelerated})^{-n} \exp\left(\frac{Q}{K_B T}\right)$$

$$AF = \frac{TTF_{normal}}{TTF_{accelerated}} = \left(\frac{\xi_{Accelerated}}{\xi_{Normal}}\right)^n = \left(\frac{100}{12}\right)^2 = 69.4$$

The AF is 69.4 for this experiment.

12.2.4 Exponential Model

The exponential model is a simple model with a form as

$$TTF = B_0 \exp(-\gamma\xi) \tag{12.10}$$

B_0: Constant
ξ: Stress (for example humidity)
γ: parameter

In ALT, the common factor humidity, an environmental condition considered in many ALT test chambers, is of the exponential form [3].

$$TTF = A_0 \exp(-\alpha.RH) \tag{12.11}$$

Example 12.4
It has been known a processing time is 5 hours at 50% relative humidity, how much longer would the processing time be at 35%RH? Assume the parameter $a = 0.2/\ \%\ RH$.

Solution:

$$TTF_1 = A_0\, exp\left[-a.(\%RH_1) + \frac{Q}{K_B T}\right]$$

$$TTF_2 = A_0\, exp\left[-a.(\%RH_2) + \frac{Q}{K_B T}\right]$$

$$AF = TTF_2/TTF_1 = exp[a(\%RH_1) - a(\%RH_2)] = exp[a * 50 - a * 35]$$

$$AF = 20.0855$$

Therefore, the safe time window at 35%RH is

$$Time_{\%35RH} = AF.Time_{\%50RH} = 20.0855 \times 5 = 100.4\ \text{h}$$

12.3 More General ALT Model

The ALT model is not restricted to the form mentioned above; in fact, it can take any form. The main challenge with ALT models is the difficulty of parameter estimation when the model becomes complex. Therefore, in practice, it is advisable to use a standard model whenever it is suitable. This section introduces another general model, called the general log-linear model, which can also be used as an ALT model.

12.3.1 General Log-Linear Model

In Chap. 9, we have intensively discussed the Proportional Hazard Model (PHM). As we can find from this model, the failure rate depends on the covariates exponentially. As the popular use of proportional hazard model in reliability data analysis, the PHM is also used in the state of art research as an ALT model. Strictly speaking, the PHM is also a kind of exponential model, referring to Eq. (12.10). Very similar to PHM, another general model, named general log-linear model (GLL) can also be used as ALT model [4, 5].

Suppose the r factors, denoted as vector $\boldsymbol{X} = (x_1, x_2, x_3, \ldots, x_r)$, are linearly combined as

$$L(\boldsymbol{X}) = e^{\alpha_0 + \sum_{i=1}^{r} \alpha_i x_i} \tag{12.12}$$

Each x_i can be a factor such as temperature, humidity, stress, vibration level, voltage etc. The GLL model is a general exponential model, and the Arrhenius model, the Power-law model can be transformed from GLL by transforming variable x_i. For Arrhenius model given as $TTF = Ae^{\frac{Q}{k_b T}}$, it is easy to find

$$r = 1; L(\boldsymbol{X}) = TTF; A = e^{\alpha_0}; \alpha_1 = \frac{Q}{k_b}; x_1 = \frac{1}{T}$$

For the Power-Law model written by $TTF = A_0(\xi)^{-n}$, it corresponds to GLL by

$$r = 1; L(\boldsymbol{X}) = TTF; A_0 = e^{\alpha_0}; \alpha_1 = n; x_1 = -\ln \xi$$

The general life model can be linearized by taking logarithm at both side of Eq. (12.12). We obtain

$$\ln L(\boldsymbol{X}) = \alpha_0 + \sum_{i=1}^{r} \alpha_i x_i \tag{12.13}$$

For such linear expression, the unknown parameters α_i can readily be estimated by linear regression by using the observed $\boldsymbol{X}$ and lifetime L.

12.3.2 Considering Uncertainty in the ALT

In practice, the observed lifetime, whether under normal or accelerated conditions, exhibits some degree of uncertainty. When using classical linear regression to estimate unknown parameters, such as the α_i in Eq. (12.13), the noise of the lifetime regression is implicitly assumed to follow normal distribution with a mean 0. Another method to accommodate the uncertainty assumes that the lifetime follows a specific statistical distribution, such as the Exponential, Weibull distribution, when the stress is fixed at a certain level, for example, $\boldsymbol{X_0}$.

Suppose at stress level $\boldsymbol{X_0}$, the observed lifetimes are given by

$$x_1, x_2, x_3, x_4, \ldots . x_n \tag{12.14}$$

If Eq. (12.12) holds, the life x_i in Eq. (12.14) and the life y_i which is under another stress $\boldsymbol{X}$ is of relationship

$$\frac{L(\boldsymbol{X_0})}{L(\boldsymbol{X})} = \frac{x_i}{y_i} \tag{12.15}$$

Rearrange Eq. (12.15) as

$$y_i = \frac{L(\boldsymbol{X})}{L(\boldsymbol{X_0})} x_i \tag{12.16}$$

For simplicity of notation, let the $A = \frac{L(\boldsymbol{X})}{L(\boldsymbol{X_0})}$. The A is similar to the acceleration factor discussed in the previous section, then the lifetimes at new stress level are

$$y_i = Ax_i; i = 1, 2, \ldots, n \tag{12.17}$$

Suppose the PDF of the x is $f(x)$. Refer to the Appendix 4A in Chap. 4, the PDF of the y is then

$$g(y) = f\left(T^{-1}(y)\right) . \left|\frac{dx}{dy}\right| = f(y/A)/A \tag{12.18}$$

Exponential Distribution
Give the $f(x)$ as the Exponential distribution $f(x) = \lambda e^{-\lambda x}$, the

$$f(y) = \frac{f\left(\frac{y}{A}\right)}{A} = \frac{\lambda e^{-\lambda x/A}}{A} = \frac{\lambda}{A} e^{-\lambda x/A} \tag{12.19}$$

As the $L(\boldsymbol{X})$ is the general log-linear model specified in Eq. (12.12) and the

$$A = \frac{L(\boldsymbol{X})}{L(\boldsymbol{X_0})} \tag{12.20}$$

Equation (12.19) is then

$$f(y) = \frac{\lambda}{e^{\alpha_0 + \sum_{i=1}^{r} \alpha_i x_i}} e^{-\lambda y / e^{\alpha_0 + \sum_{i=1}^{r} \alpha_i x_i}} \tag{12.21}$$

The parameter α_0 and λ can be combined since both are unknown parameters, Eq. (12.21) is then written as

$$f(y) = e^{-\left(\alpha_0 + \sum_{i=1}^{r} \alpha_i x_i\right)} e^{-\left(e^{-\alpha_0 - \sum_{i=1}^{r} \alpha_i x_i}\right) y} \tag{12.22}$$

which is General log-linear Exponential distribution.

Weibull Distribution
Given the failure times follows Weibull distribution as

$$f(x) = \frac{\beta}{\alpha}\left(\frac{x}{\alpha}\right)^{\beta-1} e^{-(x/\alpha)^{\beta}} \tag{12.23}$$

Combining the general log-linear ALT model with the Weibull model given in Eq. (12.23), according to Eq. (12.18), it gives

$$g(y) = \frac{\beta}{A\alpha}\left(\frac{y}{A\alpha}\right)^{\beta-1} e^{-(y/A\alpha)^{\beta}} \tag{12.24}$$

It is equivalent to substitute the scale parameter α in the original Weibull distribution to $A\alpha$. Since the $A = \frac{L(\boldsymbol{X})}{L(\boldsymbol{X_0})}$, and letting the α combines with α_0 in the general log-linear model, the distribution (12.24) is then

$$g(y) = \frac{\beta}{L(\boldsymbol{X})}\left(\frac{y}{L(\boldsymbol{X})}\right)^{\beta-1} e^{-(y/L(\boldsymbol{X}))^{\beta}} \tag{12.25}$$

which is the general log-linear Weibull distribution. A similar derivation can be performed for other distribution, allowing the combination of statistical distribution with the ALT model. The model (12.25) is named as Weibull Generalized Log Linear (WGLL) model.

12.3.3 Weibull GLL Model and Proportional Hazard Model

Both the proportional hazard model and the GLL model contains an exponential component with linear combination of factors, named covariates in proportional hazard mode.

$$L(\boldsymbol{X}) = e^{\left(\alpha_0 + \sum_{i=1}^{r} \alpha_i x_i\right)}$$

We can investigate whether there is any connection between the two models. The failure rate of Weibull distribution is

$$\lambda(t) = \frac{\beta}{\alpha}\left(\frac{t}{\alpha}\right)^{\beta-1} = \frac{\beta}{\alpha}\left(\frac{t}{\alpha}\right)^{\beta-1} \quad (12.26)$$

For the GLL model, the failure rate is then

$$\lambda(t) = \frac{\beta}{L}\left(\frac{t}{L}\right)^{\beta-1} = \beta t^{\beta-1} L^{-\beta} \quad (12.27)$$

Substitute the $L(\boldsymbol{X}) = e^{\left(\alpha_0 + \sum_{i=1}^{r} \alpha_i x_i\right)}$ into Eq. (12.27). The failure rate function is given by

$$\lambda(t) = \beta t^{\beta-1} e^{\left(-\beta\alpha_0 + \sum_{i=1}^{r} \beta\alpha_i x_i\right)} = \frac{\beta}{e^{\beta\alpha_0}} t^{\beta-1} e^{\left(-\sum_{i=1}^{r} \beta\alpha_i x_i\right)} \quad (12.28)$$

Equation (12.28) is obviously a proportional hazard model with baseline function of Weibull distribution. The scale parameter of the Weibull distribution is e^{a_0}, the coefficients of covariates of PHM corresponds to the $\beta\alpha_i$. The connection between the PHM and the WGLL suggests the parameter estimation of WGLL can be used from PHM.

12.3.4 Some Issues of WGLL

The WGLL can be derived from the Weibull distribution by changing the scale parameter α. The scale parameter characterizes the life of the system. For example, the reliability at a life equal to α is always constant as e^{-1}, regardless of its shape parameter, i.e.

$$R(t = \alpha) = e^{-(x/\alpha)^{\beta}} = e^{-1} \approx 0.62$$

To combine the Weibull distribution with the life model, we can use the life model as the scale parameter α, since α characterizes life, which is

$$\alpha = L(\boldsymbol{X}) = e^{\alpha_0 + \sum_{i=1}^{r} \alpha_i x_i}$$

Then a new model, which is named log-linear Weibull model, is given by

$$f(x) = \frac{\beta}{L(\boldsymbol{X})}\left(\frac{x}{L(\boldsymbol{X})}\right)^{\beta-1} e^{-(x/L(\boldsymbol{X}))^{\beta}} \tag{12.29}$$

which is exactly same as the WGLL given by Eq. (12.25).

12.3.5 Parameter Estimation for WGLL Model

Using the least squares method to estimate the WGLL requires transformation and relies on the Type I (Gumbel) Extreme Value distribution. It transforms the Weibull distribution into an Extreme Value Distribution for convenience. The cumulative probability function of Weibull distribution is

$$F(t) = 1 - \exp\left(-(t/\alpha)^{\beta}\right) \tag{12.30}$$

In terms of Extreme Value distribution, we just introduce a new variable $t' = \ln t$ with two parameters μ and σ. The CDF for Extreme Value distribution is as follows.

$$F(t') = 1 - \exp\left(-\exp\left(\frac{t' - \mu}{\sigma}\right)\right) \tag{12.31}$$

The CDF looks similar to the CDF of Normal distribution with the same $\exp\left(\frac{t'-\mu}{\sigma}\right)$. But the σ in Extreme Value distribution is not the standard deviation such as in the Normal distribution. The parameters of the two distributions Eqs. (12.30) and (12.31) correspond as

$$\alpha = e^{\mu}; \quad \beta = 1/\sigma \tag{12.32}$$

The mean and variance of the Extreme distribution is as in Table 12.2.

For the log-linear Weibull GLL, the use of extreme distribution is natural. We can let the μ of the Extreme distribution be the linear function of the factors as proposed in Eq. (12.29).

$$\mu = \alpha_0 + \sum_{i=1}^{r} \alpha_i x_i \tag{12.33}$$

Let z_j denotes the mean of the distribution, then

Table 12.2 Properties of extreme distribution [6]

Expectation	$\mu + 0.5772\sigma$
Variance	$\frac{\pi^2}{6}\sigma^2$

$$z_j = \alpha_0 + \sum_{i=1}^{r} \alpha_i x_{ji} + 0.5772\sigma; j = 1, 2, 3, , \ldots n \tag{12.34}$$

Trickly, let the α_0 + 0.5772σ be one parameter $\alpha_0{}'$ to simplify the computation. A regression model can be written as

$$\begin{bmatrix} \ln y_1 \\ \ln y_2 \\ \cdots \\ \ln y_n \end{bmatrix} = \begin{bmatrix} z_1 \\ z_2 \\ \cdots \\ z_n \end{bmatrix} + \begin{bmatrix} \varepsilon_1 \\ \varepsilon_2 \\ \cdots \\ \varepsilon_n \end{bmatrix} \tag{12.35}$$

The ε_i are identical and independent random variables following same Extreme distribution with zero mean.

Assume the data are grouped by its stress level if data is grouped, the mean for group m is

$$\overline{Y_{m.}} = \frac{\sum_{j}^{n_m} Y_{m,j}}{n_m} \tag{12.36}$$

where $Y_{m,\,j} =$ ln $y_{m,\,j}$. For stress level m, the variance is written as

$$s_m^2 = \frac{\sum_{j}^{n_m} \left(Y_{m,j} - \overline{Y_{m.}}\right)^2}{n_m - 1} \tag{12.37}$$

The estimate of variance for the whole data is then

$$s^2 = \frac{\sum_{m=1}^{r} (n_m - 1) s_m^2}{\sum_{m=1}^{r} (n_m - 1)} \tag{12.38}$$

Or alternatively using

$$s^2 = \frac{\sum_{j=1}^{n} \left(Y_j - \widehat{Y}_j\right)^2}{n - (r + 1)} \tag{12.39}$$

The r is the number of factors without considering the constant threshold. The $\widehat{Y}_j$ is the estimated valued by using the $\widehat{Y}_j = \alpha_0{}' + \sum_{i=1}^{r} \alpha_i x_{ji}$ when the $\alpha_0{}'$ and α_i have been estimated from the linear regression.

Estimate the Linear Coefficient in *L*(*X*)

Referring to Eq. (12.35), $\begin{bmatrix} z_1 \\ z_2 \\ \cdots \\ z_n \end{bmatrix} = \begin{bmatrix} 1 & x_{11} & x_{12} & \cdots & x_{1r} \\ 1 & x_{21} & x_{12} & \cdots & x_{2r} \\ \cdots & \cdots & \cdots & \cdots & \cdots \\ 1 & x_{n1} & x_{n2} & \cdots & x_{nr} \end{bmatrix} \begin{bmatrix} \alpha_0{}' \\ \alpha_1 \\ \cdots \\ \alpha_r \end{bmatrix}$. The Estimates of $\boldsymbol{\alpha}$ is then [7].

$$\widehat{\boldsymbol{\alpha}} = (\boldsymbol{X}'\boldsymbol{X})^{-1}\boldsymbol{X}'\boldsymbol{Y} \tag{12.40}$$

The $\boldsymbol{Y} = \begin{bmatrix} \ln y_1 \\ \ln y_2 \\ \cdots \\ \ln y_n \end{bmatrix}$, $\boldsymbol{X} = \begin{bmatrix} 1 & x_{11} & x_{12} & \cdots & x_{1r} \\ 1 & x_{21} & x_{12} & \cdots & x_{2r} \\ \cdots & \cdots & \cdots & \cdots & \cdots \\ 1 & x_{n1} & x_{n2} & \cdots & x_{nr} \end{bmatrix}_{n\times(r+1)}$.

After obtaining the parameters. The empirical variance s^2 can then be computed and use it to estimate the variance of Extreme distribution. Let the empirical variance equal to the expected variance shown in Table 12.1, which is given by

$$s^2 = \frac{\pi^2}{6}\sigma^2$$

We then obtain

$$\sigma^2 = 6s^2/\pi^2 \quad \text{or equivalently} \quad \sigma = \frac{\sqrt{6}}{\pi}s \tag{12.41}$$

As $\alpha = e^{\mu}$; $\beta = 1/\sigma$, then the

$$\beta = \frac{\pi}{\sqrt{6}s} \tag{12.42}$$

The scale parameter for the Weibull distribution is

$$\alpha = exp(\widehat{\boldsymbol{\alpha}}\boldsymbol{X} - 0.5772/\beta) \tag{12.43}$$

The 0.5772 is a constant named Euler's constant.

Example 12.5
The example uses the data from book by Nelson [4]. The data are shown in Appendix 12A of this chapter. The time data in Table 12A.1 is already transformed by ln(.).

Solution: According to Eq. (12.35), by using classical linear regression approach, the $\widehat{\boldsymbol{\alpha}} = [18.8012, -0.5031]$. According to Eq. (12.39), the

(continued)

Example 12.5 (continued)

$$s = 1.5701$$

According to Eq. (12.42), the shape parameter for Weibull distribution is then

$$\beta = \frac{\pi}{\sqrt{6}s} = \frac{\pi}{\sqrt{6} * 1.5701} = 0.8165$$

According to Eq. (12.43), the scale parameters for the Weibull is

$$\alpha = L(\boldsymbol{X}) = e^{\alpha_0 + \sum_{i=1}^{r} \alpha_i x_i}$$

$$\alpha = exp\left(\widehat{\boldsymbol{\alpha}}\boldsymbol{X} - \frac{0.5772}{\beta}\right) = exp\left(18.8012 - 0.5031x_1 - \frac{0.5772}{0.8165}\right)$$

$$= e^{18.8012 - 0.5031x_1 - 0.7069} = e^{18.0943 - 0.5031x_1}$$

12.3.6 Simulation Verification on Least Squares Method

A simulation study was carried out to verify the least squares method introduced in the previous section. The model used was

$$f(x) = \frac{\beta}{L(\boldsymbol{X})}\left(\frac{x}{L(\boldsymbol{X})}\right)^{\beta-1} e^{-(x/L(\boldsymbol{X}))^{\beta}}; \alpha = L(\boldsymbol{X}) = e^{\alpha_0 + \alpha_1 x_1} \tag{12.44}$$

The simulation carried out for $\beta = 2$; $\alpha_0 = 10$; $\alpha_1 = 1$. The stress level x_1 was used as 10, 20, 30, 40, 50, 60. The number of repeats for each stress level is fixed as 5. The simulation generated the random data x by Eq. (12.44) for each run. The data used for each run differs as the data for each run is random. The results for the first 5 iterations are shown as follows:

$$\widehat{\alpha_0} = 0.95; \widehat{\alpha_1} = 10.02; \widehat{\beta} = 0.403$$

$$\widehat{\alpha_0} = 1.81; \widehat{\alpha_1} = 10.05; \widehat{\beta} = 1.33$$

$$\widehat{\alpha_0} = 2.08; \widehat{\alpha_1} = 10.00; \widehat{\beta} = 1.49$$

$$\widehat{\alpha_0} = 3.10; \widehat{\alpha_1} = 9.97; \widehat{\beta} = 2.411$$

$$\widehat{\alpha_0} = 2.44; \widehat{\alpha_1} = 9.98; \widehat{\beta} = 1.19$$

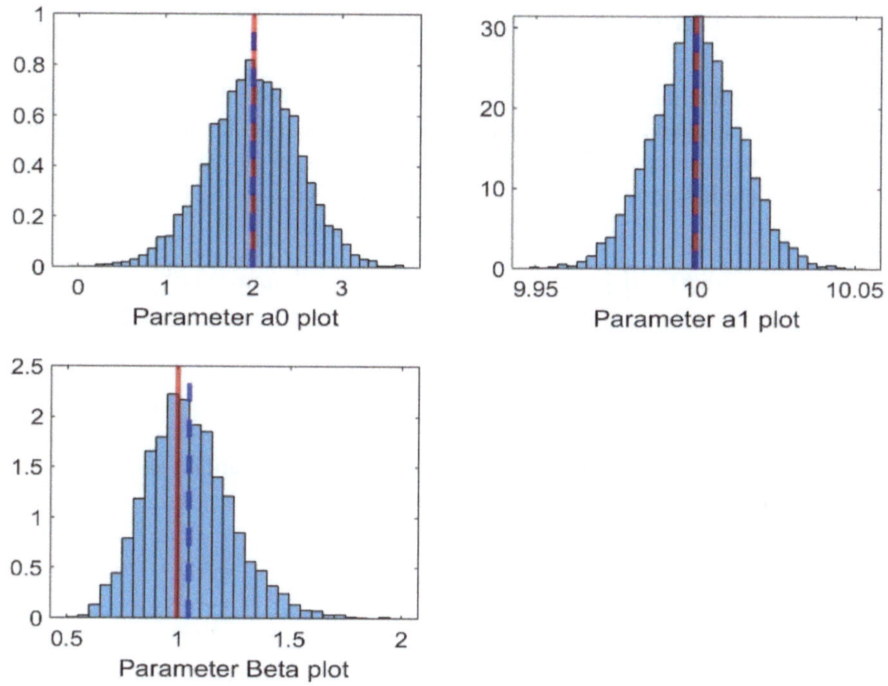

Fig. 12.3 Histogram of parameter estimates

Table 12.3 Simulation results

True Value	Mean Estimated	True Value	Mean Estimated	True Value	Mean Estimated
$\alpha_0 = 10$	9.9850	$\alpha_0 = 10$	10.000	$\alpha_0 = 10$	9.9932
$\alpha_1 = 1$	1.0001	$\alpha_1 = 1$	0.9999	$\alpha_1 = 2$	2.0001
$\beta = 1$	1.0534	$\beta = 2$	2.1150	$\beta = 4$	4.2058

The uncertainty in the estimates shown above is expected since the data size is small, only 5 for each run. If we run the simulation for an iteration of 5000. The mean of the $\widehat{\alpha_0}$ is 1.9914 with real parameter 2, mean of $\widehat{\alpha_1}$ is 10 with real parameter value 10, mean of $\widehat{\beta}$ is 1.0487 with real value 1. It shows only $\widehat{\beta}$ is roughly unbiased. The histograms of the three parameter estimates are shown in Fig. 12.3, where we can see the $\widehat{\beta}$ is biased. It is a drawback of the least squares method for small sample situation.

Further simulation was run for other parameters. The stress level x_1 was used from 10, 20, 30, 40, 50. The number of iterations for each stress is fixed as 5. The estimating results are shown in Table 12.3, where we can see the estimation methods work well as all of them are near the true parameter values.

12.3.7 Maximum Likelihood Function of WGLL

The maximum likelihood estimation is feasible for the WGLL since the analytical PDF of WGLL is available. Deriving the likelihood function can one way to estimate the parameters. It can be also used for confidence interval derivation. According to Eq. (12.44), the likelihood function for PDF of the WGLL is [8].

$$\mathcal{L}(\boldsymbol{\alpha},\beta)=\prod_{i=1}^{n}\frac{\beta}{L(\boldsymbol{x_i})}\left(\frac{t_i}{L(\boldsymbol{x_i})}\right)^{\beta-1}e^{-(t_i/L(\boldsymbol{x_i}))^{\beta}};L(\boldsymbol{x_i})=e^{\boldsymbol{\alpha}\boldsymbol{x_i}'}$$

$$\boldsymbol{\alpha}=(\alpha_1,\alpha_2,\ldots,\alpha_r)_{1\times r}$$

$$\boldsymbol{x_i}=(x_{i1},x_{i2},\ldots,x_{ir})_{1\times r}$$

The row vector $\boldsymbol{\alpha}$ is coefficient and row vector $\boldsymbol{x_i}$ is the observation value. If constant threshold such in $\alpha_0 + \alpha_1 x_1$ exists, the corresponding x_{i*} is set as 1.

$$-\ln\mathcal{L}=\beta\sum\ln L_i-n\ln\beta-(\beta-1)\sum\ln t_i+\sum\left(\frac{t_i}{L_i}\right)^{\beta}$$

Substitute $L(\boldsymbol{x_i})=e^{\boldsymbol{\alpha}\boldsymbol{x_i}'}$ into Eq. (12.45).

$$-\ln\mathcal{L}=\beta\sum\boldsymbol{\alpha}\boldsymbol{x_i}'-n\ln\beta-(\beta-1)\sum\ln t_i+\sum\left(\frac{t_i}{e^{\boldsymbol{\alpha}\boldsymbol{x_i}'}}\right)^{\beta}\tag{12.45}$$

According to Eq. (12.45), the first derivative of the likelihood function $-\ln\mathcal{L}$ is as follows

$$\begin{cases}\dfrac{\partial(-\ln\mathcal{L})}{\partial\alpha_k}=\dfrac{\partial(-\ln\mathcal{L})}{\partial L}\dfrac{\partial L}{\partial\alpha_k}=\beta\sum x_{ik}-\beta\sum\left(t_i^{\beta}L_i^{-\beta}x_{ik}\right);k=1,2,\ldots r\\ \dfrac{\partial(-\ln\mathcal{L})}{\partial\beta}=-\dfrac{n}{\beta}+\sum\ln L_i-\sum\ln t_i+\sum\left\{\left(\dfrac{t_i}{L_i}\right)^{\beta}\ln\left(\dfrac{t_i}{L_i}\right)\right\}\end{cases}\tag{12.46}$$

By setting the first derivatives of Eq. (12.46) to zero, the maximum likelihood estimates are obtained by performing iterative numerical method, since the analytical solutions cannot be obtained. The algorithm used for the numerical method is the trust-region-dogleg for a simulation. The simulation was run to check the performance of MLE using Matlab. The sample size for each stress level is 5. Total 5000 iterations have run. The means of the estimated values are tabulated in Table 12.4.

The histogram of the estimates of the 5000 iterations are shown in Fig. 12.4. From Table 12.4 and Fig. 12.4, the estimates for the β is obviously biased for MLE method.

Table 12.4 Simulation results for MLE

Real Value	Mean Estimated	Real Value	Mean Estimated	Real Value	Mean Estimated
$\alpha_0 = 10$	9.9652	$\alpha_0 = 10$	9.9934	$\alpha_0 = 10$	9.9917
$\alpha_1 = 1$	0.9953	$\alpha_1 = 1$	0.9999	$\alpha_1 = 2$	1.0001
$\beta = 1$	1.0599	$\beta = 2$	2.1564	$\beta = 4$	4.2870

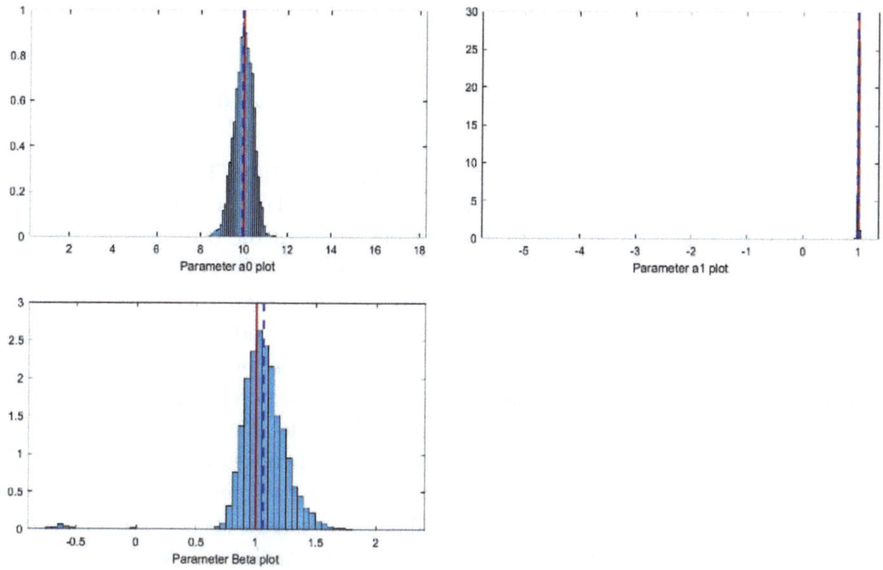

Fig. 12.4 Histogram of MLE parameter estimates

The variance of the estimators can be approximated from Fisher information matrix. The Fisher matrix requires the 2nd derivative of the likelihood function. Differentiate Eq. (12.46) in terms of α_k and β respectively again. We obtain the 2nd derivative as follows

$$\begin{cases} \dfrac{\partial^2(-\ln \mathcal{L})}{\partial \alpha_k \partial \alpha_j} = \beta^2 \sum \left(t_i^{\beta} L_i{}^{-\beta} . x_{ik} . x_{ij} \right) \\ \dfrac{\partial^2(-\ln \mathcal{L})}{\partial \alpha_k \partial \beta} = \sum x_{ik} - \sum \left(t_i^{\beta} L_i^{-\beta} x_{ik} \right) - \beta \sum \left\{ \left(\dfrac{t_i}{L_i} \right)^{\beta} . \ln \left(\dfrac{t_i}{L_i} \right) . x_{ik} \right\}; k = 1, 2, \ldots r \\ \dfrac{\partial^2(-\ln L)}{\partial \beta^2} = \dfrac{n}{\beta^2} + \sum \left\{ \ln^2 \left(\dfrac{t_i}{L_i} \right) . (t_i / L_i)^{\beta} \right\} \end{cases} \tag{12.47}$$

The Fisher information matrix is then given by [8]

$$
I(\boldsymbol{\alpha},\beta)=\begin{pmatrix}
\frac{\partial^2(-\ln L)}{\partial\alpha_1\partial\alpha_1} & \cdots & \frac{\partial^2(-\ln L)}{\partial\alpha_1\partial\beta} \\
\frac{\partial^2(-\ln L)}{\partial\alpha_2\partial\alpha_1} & \cdots & \frac{\partial^2(-\ln L)}{\partial\alpha_2\partial\beta} \\
\vdots & & \vdots \\
\frac{\partial^2(-\ln L)}{\partial\alpha_r\partial\alpha_1} & \cdots & \frac{\partial^2(-\ln L)}{\partial\alpha_r\partial\beta} \\
\frac{\partial^2(-\ln L)}{\partial\beta\partial\alpha_1} & \cdots & \frac{\partial^2(-\ln L)}{\partial\beta\partial\beta}
\end{pmatrix} \tag{12.48}
$$

12.3.8 Performance Comparison Between MLE and LSE

The performance comparison was conducted by simulation. The pseudo random numbers obtained from the WGLL model of Eq. (12.44). Total 5000 simulations have been run. The simulation results show the mean of the estimates of the two methods does not differ significantly, as shown in Table 12.5. For the STD, no one can dominate the other for all three parameters.

As stated by other researchers [5], the MLE highly depends on the start points of the iteration in the numerical method. If starting points are not close to true values, the estimated values from the numerical method would be nonsense. This simulation uses the LSE estimates as the starting point for MSE to avoid arbitrary guessing. Nevertheless, incorrect estimates from the MLE still appear in the histogram plot for some iterations. As we can see from Fig. 12.5a, for the parameter α_0, the first figure in Fig. 12.5a, most values are around true value 10, but some spread to even around 30, which goes far beyond the true values. It is a major drawback of MLE. Moreover, for the case with true β around 4, the MLE shows larger deviation from the true value 4 than LSE.

As shown in Fig. 12.5b, the LSE shows single peak, which is desirable, while the MSE unexpectedly shows multiple peaks. The multiple peaks from MSE are not visual in Fig. 12.5a as the small peaks are not dominant. But it is clearly shown in the last plot of Fig. 12.4 where we can see a clear multiple-peak pattern for β. From this perspective, the LSE method exhibits more stable performance.

A two-step solution to solve the starting point problem for MLE is a practical approach as did in the simulation cases, as shown in Fig. 12.6. LSE can be used to obtain approximate values, then the MLE elaborates the values further. Use the estimates from the LSE as the starting point for MLE does not increase so much computing time, since the LSE is computational efficiency.

Table 12.5 Simulation results for MLE

True Value	Estimated Mean (LSE)	Estimated Mean (MLE)	STD (LSE)	STD (MLE)
$\alpha_0 = 10$	9.9965	9.9990	0.1312	0.3499
$\alpha_1 = 1$	1.0000	0.9999	0.0035	0.0074
$\beta = 4$	4.1989	4.2798	0.7753	0.6642

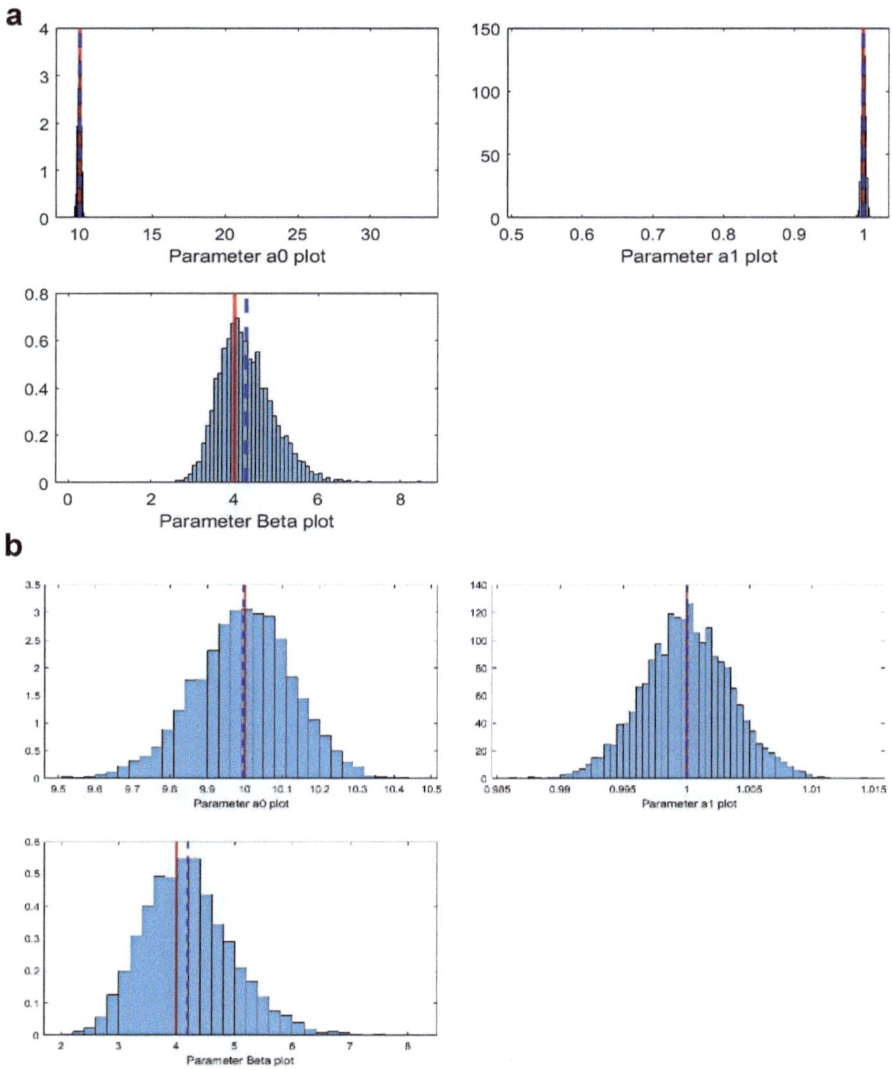

Fig. 12.5 (**a**) Histogram of MSE; (**b**) Histogram of LSE

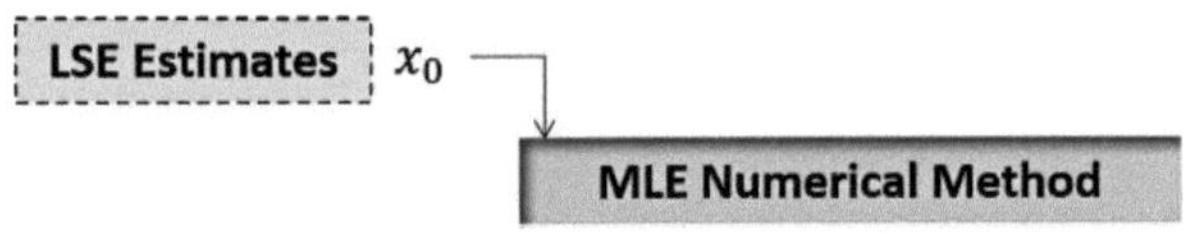

Fig. 12.6 Use LSE estimates as starting point for MLE

12.4 Generalized Linear Model for ALT

The Weibull GLL discussed in the preceding sections is conceptually very similar to another well-known framework: the Generalized Linear Model (GLM). The concepts and some examples of the GLM are presented in detail in Appendix 12B. In essence, a GLM links a linear predictor function to the *mean* of a response variable's probability distribution through a specified link function, as follows:

$$g(u_i) = \eta_i = \boldsymbol{x}_i.\boldsymbol{a} \tag{12.49}$$

where the definition of $\eta_i = \boldsymbol{x}_i.\boldsymbol{a}$ can be found in Appendix 12B. The factors in $\boldsymbol{x}_i$ for example can be humidity, temperature, stress. The

$$u_i = E(y_i) \tag{12.50}$$

According to Eq. (12.49), the

$$u_i = g^{-1}(\boldsymbol{x}_i.\boldsymbol{a}) \tag{12.51}$$

It tells that the GLM links a function of the linear formula $\boldsymbol{x}_i.\boldsymbol{a}$ with the expectation of a distribution u_i.

We can check if the Weibull GLL discussed in Eq. (12.29) is a GLM. The Weibull GLL defines the scale parameter of Weibull distribution as

$$\alpha = L(\boldsymbol{x}_i) = e^{\alpha_0 + \sum_{j=1}^{r} \alpha_j x_{ij}}$$

Therefore,

$$\alpha_0 + \sum_{j=1}^{r} \alpha_j x_j = \boldsymbol{x}_i.\boldsymbol{a} = \ln \alpha \tag{12.52}$$

According to Table 2.9 in Chap. 2, the expectation of Weibull distribution is $u_i = \alpha\Gamma(1 + 1/\beta)$. The linear formula (12.52) cannot be written in a form of $\alpha\Gamma(1 + 1/\beta)$ since β does not contain any term of $\alpha_0 + \sum_{j=1}^{r} \alpha_j x_j$ in our Weibull GLL. Therefore, the Weibull GLL discussed before is not within the GLM framework. However, as a special case, if the β for the Weibull distribution is known, then the mean of Weibull distribution can be written as $k\alpha$, where the constant $k = \Gamma(1 + 1/\beta)$. Then

$$g(u_i) = \boldsymbol{x}_i.\boldsymbol{a} = \ln \alpha = \ln k\alpha/k = \ln \frac{u_i}{k} = \ln u_i - \ln k$$

Therefore, this special Weibull GLL is a GLM with link function $g(u_i) = \ln u_i - \ln k$.

Back to the GLM, here we focus on GLMs within the exponential family of distributions. The discussion of the exponential family distribution is present in Appendix 12B. Suppose the parameter ϕ are known, then one parameter exponential family PDF can be written as:

$$f(y;\theta,\phi) = exp(a(y)b(\theta) + c(\theta) + d(y)) \tag{12.53}$$

The likelihood function $l = \ln f$ for the exponential family distribution (12.53) is then

$$l = a(y)b(\theta) + c(\theta) + d(y) \tag{12.54}$$

Its first derivative is

$$\frac{\partial l}{\partial \theta} = U = a(y)b'(\theta) + c'(\theta) \tag{12.55}$$

For independent observation y_1, y_2, $\ldots y_i \ldots$, y_n, the total likelihood function is as follows and we let it equal to zero so it can be used for parameter estimation.

$$\sum_{i=1}^{n} \{a(y_i)b'(\theta) + c'(\theta)\} = 0 \tag{12.56}$$

We can have a linear factor function $\alpha_0 + \alpha_1 x_1 + \alpha_2 x_2 + \ldots$ equal to function $g(u)$, where the u is the *mean* of this distribution. From Appendix 12B, it is $\mu = b'(\theta)$. Thereafter, by using Eq. (12.56), we can obtain the parameter α_j in the linear function. There are mature algorithms for solving Eq. (12.56); one such algorithm is described in Reference [9].

For ALT data analysis, the linear function as $\alpha_0 + \alpha_1 x_1 + \alpha_2 x_2 + \ldots$ finds widely used to link several factors in one model because of its simplicity and the clear advantages it offers in parameter estimation through linear regression in some cases [4]. However, in the case of GLM, parameter estimation typically relies on maximum likelihood estimation as shown in Eq. (12.56) [9]. Using likelihood estimation means the main advantage of linear regression, its straightforward and efficient parameter estimation by regression, has not been fully utilized. Future research could focus on developing parameter estimation approaches that better utilize linear regression. Conclusively, for ALT data analysis, GLM are promising because they can incorporate multiple influencing factors while also capturing uncertainty, as it is hybrid model of linear regression and statistical distributions. Therefore, GLMs are highly recommended as models for ALT analysis.

12.5 Some TTF Models

The models discussed above are very general and less attention has been paid to their physical background. This section introduces some time-to-failure (TTF) models, which are also ALT models, but they take the physical background into account.

12.5.1 Electromigration (EM)

Electromigration is a failure mechanism in wire connection of electronic devices. When the electron wind is sufficiently strong, the metal ions, which are confined by the crystalline structure normally will be moved by the electron wind, as shown in Fig. 12.7. The disconnection, displacement thus will occur from the moving of the metal ions [10].

The TTF from EM can be modelled by Arrhenius model as follows [3]:

$$TTF = A_0.(\mathrm{J} - \mathrm{J}_{Crit})^{-n} \exp\left(\frac{E_\alpha}{k_b T}\right) \tag{12.57}$$

E_α: Activation energy of the failure
k_b: Boltzmann constant $8.61 \times 10^{-5} eV/k$.
T: Temperature in Kelvin
J: Electron current density
J_{Crit}: Threshold current density

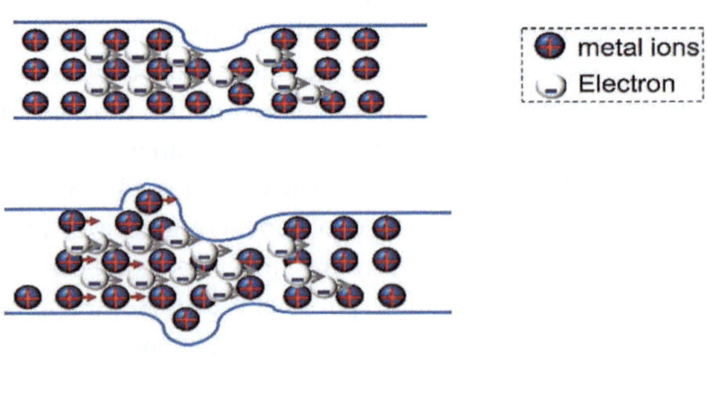

Fig. 12.7 Electron wind on metal atoms

Fig. 12.8 Stressed connection marked with circles

12.5.2 Stress Migration (SM)

The continuous mechanical stress applied on a connection, typical on metal connection of electronic units can induce failure. Such a failure is not immediate, but a gradual developing process. Figure 12.8 shows a connection of solder joints on an electronic board. The connection marked by circled line is under stress.

One TTF model considered stress is as follows [3]

$$TTF = A_0 \sigma^{-n} \exp\left(\frac{Q}{K_B T}\right) \tag{12.58}$$

A_0: Coefficient

n: stress migration exponent. n = 2–4 for soft metals such as aluminium and copper, n = 4–6 for mild steels, and n = 6–9 for strong metals

Q: activation energy. Aluminum is from 0.6 to 1 eV.

σ: is the tensile stress in the metal.

12.5.3 Corrosion

Corrosion is a significant challenge for engineering systems that operate outdoors. The moisture and contaminants nearby affect the function of both electronic and mechanical units. The cause of corrosion depends on the materials and the environment. One of the primary causes of corrosion is the electrochemical process. In the process, for example, metal chemically reacts with particles in ambient humidity or pollutants, leading to permanent altering of the material structure. This process can be time-dependent, and temperature is one of the factors influencing it. One corrosion model considers the humidity and temperature are as follows [3].

$$TTF = A_0\, exp\left[-a.(\%RH) + \frac{Q}{K_B T}\right] \tag{12.59}$$

A_0: Coefficient

$\%RH$: Humidity

a and Q: Kinetics of corrosion. a is humidity acceleration parameter 0.10–0.15 $(\% RH)^{-1}$. Q is around 0.7–0.8 eV for chloride-induced corrosion of aluminum in plastic packages.

Example 12.6

Suppose a corrosion with parameter $A_0 = 100$ hours; $Q = 0.2$ eV; $a = 0.12/\%$ RH. What is the TTF at temperature 22 °C and humidity 50%. ($K_B = 8.61 \times 10^{-5}$)

Solution:

$$TTF = A_0\, exp\left[-a.(\%RH) + \frac{Q}{K_B T}\right]$$

$$TTF = 100 * exp\left[-0.12 * 50 + \frac{0.2}{8.61 * 10^{-5} * (273 + 22)}\right] = 651.5 \text{ h}$$

12.6 Practices of ALT in Industries

One of the major activities of reliability engineering in industry is designing and conducting ALT. This section presents some examples conducted in IT industry.

Case 1: Speaker Durability Test

Speaker is a component often used in IT devices. The IT manufacturer requires the supplier to perform ALT test to validate if the speaker can work for a period of 96 hours at temperature of 28 °C and humidity 61%. One supplier thus set up test experiments at that condition and measures the performance of the speaker including impedance and dB value. They selected 4 speakers for test. During test, all speakers function without notable defect. When checking the measurement data, it found the impedance changed during the test, but fortunately it is still acceptable. The dB level also changes but not significantly. The conclusion thus asserts that the speaker passes the ALT test. Figure 12.9 demonstrates a speaker used in a computer. Usually, the speaker used on the computer already passed the ALT test during the design.

Case 2: DC Fan Accelerated Life Test

The fan is one frequently failed device in the modern electronic system such as in the desktop and laptop. To assess the fan life, ALT was performed. The fan has

Fig. 12.9 Speaker used in a computer

relatively long life for which the life test at normal condition is infeasible. The manufacturing company performed a test by selecting the elevated temperature to be 70 °C, the normal operating temperature is assumed to be 40 °C. The acceleration factor is 8 by using following acceleration factor

$$AF = 2^{(Acclelerated\ Temp\ - Normal\ Temprature)/10} \tag{12.60}$$

During test, the parameter of fan speed, electric DC current, and noise level are measured. The failure for this case was defined as the fan speed, DC current deviating over 7% than its initial values, the noise level cannot be over 50 dB, and it cannot be 3% higher than its original value. The requirement specifies:

$$\text{MTTF as 350,000 hours at } 40\,^{\circ}\text{C}$$

Total 100 fans were put on test. The experiment was run up to 600 hours at an accelerated temperature 70 °C. No failure was detected. Suppose the fan failure time follows Exponential distribution, the MTTF is then at least.

$$600 \times 100 = 60{,}000 \text{ hours}$$

Since the AF is 8, the MTTF so far is at least 60,000 × 8 = 480,000 hours at 40 °C, which is higher than the requirement of MTTF as 350,000 hours at 40 °C. The Fan can meet the requirement. Figure 12.10 shows a CPU cooling fan installed on a desktop. The fan helps to dissipate the heat generated from CPU. As the PC often runs continuously for a long period. The fan was a vulnerable component, especially in a dusty environment.

Case 3: Electronic Mainboard Survivability Test

Test the survivability of the electronic mainboard under extreme conditions. Aim was to test if the device can survive at extreme voltage and temperature level. Voltage was selected at level 3.3 V, 5 V, 12 V. To test its survivability on low

Fig. 12.10 CPU Fan installed in a desktop

Fig. 12.11 A mini mainboard for a computing device

temperature, temperature was selected as 0 °C, −5 °C, −10 °C, −15 °C, −20 °C, −30 °C, −40 °C. The results show the mainboard can function well only until −30 °C. Failure occurred when temperature was lower than −40 °C. Failure occurred at all the three voltage levels. To test its survivability on high temperature, it has tested the board under 60 °C, 70 °C, 80 °C, 85 °C. Results find the mainboard can function only lower than 85 °C. The 85 °C is the limit. Conclusively, for all three voltage levels, the work range of the mainboard is from −30 °C to 85 °C. This survivability test is simple but useful for the designer to select the best supplier, or to give accurate specifications to inform the customer. Some industrial computers used in the harsh environment require that the computer can operate in extremely cold or hot conditions. This test can provide credible information for the selection. Figure 12.11 demonstrates the mainboard of a minicomputer.

Appendix 12A: ALT Data

Table 12A.1 Accelerated life data (Data source: [4])

V =26 kV	28 kV	30 kV	32 kV	34 kV	36 kV	38 kV
1.7561	4.2319	2.0466	−1.3094	−1.6608	−1.0499	−2.4080
7.3648	4.6848	2.8361	−0.9163	−0.2485	−0.5277	−0.9417
7.7509	4.7031	3.0184	−0.3711	−0.0409	−0.0409	−0.7551
	6.0516	2.0454	−0.2358	0.2700	−0.0101	−0.3148
	6.9731	3.1206	1.0116	1.0224	0.5247	−0.3012
		3.7704	1.3635	1.1505	0.6780	0.1222
		3.8565	2.2905	1.4231	0.7275	0.3364
		4.9349	2.6354	1.5411	0.9477	0.8671
		4.9706	2.7682	1.5789	0.9969	
		5.1698	3.3250	1.8718	1.0647	
		5.2724	3.9748	1.9947	1.3001	
			4.4170	2.0806	1.3837	
			4.4918	2.1126	1.6770	
			4.6109	2.4898	2.6224	
			5.3711	3.4578	3.2386	
				3.4818		
				3.5237		
				3.6030		
				4.2889		

Appendix 12B: Generalized Linear Model [9, 11, 12]

The Generalized Linear Model (GLM) is another domain of regression-based model that can be used as ALT model. Suppose the response of factors vectors $\boldsymbol{X}$ is $\boldsymbol{\eta}$, i.e. $\boldsymbol{\eta} = \boldsymbol{X}.\boldsymbol{a}$, which is written as

$$\begin{bmatrix} \eta_1 \\ \eta_2 \\ \cdots \\ \eta_n \end{bmatrix} = \begin{bmatrix} 1 & x_{11} & x_{12} & \ldots & x_{1r} \\ 1 & x_{21} & x_{12} & \ldots & x_{2r} \\ \cdots & \cdots & \cdots & \cdots & \cdots \\ 1 & x_{n1} & x_{n2} & \ldots & x_{nr} \end{bmatrix} \begin{bmatrix} a_0 \\ a_1 \\ \cdots \\ a_r \end{bmatrix} \tag{12B.1}$$

where $\boldsymbol{X} = \begin{bmatrix} 1 & x_{11} & x_{12} & \ldots & x_{1r} \\ 1 & x_{21} & x_{12} & \ldots & x_{2r} \\ \cdots & \cdots & \cdots & \cdots & \cdots \\ 1 & x_{n1} & x_{n2} & \ldots & x_{nr} \end{bmatrix}_{n\times(r+1)}$, $\boldsymbol{a} = \begin{bmatrix} a_0 \\ a_1 \\ \cdots \\ a_r \end{bmatrix}$.

An observation y_i corresponding to a factor vector $\boldsymbol{x}_i = [1 \quad x_{i1} \quad \ldots \quad x_{ir}]$. The $\boldsymbol{Y}$ is written as vector as follows

$$\boldsymbol{Y} = \begin{bmatrix} y_1 \\ y_2 \\ \ldots \\ y_n \end{bmatrix}$$

The expected value of $\boldsymbol{Y}$ is $\boldsymbol{\mu}$, i.e. $E(y_i) = \mu_i$. Let $= \begin{bmatrix} \mu_1 \\ \mu_2 \\ \ldots \\ \mu_n \end{bmatrix}$.

The y_i can be considered as a random variable. And write

$$y_i = \mu_i + \varepsilon_i \qquad (12\text{B}.2)$$

The ε_i is a random number with mean 0. We establish a function as follows to link the linear response η_i with the mean of y_i which is μ_i as follows

$$g(\mu_i) = \eta_i \qquad (12\text{B}.3)$$

If the $g(\mu_i) = \mu_i$ and assume the ε_i follows Normal distribution, the case is the classical linear regression. The $g(\mu_i)$ can be any monotone and continuous function, the ε_i can be any statistical distribution. For this general case, the model is named the Generalized Linear Model (GLM). The function $g(\mu_i)$ is called link function.

Exponential Family GLM

The exponential family GLM assumes the distribution of ε_i follows a distribution from the exponential family of the form as [12, 13].

$$f(y;\theta,\phi) = exp\left(\frac{y\theta - b(\theta)}{a(\phi)} + c(y,\phi)\right) \qquad (12\text{B}.4)$$

where θ, ϕ are parameters, the function $a(.)$, $b(.)$, $c(.)$ are distribution specific. Equation (12B.4) is just one form of exponential family distribution function. It has other equivalent functions, such the one discussed in Sect. 3.1.1 in Chap. 3. For the Normal distribution with PDF as

$$f(y) = \frac{1}{\sigma\sqrt{2\pi}} e^{-\frac{1}{2}\left(\frac{y-\mu}{\sigma}\right)^2} = exp\left\{\frac{y\mu - \mu^2/2}{\sigma^2} - \frac{1}{2}\left(\frac{y^2}{\sigma^2} + \ln\left(2\pi\sigma^2\right)\right)\right\}$$

It is a distribution from exponential family. In this case, the $\theta = \mu$, $a(\phi) = \phi = \sigma^2$, $b(\theta) = \frac{\theta^2}{2}, c(y,\phi) = -\frac{1}{2}\left(\frac{y^2}{\sigma^2} + \ln(2\pi\sigma^2)\right) = -\frac{1}{2}\left(\frac{y^2}{\phi} + \ln(2\pi\phi)\right)$.

GLM Exponential Distribution
For Exponential distribution, the

$$f(y) = \lambda e^{-\lambda y} = exp\{-\lambda y + \ln(\lambda))\}$$

where $\theta = -\lambda$, $a(\phi) = \phi = 1$, $b(\theta) = -\ln(-\theta)$, $c(y, \phi) = 0$. The Exponential distribution is from exponential family.

GLM Gamma Distribution
A Gamma distribution has PDF as follows

$$f(y) = \frac{1}{\gamma^k \Gamma(k)} y^{k-1} e^{-\frac{y}{\gamma}} = exp\left\{\frac{-\frac{y}{k\gamma} - \ln k\gamma}{1/k} + k \ln k + (k-1)\ln y - \ln \Gamma(k)\right\}$$

Evidently, $\theta = -\frac{1}{k\gamma}$, $a(\phi) = \phi = \frac{1}{k}$, $b(\theta) = -\ln(-\theta)$,

$$c(y, \phi) = k \ln k + (k-1)\ln y - \ln \Gamma(k) = \frac{-\ln \phi}{\phi} + \left(\frac{1}{\phi} - 1\right)\ln y - \ln \Gamma\left(\frac{1}{\phi}\right).$$

The Gamma distribution is from exponential family.

For the exponential family distribution, it has

$$E(y) = \mu = b'(\theta);$$
$$Var(y) = b''(\theta)a(\phi) \tag{12B.5}$$

Link Function for Exponential Family
We present some examples to find the link function. For the exponential family, suppose we let

$$\theta_i = \eta_i. \tag{12B.6}$$

Then for the Normal distribution, the

$$E(y_i) = \mu_i = b'(\theta_i) = \theta_i$$

The link function is thus $\mu_i = \eta_i$, i.e. $g(\mu_i) = \mu_i$.

For the Gamma distribution, the

$$E(y_i) = \mu_i = b'(\theta_i) = \frac{1}{\theta_i}$$

The link function for Gamma distribution is then

$$\mu_i = \frac{1}{\theta_i} = \frac{1}{\eta_i}.$$

According to Eq. (12B.3), the link function is then $g(\mu_i) = 1/\mu_i$.

Problems and Questions

1. The Arrhenius model used in reliability test can estimate

   a. the position where a corrosion begins
   b. the amount of electrons at a particular time
   c. the air pressure in a vessel being tested
   d. the acceleration factor for a test

2. At lab with temperature 100 °C, an electronic device is with time to failures as

$$1, 2, 1, 4, 5 \text{ hours}$$

   Arrhenius models can be used as the ALT model. The activation energy is 0.6 eV. What is the reliability of the electronic device for 100 hours at room temperature 22 °C?

3. A device is supposed to be installed in the arctic area. The temperature in the arctic area is supposed to be −5 °C. An accelerated life test in the lab shows that: at temperature 200 °C, the device has a mean life of 20 hours. The acceleration life model is

$$TTF = A_0 \, exp\left(-\alpha.RH + \frac{Q}{K_B T} \right)$$

   $K_B = 8.61 \times 10^{-5}$
   $T =$ Celsius Degree + 273.15
   $Q = 0.75$ eV

   a. What is the acceleration factor in this accelerated life test?
   b. What is the mean life expected in the arctic area?
   c. If the life of the device follows exponential distribution, oil company requires the reliability to be at least 95% for 10 hours in the arctic area. Can the device meet the reliability requirement?

4. The expected life of a device is 1 year at temperature 20 °C. We are planning to conduct accelerated life test on it. If the accelerated life test duration is set as 1 day, the accelerated life test model used is

$$TTF = A_0 . \exp\left(\frac{Q_\alpha}{k_b T}\right)$$

$K_B = 8.61 \times 10^{-5}$
$T =$ Celsius Degree + 273.15
$Q = 0.75$ eV

a. What is the acceleration factor in this accelerated life test?
b. What should the temperature for the accelerated life test be in the lab?

5. In order to evaluate the life of a capacitor (electronic device), lab engineer has carried an acceleration life test. The ALT test is at a temperature of 120 °C and the time to failure model is

$$TTF = A_0 . \exp\left(\frac{Q_\alpha}{k_b T}\right)$$

Q_α: Activation energy of the failure
k_b: Boltzmann constant $8.61 \times 10^{-5} eV/k$.

6. For this failure mechanism, the activation energy of failure is 0.5 eV. The normal operation condition for the device is at temperature 20 °C. At the normal temperature, the time to failure was observed as (in hours):

$$120, 11, 14, 71, 90, 1, 52$$

a. What is the acceleration factor for this ALT experiment?
b. Assume the time to failure follows Exponential distribution. What is the failure rate at the elevated temperature 120 °C? What is the reliability of the device at 0.1 hours at the elevated temperature?
c. Under the elevated temperature 120 °C, if we put a total of 100 capacitors in the chamber to test their lives, how many capacitors are expected to fail within 0.1 hours?

7. Early knowledge tell us the time to failure of an electronic device follows Arrhenius model as

$$TTF = A_0 . \exp\left(\frac{Q_\alpha}{k_b T}\right)$$

Q_α: Activation energy of the failure
k_b: Boltzmann constant $8.61 \times 10^{-5} eV/k$

The activation energy of the failure is 0.2 eV. In the field at temperature 22 °C, we collected the time to failures data as (in years)

$$1; 5; 2; 7; 9$$

a. What is the acceleration factor if temperature is set on 200 °C for this ALT experiment?
b. What is the failure rate at the field, and what is the failure rate at temperature 200 °C?
c. In the lab, how long should the experiment run to expect the probability of failure of the device is 99%?

8. At room temperature of Celsius degree 20 °C, engineer found the life of an electronic device is 1.5 years. They plan to conduct acceleration life test on this product. The time duration for the test is 10 days. The failure model is

$$TTF = A_0 . \exp\left(\frac{Q_\alpha}{k_b T}\right)$$

Q_α: Activation energy of the failure
k_b: Boltzmann constant $8.61 \times 10^{-5} eV/k$

For the failure of this device, the activation energy of failure is 0.6 eV (Hint: Kelvin = [°C] + 273.15).

a. What is the acceleration factor for this ALT?
b. What is the temperature should the engineer set for this ALT test?
c. If the failure rate at the accelerated temperature is 0.001 per day, what is the failure rate at the normal temperature of the Celsius degree 20 °C?

References

1. Fuqing Y (2024) Reliability evaluation and reliability prediction for internet of things (IOT) appliance. In: Frontiers of performability engineering: In Honor of Prof. KB Misra. Springer, pp 69–101
2. Evseev S (2007) Eyring acceleration model in thick nitride/oxide dielectrics. Microelectron Reliab 47(4–5):748–751
3. J. W. McPherson, Reliability physics and engineering time-to-failure modeling, 2 New York: Springer,2013
4. W. B.Nelson, Accelerated testing statistical models, test plans, and data analysis. Wiley, 2004
5. Wang W, Kececioglu DB (2002) Fitting the Weibull log-linear model to accelerated life-test data. IEEE Trans Reliab 49(2):217–223

6. Krishnamoorthy K (2006) Handbook of statistical distributions with applications, Statistics, textbooks and monographs. Chapman & Hall/CRC, Boca Raton
7. Myers RH (2000) Classical and modern regression with applications. Duxbury Press
8. Yuan F, Lu J (2025) Parameter estimation for the weibull general log-linear model in accelerated life testing. J Phys Conf Ser 2964(1):012047
9. A. J.Dobson and A. G.Barnett, An introduction to generalized linear models, 3 CRC Press, 2008
10. Tan CM, Roy A (2007) Electromigration in ULSI interconnects. Mater Sci Eng 58:1–75
11. McCullagh P, Nelder JA (1983) Generalized linear models, 3rd edn. Chapman and Hall, London
12. J. J.Faraway, Extending the linear model with R. New York: CRC Taylor & Francis, 2006
13. Rodríguez G (2007) Appendix B generalized linear model theory, Lecture notes on generalized linear models. Princeton University

MIX
Papier aus verantwortungsvollen Quellen
Paper from responsible sources
FSC® C105338

If you have any concerns about our products,
you can contact us on
ProductSafety@springernature.com

In case Publisher is established outside the EU,
the EU authorized representative is:
Springer Nature Customer Service Center GmbH
Europaplatz 3, 69115 Heidelberg, Germany

Printed by Libri Plureos GmbH
in Hamburg, Germany